ADAPTIVE CONTROL

──────

Second Edition

ADAPTIVE CONTROL

──

Second Edition

Karl Johan Åström
Björn Wittenmark

Lund Institute of Technology

▲▼

Addison-Wesley Publishing Company, Inc.

Reading, Massachusetts   Menlo Park, California   New York
Don Mills, Ontario   Wokingham, England   Amsterdam
Bonn   Sydney   Singapore   Tokyo   Madrid   San Juan
Milan   Paris

LIBRARY OF CONGRESS CATALOGING-IN-PUBLICATION DATA

Åström, Karl J. (Karl Johan), 1934–
    Adaptive control / Karl Johan Åström, Björn Wittenmark.—2nd ed.
        p. cm.
    Includes index.
    ISBN 0-201-55866-1
    1. Adaptive control systems.   I. Wittenmark, Björn.   II. Title.
TJ217.A67 1995
629.8'36–dc20   94-12682
                                                                              CIP

This book is in the *Addison-Wesley Series in Electrical Engineering: Control Engineering*

Many of the designations used by manufacturers and sellers to distinguish their products are claimed as trademarks. Where those designations appear in this book, and Addison-Wesley was aware of a trademark claim, the designations have been printed in initial caps or all caps.

    4 5 6 7 8 9 10 MA 989796

To Bia, Emma, Ida, Johanna, Kalle, Kalle, Karin, and Karin

PREFACE

Adaptive control is a fascinating field for study and research. It is also of increasing practical importance, since adaptive techniques are being used more and more in industrial control systems. However, there are still many unsolved theoretical and practical issues.

• *Goal of the book*   Our goal is to give an introduction and an overview of the theoretical and practical aspects of adaptive control.

Since knowledge about adaptive techniques is widely scattered in the literature, it is difficult for a newcomer to get a good grasp of the field. In the book we introduce the basic ideas of adaptive control and compare different approaches. Practical aspects such as implementation and applications are presented in depth. These are very important for the understanding of the advantages and shortcomings of adaptive control. This book has evolved from many years of research and teaching in the field.

After learning the material in the book a reader should have a good perspective of adaptive techniques, an active knowledge of the key approaches, and a good sense of when adaptive techniques can be used and when other methods are more appropriate.

• *The new edition*   Adaptive control is a dynamic field of research and industrial applications. Much new knowledge has appeared which by itself motivates a new edition.

We have used the first edition of the book to teach to a wide variety of audiences, in regular university courses, courses to engineers in industry, and short courses at conferences. These experiences combined with advances in research, have shaped the major revisions made in the new edition. We have also benefited from feedback of students and colleagues in industry and universities, who have used the first edition of the book.

New chapters have been added, and the material has been reorganized. Most of the chapters have been substantially revised. In the revision we have also given more emphasis to the connection between different design methods in adaptive control. There is a major change in the way we deal with the theory. In the first edition we relied on mathematics from a wide variety of sources. In the new edition we have to a large extent developed the results from first principles. To make this possible we have made stronger assumptions in a few cases, but the material is now much easier to teach. The reorganization of the material also makes it easier to use the book for different audiences.

The first edition had two introductory chapters; they have now been compressed to one. In the first edition we started with model-reference adaptive systems following the historical tradition. In the second edition we start with parameter estimation and the self-tuning regulator. This has several advantages, one is that students can start to simulate and experiment with computer-based adaptive control at a much earlier state, another is that system identification gives the natural background and the key concepts required to understand many aspects of adaptive control.

The material on self-tuning control has been expanded substantially by introducing an extra chapter. This has made it possible to give a strict separation between deterministic and stochastic self-tuners. This is advantageous in courses which are restricted to the deterministic case.

The chapter on model-reference adaptive control has been expanded substantially. The key results on stability theory are now derived from first principles. This makes it much easier to teach to students who lack a background in stability theory. A new section on adaptive control of nonlinear systems has also been added.

The reorganization makes the transformation from algorithms to theory much smoother. The chapter on theory now naturally follows the development of nonlinear stability theory. The presentation of the theory has been modified substantially. A new section on stability of time-varying systems has been added. This makes it possible to get a much better understanding of adaptation of feedforward gains. It also is a good transition to the nonlinear case. Material on the nonlinear behavior of adaptive systems has also been added. This adds substantially to the understanding of the behavior of adaptive systems.

The chapter on practical aspects and implementation has been rewritten completely to reflect the increased experience of practical use of adaptive control. It has been very rewarding to observe the drastically increased industrial use of adaptive control. This has influenced the revision of the chapter on applications. For example, adaptive control is now used extensively in automobiles.

Many examples and simulations are given throughout the book to illustrate ideas and theory. Numerous problems are also given. There are theoretical problems as well as problems in which computers must be used for analysis and simulations. The examples and problems give the reader good insight into properties, design procedures, and implementation of adaptive controllers. To maintain a reasonable size of the book we have also done careful pruning.

To summarize, new research and new experiences have made it possible to present the field of adaptive control in what we hope is a better way.

Outline of the Book

• *Background Material*   The first chapter gives a broad presentation of adaptive control and background for its use. Real-time estimation, which is an essential part of adaptive control, is introduced in Chapter 2. Both discrete-time and continuous-time estimation are covered.

• *Self-tuning Regulators and Model-reference Adaptive Systems*   Chapters 3, 4, and 5 give two basic developments of adaptive control: self-tuning regulators (STR) and model-reference adaptive systems (MRAS). Today we do not make a distinction between these two approaches, since they are actually equivalent. We have tried to follow the historical development by mainly treating MRAS in continuous time and STR in discrete time. By doing so it is possible to cover many aspects of adaptive regulators. These chapters mainly cover the ideas and basic properties of the controllers. They also serve as a source of algorithms for adaptive control.

• *Theory*   Chapter 6 gives deeper coverage of the theory of adaptive control. Questions such as stability, convergence, and robustness are discussed. Stochastic adaptive control is treated in Chapter 7. Depending on the background of the students, some of the material in Chapters 6 and 7 can be omitted in an introductory course.

• *Broadening the View*   Automatic tuning of regulators, which is rapidly gaining industrial acceptance, is presented in Chapter 8. Gain scheduling is discussed in Chapter 9. Even though adaptive controllers are very useful tools, they are not the only ways to deal with systems that have varying parameters. Since we believe that it is useful for an engineer to have several ways of solving a problem, alternatives to adaptive control are also included. Robust high-gain control and self-oscillating controllers are presented in Chapter 10.

• *Practical Aspects and Applications*   Chapter 11 gives suggestions for the implementation of adaptive controllers. The guidelines are based on practical experience in using adaptive controllers on real processes. Chapter 12 is a summary of applications and description of some commercial adaptive controllers. The applications show that adaptive control can be used in many different types of processes, but also that all applications have special features that must be considered to obtain a good control system.

• *Perspectives*   Finally, Chapter 13 contains a brief review of some areas closely related to adaptive control that we have not been able to cover in the book. Connections to adaptive signal processing, expert systems, and neural networks are given.

Prerequisites

The book is for a course at the graduate level for engineering majors. It is assumed that the reader already has good knowledge in automatic control and a basic knowledge in sampled data systems. At our university the course can be taken after an introductory course in feedback control and a course in digital control. The intent is also that the book should be useful for an industrial audience.

Course Configurations

The book has been organized so that it can be used in different ways. An introductory course in adaptive control could cover Chapters 1, 2, 3, 4, 5, 8, 11, 12, and 13. A more advanced course might include all chapters in the book. A course for an industrial audience could contain Chapters 1, parts of Chapters 2, 3, 4, and 5, and Chapters 8, 9, 11, and 12. To get the full benefit of a course, it is important to supplement lectures with problem-solving sessions, simulation exercises, and laboratory experiments.

Simulation Tools

Computer simulation is an indispensible tool for understanding the behavior of adaptive systems. Most of the simulations in the book are done by using the interactive simulation package Simnon, which has been developed at our department. Simnon is available for IBM-PC compatible computers and also for several workstations and mainframe computers. Further information can be obtained from SSPA Systems, Box 24001, S-400 22 Göteborg, Sweden, e-mail: simnon@sspa.se. The macros used in the simulations are available for anonymous FTP from `ftp.control.lth.se`, directory `/pub/books/adaptive_control`. Adaptive systems can of course also be simulated using other tools.

Supplements

An instructor's manual with solutions to the problems is available through the publisher.

Wanted: Feedback

As teachers and researchers in automatic control, we know the importance of feedback. We therefore encourage all readers to write to us about errors, misunderstandings, suggestions for improvements, and also about what may be valuable in the material we have presented.

Acknowledgments

During the years we have done research in adaptive control and written the book, we have had the pleasure and privilege of interacting with many colleagues throughout the world. Consciously and subconsciously, we have picked up material from the knowledge base called adaptive control. It is impossible to mention everyone who has contributed ideas, suggestions, concepts, and examples, but we owe you all our deepest thanks. The long-term support of our research on adaptive control by the Swedish Board of Industrial and Technical Development (NUTEK) and by the Swedish Research Council for Engineering Sciences (TFR) are gratefully acknowledged.

For the second edition we want to thank Petar V. Kokotovic, P. R. Kumar, David G. Taylor, A. Galip Ulsoy, and Baxter F. Womack, who have reviewed the manuscript and given us very valuable feedback.

Finally, we want to thank some people who, more than others, have made it possible for us to write this book. Leif Andersson has been our TEXpert. He and Eva Dagnegård have been invaluable when solving many of the TEX problems. Eva Dagnegård and Agneta Tuszynski have done an excellent job of typing many versions of the manuscript. Most of the illustrations have been done by Britt-Marie Carlsson and Doris Nilsson. Without all their patience and understanding of our whims, there would never have been a final book. We also want to thank the staff at Addison-Wesley for their support and professionalism in bookmaking.

Karl Johan Åström
Björn Wittenmark

Department of Automatic Control
Lund Institute of Technology
Box 118, S-221 00 Lund, Sweden

karl_johan.astrom@control.lth.se
bjorn.wittenmark@control.lth.se

CONTENTS

CHAPTER 1

WHAT IS
ADAPTIVE CONTROL?

1.1 INTRODUCTION

In everyday language, "to adapt" means to change a behavior to conform to
new circumstances. Intuitively, an adaptive controller is thus a controller that
can modify its behavior in response to changes in the dynamics of the process
and the character of the disturbances. Since ordinary feedback also attempts
to reduce the effects of disturbances and plant uncertainty, the question of the
difference between feedback control and adaptive control immediately arises.
Over the years there have been many attempts to define adaptive control
formally. At an early symposium in 1961 a long discussion ended with the
following suggestion: "An adaptive system is any physical system that has
been designed with an adaptive viewpoint." A renewed attempt was made by
an IEEE committee in 1973. It proposed a new vocabulary based on notions like
self-organizing control (SOC) system, parameter-adaptive SOC, performance-
adaptive SOC, and learning control system. However, these efforts were not
widely accepted. A meaningful definition of adaptive control, which would make
it possible to look at a controller hardware and software and decide whether
or not it is adaptive, is still lacking. However, there appears to be a consensus
that a constant-gain feedback system is not an adaptive system.

   In this book we take the pragmatic attitude that *an adaptive controller
is a controller with adjustable parameters and a mechanism for adjusting
the parameters*. The controller becomes nonlinear because of the parameter
adjustment mechanism. It has, however, a very special structure. Since general
nonlinear systems are difficult to deal with, it makes sense to consider special
classes of nonlinear systems. An adaptive control system can be thought of
as having two loops. One loop is a normal feedback with the process and the
controller. The other loop is the parameter adjustment loop. A block diagram

1

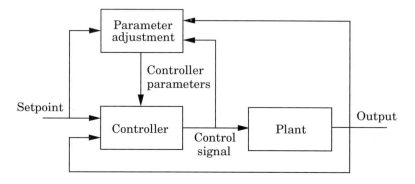

Figure 1.1   Block diagram of an adaptive system.

of an adaptive system is shown in Fig. 1.1. The parameter adjustment loop is often slower than the normal feedback loop.

A control engineer should know about adaptive systems because they have useful properties, which can be profitably used to design control systems with improved performance and functionality.

A Brief History

In the early 1950s there was extensive research on adaptive control in connection with the design of autopilots for high-performance aircraft (see Fig. 1.2). Such aircraft operate over a wide range of speeds and altitudes. It was found that ordinary constant-gain, linear feedback control could work well in one operating condition but not over the whole flight regime. A more sophisticated controller that could work well over a wide range of operating conditions was therefore needed. After a significant development effort it was found that gain scheduling was a suitable technique for flight control systems. The interest in adaptive control diminished partly because the adaptive control problem was too hard to deal with using the techniques that were available at the time.

In the 1960s there were much research in control theory that contributed to the development of adaptive control. State space and stability theory were introduced. There were also important results in stochastic control theory. Dynamic programming, introduced by Bellman, increased the understanding of adaptive processes. Fundamental contributions were also made by Tsypkin, who showed that many schemes for learning and adaptive control could be described in a common framework. There were also major developments in system identification. A renaissance of adaptive control occurred in the 1970s, when different estimation schemes were combined with various design methods. Many applications were reported, but theoretical results were very limited.

In the late 1970s and early 1980s, proofs for stability of adaptive systems appeared, albeit under very restrictive assumptions. The efforts to merge ideas

Figure 1.2   Several advanced flight control systems were tested on the X-15 experimental aircraft. (By courtesy of Smithsonian Institution.)

of robust control and system identification are of particular relevance. Investigation of the necessity of those assumptions sparked new and interesting research into the robustness of adaptive control, as well as into controllers that are universally stabilizing. Research in the late 1980s and early 1990s gave new insights into the robustness of adaptive controllers. Investigations of nonlinear systems led to significantly increased understanding of adaptive control. Lately, it has also been established that adaptive control has strong relations to ideas on learning that are emerging in the field of computer science.

There have been many experiments on adaptive control in laboratories and industry. The rapid progress in microelectronics was a strong stimulation. Interaction between theory and experimentation resulted in a vigorous development of the field. As a result, adaptive controllers started to appear commercially in the early 1980s. This development is now accelerating. One result is that virtually all single-loop controllers that are commercially available today allow adaptive techniques of some form. The primary reason for introducing adaptive control was to obtain controllers that could adapt to changes in process dynamics and disturbance characteristics. It has been found that adaptive techniques can also be used to provide automatic tuning of controllers.

1.2   LINEAR FEEDBACK

Feedback by itself has the ability to cope with parameter changes. The search for ways to design a system that are insensitive to process variations was in fact one of the driving forces for inventing feedback. Therefore it is of interest to know the extent to which process variations can be dealt with by using linear feedback. In this section we discuss how a linear controller can deal with variations in process dynamics.

Robust High-Gain Control

A linear feedback controller can be represented by the block diagram in Fig. 1.3. The feedback transfer function G_{fb} is typically chosen so that disturbances acting on the process are attenuated and the closed-loop system is insensitive to process variations. The feedforward transfer function G_{ff} is then chosen to give the desired response to command signals. The system is called a *two-degree-of-freedom system* because the controller has two transfer functions that can be chosen independently. The fact that linear feedback can cope with significant variations in process dynamics can be seen from the following intuitive argument. Consider the system in Fig. 1.3. The transfer function from y_m to y is

$$T = \frac{G_p G_{fb}}{1 + G_p G_{fb}}$$

Taking derivatives with respect to G_p, we get

$$\frac{dT}{T} = \frac{1}{1 + G_p G_{fb}} \frac{dG_p}{G_p}$$

The closed-loop transfer function T is thus insensitive to variations in the process transfer function for those frequencies at which the loop transfer function

$$L = G_p G_{fb} \tag{1.1}$$

is large. To design a robust controller, it is thus attempted to find G_{fb} such that the loop transfer function is large for those frequencies at which there are large variations in the process transfer function. For those frequencies where $L(i\omega) \approx 1$, however, it is necessary that the variations be moderate for the system to have sufficient robustness properties.

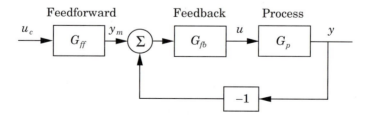

Figure 1.3   Block diagram of a robust high-gain system.

Judging Criticality of Process Variations

We now consider some specific examples to develop some intuition for judging the effects of parameter variations. The following example illustrates that significant variations in open-loop step responses may have little effect on the closed-loop performance.

EXAMPLE 1.1 **Different open-loop responses**

Consider systems with the open-loop transfer functions

$$G_0(s) = \frac{1}{(s+1)(s+a)}$$

where $a = -0.01$, 0, and 0.01. The dynamics of these processes are quite different, as is illustrated in Fig. 1.4(a). Notice that the responses are significantly different. The system with $a = 0.01$ is stable; the others are unstable. The initial parts of the step responses, however, are very similar for all systems. The closed-loop systems obtained by introducing the proportional feedback with unit gain, that is, $u = u_c - y$, give the step responses shown in Fig. 1.4(b). Notice that the responses of the closed-loop systems are virtually identical. Some insight is obtained from the frequency responses. Bode diagrams for the

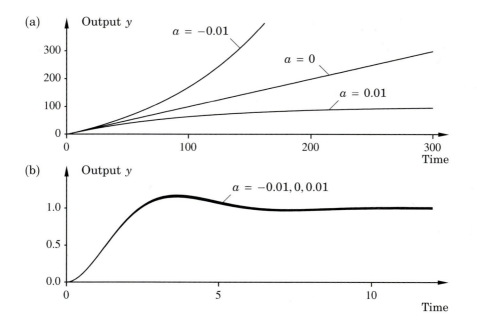

Figure 1.4 (a) Open-loop unit step responses for the process in Example 1.1 with $a = -0.01$, 0, and 0.01. (b) Closed-loop step responses for the same system, with the feedback $u = u_c - y$. Notice the difference in time scales.

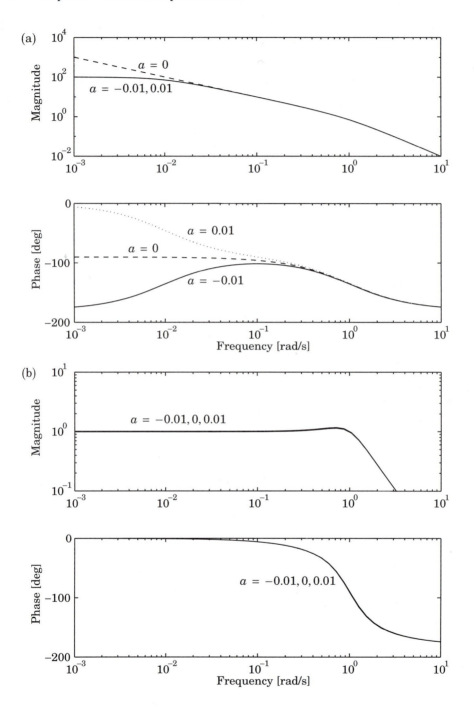

Figure 1.5   (a) Open-loop and (b) closed-loop Bode diagrams for the process in Example 1.1.

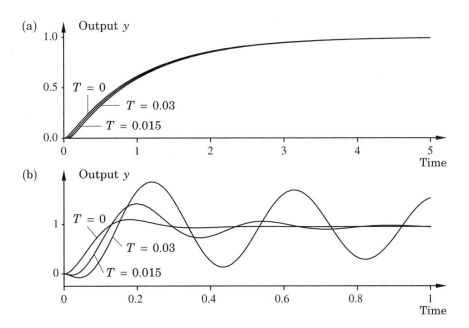

Figure 1.6   (a) Open-loop unit step responses for the process in Example 1.2 with $T = 0$, 0.015, and 0.03. (b) Closed-loop step responses for the same system, with the feedback $u = u_c - y$. Notice the difference in time scales.

open and closed loops are shown in Fig. 1.5. Notice that the Bode diagrams for the open-loop systems differ significantly at low frequencies but are virtually identical for high frequencies. Intuitively, it thus appears that there is no problem in designing a controller that will work well for all systems, provided that the closed-loop bandwidth is chosen to be sufficiently high. This is also verified by the Bode diagrams for the closed-loop systems shown in Fig. 1.5(b), which are practically identical. Also compare the step responses of the closed-loop systems in Fig. 1.4(b).   □

The next example illustrates that process variations may be significant even if changes in the open-loop step responses are small.

EXAMPLE 1.2   Similar open-loop responses

Consider systems with the open-loop transfer functions

$$G_0(s) = \frac{400(1 - sT)}{(s + 1)(s + 20)(1 + Ts)}$$

with $T = 0$, 0.015, and 0.03. The open-loop step responses are shown in Fig. 1.6(a). Figure 1.6(b) shows the step responses for the closed-loop systems obtained with the feedback $u = u_c - y$. Notice that the open-loop responses

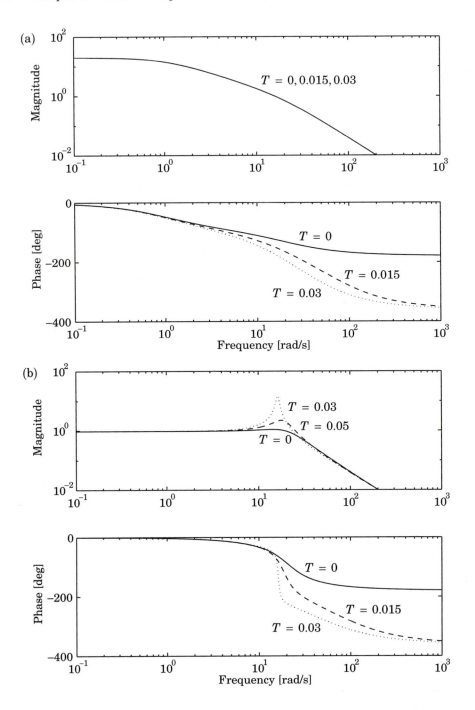

Figure 1.7   Bode diagrams for the process in Example 1.2. (a) The open-loop system; (b) The closed-loop system.

are very similar but that the closed-loop responses differ considerably. The frequency responses give some insight. The Bode diagrams for the open- and closed-loop systems are shown in Fig. 1.7. Notice that the frequency responses of the open-loop systems are very close for low frequencies but differ considerably in the phase at high frequencies. It is thus possible to design a controller that works well for all systems provided that the closed-loop bandwidth is chosen to be sufficiently small. At the crossover frequency chosen in the example there are, however, significant variations that show up in the Bode diagrams of the closed-loop systems in Fig. 1.7(b) and in the step responses of the closed-loop system in Fig. 1.6(b).   □

The examples discussed show that to judge the consequences of process variations from open-loop dynamics, it is better to use frequency responses than time responses. It is also necessary to have some information about the desired crossover frequency of the closed-loop system. Intuitively, it may be expected that a process variation that changes dynamics from unstable to stable is very severe. Example 1.1 shows that this is not necessarily the case.

EXAMPLE 1.3   Integrator with unknown sign

Consider a process whose dynamics is described by

$$G_0(s) = \frac{k_p}{s} \qquad (1.2)$$

where the gain k_p can assume both positive and negative values. This is a very severe variation because the phase of the system can change by $180°$. This process cannot be controlled by a linear controller with a rational transfer function. This can be seen as follows. Let the controller transfer function be $S(s)/R(s)$, where $R(s)$ and $S(s)$ are polynomials. Assume that $\deg R \geq \deg S$. The characteristic polynomial of the closed-loop system is then

$$P(s) = sR(s) + k_p S(s)$$

Without lack of generality it can be assumed that the coefficient of the highest power of s in the polynomial $R(s)$ is 1. The coefficient of the highest power of s of $P(s)$ is thus also 1. The constant coefficient of polynomial $k_p S(s)$ is proportional to k_p and can thus be either positive or negative. A necessary condition for $P(s)$ to have all roots in the left half-plane is that all coefficients are positive. Since k_p can be both positive and negative, the polynomial $P(s)$ will always have a zero in the right half-plane for some value of k_p.   □

1.3   EFFECTS OF PROCESS VARIATIONS

The standard approach to control system design is to develop a linear model for the process for some operating condition and to design a controller having

constant parameters. This approach has been remarkably successful. A fundamental property is also that feedback systems are intrinsically insensitive to
modeling errors and disturbances. In this section we illustrate some mechanisms that give rise to variations in process dynamics. We also show the effects
of process variations on the performance of a control system.

The examples are simplified to the extent that they do not create significant
control problems but do illustrate some of the difficulties that might occur in
real systems.

Nonlinear Actuators

A very common source of variations is that actuators, like valves, have a
nonlinear characteristic. This may create difficulties, which are illustrated by
the following example.

EXAMPLE 1.4   Nonlinear valve

A simple feedback loop with a Proportional and Integrating (PI) controller,
a nonlinear valve, and a process is shown in Fig. 1.8. Let the static valve
characteristic be

$$v = f(u) = u^4 \qquad u \geq 0$$

Linearizing the system around a steady-state operating point shows that the
incremental gain of the valve is $f'(u)$, and hence the loop gain is proportional
to $f'(u)$. The system can perform well at one operating level and poorly at
another. This is illustrated by the step responses in Fig. 1.9. The controller is
tuned to give a good response at low values of the operating level. For higer
values of the operating level the closed-loop system even becomes unstable.
One way to handle this type of problem is to feed the control signal u through
an inverse of the nonlinearity of the valve. It is often sufficient to use a fairly
crude approximation (see Example 9.1). This can be interpreted as a special
case of gain scheduling, which is treated in detail in Chapter 9.   □

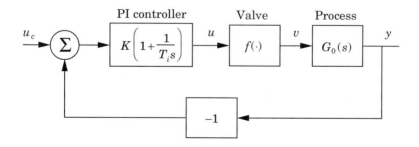

Figure 1.8   Block diagram of a flow control loop with a PI controller and a
nonlinear valve.

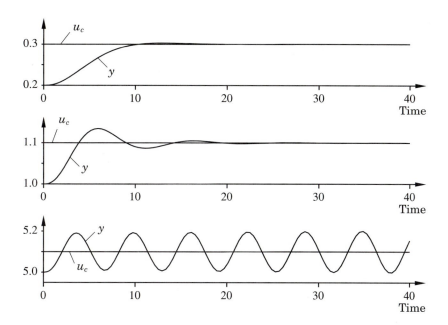

Figure 1.9 Step responses for PI control of the simple flow loop in Example 1.4 at different operating levels. The parameters of the PI controller are $K = 0.15$, $T_i = 1$. The process characteristics are $f(u) = u^4$ and $G_0(s) = 1/(s + 1)^3$.

Flow and Speed Variations

Systems with flows through pipes and tanks are common in process control. The flows are often closely related to the production rate. Process dynamics thus change when the production rate changes, and a controller that is well tuned for one production rate will not necessarily work well for other rates. A simple example illustrates what may happen.

EXAMPLE 1.5   Concentration control

Consider concentration control for a fluid that flows through a pipe, with no mixing, and through a tank, with perfect mixing. A schematic diagram of the process is shown in Fig. 1.10. The concentration at the inlet of the pipe is c_{in}. Let the pipe volume be V_d and let the tank volume be V_m. Furthermore, let the flow be q and let the concentration in the tank and at the outlet be c. A mass balance gives

$$V_m \frac{dc(t)}{dt} = q(t)\,(c_{in}(t - \tau) - c(t)) \tag{1.3}$$

where

$$\tau = V_d/q(t)$$

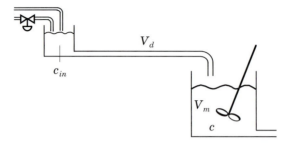

Figure 1.10   Schematic diagram of a concentration control system.

Introduce

$$T = V_m/q(t) \tag{1.4}$$

For a fixed flow, that is, when $q(t)$ is constant, the process has the transfer function

$$G_0(s) = \frac{e^{-s\tau}}{1 + sT} \tag{1.5}$$

The dynamics are characterized by a time delay and first-order dynamics. The time constant T and the time delay τ are inversely proportional to the flow q.

The closed-loop system is as in Fig. 1.8 with $f(\cdot) = 1$ and $G_0(s)$ given by Eq. (1.5). A controller will first be designed for the nominal case, which corresponds to $q = 1$, $T = 1$, and $\tau = 1$. A PI controller with gain $K = 0.5$ and integration time $T_i = 1.1$ gives a closed-loop system with good performance in this case. Figure 1.11 shows the step responses of the closed-loop system for different flows and the corresponding control actions. The overshoot will increase with decreasing flows, and the system will become sluggish when the flow increases. For safe operation it is thus good practice to tune the controller at the lowest flow. Figure 1.11 shows that the system can easily cope with a flow change of ±10% but that the performance deteriorates severely when the flow changes by a factor of 2.   □

Variations in speed give rise to similar problems. This happens for example in rolling mills and paper machines.

Flight Control

The dynamics of an airplane change significantly with speed, altitude, angle of attack, and so on. Control systems such as autopilots and stability augmentation systems were used early. These systems were based on linear feedback with constant coefficients. This worked well when speeds and altitudes were low, but difficulties were encountered with increasing speed and altitude. The problems became very pronounced at supersonic flight. Flight control was one of the strong driving forces for the early development of adaptive control. The

(a)

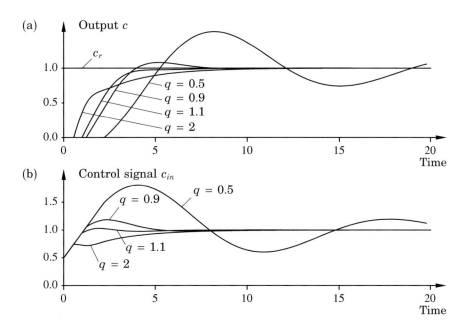

(b)

Figure 1.11   Change in reference value for different flows for the system in Example 1.5. (a) Output c and reference c_r concentration, (b) control signal.

following example from Ackermann (1983) illustrates the variations in dynamics that can be encountered. The variations can be even larger for aircraft with larger variations in flight regimes.

EXAMPLE 1.6   **Short-period aircraft dynamics**

A schematic diagram of an airplane is given in Fig. 1.12. To illustrate the effect of parameter variations, we consider the pitching motion of the aircraft. Introduce the pitch angle θ. Choose normal acceleration N_z, pitch rate $q = \dot{\theta}$,

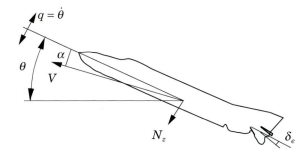

Figure 1.12   Schematic diagram of the aircraft in Example 1.6.

and elevon angle δ_e as state variables and the input to the elevon servo as the input signal u. The following model is obtained if we assume that the aircraft is a rigid body:

$$\frac{dx}{dt} = \begin{pmatrix} a_{11} & a_{12} & a_{13} \\ a_{21} & a_{22} & a_{23} \\ 0 & 0 & -a \end{pmatrix} x + \begin{pmatrix} b_1 \\ 0 \\ a \end{pmatrix} u \qquad (1.6)$$

where $x^T = \begin{pmatrix} N_z & \dot{\theta} & \delta_e \end{pmatrix}$. This model is called short-period dynamics. The parameters of the model given depend on the operating conditions, which can be described in terms of Mach number and altitude; see Fig. 1.13, which shows the flight envelope.

Table 1.1 shows the parameters for the four flight conditions (FC) indicated in Fig. 1.13. The data applies to the supersonic aircraft F4-E. The system has three eigenvalues. One eigenvalue, $-a = -14$, which is due to the elevon servo, is constant. The other eigenvalues, λ_1 and λ_2, depend on the flight conditions. Table 1.1 shows that the system is unstable for subsonic speeds (FC 1, 2, and 3) and stable but poorly damped for the supersonic condition FC 4. Because of these variations it is not possible to use a controller with the same parameters for all flight conditions. The operating condition is determined from air data sensors that measure altitude and Mach number. The controller parameters are then changed as a function of these parameters. How this is done is discussed in Chapter 9.

Much more complicated models will have to be considered in practice because the airframe is elastic and will bend. Notch prefilters on the command signal from the pilot are also used so that the control actions will not excite the bending modes of the airplane. □

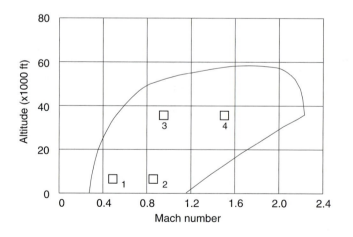

Figure 1.13   Flight envelope of the F4-E. Four different flight conditions are indicated. (From Ackermann (1983), courtesy of Springer-Verlag.)

Table 1.1   Parameters of the airplane state model of Eq. (1.6) for different flight conditions (FC).

	FC 1	FC 2	FC 3	FC 4
Mach	0.5	0.85	0.9	1.5
Altitude (feet)	5000	5000	35000	35000
a_{11}	−0.9896	−1.702	−0.667	−0.5162
a_{12}	17.41	50.72	18.11	26.96
a_{13}	96.15	263.5	84.34	178.9
a_{21}	0.2648	0.2201	0.08201	−0.6896
a_{22}	−0.8512	−1.418	−0.6587	−1.225
a_{23}	−11.39	−31.99	−10.81	−30.38
b_1	−97.78	−272.2	−85.09	−175.6
λ_1	−3.07	−4.90	−1.87	−0.87 ± 4.3i
λ_2	1.23	1.78	0.56	

Variations in Disturbance Characteristics

So far, we have discussed effects of variations in process dynamics. There are also situations in which the key issue is variations in disturbance characteristics. Two examples follow.

EXAMPLE 1.7   Ship steering

A key problem in the design of an autopilot for ship steering is to compensate for the disturbing forces that act on the ship because of wind, waves, and current. The wave-generated forces are often the dominating forces. Waves have strong periodic components. The dominating wave frequency may change by a factor of 3 when the weather conditions change from light breeze to fresh gale. The frequency of the forces generated by the waves will change much more because it is also influenced by the velocity and heading of the ship. Examples of wave height and spectra for two weather conditions are shown in Fig. 1.14. It seems natural to take the nature of the wave disturbances into account in designing autopilots and roll dampers. Since the wave-induced forces change so much, it seems natural to adjust the controller parameters to cope with the disturbance characteristics.   □

Positioning of ships and platforms is another example that is similar to ship steering. In this case the control system will typically have less control authority. This means that the platform to a greater extent has to "ride the waves" and can compensate only for a low-frequency component of the disturbances. This makes it even more critical to have a model for the disturbance pattern.

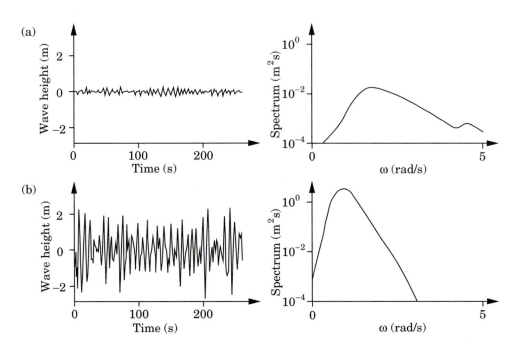

Figure 1.14 Measurements and spectra of waves at different weather conditions at Hoburgen, Sweden. (a) Wind speed 3–4 m/s. (b) Wind speed 18–20 m/s. (Courtesy of SSPA Maritime Consulting AB, Sweden.)

In process control the key issue is often to perform accurate regulation. For important quality variables, even moderate reductions in the fluctuation of a quality variable can give substantial savings. If the disturbances have some statistical regularity, it is possible to obtain significant improvements in control quality by having a controller that is tuned to the particular character of the disturbance. Such controllers can give much better performance than standard PI controllers. The consequences of compensating for disturbances are illustrated by an example.

EXAMPLE 1.8   **Regulation of a quality variable in process control**

Consider regulation of a quality variable of an industrial process in which there are disturbances whose characteristics are changing. A block diagram of the system is shown in Fig. 1.15. In the experiment it is assumed that the process dynamics are first order with time constant $T = 1$. It is assumed that the disturbance acts on the process input. The disturbance is simulated by sending white noise through a band-pass filter. The process dynamics are constant, but the frequency of the band-pass filter changes. Regulation can be done by a PI controller, but performance can be improved significantly by using a more complex controller that is tuned to the disturbance character. Such a

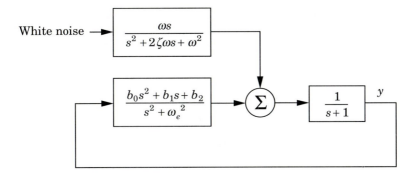

Figure 1.15   Block diagram of the system with disturbances used in Example 1.8.

controller has a very high gain at the center frequency of the disturbance. Figure 1.16 shows the control error under different conditions. The center frequency of the band-pass filter used to generate the disturbance is ω, and the corresponding value used in the design of the controller is ω_e. In Fig. 1.16(a) we show the control error obtained when the controller is tuned to the disturbance,

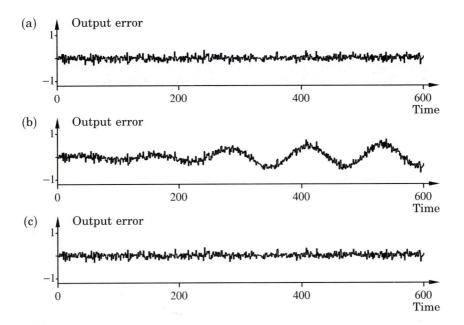

Figure 1.16   Illustrates performance of controllers that are tuned to the disturbance characteristics. Output error when (a) $\omega = \omega_e = 0.1$; (b) $\omega = 0.05$, $\omega_e = 0.1$; (c) $\omega = \omega_e = 0.05$.

that is, $\omega_e = \omega = 0.1$. In Fig. 1.16(b) we illustrate what happens when the disturbance properties change. Parameter ω is changed to 0.05, while $\omega_e = 0.1$. The performance of the control system now deteriorates significantly. In Fig. 1.16(c) we show the improvement obtained by tuning the controller to the new conditions, that is, $\omega = \omega_e = 0.05$. □

There are many other practical problems of a similar type in which there are significant variations in the disturbance characteristics. Having a controller that can adapt to changing disturbance patterns is particularly important when there is limited control authority or dead time in the process dynamics.

Summary

The examples in this section illustrate some mechanisms that can create variations in process dynamics. The examples have of necessity been very simple to show some of the difficulties that may occur. In some cases it is straightforward to reduce the variations by introducing nonlinear compensations in the controllers. For the nonlinear valve in Example 1.4 it is natural to introduce a nonlinear compensator at the controller output that is the inverse of the valve characteristics. This modification is done in Example 9.1. The variations in flow rate in Example 1.5 can be dealt with in a similar way by measuring the flow and changing the controller parameters accordingly. To compensate for the variations in dynamics in Example 1.6, it is necessary to measure the flight conditions. In Examples 1.7 and 1.8, in which the variations are due to changes in the disturbances, it is not possible to directly relate the variation to a measurable quantity. In these cases it may be very advantageous to use adaptive control.

In practice there are many different sources of variations, and there is usually a mixture of different phenomena. The underlying reasons for the variations are in most cases not fully understood. When the physics of the process is reasonably well known (as for airplanes), it is possible to determine suitable controller parameters for different operating conditions by linearizing the models and using some method for control design. This is the common way to design autopilots for airplanes. System identification is an alternative to physical modeling. Both approaches do, however, require a significant engineering effort.

Most industrial processes are very complex and not well understood; it is neither possible nor economical to make a thorough investigation of the causes of the process variations. Adaptive controllers can be a good alternative in such cases. In other situations, some of the dynamics may be well understood, but other parts are unknown. A typical example is robots, for which the geometry, motors, and gearboxes do not change but the load does change. In such cases it is of great importance to use the available *a priori* knowledge and estimate and adapt only to the unknown part of the process.

1.4 ADAPTIVE SCHEMES

In this section we describe four types of adaptive systems: gain scheduling, model-reference adaptive control, self-tuning regulators, and dual control.

Gain Scheduling

In many cases it is possible to find measurable variables that correlate well with changes in process dynamics. A typical case is given in Example 1.4. These variables can then be used to change the controller parameters. This approach is called *gain scheduling* because the scheme was originally used to measure the gain and then change, that is, schedule, the controller to compensate for changes in the process gain. A block diagram of a system with gain scheduling is shown in Fig. 1.17. The system can be viewed as having two loops. There is an inner loop composed of the process and the controller and an outer loop that adjusts the controller parameters on the basis of the operating conditions. Gain scheduling can be regarded as a mapping from process parameters to controller parameters. It can be implemented as a function or a table lookup.

The concept of gain scheduling originated in connection with the development of flight control systems. In this application the Mach number and the altitude are measured by air data sensors and used as scheduling variables. This was used, for instance, in the X-15 in Fig. 1.2. In process control the production rate can often be chosen as a scheduling variable, since time constants and time delays are often inversely proportional to production rate. Gain scheduling is thus a very useful technique for reducing the effects of parameter variations. Historically, it has been a matter of controversy whether gain scheduling should be considered an adaptive system or not. If we use the informal definition in Section 1.1 that an adaptive system is a controller with

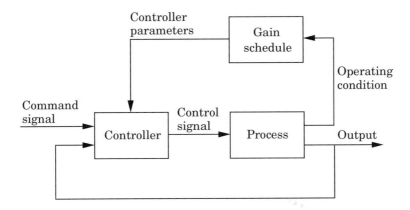

Figure 1.17 Block diagram of a system with gain scheduling.

adjustable parameters and an adjustment mechanism, it is clearly adaptive. An in-depth discussion of gain scheduling is given in Chapter 9.

Model-Reference Adaptive Systems (MRAS)

The *model-reference adaptive system* (MRAS) was originally proposed to solve a problem in which the performance specifications are given in terms of a reference model. This model tells how the process output ideally should respond to the command signal. A block diagram of the system is shown in Fig. 1.18. The controller can be thought of as consisting of two loops. The inner loop is an ordinary feedback loop composed of the process and the controller. The outer loop adjusts the controller parameters in such a way that the error, which is the difference between process output y and model output y_m, is small. The MRAS was originally introduced for flight control. In this case the reference model describes the desired response of the aircraft to joystick motions.

The key problem with MRAS is to determine the adjustment mechanism so that a stable system, which brings the error to zero, is obtained. This problem is nontrivial. The following parameter adjustment mechanism, called the *MIT rule*, was used in the original MRAS:

$$\frac{d\theta}{dt} = -\gamma e \, \frac{\partial e}{\partial \theta} \tag{1.7}$$

In this equation, $e = y - y_m$ denotes the model error and θ is a controller parameter. The quantity $\partial e / \partial \theta$ is the sensitivity derivative of the error with respect to parameter θ. The parameter γ determines the adaptation rate. In practice it is necessary to make approximations to obtain the sensitivity derivative. The MIT rule can be regarded as a gradient scheme to minimize the squared error e^2.

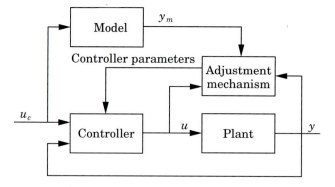

Figure 1.18   Block diagram of a model-reference adaptive system (MRAS).

Self-tuning Regulators (STR)

The adaptive schemes discussed so far are called *direct* methods, because the adjustment rules tell directly how the *controller* parameters should be updated. A different scheme is obtained if the estimates of the *process* parameters are updated and the controller parameters are obtained from the solution of a design problem using the estimated parameters. A block diagram of such a system is shown in Fig. 1.19. The adaptive controller can be thought of as being composed of two loops. The inner loop consists of the process and an ordinary feedback controller. The parameters of the controller are adjusted by the outer loop, which is composed of a recursive parameter estimator and a design calculation. It is sometimes not possible to estimate the process parameters without introducing probing control signals or perturbations. Notice that the system may be viewed as an automation of process modeling and design, in which the process model and the control design are updated at each sampling period. A controller of this construction is called a *self-tuning regulator (STR)* to emphasize that the controller automatically tunes its parameters to obtain the desired properties of the closed-loop system. Self-tuning regulators are discussed in detail in Chapters 3 and 4.

The block labeled "Controller design" in Fig. 1.19 represents an on-line solution to a design problem for a system with known parameters. This is the *underlying design problem.* Such a problem can be associated with most adaptive control schemes, but it is often given indirectly. To evaluate adaptive control schemes, it is often useful to find the underlying design problem, because it will give the characteristics of the system under the ideal conditions when the parameters are known exactly.

The STR scheme is very flexible with respect to the choice of the underlying design and estimation methods. Many different combinations have been

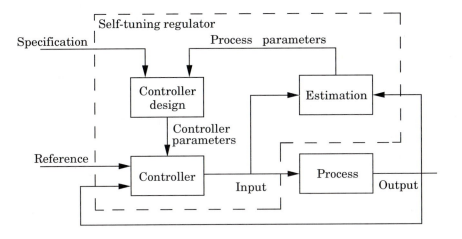

Figure 1.19   Block diagram of a self-tuning regulator (STR).

explored. The controller parameters are updated indirectly via the design calculations in the self-tuner shown in Fig. 1.19. It is sometimes possible to reparameterize the process so that the model can be expressed in terms of the controller parameters. This gives a significant simplification of the algorithm because the design calculations are eliminated. In terms of Fig. 1.19 the block labeled "Controller design" disappears, and the controller parameters are updated directly.

In the STR the controller parameters or the process parameters are estimated in real time. The estimates are then used as if they are equal to the true parameters (i.e., the uncertainties of the estimates are not considered). This is called the *certainty equivalence principle*. In many estimation schemes it is also possible to get a measure of the quality of the estimates. This uncertainty may then be used in the design of the controller. For example, if there is a large uncertainty, one may choose a conservative design. This is discussed in Chapter 7.

Dual Control

The schemes for adaptive control described so far look like reasonable heuristic approaches. Already from their description it appears that they have some limitations. For example, parameter uncertainties are not taken into account in the design of the controller. It is then natural to ask whether there are better approaches than the certainty equivalence scheme. We may also ask whether adaptive controllers can be obtained from some general principles. It is possible to obtain a solution that follows from an abstract problem formulation and use of optimization theory. The particular tool one could use is nonlinear stochastic control theory. This will lead to the notion of *dual control*. The approach will give a controller structure with interesting properties. A major consequence is that the uncertainties in the estimated parameters will be taken into account in the controller. The controller will also take special actions when it has poor knowledge about the process. The approach is so complicated, however, that so far it has not been possible to use it for practical problems. Since the ideas are conceptually useful, we will discuss them briefly in this section.

The first problem that we are faced with is to describe mathematically the idea that a constant or slowly varying parameter is unknown. An unknown constant can be modeled by the differential equation

$$\frac{d\theta}{dt} = 0 \tag{1.8}$$

with an initial distribution that reflects the parameter uncertainty. Parameter drift can be described by adding random variables to the right-hand side of Eq. (1.8). A model of a plant with uncertain parameters is thus obtained by augmenting the state variables of the plant and its environment by the parameter vector whose dynamics is given by Eq. (1.8). Notice that with this

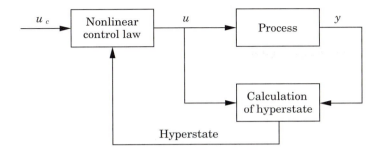

Figure 1.20 Block diagram of a dual controller.

formulation there is no distinction between these parameters and the other state variables. This means that the resulting controller can handle very rapid parameter variations. An augmented state $z = \left(\begin{array}{cc} x^T & \theta^T \end{array} \right)^T$ consisting of the state of the process and the parameters can now be introduced. The goal of the control is then formulated to minimize a loss function

$$V = E \left(G\left(z(T), u(T)\right) + \int_0^T g(z, u)\, dt \right)$$

where E denotes mathematical expectation, u is the control variable, and G and g are scalar functions of z and u. The expectation is taken with respect to the distribution of all initial values and all disturbances appearing in the models of the system. The criterion V should be minimized with respect to admissible controls that are such that $u(t)$ is a function of past and present measurements and the prior distributions. The problem of finding a controller that minimizes the loss function is difficult. By making sufficient assumptions a solution can be obtained by using dynamic programming. The solution is then given in terms of a functional equation that is called the *Bellman equation*. This equation is an extension of the Hamilton-Jacobi equation in the calculus of variations. It is very difficult and time-consuming, if at all possible, to solve the Bellman equation numerically.

Some structural properties are shown in Fig. 1.20. The controller can be regarded as being composed of two parts: a nonlinear estimator and a feedback controller. The estimator generates the conditional probability distribution of the state from the measurements, $p(z|y, u)$. This distribution is called the *hyperstate* of the problem. The feedback controller is a nonlinear function that maps the hyperstate into the space of control variables. This function could be computed off-line. The hyperstate must, however, be updated on-line. The structural simplicity of the solution is obtained at the price of introducing the hyperstate, which is a quantity of very high dimension. Updating of the hyperstate generally requires solution of a complicated nonlinear filtering problem. In simple cases the distribution can be characterized by its mean and covariance, as will be shown in Chapter 7.

The optimal controller sometimes has some interesting properties, which have been found by solving a number of specific problems. It attempts to drive the output to its desired value, but it will also introduce perturbations (probing) when the parameters are uncertain. This improves the quality of the estimates and the future performance of the closed-loop system. The optimal control gives the correct balance between maintaining good control and small estimation errors. The name *dual control* was coined to express this property.

It is interesting to compare the controller in Fig. 1.20 with the self-tuning regulator in Fig. 1.19. In the STR the states are separated into two groups: the ordinary state variables of the underlying constant parameter model and the parameters, which are assumed to vary slowly. The parameter estimator may be considered as an observer for the parameters. Notice that many estimators will also provide estimates of the uncertainties, although this is not used in calculating the control signal. The calculation of the hyperstate in the dual controller gives the conditional distribution of all states and all parameters of the process. The conditional mean value represents estimates, and the conditional covariances give the uncertainties of the estimates. Uncertainties are not used in computing the control signal in the self-tuning regulator. They are important for the dual controller because it may automatically introduce perturbations when the estimates are poor. Dual control is discussed in more detail in Chapter 7.

1.5   THE ADAPTIVE CONTROL PROBLEM

In this section we formulate the adaptive control problem. We do this by giving examples of process models, controller structures, and ways to adapt the controller parameters.

Process Descriptions

In this book the processes will mainly be described by linear single-input, single-output systems. In continuous time the process can be in state space form:

$$\frac{dx}{dt} = Ax + Bu$$
$$y = Cx \tag{1.9}$$

or in transfer function form:

$$G_p(s) = \frac{B(s)}{A(s)} = \frac{b_0 s^m + b_1 s^{m-1} + \cdots b_m}{s^n + a_1 s^{n-1} + \cdots a_n} \tag{1.10}$$

where s is the Laplace transform variable. Notice that A, B, and C are used for matrices as well as polynomials. In normal cases this will not cause any misunderstanding. In ambiguous cases the argument will be used in the poly-nomials.

In discrete time the process can be described in state space form:

$$x(t + 1) = \Phi x(t) + \Gamma u(t)$$
$$y(t) = C x(t)$$

where the sampling interval is taken as the time unit. The discrete time system can also be represented by the pulse transfer function

$$H_p(z) = \frac{B(z)}{A(z)} = \frac{b_0 z^m + b_1 z^{m-1} + \cdots b_m}{z^n + a_1 z^{n-1} + \cdots a_n} \tag{1.11}$$

where z is the z-transform variable.

The parameters, $b_0, b_1, \ldots, b_m, a_1, \ldots, a_n$ of systems (1.10) and (1.11) as well as the orders m, n are often assumed to be unknown or partly unknown.

A Remark on Notation

Throughout this book we need a convenient notation for the time functions obtained in passing signals through linear systems. For this purpose we will use the *differential operator p* $= d/dt$. The output of the system with the transfer function $G(s)$ when the input signal is $u(t)$ will then be denoted by

$$y(t) = G(p)u(t)$$

The output will also depend on the initial conditions. In using the above notation it is assumed that all initial conditions are zero. To deal with discrete time systems, we introduce the *forward shift operator q* defined by

$$qy(t) = y(t + 1)$$

The output of a system with input u and pulse transfer function $H(z)$ is denoted by

$$y(t) = H(q)u(t)$$

In this case it is also assumed that all initial conditions are zero.

Controller Structures

The process is controlled by a controller that has adjustable parameters. It is assumed that there exists some kind of design procedure that makes it possible to determine a controller that satisfies some design criteria if the process and its environment are known. This is called the *underlying design problem*. The adaptive control problem is then to find a method of adjusting the controller when the characteristics of the process and its environment are unknown or changing. In *direct adaptive control* the controller parameters are changed directly without the characteristics of the process and its disturbances first being determined. In *indirect adaptive methods* the process model and possibly the disturbance characteristics are first determined. The controller parameters are designed on the basis of this information.

One key problem is the parameterization of the controller. A few examples are given to illustrate this.

EXAMPLE 1.9   **Adjustment of gains in a state feedback**

Consider a single-input, single-output process described by Eq. (1.9). Assume that the order n of the process is known and that the controller is described by

$$u = -Lx$$

In this case the controller is parameterized by the elements of the matrix L.

□

EXAMPLE 1.10   **A general linear controller**

A general linear controller can be described by

$$R(s)U(s) = -S(s)Y(s) + T(s)U_c(s)$$

where R, S, and T are polynomials and U, Y, and U_c are the Laplace transform of the control signal, the process output, and the reference value, respectively. Several design methods are available to determine the parameters in the controller when the system is known.

□

In Examples 1.9 and 1.10 the controller is linear. Of course, parameters can also be adjusted in nonlinear controllers. A common example is given next.

EXAMPLE 1.11   **Adjustment of a friction compensator**

Friction is common in all mechanical systems. Consider a simple servo drive. Friction can to some extent be compensated for by adding the signal u_{fc} to a controller, where

$$u_{fc} = \begin{cases} u_+ & \text{if } v > 0 \\ -u_- & \text{if } v < 0 \end{cases}$$

where v is the velocity. The signal attempts to compensate for Coulomb friction by adding a positive control signal u_+ when the velocity is positive and subtracting u_- when the velocity is negative. The reason for having two parameters is that the friction forces are typically not symmetrical. Since there are so many factors that influence friction, it is natural to try to find a mechanism that can adjust the parameters u_+ and u_- automatically.

□

The Adaptive Control Problem

An adaptive controller has been defined as a controller with adjustable parameters and a mechanism for adjusting the parameters. The construction of an adaptive controller thus contains the following steps:

- Characterize the desired behavior of the closed-loop system.
- Determine a suitable control law with adjustable parameters.
- Find a mechanism for adjusting the parameters.
- Implement the control law.

In this book, different ways to derive the adjustment rule will be discussed.

1.6   APPLICATIONS

There have been a number of applications of adaptive feedback control since the mid-1950s. The early experiments, which used analog implementations, were plagued by hardware problems. Systems implemented by using minicomputers appeared in the early 1970s. The number of applications has increased drastically with the advent of the microprocessor, which has made the technology cost-effective. Adaptive techniques have been used in regular industrial controllers since the early 1980s. Today, a large number of industrial control loops are under adaptive control. These include a wide range of applications in aerospace, process control, ship steering, robotics, and automotive and biomedical systems. The applications have shown that there are many cases in which adaptive control is very useful, others in which the benefits are marginal, and yet others in which it is inappropriate. On the basis of the products and their uses, it is clear that adaptive techniques can be used in many different ways. In this section we give a brief discussion of some applications. More details are given in Chapter 12.

Automatic Tuning

The most widespread applications are in automatic tuning of controllers. By automatic tuning we mean that the parameters of a standard controller, for instance a PID controller, are tuned automatically at the demand of the operator. After the tuning, the parameters are kept constant. Practically all controllers can benefit from tools for automatic tuning. This will drastically simplify the use of controllers. Practically all adaptive techniques can be used for automatic tuning. There are also many special techniques that can be used for this purpose. Single-loop controllers and distributed systems for process control are important application areas. Most of these controllers are of the PID type. This is a vast application area because there are millions of controllers of this type in use. Many of them are poorly tuned.

Although automatic tuning is currently widely used in simple controllers, it is also beneficial for more complicated controllers. It is in fact a prerequisite for the widespread use of more advanced control algorithms. A mechanism for automatic tuning is often necessary to get the correct time scale and to find a starting value for a more complex adaptive controller. The main advantage of using an automatic tuner is that it simplifies tuning drastically and thus contributes to improved control quality. Tuners have also been developed for other standard applications such as motor control. This is also a case in which a fairly standardized system has to be applied to a wide variety of applications.

Gain Scheduling

Gain scheduling is a powerful technique that is straightforward and easy to use. The key problem is to find suitable scheduling variables, that is, variables that characterize the operating conditions (see Fig. 1.17). It may also be a significant engineering effort to determine the schedules. This effort can be reduced significantly by using automatic tuning because the schedules can then be determined experimentally. Auto-tuning or adaptive algorithms may be used to build gain schedules. A scheduling variable is first determined. Its range is quantized into a number of discrete operating conditions. The controller parameters are determined by automatic tuning when the system is running in one operating condition. The parameter values are stored in a table. The procedure is repeated until all operating conditions are covered. In this way it is easy to install and tune gain scheduling into a computer-controlled system. The only facility required is a table for storing and recalling controller parameters.

Gain scheduling is the standard technique used in flight control systems for high-performance aircrafts. An example is given in Fig. 1.21. A massive engineering effort is required to develop such systems. Gain scheduling is increasingly being used for industrial process control. A combination with automatic tuning makes it possible to significantly reduce the engineering effort in developing the systems.

Continuous Adaptation

There are several cases in which the process or the disturbance characteristics are changing continuously. Continuous adaptation of controller parameters is then needed. The MRAS and the STR are the most common approaches for parameter adjustment. There are many different ways to use the techniques. In some cases, it is natural to assume that the process is described by a general linear model. In other cases, parts of the model are known and only a few parameters are adjusted. In many situations it is possible to measure the disturbances acting on a system. A typical example is climate control in houses in which the outdoor temperature can be measured. The process of using the

Figure 1.21 Gain scheduling is an important ingredient in modern flight control systems. (By courtesy of Nawrocki Stock Photo, Inc., Neil Hargreave.)

measurable disturbance and compensating for its influence is called *feedforward*. Adaptation of feedforward compensators has been found particularly beneficial. One reason for this is that feedforward control requires good models. Another is that it is difficult and time consuming to tune feedforward loops because it is necessary to wait for a proper disturbance to appear. Adaptation is thus almost a prerequisite for using feedforward control.

Since adaptive control is a relatively new technology, there is limited experience of its use in products. One observation that has been made is that the human-machine interface is very important. Adaptive controllers also have their own parameters, which must be chosen. It has been our experience that controllers without any externally adjusted parameters can be designed for specific applications in which the purpose of control can be stated *a priori*. Autopilots for missiles and ships are typical examples. However, in many cases it is not possible to specify the purpose of control *a priori*. It is at least necessary to tell the controller what it is expected to do. This can be done by introducing dials that give the desired properties of the closed-loop system. Such dials are *performance-related*. New types of controllers can be designed by using this concept. For example, it is possible to have a controller with one dial, labeled with the desired closed-loop bandwidth. This is very convenient for applications to motor control. Another possibility would be to have a controller with a dial labeled with the weighting between state deviation and control action in a quadratic optimization problem. Adaptation can also be combined with gain scheduling. A gain schedule can be used to get the parameters quickly into the correct region, and adaptation can then be used for fine-tuning. On the whole it appears that there is significant room for engineering ingenuity in the packaging of adaptive techniques.

Abuses of Adaptive Control

An adaptive controller, being inherently nonlinear, is more complicated than a fixed-gain controller. Before attempting to use adaptive control, it is therefore important to investigate whether the control problem might be solved by constant-gain feedback. In the literature on adaptive control there are many cases in which constant-gain feedback can do as well as an adaptive controller. This is one reason why we are discussing alternatives to adaptive control in this book. One way to proceed in deciding whether adaptive control should be used is sketched in Fig. 1.22.

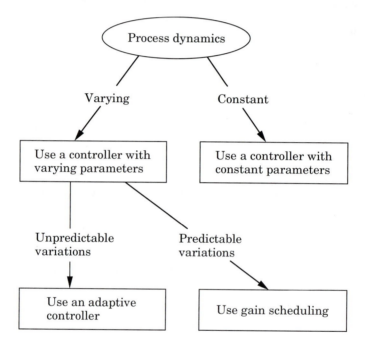

Figure 1.22   Procedure to decide what type of controller to use.

Industrial Products

The industrial products can, broadly speaking, be divided into three different categories: standard controllers, distributed control systems, and dedicated special-purpose systems.

Standard controllers form the largest category. They are typically based on some version of the PID algorithm. Currently, there is very vigorous development of these systems, which are manufactured in large quantities. Practically all new single-loop controllers introduced use some form of adaptation. Many different schemes are used. The single-loop controller is in fact becoming a proving ground for adaptive control. One example is shown in Fig. 1.23. This

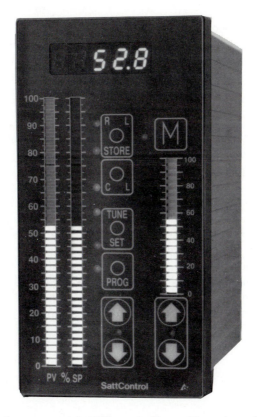

Figure 1.23 A commercial PID controller with automatic tuning, gain scheduling, and feedforward (SattControl Instruments ECA50). Tuning is performed on operator demand when the tune button is pushed. (By courtesy of SattControl Instrument.)

system has automatic tuning of the PID controller. The controller also has feedforward and gain scheduling. The automatic tuning is implemented in such a way that the user only has to push a button to execute the tuning.

A standard controller may be regarded as automation of the actions of a process operator. The controller shown in Fig. 1.23 may be viewed as the next level of automation, in which the actions of an instrument engineer are automated.

Distributed control systems are general-purpose systems primarily for process control applications. These systems may be viewed as a toolbox for implementing a wide variety of control systems. Typically, in addition to tools for PID control, alarm, and startup, more advanced control schemes are also incorporated. Adaptive techniques are now being introduced in the distributed systems, although the rate of development is not as rapid as for single-loop controllers.

There are many special-purpose systems for adaptive control. The applications range from space vehicles to automobiles and consumer electronics. The spacecraft Gemini, for example, has an adaptive notch filter and adaptive friction compensation. The following is another example of an adaptive controller.

EXAMPLE 1.12   An adaptive autopilot for ship steering

This is an example of a dedicated system for a special application. The adaptive autopilot is superior to a conventional autopilot for two reasons: It gives better performance, and it is easier to operate. A conventional autopilot has three dials, which have to be adjusted over a continuous scale. The adaptive autopilot has a performance-related switch with two positions (tight steering and economic propulsion). In the tight steering mode the autopilot gives good, fast response to commands with no consideration for propulsion efficiency. In the economic propulsion mode the autopilot attempts to minimize the steering loss. The control performance is significantly better than that of a well-adjusted conventional autopilot, as shown in Fig. 1.24. The figure shows heading deviations and rudder motions for an adaptive autopilot and a conventional autopilot. The experiments were performed under the same weather conditions. Notice that the heading deviations for the adaptive autopilot are much smaller than those for the conventional autopilot but that the rudder motions are of the same magnitude. The adaptive autopilot is better because

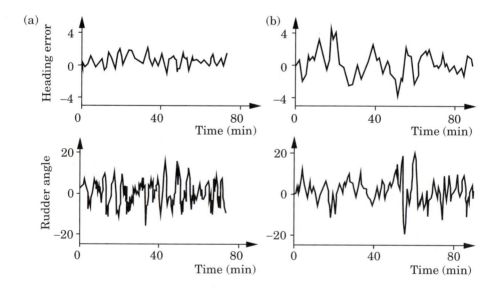

Figure 1.24   The figure shows the variations in heading and the corresponding rudder motions of a ship. (a) Adaptive autopilot. (b) Conventional autopilot based on a PID-like algorithm.

it uses a more complicated control law, which has eight parameters instead of three for the conventional autopilot. For example, the adaptive autopilot has an internal model of the wave motion. If the adaptation mechanism is switched off, the constant parameter controller obtained will perform well for a while, but its performance will deteriorate as the conditions change. Since it is virtually impossible to adjust eight parameters manually, adaptation is a necessity for using such a controller. The adaptive autopilot is discussed in more detail in Chapter 12.   □

The next example illustrates a general-purpose adaptive system.

EXAMPLE 1.13   **Novatune**

The first general-purpose adaptive system was Novatune, announced by the Swedish company Asea in 1982. The system can be regarded as a software-configured toolbox for solving control problems. It broke with conventional process control by using a general-purpose discrete-time pulse transfer function as the building block. The system also has elements for conventional PI and PID control, lead-lag filter, logic, sequencing, and three modules for adaptive control. It has been used to implement control systems for a wide range of process control problems. The advantage of the system is that the control system designer has a simple means of introducing adaptation. The adaptive controller is now incorporated in ABB Master (see Chapter 12).   □

1.7   CONCLUSIONS

The purpose of this chapter has been to introduce the notion of adaptive control, to describe some adaptive systems, and to indicate why adaptation is useful. An adaptive controller was defined as a controller with adjustable parameters and a mechanism for adjusting the parameters.

The key new element is the parameter adjustment mechanisms. Five ways of doing this were discussed: gain scheduling, auto tuning, model-reference adaptive control, self-tuning control, and dual control. To present a balanced account and to give the knowledge required to make complete systems, all aspects of the adaptive problem will be discussed in the book.

Some reasons for using adaptive control have also been discussed in this chapter. The key factors are

- variations in process dynamics,
- variations in the character of the disturbances, and
- engineering efficiency and ease of use.

Examples of mechanisms that cause variations in process dynamics have been given. The examples are simplistic; in many real-life problems it is difficult

to describe the mechanisms analytically. Variations in the character of disturbances is another strong reason for using adaptation.

Adaptive control is not the only way to deal with parameter variations.
Robust control is an alternative. A robust controller is a controller that can
satisfactorily control a class of system with specified uncertainties in the process model. To have a balanced view of adaptive techniques, it is therefore
necessary to know these methods as well (see Chapter 10). Notice particularly that there are few alternatives to adaptation for feedforward control of
processes with varying dynamics.

Engineering efficiency is an often overlooked argument in the choice between different techniques. It may be advantageous to trade engineering efforts against more "intelligence" in the controller. This tradeoff is one reason
for the success of automatic tuning. When a control loop can be tuned simply
by pushing a button, it is easy to commission control systems and to keep them
running well. This also makes it possible to use a more complex controller like
feedforward. With toolboxes for adaptive control (such as ABB Master) it is
often a simple matter to configure an adaptive control system and to try it experimentally. This can be much less time-consuming than the alternative path
of modeling, design, and implementation of a conventional control system. The
knowledge required to build and use toolboxes for adaptive control is given
in the chapters that follow. It should be emphasized that typical industrial
processes are so complex that the parameter variations cannot be determined
from first principles.

A more complex controller may be used on different processes, and the development expenses can be shared by many applications. However, it should be
pointed out that the use of an adaptive controller will not replace good process
knowledge, which is still needed to choose the specifications, the structure of
the controller, and the design method.

PROBLEMS

1.1 Look up the definitions of "adaptive" and "learning" in a good dictionary.
Compare the uses of the words in different fields.

1.2 Find descriptions of adaptive controllers from some manufacturers and
browse through them.

1.3 Give some situations in which adaptive control may be useful. What
factors would you consider when judging the need for adaptive control?

1.4 Make an assessment of the field of adaptive control by making a literature
search. Look for the distribution of publications on adaptive control over
the years. Can you see some pattern in the publications concerning uses
of different methods, emphasis on theory and applications, and so on?

1.5 The system in Example 1.4 has the following characteristics:

$$G_0(s) = \frac{1}{(s+1)^3}$$

$$f(u) = u^4$$

The PI controller has the gain $K = 0.15$ and the reset time $T_i = 1$. Linearize the equations when the reference values are $u_c = 0.3, 1.1,$ and 5.1. Determine the roots of the characteristic equation in the different cases. Determine a reference value such that the linearized equations just become unstable.

1.6 Consider the concentration control system in Example 1.5. Assume that $V_d = V_m = 1$ and that the nominal flow is $q = 1$. Determine PI controllers with the transfer function

$$K_c \left(1 + \frac{1}{T_i s} \right)$$

that give good closed-loop performance for the flows $q = 0.5, 1,$ and 2. Test the controllers for the nominal flow.

1.7 Consider the following system with two inputs and two outputs:

$$\frac{dx}{dt} = \begin{pmatrix} -1 & 0 & 0 \\ 0 & -3 & 0 \\ 0 & 0 & -1 \end{pmatrix} x + \begin{pmatrix} 1 & 0 \\ 0 & 2 \\ 0 & 1 \end{pmatrix} u$$

$$y = \begin{pmatrix} 1 & 1 & 0 \\ 1 & 0 & 1 \end{pmatrix} x$$

Assume that proportional feedback is introduced around the second loop:

$$u_2 = -k_2 y_2$$

(a) Determine the transfer function from u_1 to y_1, and determine how the steady-state gain depends on k_2.

(b) Simulate the response of y_1 and y_2 when u_1 is a step for different values of k_2.

1.8 A block diagram of a system used for metal cutting on a numerically controlled machine is shown in Fig. 1.25. The machine is equipped with a force sensor, which measures the cutting force. A controller adjusts the feedback to maintain a constant cutting force. The cutting force is approximately given by

$$F = k\,a \left(\frac{v}{N} \right)^{\alpha}$$

where a is the depth of the cut, v is the feed rate, N is the spindle speed, α is a parameter in the range $0.5 < \alpha < 1$, and k is a positive parameter.

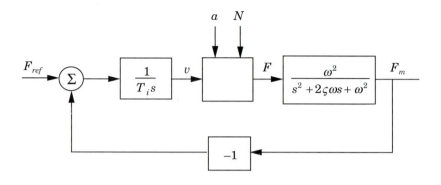

Figure 1.25   Block diagram of a control system for metal cutting.

The steady-state gain from feed rate to force is

$$K = k \alpha a v^{\alpha-1} N^{-\alpha}$$

The gain increases with increasing depth a, decreasing feed rate v, and decreasing spindle speed N. Assume that $\alpha = 0.7$, $k = 1$, $a = 1$, $\zeta = 0.7$, and $\omega = 5$. Determine T_i such that the closed-loop system shows good closed-loop behavior for $N = 1$ and $a = 1$.

(a) Investigate the performance of the closed-loop system when N varies between 0.2 and 2 and $a = 1$.

(b) Repeat part (a) but for a varying between 0.5 and 4 and $N = 1$.

1.9   Consider the system in Fig. 1.26. Let the process be

$$G_0(s) = \frac{K}{s + a}$$

where

$$K = K_0 + \Delta K \qquad K_0 = 1$$
$$a = a_0 + \Delta a \qquad a_0 = 1$$

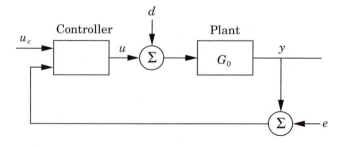

Figure 1.26   Block diagram for Problems 1.9 and 1.10.

and

$$-0.5 \le \Delta K \le 2.0$$
$$-2.0 \le \Delta a \le 2.0$$

Let the ideal closed-loop response be given by

$$Y_m(s) = \frac{1}{s+1} U_c(s)$$

(a) Simulate the open-loop responses for some values of K and a.

(b) Determine a controller for the nominal system such that the difference between step responses of the closed-loop system and of the desired system is less than 1% of the magnitude of the step.

(c) Use the controller from part (b) and investigate the sensitivity to parameter changes.

(d) Use the controller from part (b) and investigate the sensitivity to the disturbance $d(t)$ when

$$d(t) = \begin{cases} -1 & 0 \le t < 6 \\ 2 & 6 \le t < 15 \\ 1 & 15 \le t \end{cases}$$

(e) Use the controller from part (b) and investigate the influence of measurement noise, $e(t)$. Let $e(t)$ be zero mean white noise.

This problem and the next example are based on a special session at the 1988 American Control Conference in Atlanta, Georgia. A detailed discussion of the problem is found in *International Journal of Adaptive Control and Signal Processing*, No. 2, June 1989, which is entirely devoted to the problem.

1.10 Make the same investigation as in Problem 1.9 when the process is

$$G_0(s) = \frac{K}{s^2 + a_1 s + a_2}$$

where

$$K = K_0 + \Delta K \qquad K_0 = 1$$
$$a_1 = a_{10} + \Delta a_1 \qquad a_{10} = 1.4$$
$$a_2 = a_{20} + \Delta a_2 \qquad a_{20} = 1$$

and

$$-0.5 \le \Delta K \le 2.0$$
$$-2.0 \le \Delta a_1 \le 2.0$$
$$-3.0 \le \Delta a_2 \le 3.0$$

Let the desired closed-loop response be given by

$$Y_m(s) = \frac{1}{s^2 + 1.4s + 1} U_c(s)$$

REFERENCES

Many papers, books, and reports have been written on adaptive control. Some of the earlier developments are found in:

Kalman, R. E., 1958. "Design of Self-optimizing Control Systems." *ASME Transactions* **80**: 468–478.

Gregory, P. C., ed., 1959. *Proc. Self Adaptive Flight Control Symposium.* Wright-Patterson Air Force Base, Ohio: Wright Air Development Center.

Bellman, R., 1961. *Adaptive Control—A Guided Tour.* Princeton, N.J.: Princeton University Press.

Mishkin, E., and L. Braun, 1961. *Adaptive Control Systems.* New York: Mc-Graw-Hill.

Tsypkin, Y. Z., 1971. *Adaptation and Learning in Automatic Systems.* New York: Academic Press.

The conference proceedings edited by Gregory is an interesting historical document. The papers and the discussions quoted give a good perspective on early research on adaptive control. Most schemes in the conference are also found in the book by Mishkin and Braun. Bellman's book is still interesting reading. The relation to learning is emphasized both in this book and in the book by Tsypkin. Reprints of many original papers are found in:

Gupta, M. M., ed., 1986. *Adaptive Methods for Control System Design.* New York: IEEE Press.

Narendra, K. S., R. Ortega, and P. Dorato, eds., 1991. *Advances in Adaptive Control.* New York: IEEE Press.

There are several good survey papers on adaptive control:

Åström, K. J., 1983. "Theory and applications of adaptive control—A survey." *Automatica* **19**: 471–486.

Kumar, P. R., 1985. "A survey of some results in stochastic adaptive control." *SIAM J. Control and Opt.* **23**: 329–380.

Seborg, D. E., T. F. Edgar, and S. L. Shah, 1986. "Adaptive control strategies for process control: A survey." *AIChE Journal* **32**: 881–913.

Åström, K. J., 1987. "Adaptive feedback control." *Proc. IEEE* **75**: 185–217.

Ioannou, P. A., and A. Datta, 1991. "Robust Adaptive Control: A Unified Approach." *Proc. IEEE* **79**: 1736–1768.

Among the textbooks in adaptive control we can mention:

Narendra, K. S., and R. V. Monopoli, eds., 1980. *Applications of Adaptive Control.* New York: Academic Press.

Unbehauen, H., ed., 1980. *Methods and Applications in Adaptive Control.* Berlin: Springer-Verlag.

Harris, C. J., and S. A. Billings, 1981. *Self-tuning and Adaptive Control: Theory and Applications.* London: Peter Peregrinus.

Goodwin, G. C., and K. S. Sin, 1984. *Adaptive Filtering Prediction and Control.* Englewood Cliffs, N.J.: Prentice-Hall.

Anderson, B. D. O., R. R. Bitmead, C. R. Johnson, P. V. Kokotovic, R. L. Kosut, I. M.Y. Mareels, L. Praly, and B. D. Riedle, 1986. *Stability of Adaptive Systems: Passivity and Averaging Analysis.* Cambridge, Mass.: MIT Press.

Gawthrop, P. J., 1986. *Continuous Time Self-Tuning Control.* Letchworth, U.K.: Research Studies Press.

Narendra, K. S., and A. M. Annaswamy, 1989. *Stable Adaptive Systems.* Englewood Cliffs, N.J.: Prentice-Hall.

Sastry, S., and M. Bodson, 1989. *Adaptive Control: Stability, Convergence and Robustness.* Englewood Cliffs, N.J.: Prentice-Hall.

Wellstead, P. E., and M. B. Zarrop, 1991. *Selftuning Systems: Control and Signal Processing.* Chichester, U.K.: John Wiley & Sons.

Isermann, R., K.-H. Lachmann, and D. Matko, 1992. *Adaptive Control Systems.* Hemel Hempstead, U.K.: Prentice-Hall International.

Recent developments with particular emphasis on nonlinear systems are discussed in:

Kokotovic, P. V., ed., 1991. *Foundations of Adaptive Control.* Berlin: Springer-Verlag.

There are normally sessions on adaptive control at the major control conferences. The International Federation of Automatic Control is responsible for the Symposium on Adaptive Systems in Control and Signal Processing (ACASP), which is held every third year. The first symposium was held in San Francisco in 1983. These symposia provide up-to-date information about progess in the field. There are few discussions of when to use adaptive control in the literature. Some papers in which this is discussed are:

Åström, K. J., 1980. "Why use adaptive techniques for steering large tankers?" *Int. J. Control* **32**: 689–708.

Jacobs, O. L. R., 1981. "When is adaptive control useful?" *Proceedings Third IMA Conference on Control Theory.* New York: Academic Press.

Flight control systems are usually based on gain scheduling. Feasibility studies of using adaptive control for airplane control are reported in:

IEEE, 1977. "Mini-issue on NASA's advanced control law program for the F-8 DFBW aircraft." *IEEE Trans. Automat. Contr.* **AC-22**: 752–806.

A discussion of adaptive flight control is found in:

Stein, G., 1980. "Adaptive flight control: A pragmatic view." In *Applications of Adaptive Control*, eds. K. S. Narendra and R. V. Monopoli. New York: Academic Press.

The airplane problem in Example 1.6 is taken from:

Ackermann, J., 1983. *Abtastregelung Band II: Entwurf robuster Systeme.* Berlin: Springer-Verlag.

Robust high-gain control is thoroughly discussed in:

Horowitz, I. M., 1963. *Synthesis of Feedback Systems.* New York: Academic Press.

Horowitz's book contains the foundation of feedback control systems synthesis in the frequency domain, including benefits and disadvantages of feedback, parameter-uncertain systems, tolerances and specification, and reasoning about slowly varying parameters. Basic background material for feedback and sensitivity is found in:

Bode, H. W., 1945. *Network Analysis and Feedback Amplifier Design.* New York: Van Nostrand.

Unstructured perturbations are discussed in:

Doyle, J. C., and G. Stein, 1981. "Multivariable feedback design: Concepts for a classical/modern synthesis." *IEEE Trans. Automat. Contr.* **AC-26**: 4–16.

A survey of linear quadratic Gaussian design and its robustness properties is found in:

Stein, G., and M. Athans, 1987. "The LQG/LTR procedure for multivariable feedback control design." *IEEE Trans. Automat. Contr.* **AC-32**: 105–114.

Other references on robustness and sensitivity are:

Zames, G., 1981. "Feedback and optimal sensitivity: Model reference transformations, multiplicative seminorms and approximate inverses." *IEEE Trans. Automat. Contr.* **AC-26**: 301–320.

Zames, G., and B. A. Francis, 1983. "Feedback, minimax sensitivity and optimal robustness." *IEEE Trans. Automat. Contr.* **AC-28**: 585–601.

Morari, M., and E. Zafiriou, 1989. *Robust Process Control.* Englewood Cliffs, N.J.: Prentice-Hall.

Doyle, J. C., B. A. Francis, and A. R. Tannenbaum, 1992. *Feedback Control Theory.* New York: Macmillan.

Doug Camomile
610-323-4829

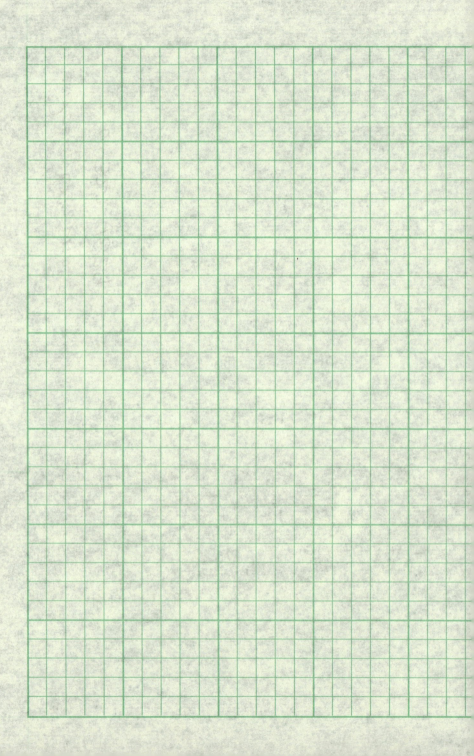

CHAPTER 2

REAL-TIME
PARAMETER ESTIMATION

——————

2.1 INTRODUCTION

On-line determination of process parameters is a key element in adaptive control. A recursive parameter estimator appears explicitly as a component of a self-tuning regulator (see Fig. 1.19). Parameter estimation also occurs implicitly in a model-reference adaptive controller (see Fig. 1.18). This chapter presents some methods for real-time parameter estimation. It is useful to view parameter estimation in the broader context of system identification. The key elements of system identification are selection of model structure, experiment design, parameter estimation, and validation. Since system identification is executed automatically in adaptive systems, it is essential to have a good understanding of all aspects of the problem. Selection of model structure and parameterization are fundamental issues. Simple transfer function models will be used in this chapter. The identification problems are simplified significantly if the models are linear in the parameters.

The experiment design is crucial for successful system identification. In control problems this boils down to selection of the input signal. Choosing an input signal requires some knowledge of the process and the intended use of the model. In adaptive systems there is an additional complication because the input signal to the plant is generated by feedback. In certain cases this does not permit the parameters to be determined uniquely, a situation that has far-reaching consequences. In some cases it may be necessary to introduce perturbation signals, as discussed in more detail in Chapters 6 and 7. In adaptive control the parameters of a process change continuously, so it is necessary to have estimation methods that update the parameters recursively.

In solving identification problems it is very important to validate the results. This is especially important for adaptive systems, in which identification is done automatically. Some validation techniques will therefore be discussed.

The least-squares method is a basic technique for parameter estimation. The method is particularly simple if the model has the property of being *linear in the parameters*. In this case the least-squares estimate can be calculated analytically. A compact presentation of the method of least squares is given in Section 2.2. The formulas for the estimate are derived, and geometric and statistical interpretations are given. It is shown how the computations can be done recursively. In Section 2.3 it is shown how the least-squares method can be used to estimate parameters in dynamical systems. Experimental conditions are discussed in Section 2.4. In particular we introduce the notion of persistent excitation. In using parameter estimation in adaptive control it is useful to have an intuitive insight into the properties of parameter estimators. To start to develop this, we give a number of simulations that illustrate the properties of the different algorithms in Section 2.5. More properties of different estimation schemes are given in Chapter 6 in connection with convergence and stability analysis of adaptive controllers.

2.2   LEAST SQUARES AND REGRESSION MODELS

Karl Friedrich Gauss formulated the principle of least squares at the end of the eighteenth century and used it to determine the orbits of planets and asteroids. Gauss stated that, according to this principle, the unknown parameters of a mathematical model should be chosen in such a way that

> the sum of the squares of the differences between the actually observed and the computed values, multiplied by numbers that measure the degree of precision, is a minimum.

The least-squares method can be applied to a large variety of problems. It is particularly simple for a mathematical model that can be written in the form

$$y(i) = \varphi_1(i)\theta_1^0 + \varphi_2(i)\theta_2^0 + \cdots + \varphi_n(i)\theta_n^0 = \varphi^T(i)\theta^0 \qquad (2.1)$$

where y is the observed variable, $\theta_1^0, \theta_2^0, \ldots, \theta_n^0$ are parameters of the model to be determined, and $\varphi_1, \varphi_2, \ldots, \varphi_n$ are known functions that may depend on other known variables. The vectors

$$\varphi^T(i) = \begin{pmatrix} \varphi_1(i) & \varphi_2(i) & \cdots & \varphi_n(i) \end{pmatrix}$$

$$\theta^0 = \begin{pmatrix} \theta_1^0 & \theta_2^0 & \cdots & \theta_n^0 \end{pmatrix}^T$$

have also been introduced. The model is indexed by the variable i, which often denotes time. It will be assumed initially that the index set is a discrete set. The variables φ_i are called the *regression variables* or the *regressors*, and the model in Eq. (2.1) is also called a *regression model*. Pairs of observations and regressors $\{(y(i), \varphi(i)), i = 1, 2, \ldots, t\}$ are obtained from an experiment. The problem is to determine the parameters in such a way that the outputs computed from the model in Eq. (2.1) agree as closely as possible with the measured variables $y(i)$ in the sense of least squares. That is, the parameter θ should be chosen to minimize the least-squares loss function

$$V(\theta, t) = \frac{1}{2} \sum_{i=1}^{t} \left(y(i) - \varphi^T(i)\theta \right)^2 \tag{2.2}$$

Since the measured variable y is linear in parameters θ^0 and the least-squares criterion is quadratic, the problem admits an analytical solution. Introduce the notations

$$Y(t) = \left(\begin{array}{cccc} y(1) & y(2) & \ldots & y(t) \end{array} \right)^T$$

$$\mathcal{E}(t) = \left(\begin{array}{cccc} \varepsilon(1) & \varepsilon(2) & \ldots & \varepsilon(t) \end{array} \right)^T$$

$$\Phi(t) = \left(\begin{array}{c} \varphi^T(1) \\ \vdots \\ \varphi^T(t) \end{array} \right)$$

$$P(t) = \left(\Phi^T(t)\Phi(t) \right)^{-1} = \left(\sum_{i=1}^{t} \varphi(i)\varphi^T(i) \right)^{-1} \tag{2.3}$$

where the *residuals* $\varepsilon(i)$ are defined by

$$\varepsilon(i) = y(i) - \hat{y}(i) = y(i) - \varphi^T(i)\theta$$

With these notations the loss function (2.2) can be written as

$$V(\theta, t) = \frac{1}{2} \sum_{i=1}^{t} \varepsilon^2(i) = \frac{1}{2} \mathcal{E}^T \mathcal{E} = \frac{1}{2} \|\mathcal{E}\|^2$$

where $\mathcal{E}$ can be written as

$$\mathcal{E} = Y - \hat{Y} = Y - \Phi\theta \tag{2.4}$$

The solution to the least-squares problem is given by the following theorem.

THEOREM 2.1 Least-squares estimation

The function of Eq. (2.2) is minimal for parameters $\hat{\theta}$ such that

$$\Phi^T \Phi \hat{\theta} = \Phi^T Y \tag{2.5}$$

If the matrix $\Phi^T \Phi$ is nonsingular, the minimum is unique and given by

$$\hat{\theta} = (\Phi^T \Phi)^{-1} \Phi^T Y \tag{2.6}$$

Proof: The loss function of Eq. (2.2) can be written as

$$2V(\theta, t) = \mathcal{E}^T \mathcal{E} = (Y - \Phi\theta)^T (Y - \Phi\theta)$$
$$= Y^T Y - Y^T \Phi\theta - \theta^T \Phi^T Y + \theta^T \Phi^T \Phi\theta \qquad (2.7)$$

Since the matrix $\Phi^T \Phi$ is always nonnegative definite, the function V has a minimum. The loss function is quadratic in θ. The minimum can be found in many ways. One way is to determine the gradient of Eq. (2.7) with respect to θ. (See Problem 2.1 at the end of the chapter). The gradient is zero when Eq. (2.5) is satisfied. Another way to find the minimum is by completing the square. We get

$$2V(\theta, t) = Y^T Y - Y^T \Phi\theta - \theta^T \Phi^T Y + \theta^T \Phi^T \Phi\theta$$

$$+ Y^T \Phi(\Phi^T \Phi)^{-1} \Phi^T Y - Y^T \Phi(\Phi^T \Phi)^{-1} \Phi^T Y$$

$$= Y^T \left(I - \Phi(\Phi^T \Phi)^{-1} \Phi^T \right) Y$$

$$+ \left(\theta - (\Phi^T \Phi)^{-1} \Phi^T Y \right)^T \Phi^T \Phi \left(\theta - (\Phi^T \Phi)^{-1} \Phi^T Y \right) \qquad (2.8)$$

The first term on the right-hand side is independent of θ. The second term is always positive. The minimum is obtained for

$$\theta = \hat{\theta} = (\Phi^T \Phi)^{-1} \Phi^T Y$$

and the theorem is proven.   □

Remark 1.   Equation (2.5) is called the *normal equation*. Equation (2.6) can be written as

$$\hat{\theta}(t) = \left(\sum_{i=1}^{t} \varphi(i) \varphi^T(i) \right)^{-1} \left(\sum_{i=1}^{t} \varphi(i) y(i) \right) = P(t) \left(\sum_{i=1}^{t} \varphi(i) y(i) \right) \qquad (2.9)$$

Remark 2.   The condition that the matrix $\Phi^T \Phi$ is invertible is called an *excitation* condition.

Remark 3.   The least-squares criterion weights all errors $\varepsilon(i)$ equally, and this corresponds to the assumption that all measurements have the same precision.   □

Different weighting of the errors can be accounted for by changing the loss function (2.2) to

$$V = \frac{1}{2} \mathcal{E}^T W \mathcal{E} \qquad (2.10)$$

where W is a diagonal matrix with the weights in the diagonal. The least-squares estimate is then given by

$$\hat{\theta} = \left(\Phi^T W \Phi \right)^{-1} \Phi^T W Y \qquad (2.11)$$

EXAMPLE 2.1   **Least-squares estimation of static system**

Consider the system

$$y(i) = b_0 + b_1 u(i) + b_2 u^2(i) + e(i)$$

where $e(i)$ is zero mean Gaussian noise with standard deviation 0.1. The system is linear in the parameters and can be written in the form (2.1) with

$$\varphi^T(i) = \begin{pmatrix} 1 & u(i) & u^2(i) \end{pmatrix}$$
$$\theta^T = \begin{pmatrix} b_0 & b_1 & b_2 \end{pmatrix}$$

The output is measured for the seven different inputs shown by the dots in Fig. 2.1. In practice the model structure is usually unknown, and the user must decide on an appropriate model. We illustrate this by estimating parameters of the following models:

Model 1 : $y(i) \;=\; b_0$

Model 2 : $y(i) \;=\; b_0 + b_1 u$

Model 3 : $y(i) \;=\; b_0 + b_1 u + b_2 u^2$

Model 4 : $y(i) \;=\; b_0 + b_1 u + b_2 u^2 + b_3 u^3$

The different models give a polynomial dependence of different orders between y and u.

Table 2.1 shows the least-squares estimates of the different models together with the resulting loss function. Figure 2.1 also shows the estimated relation between u and y for the different models. From the table it is seen that about the same losses are obtained for Models 3 and 4. The fit to the data points is almost the same for these two models, as is seen in Fig. 2.1.   □

The example shows that it is important to choose the correct model structure to get a good model. With few parameters it is not possible to get a good fit to the data. If too many parameters are used, the fit to the measured data will be very good but the fit to another data set may be very poor. This latter situation is called *overfitting*.

Table 2.1   Least-squares estimates and loss functions for the system in Example 2.1 using different model structures.

Model	$\hat{b}_0$	$\hat{b}_1$	$\hat{b}_2$	$\hat{b}_3$	V
1	3.85				34.46
2	0.57	1.09			1.01
3	1.11	0.45	0.11		0.031
4	1.13	0.37	0.14	-0.003	0.027

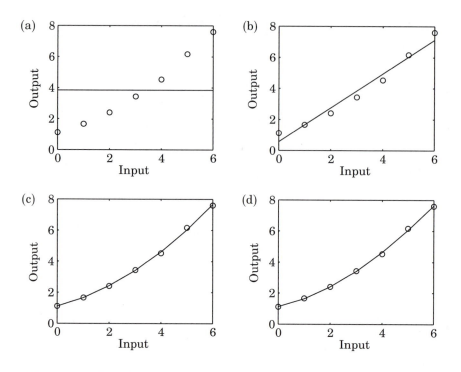

Figure 2.1   The dots represent the measured data points. Resulting models, indicated by the solid lines, based on the least-squares estimates are also given for (a) Model 1, (b) Model 2, (c) Model 3, (d) Model 4.

Geometric Interpretation

The least-squares problem can be interpreted as a geometric problem in R^t, where t is the number of observations. Figure 2.2 illustrates the situation with two parameters and three observations. The vectors φ^1 and φ^2 spans a plane if they are linearly independent. The predicted output $\hat{Y}$ lies in the plan spanned by φ^1 and φ^2. The error $\mathcal{E} = Y - \hat{Y}$ is smallest when $\mathcal{E}$ is orthogonal to this plane. In the general case, Eq. (2.4) can be written as

$$\begin{pmatrix} \varepsilon(1) \\ \varepsilon(2) \\ \vdots \\ \varepsilon(t) \end{pmatrix} = \begin{pmatrix} y(1) \\ y(2) \\ \vdots \\ y(t) \end{pmatrix} - \begin{pmatrix} \varphi_1(1) \\ \varphi_1(2) \\ \vdots \\ \varphi_1(t) \end{pmatrix} \theta_1 - \cdots - \begin{pmatrix} \varphi_n(1) \\ \varphi_n(2) \\ \vdots \\ \varphi_n(t) \end{pmatrix} \theta_n$$

or

$$\mathcal{E} = Y - \varphi^1 \theta_1 - \varphi^2 \theta_2 - \cdots - \varphi^n \theta_n$$

where φ^i are the columns of the matrix Φ. The least-squares problem can thus be interpreted as the problem of finding constants $\theta_1, \dots, \theta_n$ such that

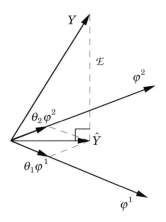

Figure 2.2   Geometric interpretation of the least-squares estimate.

the vector Y is approximated as well as possible by a linear combination of the vectors $\varphi^1, \varphi^2, \ldots, \varphi^n$. Let $\hat{Y}$ be the vector in the span of $\varphi^1, \varphi^2, \ldots, \varphi^n$, which is the best approximation, and let $\mathcal{E} = Y - \hat{Y}$. The vector $\mathcal{E}$ is smallest when it is orthogonal to all vectors φ^i. This gives

$$(\varphi^i)^T \left(Y - \theta_1 \varphi^1 - \theta_2 \varphi^2 - \cdots - \theta_n \varphi^n \right) = 0 \qquad i = 1, \ldots, t$$

which is identical to the normal equation (2.5). The vector θ is unique if the vectors $\varphi^1, \varphi^2, \ldots, \varphi^n$ are linearly independent.

Statistical Interpretation

The least-squares method can be interpreted in statistical terms. It is then necessary to make assumptions about how the data has been generated. Assume that the process is

$$y(i) = \varphi^T(i)\theta^0 + e(i) \tag{2.12}$$

where θ^0 is the vector of "true" parameters and $\{e(i), i = 1, 2, \ldots\}$ is a sequence of independent, equally distributed random variables with zero mean. It is also assumed that e is independent of φ. Equation (2.4) can be written as

$$Y = \Phi\theta^0 + \mathcal{E}$$

Multiplying by $(\Phi^T\Phi)^{-1}\Phi^T$ gives

$$\left(\Phi^T\Phi\right)^{-1}\Phi^T Y = \hat{\theta} = \hat{\theta}^0 + \left(\Phi^T\Phi\right)^{-1}\Phi^T\mathcal{E} \tag{2.13}$$

Provided that $\mathcal{E}$ is independent of Φ^T, which is equivalent to saying that $e(i)$ is independent of $\varphi(i)$, the mathematical expectation of $\hat{\theta}$ is equal to θ^0. An estimate with this property is called unbiased. The following theorem is given without proof.

T H E O R E M   2.2   Statistical properties of least-squares estimation

Consider the estimate in Eq. (2.6) and assume that data is generated from Eq. (2.12), where $\{e(i), i = 1, 2, \ldots\}$ is a sequence of independent random variables with zero mean and variance σ^2. Let E denote mathematical expectation and cov the covariance of a random variable.

   If $\Phi^T \Phi$ is nonsingular, then

(i) $E\hat{\theta}(t) = \theta^0$

(ii) $\text{cov } \hat{\theta}(t) = \sigma^2 (\Phi^T \Phi)^{-1}$

(iii) $\hat{\sigma}^2(t) = 2V(\hat{\theta}, t)/(t - n)$ is an unbiased estimate of σ^2

where n is the number of parameters in θ^0 and $\hat{\theta}$ and t is the number of data points.   ☐

   The theorem states that the estimates are unbiased, that is, $E\hat{\theta}(t) = \theta^0$. Further, it is desirable that an estimate converge to the true parameter value as the number of observations increases toward infinity. This property is called *consistency*. There are several notions of consistency corresponding to different convergence concepts for random variables. Mean square convergence is one possibility, which can be investigated simply by analyzing the variance of the estimate. The result (ii) can be used to determine how the variance of the estimate decreases with the number of observations. This is illustrated by an example.

EXAMPLE 2.2   Decrease of variance

Consider the case in which the model in Eq. (2.12) has only one parameter. Let t be the number of observations. It follows from (ii) of Theorem 2.2 that the variance of the estimate is given by

$$\text{cov } \hat{\theta} = \frac{\sigma^2}{\displaystyle\sum_{k=1}^{t} \varphi^2(k)}$$

Several different cases can now be considered, depending on the asymptotic behavior of $\varphi(k)$ for large k. Introduce the notation $a \sim b$ to indicate that a and b are proportional.

(a) Assume that $\varphi(k) \sim e^{-\alpha k}, \alpha > 0$. The sum in the denominator above then converges, and the variance goes to a constant.

(b) Assume that $\varphi(k) \sim k^{-a}, a > 0$. Then

$$\sum_{k=1}^{t} \varphi^2(k) \sim \begin{cases} \text{const} & a > 0.5 \\ \log t & a = 0.5 \\ t^{1-2a} & a < 0.5 \end{cases}$$

The variance goes to zero if $a \leq 0.5$.

(c) Assume that $\varphi(k) \sim 1$. The variance then goes to zero as $1/t$.

(d) Assume that $\varphi(k) \sim k^a, a > 0$. The variance then goes to zero as $t^{-(1+2a)}$.

(e) Assume that $\varphi(k) \sim e^{\alpha k}, \alpha > 0$. The variance then goes to zero as $e^{-2\alpha t}$.

□

The example shows clearly how the precision of the estimate depends on the rate of growth of the regression vector. The variance does not go to zero with increasing number of observations if the regression variable decreases faster than $1/\sqrt{t}$. In the normal situation, when the regressors are of the same order of magnitude, the variance decreases as $1/t$. The variance decreases more rapidly if the regression variables increase with time.

When several parameters are estimated, the convergence rates may be different for different parameters. This is related to the structure of the matrix $(\Phi^T \Phi)^{-1}$ in Eq. (2.6).

Recursive Computations

In adaptive controllers the observations are obtained sequentially in real time. It is then desirable to make the computations recursively to save computation time. Computation of the least-squares estimate can be arranged in such a way that the results obtained at time $t - 1$ can be used to get the estimates at time t. The solution in Eq. (2.6) to the least-squares problem will be rewritten in a recursive form. Let $\hat{\theta}(t - 1)$ denote the least-squares estimate based on $t - 1$ measurements. Assume that the matrix $\Phi^T \Phi$ is nonsingular for all t. It follows from the definition of $P(t)$ in Eq. (2.3) that

$$
\begin{aligned}
P^{-1}(t) = \Phi^T(t)\Phi(t) &= \sum_{i=1}^{t} \varphi(i)\varphi^T(i) \\
&= \sum_{i=1}^{t-1} \varphi(i)\varphi^T(i) + \varphi(t)\varphi^T(t) \\
&= P^{-1}(t-1) + \varphi(t)\varphi^T(t)
\end{aligned}
\tag{2.14}
$$

The least-squares estimate $\hat{\theta}(t)$ is given by Eq. (2.9):

$$
\hat{\theta}(t) = P(t)\left(\sum_{i=1}^{t} \varphi(i)y(i) \right) = P(t)\left(\sum_{i=1}^{t-1} \varphi(i)y(i) + \varphi(t)y(t) \right)
$$

It follows from Eqs. (2.9) and (2.14) that

$$
\sum_{i=1}^{t-1} \varphi(i)y(i) = P^{-1}(t-1)\hat{\theta}(t-1) = P^{-1}(t)\hat{\theta}(t-1) - \varphi(t)\varphi^T(t)\hat{\theta}(t-1)
$$

The estimate at time t can now be written as

$$\hat{\theta}(t) = \hat{\theta}(t-1) - P(t)\varphi(t)\varphi^T(t)\hat{\theta}(t-1) + P(t)\varphi(t)y(t)$$
$$= \hat{\theta}(t-1) + P(t)\varphi(t)\left(y(t) - \varphi^T(t)\hat{\theta}(t-1)\right)$$
$$= \hat{\theta}(t-1) + K(t)\varepsilon(t)$$

where

$$K(t) = P(t)\varphi(t)$$
$$\varepsilon(t) = y(t) - \varphi^T(t)\hat{\theta}(t-1)$$

The residual $\varepsilon(t)$ can be interpreted as the error in predicting the signal $y(t)$ one step ahead based on the estimate $\hat{\theta}(t-1)$.

To proceed, it is necessary to derive a recursive equation for $P(t)$ rather than for $P(t)^{-1}$ as in Eq. (2.14). The following lemma is useful.

LEMMA 2.1   Matrix inversion lemma

Let A, C, and $C^{-1} + DA^{-1}B$ be nonsingular square matrices. Then $A + BCD$ is invertible, and

$$(A + BCD)^{-1} = A^{-1} - A^{-1}B(C^{-1} + DA^{-1}B)^{-1}DA^{-1}$$

Proof: By direct multiplication we find that

$$(A + BCD)\left(A^{-1} - A^{-1}B(C^{-1} + DA^{-1}B)^{-1}DA^{-1}\right)$$
$$= I + BCDA^{-1} - B(C^{-1} + DA^{-1}B)^{-1}DA^{-1}$$
$$\quad - BCDA^{-1}B(C^{-1} + DA^{-1}B)^{-1}DA^{-1}$$
$$= I + BCDA^{-1} - BC(C^{-1} + DA^{-1}B)(C^{-1} + DA^{-1}B)^{-1}DA^{-1}$$
$$= I \qquad\qquad \square$$

Applying Lemma 2.1 to $P(t)$ and using Eq. (2.14), we get

$$P(t) = \left(\Phi^T(t)\Phi(t)\right)^{-1} = \left(\Phi^T(t-1)\Phi(t-1) + \varphi(t)\varphi^T(t)\right)^{-1}$$
$$= \left(P(t-1)^{-1} + \varphi(t)\varphi^T(t)\right)^{-1}$$
$$= P(t-1) - P(t-1)\varphi(t)\left(I + \varphi^T(t)P(t-1)\varphi(t)\right)^{-1}\varphi^T(t)P(t-1)$$

This implies that

$$K(t) = P(t)\varphi(t) = P(t-1)\varphi(t)\left(I + \varphi^T(t)P(t-1)\varphi(t)\right)^{-1}$$

Notice that a matrix inversion is necessary to compute P. However, the matrix to be inverted is of the same dimension as the number of measurements. That is, for a single output system it is a scalar.

The recursive calculations are summarized in the following theorem.

THEOREM 2.3   Recursive least-squares estimation (RLS)

Assume that the matrix $\Phi(t)$ has full rank, that is, $\Phi^T(t)\Phi(t)$ is nonsingular, for all $t \geq t_0$. Given $\hat{\theta}(t_0)$ and $P(t_0) = (\Phi^T(t_0)\Phi(t_0))^{-1}$, the least-squares estimate $\hat{\theta}(t)$ then satisfies the recursive equations

$$\hat{\theta}(t) = \hat{\theta}(t-1) + K(t)\Big(y(t) - \varphi^T(t)\hat{\theta}(t-1)\Big) \tag{2.15}$$

$$K(t) = P(t)\varphi(t) = P(t-1)\varphi(t)\Big(I + \varphi^T(t)P(t-1)\varphi(t)\Big)^{-1} \tag{2.16}$$

$$P(t) = P(t-1) - P(t-1)\varphi(t)\Big(I + \varphi^T(t)P(t-1)\varphi(t)\Big)^{-1}\varphi^T(t)P(t-1)$$

$$= \Big(I - K(t)\varphi^T(t)\Big)P(t-1) \tag{2.17}$$

$\square$

Remark 1.   Equation (2.15) has strong intuitive appeal. The estimate $\hat{\theta}(t)$ is obtained by adding a correction to the previous estimate $\hat{\theta}(t-1)$. The correction is proportional to $y(t) - \varphi^T(t)\hat{\theta}(t-1)$, where the last term can be interpreted as the value of y at time t predicted by the model of Eq. (2.1). The correction term is thus proportional to the difference between the measured value of $y(t)$ and the prediction of $y(t)$ based on the previous parameter estimate. The components of the vector $K(t)$ are weighting factors that tell how the correction and the previous estimate should be combined.

Remark 2.   The least-squares estimate can be interpreted as a Kalman filter for the process

$$\begin{aligned} \theta(t+1) &= \theta(t) \\ y(t) &= \varphi^T(t)\theta(t) + e(t) \end{aligned} \tag{2.18}$$

Remark 3.   The recursive equations can also be derived by starting with the loss function of Eq. (2.2). Using Eqs. (2.8) and (2.6) gives

$$2V(\theta,t) = 2V(\theta,t-1) + \varepsilon^2(\theta,t)$$

$$= Y^T(t-1)\left(I - \Phi(t-1)\Big(\Phi^T(t-1)\Phi(t-1)\Big)^{-1}\Phi(t-1)\right)Y(t-1)$$

$$+ \Big(\theta - \hat{\theta}(t-1)\Big)^T\Phi^T(t-1)\Phi(t-1)\Big(\theta - \hat{\theta}(t-1)\Big)$$

$$+ \Big(y(t) - \varphi^T(t)\theta\Big)^T\Big(y(t) - \varphi^T(t)\theta\Big) \tag{2.19}$$

The first term on the right-hand side is independent of θ, and the remaining two terms are quadratic in θ. $V(\theta,t)$ can then easily be minimized with respect to θ. $\square$

Notice that the matrix $P(t)$ is defined only when the matrix $\Phi^T(t)\Phi(t)$ is nonsingular. Since

$$\Phi^T(t)\Phi(t) = \sum_{i=1}^{t}\varphi(i)\varphi^T(i)$$

it follows that $\Phi^T\Phi$ is always singular if $t < n$. To obtain an initial condition for P, it is thus necessary to choose $t = t_0$ such that $\Phi^T(t_0)\Phi(t_0)$ is nonsingular. The initial conditions are then

$$P(t_0) = \left(\Phi^T(t_0)\Phi(t_0)\right)^{-1}$$
$$\hat{\theta}(t_0) = P(t_0)\Phi^T(t_0)Y(t_0)$$

The recursive equations can then be used for $t > t_0$. It is, however, often convenient to use the recursive equations in all steps. If the recursive equations are started with the initial condition

$$P(0) = P_0$$

where P_0 is positive definite, then

$$P(t) = \left(P_0^{-1} + \Phi^T(t)\Phi(t)\right)^{-1}$$

Notice that $P(t)$ can be made arbitrarily close to $\left(\Phi^T(t)\Phi(t)\right)^{-1}$ by choosing P_0 sufficiently large.

By using the Kalman filter interpretation of the least-squares method, it may be seen that this way of starting the recursion corresponds to the situation in which the parameters have an initial distribution with mean θ_0 and covariance P_0.

Time-Varying Parameters

In the least-squares model (2.1) the parameters θ_i^0 are assumed to be constant. In several adaptive problems it is of interest to consider the situation in which the parameters are time-varying. Two cases can be covered by simple extensions of the least-squares method. In one such case parameters are assumed to change abruptly but infrequently; in the other case the parameters are changing continuously but slowly. The case of abrupt parameter changes can be covered by *resetting*. The matrix P in the least-squares algorithm (Theorem 2.3) is then periodically reset to αI, where α is a large number. This implies that the gain $K(t)$ in the estimator becomes large and the estimate can be updated with a larger step. A more sophisticated version is to run n estimators in parallel, which are reset sequentially. The estimate is then chosen by using some decision logic. (See Chapter 6.) The case of slowly time-varying parameters can be covered by relatively simple mathematical models. One pragmatic approach is simply to replace the least-squares criterion of Eq. (2.2) with

$$V(\theta, t) = \frac{1}{2}\sum_{i=1}^{t} \lambda^{t-i}\left(y(i) - \varphi^T(i)\theta\right)^2 \tag{2.20}$$

where λ is a parameter such that $0 < \lambda \le 1$. The parameter λ is called the *forgetting factor* or *discounting factor*. The loss function of Eq. (2.20) implies that a time-varying weighting of the data is introduced. The most recent data is given unit weight, but data that is n time units old is weighted by λ^n. The method is therefore called *exponential forgetting* or *exponential discounting*. By repeating the calculations leading to Theorem 2.3 for the loss function of Eq. (2.20), the following result is obtained.

T H E O R E M 2.4   Recursive least squares with exponential forgetting

Assume that the matrix $\Phi(t)$ has full rank for $t \geq t_0$. The parameter θ, which minimizes Eq. (2.20), is given recursively by

$$\hat{\theta}(t) = \hat{\theta}(t-1) + K(t)\Big(y(t) - \varphi^T(t)\hat{\theta}(t-1)\Big)$$

$$K(t) = P(t)\varphi(t) = P(t-1)\varphi(t)\left(\lambda I + \varphi^T(t)P(t-1)\varphi(t)\right)^{-1} \quad (2.21)$$

$$P(t) = \left(I - K(t)\varphi^T(t)\right)P(t-1) / \lambda \qquad \qquad \square$$

A disadvantage of exponential forgetting is that data is discounted even if $P(t)\varphi(t) = 0$. This condition implies that $y(t)$ does not contain any new information about the parameter θ. In this case it follows from Eqs. (2.21) that the matrix P increases exponentially with rate λ. Several ways to avoid this are discussed in detail in Chapter 11.

An alternative method of dealing with time-varying parameters is to assume a time-varying mathematical model. Time-varying parameters can be obtained by replacing the first equation of Eqs. (2.18) with the model

$$\theta(t+1) = \Phi_v\theta(t) + v(t)$$

where Φ_v is a known matrix and $v(t)$ is discrete-time white noise. The filtering interpretation of the least-squares problem given in Remark 2 of Theorem 2.3 can now easily be generalized. The least-squares estimator will then be the Kalman filter. The case $\Phi_v = I$ corresponds to a model in which the parameters are drifting Wiener processes.

Simplified Algorithms

The recursive least-squares algorithm given by Theorem 2.3 has two sets of state variables, $\hat{\theta}$ and P, which must be updated at each step. For large n the updating of the matrix P dominates the computing effort. There are several simplified algorithms that avoid updating the P matrix at the cost of slower convergence. Kaczmarz's *projection algorithm* is one simple solution. To describe this algorithm, consider the unknown parameter as an element of R^n. One measurement

$$y(t) = \varphi^T(t)\theta \qquad (2.22)$$

determines the projection of the parameter vector θ on the vector $\varphi(t)$. From this it is immediately clear that n measurements, where $\varphi(1), \ldots, \varphi(n)$ span

R^n, are required to determine the parameter vector θ uniquely. Assume that an estimate $\hat{\theta}(t-1)$ is available and that a new measurement such as Eq. (2.22) is obtained. Since the measurement $y(t)$ contains information only in the direction $\varphi(t)$ in parameter space, it is natural to choose as the new estimate the value $\hat{\theta}(t)$ that minimizes $\|\hat{\theta}(t) - \hat{\theta}(t-1)\|$ subject to the constraint $y(t) = \varphi^T(t)\hat{\theta}(t)$. Introducing a Lagrangian multiplier $\bar{\alpha}$ to handle the constraint, we thus have to minimize the function

$$V = \frac{1}{2}\left(\hat{\theta}(t) - \hat{\theta}(t-1)\right)^T \left(\hat{\theta}(t) - \hat{\theta}(t-1)\right) + \bar{\alpha}\left(y(t) - \varphi^T(t)\hat{\theta}(t)\right)$$

Taking derivatives with respect to $\hat{\theta}(t)$ and $\bar{\alpha}$, we get

$$\hat{\theta}(t) - \hat{\theta}(t-1) - \bar{\alpha}\varphi(t) = 0$$

$$y(t) - \varphi^T(t)\hat{\theta}(t) = 0$$

Solving these equations gives

$$\hat{\theta}(t) = \hat{\theta}(t-1) + \frac{\varphi(t)}{\varphi^T(t)\varphi(t)}\left(y(t) - \varphi^T(t)\hat{\theta}(t-1)\right) \qquad (2.23)$$

The updating formula is called *Kaczmarz's algorithm*. It is useful to be able to change the step length of the parameter adjustment by introducing a factor γ. This gives

$$\hat{\theta}(t) = \hat{\theta}(t-1) + \frac{\gamma\varphi(t)}{\varphi^T(t)\varphi(t)}\left(y(t) - \varphi^T(t)\hat{\theta}(t-1)\right)$$

To avoid a potential problem that occurs when $\varphi(t) = 0$, the denominator in the correction term is changed from $\varphi^T(t)\varphi(t)$ to $\varphi^T(t)\varphi(t) + \alpha$, where α is a positive constant. The following algorithm is then obtained.

ALGORITHM 2.1   Projection algorithm

$$\hat{\theta}(t) = \hat{\theta}(t-1) + \frac{\gamma\varphi(t)}{\alpha + \varphi^T(t)\varphi(t)}\left(y(t) - \varphi^T(t)\hat{\theta}(t-1)\right) \qquad (2.24)$$

where $\alpha \geq 0$ and $0 < \gamma < 2$.   □

Remark 1.   In some textbooks this is called the normalized projection algorithm.

Remark 2.   The bound for the parameter γ is obtained from the following analysis. Assume that data has been generated by Eq. (2.22) with parameter $\theta = \theta^0$. It then follows from Eq. (2.24) that the parameter error

$$\tilde{\theta} = \theta^0 - \hat{\theta}$$

satisfies the equation

$$\tilde{\theta}(t) = A(t)\tilde{\theta}(t-1)$$

where

$$A(t) = I - \frac{\gamma \varphi(t)\varphi^T(t)}{\alpha + \varphi^T(t)\varphi(t)}$$

The matrix $A(t)$ has one eigenvalue,

$$\lambda = \frac{\alpha + (1 - \gamma)\varphi^T \varphi}{\alpha + \varphi^T \varphi}$$

This value is less than 1 in magnitude if $0 < \gamma < 2$. The other eigenvalues of A are all equal to 1. ☐

The projection algorithm assumes that the data is generated by Eq. (2.22) with no error. When the data is generated by Eq. (2.12) with additional random error, a simplified algorithm is given by

$$\hat{\theta}(t) = \hat{\theta}(t-1) + P(t)\varphi(t)\left(y(t) - \varphi^T(t)\hat{\theta}(t-1)\right) \tag{2.25}$$

where

$$P(t) = \left(\sum_{i=1}^{t} \varphi^T(i)\varphi(i)\right)^{-1} \tag{2.26}$$

This is the *stochastic approximation* (SA) *algorithm*. Notice that $P(t) = \Phi\Phi^T$ is now a scalar when $y(t)$ is a scalar. A further simplification is the *least mean square (LMS) algorithm* in which the parameter updating is done by using

$$\hat{\theta}(t) = \hat{\theta}(t-1) + \gamma\varphi(t)\left(y(t) - \varphi^T(t)\hat{\theta}(t-1)\right)$$

where γ is a constant.

Continuous-Time Models

In the recursive schemes the variables have so far been indexed by a discrete parameter t. The notation t was chosen because in many applications it denotes time. In some cases it is natural to use continuous-time observations. It is straightforward to generalize the results to this case. Equation (2.1) is still used, but i is now assumed to be a real variable. Assuming exponential forgetting, the parameter should be determined such that the criterion

$$V(\theta) = \int_0^t e^{-\alpha(t-\tau)}\left(y(\tau) - \varphi^T(\tau)\theta\right)^2 d\tau \tag{2.27}$$

is minimized. The parameter α, where $\alpha \geq 0$, corresponds to the forgetting factor λ in Eq. (2.20). A straightforward calculation shows that the criterion is minimized if (see Problem 2.15 at the end of the chapter)

$$\left(\int_0^t e^{-\alpha(t-\tau)}\varphi(\tau)\varphi^T(\tau)\,d\tau\right)\hat{\theta}(t) = \int_0^t e^{-\alpha(t-\tau)}\varphi(\tau)y(\tau)\,d\tau \tag{2.28}$$

which is the *normal equation*. The estimate is unique if the matrix

$$R(t) = \int_0^t e^{-\alpha(t-\tau)} \varphi(\tau) \varphi^T(\tau) \, d\tau \tag{2.29}$$

is invertible. It is also possible to obtain recursive equations by differentiating Eq. (2.28). The estimate is given by the following theorem.

THEOREM 2.5   Continuous-time least-squares estimation

Assume that the matrix $R(t)$ given by Eq. (2.29) is invertible for all t. The estimate that minimizes Eq. (2.27) satisfies

$$\frac{d\hat{\theta}}{dt} = P(t)\varphi(t)e(t) \tag{2.30}$$

$$e(t) = y(t) - \varphi^T(t)\hat{\theta}(t) \tag{2.31}$$

$$\frac{dP(t)}{dt} = \alpha P(t) - P(t)\varphi(t)\varphi^T(t)P(t) \tag{2.32}$$

Proof: The theorem is proved by differentiating Eq. (2.28).   □

Remark 1.   The matrix $R(t) = P(t)^{-1}$ satisfies

$$\frac{dR}{dt} = -\alpha R + \varphi \varphi^T$$

Remark 2.   There are also continuous-time versions of the simplified algorithms. The projection algorithm corresponding to Eqs. (2.25) and (2.26) is given by Eq. (2.30) with

$$P(t) = \left(\int_0^t \varphi^T(\tau)\varphi(\tau) \, d\tau \right)^{-1}$$

where $P(t)$ is now a scalar.   □

2.3   ESTIMATING PARAMETERS IN DYNAMICAL SYSTEMS

We now show how the least-squares method can be used to estimate parameters in models of dynamical systems. The particular way to do this will depend on the character of the model and its parameterization.

Finite-Impulse Response (FIR) Models

A linear time-invariant dynamical system is uniquely characterized by its impulse response. The impulse response is in general infinite-dimensional. For stable systems the impulse response will go to zero exponentially fast and may then be truncated. Notice, however, that a large number of parameters may be required if the sampling interval is short in comparison to the slowest time constant of the system. This results in the so-called finite impulse response (FIR) model, which is also called a transversal filter. The model can be described by the equation

$$y(t) = b_1 u(t-1) + b_2 u(t-2) + \cdots + b_n u(t-n) \qquad (2.33)$$

or

$$y(t) = \varphi^T(t-1)\theta$$

where

$$\theta^T = \begin{pmatrix} b_1 & \cdots & b_n \end{pmatrix}$$
$$\varphi^T(t-1) = \begin{pmatrix} u(t-1) & \cdots & u(t-n) \end{pmatrix}$$

This model is identical to the regression model of Eq. (2.1), except for the index t of the regression vector, which is different. The reason for this change of notation is that it will be convenient to label the regression vector with the time of the most recent data that appears in the regressor. The model of Eq. (2.33) clearly fits the least-squares formulation, and the estimator is then given by Theorem 2.3.

The parameter estimator can be represented by the block diagram in Fig. 2.3. The estimator may be regarded as a system with inputs u and y and output θ. Since the signal

$$\hat{y}(t) = \hat{b}_1(t-1)u(t-1) + \cdots + \hat{b}_n(t-1)u(t-n)$$

is available in the system, we can also consider $\hat{y}(t)$ as an output. Since $\hat{y}(t)$ is a predicted estimate of y, the recursive estimator can also be interpreted as an *adaptive filter* to predict y. The use of this filter is discussed in Chapter 13.

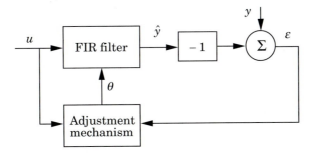

Figure 2.3   Block diagram representation of a recursive parameter estimator for an FIR model.

Transfer Function Models

The least-squares method can be used to identify parameters in dynamical systems. Let the system be described by the model

$$A(q)y(t) = B(q)u(t) \qquad (2.34)$$

where q is the forward shift operator and $A(q)$ and $B(q)$ are the polynomials

$$A(q) = q^n + a_1 q^{n-1} + \ldots + a_n$$
$$B(q) = b_1 q^{m-1} + b_2 q^{m-2} + \ldots + b_m$$

Equation (2.34) can be written as the difference equation

$$y(t) + a_1 y(t-1) + \cdots + a_n y(t-n) = b_1 u(t+m-n-1) + \cdots + b_m u(t-n)$$

Assume that the sequence of inputs $\{u(1), u(2), \ldots, u(t)\}$ has been applied to the system and the corresponding sequence of outputs $\{y(1), y(2), \ldots, y(t)\}$ has been observed. Introduce the parameter vector

$$\theta^T = \begin{pmatrix} a_1 & \ldots & a_n & b_1 & \ldots & b_m \end{pmatrix} \qquad (2.35)$$

and the regression vector

$$\varphi^T(t-1) = \begin{pmatrix} -y(t-1) & \ldots & -y(t-n) & u(t+m-n-1) & \ldots & u(t-n) \end{pmatrix}$$

Notice that the output signal appears delayed in the regression vector. The model is therefore called an *autoregressive model*. The way in which the elements are ordered in the matrix θ is, of course, arbitrary, provided that $\varphi(t-1)$ is also similarly reordered. Later, in dealing with adaptive control, it will be natural to reorder the terms. The convention that the time index of the φ vector will refer to the time when all elements in the vector are available will also be adopted. The model can formally be written as the regression model

$$y(t) = \varphi^T(t-1)\theta$$

Parameter estimates can be obtained by applying the least-squares method (Theorem 2.1). The matrix Φ is given by

$$\Phi = \begin{pmatrix} \varphi^T(n) \\ \vdots \\ \varphi^T(t-1) \end{pmatrix}$$

If we use the statistical interpretation of the least-squares estimate given by Theorem 2.2, it follows that the method described will work well when the disturbances can be described as white noise added to the right-hand side of Eq. (2.34). This leads to the least-squares model

$$A(q)y(t) = B(q)u(t) + e(t+n)$$

(Compare with Eq. (2.12).) The method is therefore called an *equation error* method. A slight variation of the method is better if the disturbances are described instead as white noise added to the system output, that is, when the model is

$$y(t) = \frac{B(q)}{A(q)} u(t) + e(t)$$

WHITE NOISE ADDED

The method obtained is then called an *output error* method. To describe such a method, let u be the input and $\hat{y}$ be the output of a system with the input-output relation

$$\hat{y}(t) + a_1\hat{y}(t-1) + \cdots + a_n\hat{y}(t-n) = b_1 u(t+m-n-1) + \cdots + b_m u(t-n)$$

that is,

$$\hat{y}(t) = \frac{B(q)}{A(q)} u(t)$$

Determine the parameters that minimize the criterion

$$\sum_{k=1}^{t} (y(k) - \hat{y}(k))^2$$

where $y(t) = \hat{y}(t) + e(t)$. This problem can be interpreted as a least-squares problem whose solution is given by

$$\hat{\theta}(t) = \hat{\theta}(t-1) + P(t)\varphi(t-1)\varepsilon(t) \qquad \# 1$$

where

$$\varphi^T(t-1) = \Big(-\hat{y}(t-1) \quad \cdots \quad -\hat{y}(t-n) \quad u(t+m-n-1) \quad \cdots \quad u(t-n) \Big)$$
$$\varepsilon(t) = y(t) - \varphi^T(t-1)\hat{\theta}(t-1) \qquad \# 2$$

Compare with Theorem 2.1. The recursive estimator obtained can be represented by the block diagram in Fig. 2.4.

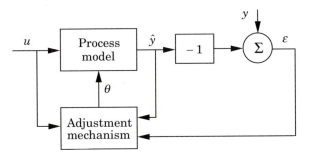

Figure 2.4 Block diagram of a least-squares estimator based on the output error.

Continuous-Time Transfer Functions

We now show that the least-squares method can also be used to estimate parameters in continuous-time transfer functions. For instance, consider a continuous-time model of the form

$$\frac{d^n y}{dt^n} + a_1 \frac{d^{n-1} y}{dt^{n-1}} + \cdots + a_n y = b_1 \frac{d^{m-1} u}{dt^{m-1}} + \cdots + b_m u$$

which can also be written as

$$A(p)y(t) = B(p)u(t) \tag{2.36}$$

where $A(p)$ and $B(p)$ are polynomials in the differential operator $p = d/dt$. In most cases we cannot conveniently compute $p^n y(t)$ because it would involve taking n derivatives of a signal. The model of Eq. (2.36) is therefore rewritten as

$$A(p)y_f(t) = B(p)u_f(t) \tag{2.37}$$

where

$$y_f(t) = H_f(p)y(t)$$
$$u_f(t) = H_f(p)u(t)$$

and $H_f(p)$ is a stable transfer function with a pole excess of n or more. See Fig. 2.5. If we introduce

$$\theta = \left(\begin{array}{ccccccc} a_1 & \cdots & a_n & b_1 & \cdots & b_m \end{array} \right)^T$$
$$\varphi^T(t) = \left(\begin{array}{ccccccc} -p^{n-1}y_f & \cdots & -y_f & p^{m-1}u_f & \cdots & u_f \end{array} \right)$$
$$= \left(\begin{array}{ccccccc} -p^{n-1}H_f(p)y & \cdots & -H_f(p)y & p^{m-1}H_f(p)u & \cdots & H_f(p)u \end{array} \right)$$

the model expressed by Eq. (2.37) can be written as

$$p^n y_f(t) = p^n H_f(p)y(t) = \varphi^T(t)\theta$$

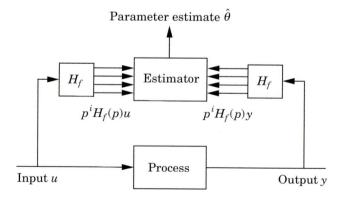

Figure 2.5   Block diagram of estimator with filters H_f.

By a proper realization of the filter H_f it is possible to use one filter to generate all the signals $p^i H_f(p)y$, $i = 0, \ldots, n$, and another filter to generate $p^i H_f(p)u$, $i = 0, \ldots, m - 1$. Standard least squares can now be applied, since this is a regression model. A recursive estimate is given by Theorem 2.5. With the restriction on H_f there will not be any pure differentiation of the output or the input to the system.

Nonlinear Models

Least squares can also be applied to certain nonlinear models. The essential restriction is that the models be linear in the parameters so that they can be written as linear regression models. Notice that the regressors do not need to be linear in the inputs and outputs. An example illustrates the idea.

EXAMPLE 2.3   **Nonlinear system**

Consider the model

$$y(t) + ay(t - 1) = b_1 u(t - 1) + b_2 \sin(u(t - 1))$$

By introducing

$$\theta = \left(\begin{array}{ccc} a & b_1 & b_2 \end{array} \right)^T$$

and

$$\varphi^T(t) = \left(\begin{array}{ccc} -y(t) & u(t) & \sin(u(t - 1)) \end{array} \right)$$

the model can be written as

$$y(t) = \varphi^T(t - 1)\theta$$

The model is linear in the parameters, and the least-squares method can be used to estimate θ.   □

Stochastic Models

The least-squares estimate is biased when it is used on data generated by Eq. (2.12), where the errors $e(i)$ are correlated. The reason is that $E\varphi^T(i)e(i) \neq 0$ (compare Eq. (2.13)). A possibility to cope with this problem is to model the correlation of the disturbances and to estimate the parameters describing the correlations. Consider the model

$$A(q)y(t) = B(q)u(t) + C(q)e(t) \tag{2.38}$$

where $A(q), B(q)$, and $C(q)$ are polynomials in the forward shift operator and $\{e(t)\}$ is white noise. The parameters of the polynomial C describe the correlation of the disturbance. The model of Eq. (2.38) cannot be converted directly to a regression model, since the variables $\{e(t)\}$ are not known. A regression model can, however, be obtained by suitable approximations. To describe these, introduce

$$\varepsilon(t) = y(t) - \varphi^T(t-1)\hat{\theta}(t-1)$$

where

$$\theta = \begin{pmatrix} a_1 & \dots & a_n & b_1 & \dots & b_n & c_1 & \dots & c_n \end{pmatrix}$$

$$\varphi^T(t-1) = \begin{pmatrix} -y(t-1) \dots -y(t-n) \; u(t-1) \dots u(t-n) \; \varepsilon(t-1) \dots \varepsilon(t-n) \end{pmatrix}$$

The variables $e(t)$ are approximated by the prediction errors $\varepsilon(t)$. The model can then be approximated by

$$y(t) = \varphi^T(t-1)\theta + e(t)$$

and standard recursive least squares can be applied. The method obtained is called *extended least squares* (ELS). The equations for updating the estimates are given by

$$\hat{\theta}(t) = \hat{\theta}(t-1) + P(t)\varphi(t-1)\varepsilon(t)$$
$$P^{-1}(t) = P^{-1}(t-1) + \varphi(t-1)\varphi^T(t-1) \tag{2.39}$$

(Compare with Theorem 2.3.) Another method of estimating the parameters in Eq. (2.38) is to use Eqs. (2.39) and let the residual be defined by

$$\hat{C}(q)\varepsilon(t) = \hat{A}(q)y(t) - \hat{B}(q)u(t) \tag{2.40}$$

and regression vector φ in Eqs. (2.39) be replaced by φ_f, where

$$\hat{C}(q)\varphi_f(t) = \varphi(t) \tag{2.41}$$

The most recent estimates should be used in these updates. The method obtained is then not truly recursive, since Eqs. (2.41) and (2.40) have to be solved from $t = 1$ for each measurement. The following approximations can be made:

$$\varepsilon(t) = y(t) - \varphi_f^T(t-1)\hat{\theta}(t-1)$$

This algorithm is called the *recursive maximum likelihood (RML) method.*

It is advantageous for both ELS and RML to replace the residual in the regression vector by the *posterior residual* defined as

$$\varepsilon_p(t) = y(t) - \varphi^T(t-1)\hat{\theta}(t)$$

that is, the latest value of $\hat{\theta}$ is used to compute ε_p.

Another possibility to model the correlated noise is to use the model

$$y(t) = \frac{B(q)}{A(q)}u(t) + \frac{C(q)}{D(q)}e(t)$$

instead of Eq. (2.38). Recursive parameter estimates for this model can be derived in the same way as for Eq. (2.38).

Details about the extended least-squares method and the recursive maximum likelihood method are found in the references at the end of the chapter.

Unification

The different recursive algorithms discussed are quite similar. They can all be described by the equations

$$\hat{\theta}(t) = \hat{\theta}(t-1) + P(t)\varphi(t-1)\varepsilon(t)$$

$$P(t) = \frac{1}{\lambda}\left(P(t-1) - \frac{P(t-1)\varphi(t-1)\varphi^T(t-1)P(t-1)}{\lambda + \varphi^T(t-1)P(t-1)\varphi(t-1)}\right)$$

where θ, φ, and ε are different for the different methods.

2.4   EXPERIMENTAL CONDITIONS

The properties of the data used in parameter estimation are crucial for the quality of the estimates. For example, it is obvious that no useful parameter estimates can be obtained if all signals are identically zero. In this section we discuss the influence of the experimental conditions on the quality of the estimates. In performing system identification automatically, as in an adaptive system, it is essential to understand these conditions, as well as the mechanisms that can interfere with proper identification. The notion of persistent excitation, which is one way to characterize process inputs, is introduced. In adaptive systems the plant input is generated by feedback. Difficulties caused by this are also discussed.

Persistent Excitation

Let us first consider estimation of parameters in a FIR model given by Eq. (2.33). The parameters of the model cannot be determined unless some conditions are imposed on the input signal. It follows from the condition for uniqueness of the least-squares estimate given by Theorem 2.1 that the minimum is unique if the matrix

$$\Phi^T\Phi = \begin{pmatrix} \sum\limits_{n+1}^{t} u^2(k-1) & \sum\limits_{n+1}^{t} u(k-1)u(k-2) & \cdots & \sum\limits_{n+1}^{t} u(k-1)u(k-n) \\[2mm] \sum\limits_{n+1}^{t} u(k-1)u(k-2) & \sum\limits_{n+1}^{t} u^2(k-2) & \cdots & \sum\limits_{n+1}^{t} u(k-2)u(k-n) \\[2mm] \vdots & & & \\[2mm] \sum\limits_{n+1}^{t} u(k-1)u(k-n) & & & \sum\limits_{n+1}^{t} u^2(k-n) \end{pmatrix}$$

$$(2.42)$$

has full rank. This condition is called an *excitation condition*. For long data sets, all sums in Eq. (2.42) can be taken from 1 to t. We then get

$$
C_n = \lim_{t\to\infty} \frac{1}{t}\, \Phi^T \Phi = \begin{pmatrix} c(0) & c(1) & \cdots & c(n-1) \\ c(1) & c(0) & \cdots & c(n-2) \\ \vdots & & & \\ c(n-1) & c(n-2) & \cdots & c(0) \end{pmatrix}
\tag{2.43}
$$

where $c(k)$ are the empirical covariances of the input, that is,

$$
c(k) = \lim_{t\to\infty} \frac{1}{t} \sum_{i=1}^{t} u(i)u(i-k)
\tag{2.44}
$$

For long data sets the condition for uniqueness can thus be expressed as the matrix in Eq. (2.43) being positive definite. This leads to the following definition.

DEFINITION 2.1   Persistent excitation

A signal u is called *persistently exciting* (PE) of order n if the limits (2.44) exist and if the matrix C_n given by Eq. (2.43) is positive definite.

Remark 1. In the adaptive control literature an alternative definition of PE is often used. The signal u is said to be persistently exciting of order n if for all t there exists an integer m such that

$$
\rho_1 I > \sum_{k=t}^{t+m} \varphi(k)\varphi^T(k) > \rho_2 I
$$

where $\rho_1, \rho_2 > 0$ and the vector $\varphi(t)$ is given by

$$
\varphi(t) = \begin{pmatrix} u(t-1) & u(t-2) & \cdots & u(t-n) \end{pmatrix}
$$

Notice that the matrix (2.43) can be written as

$$
C_n = \lim_{t\to\infty} \frac{1}{t} \sum_{k=1}^{t} \varphi(k)\varphi^T(k)
$$

Remark 2. Notice that no mean value is included in the definition of the empirical covariance $c(k)$ in Eq. (2.44).   □

The following result can be established.

THEOREM 2.6   Consistency for FIR models

Consider least-squares estimation of the parameters of a finite impulse response model with n parameters. The estimate is consistent and the variance of the estimates goes to zero as $1/t$ if the input signal is persistently exciting of order n.

Proof: The result follows from Definition 2.1 and Theorem 2.2.   □

We now introduce the following theorem.

THEOREM 2.7   Persistently exciting signals

The signal u with the property (2.44) is persistently exciting of order n if and only if

$$U = \lim_{t \to \infty} \frac{1}{t} \sum_{k=1}^{t} (A(q)u(k))^2 > 0 \tag{2.45}$$

for all nonzero polynomials A of degree $n - 1$ or less.

Proof: Let the polynomial A be

$$A(q) = a_0 q^{n-1} + a_1 q^{n-2} + \cdots + a_{n-1}$$

A straightforward calculation shows that

$$U = \lim_{t \to \infty} \frac{1}{t} \sum_{k=1}^{t} (a_0 u(k + n - 1) + \cdots + a_{n-1}u(k))^2 = a^T C_n a$$

where C_n is the matrix given by Eq. (2.43). If C_n is positive definite, the right-hand side is positive for all a, and so is the left-hand side. Conversely, if the left-hand side is positive for all a, so is the right-hand side.   □

The result is useful in investigating whether special signals are persistently exciting.

EXAMPLE 2.4   Pulse

It follows from Eq. (2.45) that $C_n \to 0$ for all n if u is a pulse. A pulse thus is not PE for any n.   □

EXAMPLE 2.5   Step

Let $u(t) = 1$ for $t > 0$ and zero otherwise. It follows that

$$(q - 1)u(t) = \begin{cases} 1 & t = 0 \\ 0 & t \neq 0 \end{cases}$$

A step can thus at most be PE of order 1. Since

$$C_1 = \frac{1}{t} \sum_{k=1}^{t} u^2(k) = 1$$

it follows that it is PE of order 1.   □

EXAMPLE 2.6   **Sinusoid**

Let $u(t) = \sin \omega t$. It follows that

$$\left(q^2 - 2q \cos \omega + 1\right) u(t) = 0$$

A sinusoid can thus at most be PE of order 2. Since

$$C_2 = \frac{1}{2} \begin{pmatrix} 1 & \cos \omega \\ \cos \omega & 1 \end{pmatrix}$$

it follows that a sinusoid is actually PE of order 2.   □

EXAMPLE 2.7   **Periodic signal**

Let $u(t)$ be periodic with period n. It then follows that

$$(q^n - 1) u(t) = 0$$

The signal can thus at most be PE of order n.   □

EXAMPLE 2.8   **Random signals**

Consider the stochastic process

$$u(t) = H(q)e(t)$$

where $e(t)$ is white noise and $H(q)$ is a pulse transfer function. It follows from the definition of white noise that Eq. (2.45) is satisfied for the signal e for any nonzero polynomial $A(q)$. This property also holds for the signal u. The signal u is thus PE of any order.   □

To be able to give a frequency domain interpretation of PE, it is useful to use the following theorem, which is given without proof.

THEOREM 2.8   **Parseval's theorem**

Let

$$H(q^{-1}) = \sum_{k=0}^{\infty} h_k q^{-k}$$

$$G(q^{-1}) = \sum_{k=0}^{\infty} g_k q^{-k}$$

be two stable transfer functions, and let $e(t)$ be white noise of zero mean and covariance σ^2. Then

$$\sigma^2 \sum_{k=0}^{\infty} h_k g_k = \frac{\sigma^2}{2\pi} \int_{-\pi}^{\pi} H(e^{i\omega})G(e^{i\omega}) \, d\omega$$

    □

Remark. The left-hand side can be interpreted as $E\left(H(q^{-1})e(t) \cdot G(q^{-1})e(t)\right)$, that is, the covariance of the two signals obtained by sending white noise through the transfer functions $H(q^{-1})$ and $G(q^{-1})$. □

EXAMPLE 2.9   **Frequency domain characterization**

Consider a quasi-stationary signal $u(t)$ with spectrum $\Phi_u(\omega)$. It follows from Parseval's theorem that

$$\lim_{t\to\infty} \frac{1}{t} \sum_{k=1}^{t} (A(q)u(k))^2 = \frac{1}{2\pi} \int_{-\pi}^{\pi} |A(e^{i\omega})|^2 \Phi_u(\omega)\, d\omega \tag{2.46}$$

This equation gives considerable insight into the notion of persistent excitation. A polynomial of degree $n-1$ can at most vanish in $n-1$ points. The right-hand side of Eq. (2.46) will thus be positive if $\Phi_u(\omega) \neq 0$ for at least n points in the interval $-\pi \leq \omega \leq \pi$. A signal whose spectrum is different from zero in an interval is thus persistently exciting of any order. □

A sinusoid has a point spectrum that differs from zero at two points. It is thus persistently exciting of second order. A signal that is a sum of k sinusoids is persistently exciting of order $2k$. The frequency domain characterization also makes it possible to derive the following result.

THEOREM 2.9   **PE of filtered signals**

Let the signal u be persistently exciting of order n. Assume that $A(q)$ is a polynomial of degree $m < n$. The signal v defined by

$$v(t) = A(q)u(t)$$

is then persistently exciting of order ℓ with $n - m \leq \ell \leq n$. Assuming that A is stable, the signal w defined by

$$w(t) = \frac{1}{A(q)} u(t)$$

is persistently exciting of order n. □

Transfer Function Models

The properties of parameter estimates for discrete-time transfer functions will now be discussed. The uniqueness of the estimates will first be explored. For this purpose it is assumed that the data is actually generated by

$$A^0(q)y(t) = B^0(q)u(t) + e(t + n) \tag{2.47}$$

where A^0 and B^0 are relatively prime. Let A and B be the estimates of A^0 and B^0, respectively. If $e = 0$, $\deg A > \deg A^0$, and $\deg B > \deg B^0$, it follows from Theorem 2.1 that the estimate is not unique because the columns of the matrix Φ are linearly dependent. However, we have the following result.

THEOREM 2.10   Transfer function estimation

Consider data generated by the model of Eq. (2.47), with A^0 stable and $e = 0$. Let the parameters of the polynomials A and B be fitted by least squares. Assume that the input u is persistently exciting of order $\deg A + \deg B + 1$. If it is further assumed that $\deg A = \deg A^0$ and $\deg B \geq B^0$, then $\lim_{t \to \infty} \Phi^T \Phi / t$ is positive definite.

Proof: Consider

$$V(\theta) = \theta^T \lim_{t \to \infty} \frac{1}{t} \Phi^T \Phi \theta = \lim_{t \to \infty} \frac{1}{t} \sum_{k=1}^{t} \left(\varphi^T(k)\theta \right)^2$$

Introduce

$$v(t) = \varphi^T(t + n - 1)\theta = B(q)u(t) - (A(q) - q^n) y(t)$$

$$= B(q)u(t) - \frac{A(q) - q^n}{A^0(q)} B^0(q)u(t)$$

$$= \left\{ A^0(q)B(q) - \left(A(q) - q^n\right) B^0(q) \right\} \frac{1}{A^0(q)} u(t)$$

Since A^0 is stable, it follows from Theorem 2.9 that the signal $1/A^0(q) \cdot u(t)$ is persistently exciting of order $\deg A + \deg B + 1$. Since the polynomial in braces has a degree lower than or equal to $\deg A + \deg B$, it follows that the signal $v(t)$ does not vanish in the mean square sense unless the polynomial is identically zero. This happens if

$$\frac{B^0(q)}{A^0(q)} = \frac{B(q)}{A(q) - q^n}$$

Since $\deg A = \deg A^0$, the denominator on the right-hand side thus has degree $\deg A - 1 = \deg A^0 - 1$. The rational functions are then not identical, and the theorem is proved.   □

Remark 1.   Notice that $\deg A + \deg B + 1$ is equal to the number of parameters in the model of Eq. (2.47). The order of PE required is thus equal to the number of estimated parameters.

Remark 2.   If the data is generated by Eq. (2.47), where $\{e(t)\}$ is white noise (i.e., a sequence of uncorrelated random variables), then the matrix

$$\lim_{t \to \infty} \frac{1}{t} \Phi^T \Phi$$

is positive definite for models of all orders provided that the input is persistently exciting of order $\deg B + 1$.   □

Theorem 2.2 does not automatically apply to estimation of parameters of a transfer function, because the output y appears in the regression vector. A consequence of this is that theoretical properties of the estimates can be established asymptotically only for large number of observations.

Identification in Closed Loop

In adaptive control, system identification is often performed under closed-loop conditions, which may give rise to certain difficulties. Consider, for example, the estimation of the coefficients of a transfer function model as in Eq. (2.34). The matrix Φ is then

$$
\Phi = \begin{pmatrix}
-y(n) & \cdots & -y(1) & u(n) & \cdots & u(1) \\
-y(n+1) & \cdots & -y(2) & u(n+1) & \cdots & u(2) \\
\vdots & & & & & \vdots \\
-y(t-1) & \cdots & -y(t-n) & u(t-1) & \cdots & u(t-n)
\end{pmatrix} \tag{2.48}
$$

A linear feedback of sufficiently low order introduces linear dependencies among the columns of the matrix Φ. This means that the parameters cannot be determined uniquely. A simple example shows what may happen.

EXAMPLE 2.10   **Loss of identifiability due to feedback**

Consider a system described by

$$
y(t+1) + ay(t) = bu(t) \tag{2.49}
$$

Assume that the parameters a and b should be estimated in the presence of the feedback

$$
u(t) = -ky(t) \tag{2.50}
$$

Multiplying Eq. (2.50) by α and adding to Eq. (2.49) give

$$
y(t+1) + (a + \alpha k)y(t) = (b - \alpha)u(t)
$$

This shows that any parameters such that

$$
\hat{a} = a + \alpha k
$$
$$
\hat{b} = b - \alpha
$$

give the same input-output relation. The above equation represents a straight line

$$
\hat{b} = b + \frac{1}{k}(a - \hat{a}) \tag{2.51}
$$

in parameter space (see Fig. 2.6). The least-squares loss function (2.2) has the same value for all parameters on this line.   □

The problem with lack of identifiability due to feedback disappears if a linear feedback of sufficiently high order is used. Then the columns of the matrix Φ given by Eq. (2.48) are no longer linearly dependent. Another possibility is to have a time-varying feedback. For example, in Example 2.10 it is sufficient to have a feedback of the form

$$
u(t) = -k_1 y(t) - k_2 y(t-1)
$$

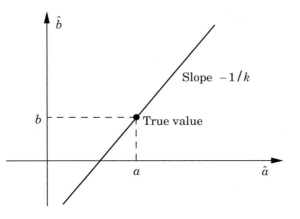

Figure 2.6   Illustration of lack of uniqueness in closed-loop identification.

with $k_2 \neq 0$. Another possibility is to use a feedback law

$$u(t) = -k(t)y(t)$$

where k varies with time. For instance, in Example 2.10 it is sufficient to use two values of the gain. Each value of the gain corresponds to a straight line with slope $-1/k$ in parameter space. Two lines give a unique intersection.

In adaptive systems there is a natural time variation in the feedback because the feedback gains are based on parameter estimates. In a typical case the variance of the parameters decreases as $1/t$, but more complex behavior is also possible. The following example shows what can happen.

EXAMPLE 2.11   Convergence rate

Consider data generated by

$$y(t) + ay(t-1) = bu(t-1) + e(t)$$

with a feedback of the form

$$u(t) = -k\left(1 + \frac{v(t)}{\sqrt{t}}\right)y(t) \tag{2.52}$$

where $\{v(t)\}$ is a sequence of independent random variables that are also independent of $\{e(t)\}$. With the feedback law of Eq. (2.52) the closed-loop system becomes

$$y(t+1) = -\left(a + bk + \frac{bkv(t)}{\sqrt{t}}\right)y(t) + e(t+1)$$

Given measurements up to $t + 1$, the matrix $\Phi^T \Phi$ of the estimation problem is

$$\Phi^T \Phi = \begin{pmatrix} \sum\limits_{j=1}^{t} y^2(j) & \sum\limits_{j=1}^{t} y(j)u(j) \\ \sum\limits_{j=1}^{t} y(j)u(j) & \sum\limits_{j=1}^{t} u^2(j) \end{pmatrix}$$

It follows that

$$\sum_{j=1}^{t} y(j)u(j) = -k \sum_{j=1}^{t} y^2(j) - k \sum_{j=1}^{t} \frac{v(j)y^2(j)}{\sqrt{j}} \approx -k \sum_{j=1}^{t} y^2(j) \approx -kt\sigma_y^2$$

$$\sum_{j=1}^{t} u^2(j) = k^2 \left(\sum_{j=1}^{t} y^2(j) + 2 \sum_{j=1}^{t} \frac{v(j)y^2(j)}{\sqrt{j}} + \sum_{j=1}^{t} \frac{v^2(j)y^2(j)}{j} \right)$$

$$\approx k^2 \left(\sum_{j=1}^{t} y^2(j) + \sum_{j=1}^{t} \frac{v^2(j)y^2(j)}{j} \right) \approx k^2 \sigma_y^2 \left(t + \sigma_v^2 \log t \right)$$

Hence for large t,

$$\Phi^T \Phi \approx \sigma_y^2 \begin{pmatrix} t & -kt \\ -kt & k^2 \left(t + \sigma_v^2 \log t \right) \end{pmatrix}$$

The covariance matrix of the estimate is thus

$$\sigma_e^2 \left(\Phi^T \Phi \right)^{-1} \approx \frac{\sigma_e^2}{\sigma_y^2 \sigma_v^2} \begin{pmatrix} \dfrac{1}{\log t} + \dfrac{\sigma_v^2}{t} & \dfrac{1}{k \log t} \\ \dfrac{1}{k \log t} & \dfrac{1}{k^2 \log t} \end{pmatrix}$$

It now follows that

$$\mathrm{cov}\left(\hat{a} - k\hat{b} \right) = \sigma_e^2 \begin{pmatrix} 1 & -k \end{pmatrix} \left(\Phi^T \Phi \right)^{-1} \begin{pmatrix} 1 & -k \end{pmatrix}^T \approx \frac{\sigma_e^2}{t\sigma_y^2}$$

$$\mathrm{cov}\left(k\hat{a} + \hat{b} \right) = \sigma_e^2 \begin{pmatrix} k & 1 \end{pmatrix} \left(\Phi^T \Phi \right)^{-1} \begin{pmatrix} k & 1 \end{pmatrix}^T \approx \frac{\sigma_e^2}{\sigma_y^2 \sigma_v^2 \log t}$$

The estimate will thus approach the line (2.51) at the rate $1/t$. The estimate will then converge toward the correct values at the rate $1/\log t$. The convergence along the line (2.51) is slower than convergence toward the line.   □

2.5   SIMULATION OF RECURSIVE ESTIMATION

In this section, different properties of the recursive least-squares (RLS) method are illustrated through simulations. Throughout the section, data is generated by

$$y(t) + ay(t-1) = bu(t-1) + e(t) + ce(t-1) \tag{2.53}$$

where $a = -0.8, b = 0.5$, and $e(t)$ is zero mean white noise with standard deviation $\sigma = 0.5$. Furthermore, $c = 0$, $P(0) = 100 \cdot I$, and $\hat{\theta}(0) = 0$ except when indicated. In most cases we use

$$\hat{\theta} = \begin{pmatrix} \hat{a} \\ \hat{b} \end{pmatrix} \qquad \varphi(t-1) = \left(-y(t-1) \quad u(t-1) \right)$$

Only in Example 2.13 is the parameter c estimated.

EXAMPLE 2.12   Excitation

The need for persistency of excitation is illustrated in this example. A simulation of the estimates when the input is a unit pulse at $t = 50$ is shown in Fig. 2.7(a). The estimate $\hat{a}$ appears to converge to the correct value, but the estimate $\hat{b}$ does not. The reason for this is that information about the a parameter is obtained through the excitation by the noise. Information about the b parameter is obtained only through the pulse that is not persistently exciting.

In Fig. 2.7(b) the experiment is repeated, but the input is now a square wave of unit amplitude and a period of 100 samples. Both $\hat{a}$ and $\hat{b}$ will converge to their true values, because the input is persistently exciting. The absolute values of the elements of $P(t)$ are decreasing with time. For the simulation in

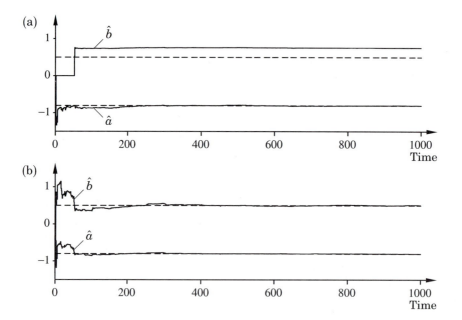

Figure 2.7 The estimated (solid line) and true (dashed line) parameter values in estimating the parameters in the model (2.53). The input signal $u(t)$ is (a) a unit pulse at $t = 50$, (b) a unit amplitude square wave with period 100.

Fig. 2.7(b) we have

$$\begin{pmatrix} \hat{a}(1000) \\ \hat{b}(1000) \end{pmatrix} = \begin{pmatrix} -0.796 \\ 0.511 \end{pmatrix} \qquad P(1000) = \begin{pmatrix} 0.550 & 1.114 \\ 1.114 & 3.258 \end{pmatrix} \cdot 10^{-3}$$

According to Theorem 2.2 this implies the following standard deviations for the estimates:

$$\sigma_{\hat{a}} = 0.5\sqrt{5.50} \cdot 10^{-2} = 0.012$$
$$\sigma_{\hat{b}} = 0.5\sqrt{32.58} \cdot 10^{-2} = 0.029$$

The estimates are thus well within one standard deviation of their true values.

□

EXAMPLE 2.13   Model structure

In this example, parameter c in Eq. (2.53) has the value -0.5. Figure 2.8(a) shows the estimates of parameters a and b. The estimates do not converge to their true values. This is because the equation error $e(t) + ce(t-1)$ is not white noise. The assumptions in Theorem 2.2 are thus violated. Figure 2.8(b) shows the estimates when the extended least squares (ELS) method is used. All three parameters a, b, and c are then estimated, and the estimates converge to the true values. When only a and b are estimated by using the least-squares

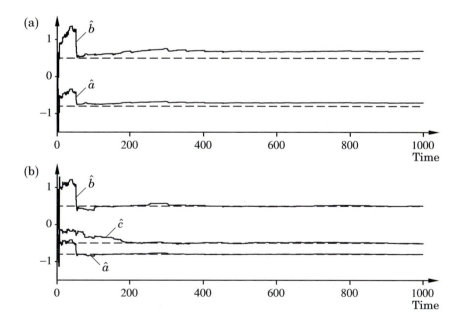

Figure 2.8   Estimated parameters when the model (2.53) is simulated with $c = -0.5$ by using (a) LS and (b) ELS.

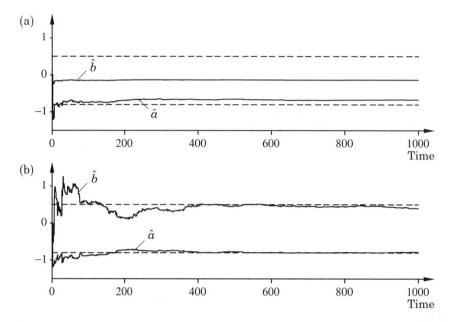

Figure 2.9 Estimates when the control signal is generated through feedback (a) $u(t) = -0.2y(t)$ and (b) $u(t) = -0.32y(t-1)$.

method, the estimates and the P-matrix at time $t = 1000$ are

$$\begin{pmatrix} \hat{a}(1000) \\ \hat{b}(1000) \end{pmatrix} = \begin{pmatrix} -0.702 \\ 0.697 \end{pmatrix} \qquad P(1000) = \begin{pmatrix} 0.710 & 1.435 \\ 1.435 & 3.903 \end{pmatrix} \cdot 10^{-3}$$

The elements in the P-matrix are small. This would indicate good accuracy if the process had fulfilled the assumptions about the noise structure. Theorem 2.2 gives the following estimates of the standard deviation of $\hat{a}$ and $\hat{b}$:

$$\sigma_{\hat{a}} = 0.5\sqrt{7.10 \cdot 10^{-2}} = 0.013$$
$$\sigma_{\hat{b}} = 0.5\sqrt{39.03 \cdot 10^{-2}} = 0.031$$

It is thus deceptive to judge the accuracy of the estimates by only looking at the P-matrix. It is necessary that the data be generated from a model of the form (2.12) to use the P-matrix for accuracy estimates.

   If we did not observe that the equation error is not white, we could thus be strongly misled. One possibility to avoid mistakes is to also compute the correlation of the equation error and check whether it is white noise.   □

EXAMPLE 2.14   Closed-loop estimation
Example 2.10 showed that identifiability can be lost owing to feedback. The

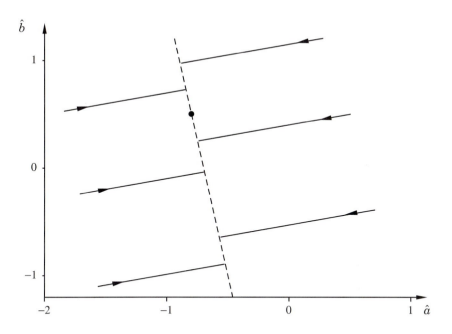

Figure 2.10 Phase plane of the estimates when the system (2.53) is simulated for different initial conditions when $u(t) = -0.2y(t)$. The dashed line shows the identifiable subspace. The dot shows the true parameter values.

estimates when the input is generated through the feedback

$$u(t) = -0.2y(t)$$

are shown in Fig. 2.9(a). The estimates converge to the wrong values. Notice, however, that the estimates are on the straight line (2.51). In Fig. 2.9(b) the feedback is more complex:

$$u(t) = -0.32y(t - 1)$$

The two control laws give approximately the same speed and output variance of the closed-loop system. Identifiability is now regained, and the estimates converge to the correct values. The phase plane, that is, $\hat{b}$ as a function of $\hat{a}$, is shown in Fig. 2.10 for different initial conditions when $u(t) = -0.2y(t)$. The initial value of the P-matrix is $P(0) = 0.01I$, and 20,000 steps have been simulated for each initial condition. The estimates converge to the identifiable subspace determined by

$$\hat{a} + 0.2\hat{b} + 0.7 = 0$$

(Compare Eq. (2.51).) This line is dashed in the phase plane. The estimates are approaching the identifiable subspace along straight lines. The same initial conditions are simulated for the control law $u(t) = -0.32y(t - 1)$ in Fig. 2.11. The estimates converge to the correct value $(-0.8, 0.5)$, independent of the initial values. □

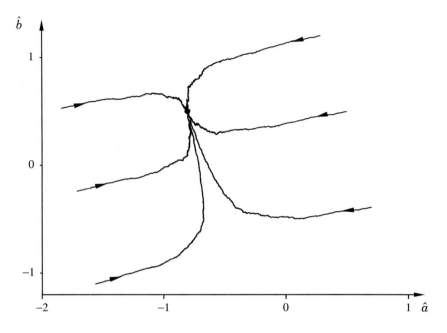

Figure 2.11   Phase plane of the estimates when the system (2.53) is simulated for different initial conditions when $u(t) = -0.32y(t-1)$. The dot shows the true parameter values.

EXAMPLE 2.15	Influence of forgetting factor

The recursive least-squares algorithm (2.21) has a forgetting factor λ. The influence of the forgetting factor is shown in Figure 2.12. When $\lambda = 1$, the estimates become smoother and smoother, since the gain $K(t)$ goes to zero. When $\lambda < 1$, the estimator gain $K(t)$ does not go to zero, and the estimates will always fluctuate. The fluctuations increase with decreasing λ. As a rule of thumb the "memory" of the estimator is

$$N = \frac{2}{1-\lambda}$$

For $\lambda = 0.99$ the estimates are based on approximately the last 200 steps. □

EXAMPLE 2.16	Different estimation methods

In the previous examples the RLS and ELS methods were used. Simplified estimation methods based on projection were discussed in Section 2.2. Three different projection algorithms will now be compared with the RLS method. All have the following form:

$$\hat{\theta}(t) = \hat{\theta}(t-1) + P(t)\varphi(t)\left(y(t) - \varphi^T(t)\hat{\theta}(t-1)\right) \tag{2.54}$$

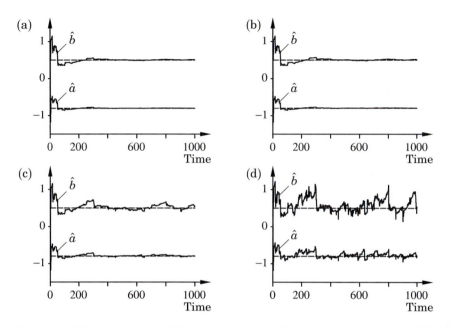

Figure 2.12   Estimates of the parameters in the process (2.53) when RLS is used and (a) $\lambda = 1$, (b) $\lambda = 0.999$, (c) $\lambda = 0.99$, and (d) $\lambda = 0.95$.

Compare with Eq. (2.24). The scalar gain $P(t)$ is given by the following algorithms.

Least mean squares (LMS):

$$P(t) = \gamma$$

Projection algorithm (PA):

$$P(t) = \frac{\gamma}{\alpha + \varphi^T(t)\varphi(t)} \qquad \alpha \geq 0, \quad 0 < \gamma < 2$$

Stochastic approximation (SA):

$$P(t) = \frac{\gamma}{\Sigma_{i=1}^{t}\varphi^T(i)\varphi(i)}$$

The convergence properties of the four algorithms RLS, LMS, PA, and SA are compared in Fig. 2.13. All algorithms are initialized with $\theta(0) = 0$. The RLS method in Fig. 2.13(a) uses $P(0) = 100I$ and $\lambda = 1$. Notice that the estimates move very quickly initially. The LMS method used in Fig. 2.13(b) has a constant gain $\gamma = 0.01$. The estimates approach values that are close to the correct ones relatively quickly, but the estimates do not converge, since the gain is not decreasing. In the PA method in Fig. 2.13(c) the gain is normalized with $\varphi^T(t)\varphi(t)$. Further, $\alpha = 0.1$ and $\gamma = 0.01$ are used. The approach toward

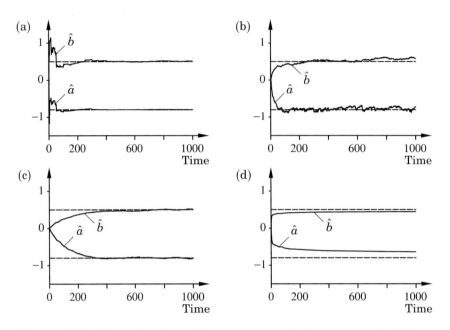

Figure 2.13   The estimates of the parameters in the process for different estimation methods. (a) Recursive least squares (RLS) with $P(0) = 100$ and $\lambda = 1$. (b) Least mean squares (LMS) with $\gamma = 0.01$. (c) Projection algorithm (PA) with $\alpha = 0.1$ and $\gamma = 0.01$. (d) Stochastic approximation (SA) with $\gamma = 0.2$.

the correct values is slower than for the LMS algorithm. However, the PA method is less sensitive than the LMS method to the size of the signals. The SA method is used in Fig. 2.13(d) with $\gamma = 0.2$, and the estimates converge to the correct values even if the convergence is very slow. About 25,000–50,000 steps are needed before the estimates are close to the correct values. From the simulations it is seen that the recursive least-squares method has superior convergence properties. The price for this is the increase in computations.   □

The examples show that there are many things that influence the performance of the estimators. In adaptive control it is important to remember that the estimation is done in closed loop.

2.6   PRIOR INFORMATION

There is a significant advantage in incorporating available prior information. It reduces the number of parameters that have to be estimated, improves the precision of the estimates, and reduces the requirements on excitation.

Prior information typically relates to properties of a model. It can, for instance, represent knowledge of time constants of an actuator. This type of knowledge is easy to incorporate in an indirect adaptive algorithm. However, it may be difficult to incorporate in a direct adaptive algorithm, since process parameters influence controller parameters in a complicated fashion. Since prior knowledge is often related to the continuous-time models it is easier to use for continuous time than for discrete time self-tuners. These properties are highlighted by a few examples.

EXAMPLE 2.17   **Prior information in continuous time**

Consider the continuous-time system with the transfer function

$$G(s) = \frac{\theta_3}{(1 + \theta_1 s)(1 + \theta_2 s)}$$

The parameter θ_1 is assumed to be known; θ_2 and θ_3 are unknown. If we introduce the filtered signal $\bar{u}$ defined by

$$\bar{u} = \frac{1}{1 + \theta_1 p} u$$

the input-output relation may be written as

$$y + \theta_2 \frac{dy}{dt} = \theta_3 \bar{u} \tag{2.55}$$

The estimation problem thus reduces to estimation of parameters θ_3 and θ_2 of the first-order system given by Eq. (2.55).   ☐

The example thus shows that it is straightforward to handle prior information for the continuous-time model. The next example illustrates some complications that occur when the model is sampled.

EXAMPLE 2.18   **Prior information in sampled models**

Consider the system in Example 2.17. Sampling the system with sampling period h gives the pulse transfer operator

$$H(q) = \frac{b_1 q + b_2}{q^2 + a_1 q + a_2}$$

where

$$b_1 = \theta_3 \frac{\theta_1 \left(1 - e^{-h/\theta_1}\right) - \theta_2 \left(1 - e^{-h/\theta_2}\right)}{\theta_1 - \theta_2}$$

$$b_2 = \theta_3 \frac{\theta_2 \left(1 - e^{-h/\theta_2}\right) e^{-h/\theta_1} - \theta_1 \left(1 - e^{-h/\theta_1}\right) e^{-h/\theta_2}}{\theta_1 - \theta_2}$$

$$a_1 = -\left(e^{-h/\theta_1} + e^{-h/\theta_2}\right)$$

$$a_2 = e^{-(1/\theta_1 + 1/\theta_2)h}$$

The pulse transfer function is nonlinear in θ_1, θ_2, and θ_3. Further, both parameters appear in *all* the coefficients of the discrete-time pulse transfer function. This implies that a change in the unknown time constant θ_2 will influence all the coefficients in the sampled data model. There is, however, some structure in the parameter dependence. The denominator polynomial ca•. be written as

$$q^2 + a_1 q + a_2 = \left(q - e^{-h/\theta_1}\right)\left(q - e^{-h/\theta_2}\right)$$
$$= \left(q - \alpha_1\right)\left(q - \alpha_2\right)$$

When θ_1 is known, one factor of $A(q)$ is thus known. By reparameterization the sampled model can be written as

$$H(q) = \frac{b_1 q + b_2}{(q - \alpha_1)(q - \alpha_2)}$$

The prior information can be used to reduce the estimated parameters from 4 to 3. Further simplifications can be made when the sampling interval is small in comparison with θ_1 and θ_2. A series approximation of b_1 and b_2 in h gives

$$b_1 \approx \frac{\theta_3}{2\theta_1\theta_2} h^2 - \frac{\theta_3(\theta_1 + \theta_2)}{60\theta_1^2\theta_2^2} h^3$$

$$b_2 \approx \frac{\theta_3}{2\theta_1\theta_2} h^2 - \frac{\theta_3(\theta_1 + \theta_2)}{\theta_1^2\theta_2^2} h^3$$

For short sampling periods we have

$$b_1 \approx b_2 \approx \frac{\theta_3}{2\theta_1\theta_2} h^2$$

The model can now be described by

$$H(q) = \frac{k(q + 1)}{(q - \alpha_1)(q - \alpha_2)}$$

where parameter α_1 is known and α_2 and $k = \theta_3/(h^2\theta_1\theta_2)$ are unknown.   □

    The observation about the structure of the sampled model for small sampling periods is a consequence of a general result about how poles and zeros are transformed by sampling. If α_i is a pole of a continuous-time system, then the sampled system has a pole at $\exp(\alpha_i h)$. There are no simple, exact formulas for transforming the zeros. For short sampling periods, however, a zero β_i is approximately transformed to $\exp(\beta_i h)$. If d is the pole excess of the continuous-time system, there will be $d - 1$ additional zeros of the sampled system. The limiting positions of these zeros as the sampling period goes to zero are given by Theorem 6.9 in Chapter 6. In this way it is possible to use prior information in terms of poles and zeros both for continuous-time self-tuners and for discrete-time self-tuners with a short sampling period.

    Example 2.18 shows that how the process model is parameterized is crucial. Different parameterizations can be attempted. This is illustrated by an example.

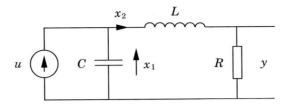

Figure 2.14   The circuit in Example 2.19.

EXAMPLE 2.19   **Reparameterization**

Consider the circuit in Fig. 2.14. The state space representation is

$$\frac{dx}{dt} = \begin{pmatrix} 0 & -1/C \\ 1/L & -R/L \end{pmatrix} x + \begin{pmatrix} 1/C \\ 0 \end{pmatrix} u$$

$$y = \begin{pmatrix} 0 & R \end{pmatrix} x$$

and the transfer function is

$$G(s) = \frac{\dfrac{R}{LC}}{s^2 + \dfrac{R}{L}s + \dfrac{1}{LC}}$$

Let $\theta_1 = R$, $\theta_2 = 1/L$, and $\theta_3 = 1/C$. Then

$$G(s) = \frac{\theta_1\theta_2\theta_3}{s^2 + \theta_1\theta_2 s + \theta_2\theta_3}$$

The coefficients are nonlinear (although of special structure) in the physical parameters R, $1/L$, and $1/C$. The system can be written as

$$G(s) = \frac{k_1}{s^2 + k_2 s + k_3} \tag{2.56}$$

and it is possible to make an estimation of θ_1, θ_2, and θ_3 by using Eq. (2.56). However, the estimates must be constrained such that the relations

$$k_1 = \theta_1\theta_2\theta_3$$
$$k_2 = \theta_1\theta_2$$
$$k_3 = \theta_2\theta_3$$

are fulfilled.   □

For indirect self-tuning regulators it is possible to estimate the continuous-time process parameters from discrete-time measurements. The model can then be sampled and the controller designed for the chosen sampling interval.

2.7   CONCLUSIONS

In this chapter we have introduced recursive parameter estimation, which is a key ingredient in adaptive control. The presentation has been focused on the least-squares method, which is a simple but useful technique. In the next chapter we will show how the method is used in adaptive systems. System identification involves several important issues that we have not discussed. One is model validation; another is computational aspects. These issues are discussed in detail in Chapter 11.

PROBLEMS

2.1   Consider the function

$$V(x) = x^T A x + b^T x + c$$

where x and b are column vectors, A is a matrix, and c is a scalar. Show that the gradient of function V with respect to x is given by

$$\text{grad}_x V = (A + A^T)x + b$$

This can be used to find the minimum of Eq. (2.7).

2.2   Consider the FIR model

$$y(t) = b_0 u(t) + b_1 u(t-1) + e(t) \qquad t = 1, 2, \ldots$$

where $\{e(t)\}$ is a sequence of independent normal $N(0, \sigma)$ random variables.

(a) Determine the least-squares estimate of the parameters b_0 and b_1 when the input signal u is a step. Analyze the covariance of the estimate when the number of observations goes to infinity. Relate the results to the notion of persistent excitation.

(b) Make the same investigation as in part (a) when the input signal is white noise with unit variance.

2.3   Consider data generated by the discrete-time system

$$y(t) = b_0 u(t) + b_1 u(t-1) + e(t)$$

where $\{e(t)\}$ is a sequence of independent $N(0, 1)$ random variables. Assume that the parameter b of the model

$$y(t) = b u(t)$$

is determined by least squares.

(a) Determine the estimates obtained for large observation sets when the input u is a step. (This is a simple illustration of the problem of fitting a low-order model to data generated by a complex model. The result obtained will critically depend on the character of the input signal.)

(b) Make the same investigation as in part (a) when the input signal is a sequence of independent $N(0, \sigma)$ random variables.

2.4 Determine which of the input signals below are persistently exciting of at least order 4.

(a)
$$u(t) = a_0 + a_1 \sin \omega t \qquad a_i \neq 0, \quad i = 0, 1$$

(b)
$$u(t) = \frac{q - 0.5}{(q - 0.4)(q - 0.6)} v(t)$$

where $v(t)$ is persistently exciting of order 5.

(c)
$$u(t) = \frac{q - 0.5}{(q - 0.4)(q - 0.6)} v(t)$$

where $v(t)$ has a spectrum $\Phi_v(\omega)$ that is not equal to zero in the interval $1 < \omega < 2$.

2.5 Consider the discrete-time system
$$y(t + 1) + ay(t) = bu(t) + e(t + 1)$$

where the input signal u and the noise e are sequences of independent random variables with zero mean values and standard deviation σ and 1. Determine the covariance of the estimates obtained for large observation sets.

2.6 Consider data generated by the least-squares model
$$y(t + 1) + ay(t) = bu(t) + e(t + 1) + ce(t) \qquad t = 1, 2, \ldots$$

where $\{u(t)\}$ and $\{e(t)\}$ are sequences of independent random variables with zero mean values and standard deviations 1 and σ. Assume that parameters a and b of the model
$$y(t + 1) + ay(t) = bu(t)$$

are estimated by least squares. Determine the asymptotic values of the estimates.

2.7 Consider least-squares estimation of the parameters b_1 and b_2 in
$$y(t) = b_1 u(t) + b_2 u(t - 1)$$

Assume that the following measurements are obtained:

t	u	y
1	1000	–
2	1001	2001
3	1000	2001

Discuss the numerical properties of computing the estimates directly and by the normal equations.

2.8 Consider the model

$$y(t) = a + b \cdot t + e(t) \qquad t = 1, 2, 3, \ldots$$

where $\{e(t)\}$ is a sequence of uncorrelated $N(0, 1)$ random variables. Determine the least-squares estimate of the parameters a and b. Also determine the covariance of the estimate. Discuss the behavior of the covariance as the number of estimates increases.

2.9 Consider the model in Problem 2.8, but assume continuous-time observation, where $e(t)$ is white noise, that is, a random function with covariance $\delta(t)$. Determine the estimate and its covariance. Analyze the behavior of the covariance for large observation intervals.

2.10 Consider data generated by

$$y(t) = b + e(t) \qquad t = 1, 2, \ldots, N$$

where $\{e(t); t = 1, 3, 4, \ldots\}$ is a sequence of independent random variables. Furthermore, assume that there is a large error at $t = 2$, that is,

$$e(2) = a$$

where a is a large number. Assume that the parameter b in the model

$$y(t) = b$$

is estimated by least squares. Determine the estimate obtained, and discuss how it depends on a. (This is a simple example that shows how sensitive the least-squares estimate is with respect to occasional large errors.)

2.11 Consider Example 2.12. Analyze the asymptotic properties of the P-matrix and explain the simulation in Figs. 2.7(a) and 2.7(b).

2.12 Show that Eq. (2.11) minimizes the weighted least-squares loss function (2.10).

2.13 Consider Eqs. (2.21) with the initial condition $\hat{\theta}(0) = \theta_0$ and $P(0) = P_0$. Show that $\hat{\theta}(t)$ minimizes the criterion

$$V(\theta, t) = \frac{1}{2} \sum_{i=1}^{t} \lambda^{t-i} \left(y(i) - \varphi^T(i)\theta\right)^2 + \frac{\lambda^t}{2} (\theta - \theta_0)^T P_0^{-1} (\theta - \theta_0)$$

Compare Theorem 2.4.

2.14 Consider the following model of time-varying parameters:

$$\theta(t) = \Phi_v \theta(t-1) + v(t)$$
$$y(t) = \varphi^T(t)\theta(t) + e(t)$$

where $\{v(t), t = 1, 2, \ldots\}$ and $\{e(t), t = 1, 2, \ldots\}$ are sequences of independent, equally distributed random vectors with zero mean values and covariances R_1 and R_2, respectively. Show that the recursive estimates of θ are given by

$$\hat{\theta}(t) = \hat{\theta}(t-1) + K(t) \left(y(t) - \varphi^T(t)\hat{\theta}(t-1)\right)$$

$$K(t) = \Phi_v P(t-1)\varphi(t-1) \left(R_2 + \varphi^T(t-1)P(t-1)\varphi(t-1)\right)^{-1}$$

$$P(t) = \Phi_v P(t-1)\Phi_v^T + R_1$$
$$\quad - \Phi_v P(t-1)\varphi(t) \left(R_2 + \varphi^T(t)P(t-1)\varphi(t)\right)^{-1} \varphi^T(t)P(t-1)\Phi_v^T$$

2.15 Show that Eq. (2.28) minimizes Eq. (2.27), and use this to prove Theorem 2.5. *Hint:* Use Remark 1 in Theorem 2.5 and that the time derivative of the identity $I = PP^{-1}$ is

$$\frac{dP}{dt} = -P \frac{d(P^{-1})}{dt} P$$

2.16 In an adaptive controller the process parameters are estimated according to the model

$$y(t) + a_1 y(t-1) + a_2 y(t-2) = b_0 u(t-1) + b_1 u(t-2) + e(t)$$

The controller has the structure

$$u(t) + r_1 u(t-1) = -s_0 y(t) - s_1 y(t-1)$$

The reference value is thus zero. Consider the case in which the controller parameters are constant.

(a) Show that the parameters a_1, a_2, b_0, and b_1 cannot be uniquely determined.

(b) Characterize the parameter combinations that can be determined.

(c) Show that with the controller structure

$$u(t) + r_1 u(t-1) + r_2 u(t-2) = -s_0 y(t) - s_1 y(t-1) - s_2 y(t-2)$$

all process parameters can be estimated uniquely.

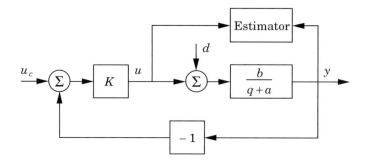

Figure 2.15   Closed-loop estimation scheme for Problem 2.17.

2.17 Figure 2.15 shows a closed-loop system for estimation of the unknown
constant b in the pulse transfer function $H(q) = b/(q+a)$. The constant
a is known and is such that $|a| > 1$. This means that the open-loop system
is unstable, and to have bounded signals for the estimation, we need to
stabilize the system with a controller. This is done with a P controller
with gain K such that $|a + Kb| < 1$. The estimator is a least-squares
(LS) estimator that is based on the regression model

$$\bar{y}(t) = \varphi^T(t-1)\theta$$

where

$$\bar{y}(t) = y(t) + ay(t-1)$$
$$\varphi(t-1) = u(t-1)$$
$$\theta = b$$

(a) Determine the asymptotic LS estimate of $\theta = b$ when $d = 0$ and $\{u_c\}$
is a sequence of independent, equally distributed random variables
with zero mean and variance σ^2 (i.e., u_c is a white noise signal).

(b) Determine the asymptotic LS estimate of b when $u_c = 0$ and $d = d_0 =$
constant.

(c) Discuss the case of least-squares estimation of b when u_c is as in part
(a) and $d = d_0 =$ constant. What should be done to avoid a biased
estimate of b?

2.18 Write a computer program to simulate the recursive least-squares estimation problem. Write the program so that arbitrary input signals can
be used. Use the program to investigate the effects of initial values on
the estimate.

2.19 Use the program from Problem 2.18 to estimate the parameters a and b
in Problem 2.6. Investigate how the bias of the estimate depends on c.

2.20 Consider the estimation problem in Problem 2.6. Use the computer program developed in Problem 2.18 to explore what happens when the con-

trol signal u is generated by the feedback

$$u(t) = -ky(t)$$

Try to support your observations by analysis.

2.21 Consider the open-loop system in Section 2.5 when $c = 0$. Let the input signal be a square wave with unit amplitude and a period of 100 samples. Investigate through simulations the convergence and behavior of the parameter estimates when varying:

(a) The initial value $\hat{\theta}(0)$.

(b) The initial value of the covariance matrix $P(0)$.

(c) The forgetting factor λ.

(d) The period of the input signal.

REFERENCES

The following textbooks can be recommended for those who would like to learn more about system identification:

Norton, J. P., 1986. *An Introduction to Identification*. London: Academic Press.

Ljung, L., 1987. *System Identification—Theory for the User*. Englewood Cliffs, N.J.: Prentice-Hall.

Söderström, T., and P. Stoica, 1988. *System Identification*. Hemel Hempstead, U.K.: Prentice-Hall International.

Johansson, R., 1992. *System Modeling and Identification*. Englewood Cliffs, N.J.: Prentice-Hall.

The regression model is commonly used in many branches of applied mathematics. See, for example:

Draper, N. R., and H. Smith, 1981. *Applied Regression Analysis*, 2nd edition. New York: John Wiley.

Recursive identification and properties of recursive estimators are treated in depth in:

Ljung, L., and T. Söderström, 1983. *Theory and Practice of Recursive Identification*. Cambridge, Mass.: MIT Press.

Goodwin, G. C., and K. S. Sin, 1984. *Adaptive Filtering, Prediction and Control*. Englewood Cliffs, N.J.: Prentice-Hall.

Properties of identification in closed-loop systems are found in:

Wellstead, P. E., and J. M. Edmunds, 1975. "Least-squares identification of closed loop systems." *Int. J. of Control* **21**: 689–699.

Gustafsson, I., L. Ljung, and T. Söderström, 1977. "Identification of processes in closed-loop—Identification and accuracy aspects." *Automatica* **13**: 59–75.

Good sources are also the proceedings of the IFAC symposia on system identification that have been held every third year since 1967.

The least-squares method was first presented in:

Gauss, K. F., 1809. *Theoria motus corposum coelestium,* (In Latin). English translation: *Theory of the Motion of the Heavenly Bodies.* New York: Dover, 1963.

The numerical solution to least-squares problems is well treated in:

Lawson, C. L., and R. J. Hanson, 1974. *Solving Least Squares Problems.* Englewood Cliffs, N.J.: Prentice-Hall.

Recursive square root algorithms are discussed in:

Bierman, G. J., 1977. *Factorization Methods for Discrete Sequential Estimation.* New York: Academic Press.

The exponential weighting of data in the least-squares estimation was first introduced in:

Plackett, R. L., 1950. "Some theorems in least squares." *Biometrika* **37**: 149–157.

Different ways to modify recursive estimators to follow time-varying parameters are suggested in:

Irving, E., 1980. "New developments in improving power network stability with adaptive generator control." In *Applications of Adaptive Control,* eds. K. S. Narendra and R. V. Monopoli. New York: Academic Press.

Fortescue, T. R., L. S. Kershenbaum, and B. E. Ydstie, 1981. "Implementation of self-tuning regulators with variable forgetting factors." *Automatica* **17**: 831–835.

Kulhavý, R., and M. Kárný, 1984. "Tracking of slowly varying parameters by directional forgetting." Paper 14.4/E-4, *9th IFAC World Congress.* Budapest.

Hägglund, T., 1985. "Recursive estimation of slowly time-varying parameters." *Preprints 7th IFAC Symposium on Identification and System Parameter Estimation,* pp. 1255–1260. York, U.K.

Kulhavý, R., 1987. "Restricted exponential forgetting in real-time identification." *Automatica* **23**: 589–600.

The Kaczmarz's algorithm was first published in German in 1937

Kaczmarz, S., 1937. "Angenäherte Auflösung von Systemen linearer Gleichunger." *Bulletin International de l'Academie Polonaise des Sciences. Lett A*: 355–357.

An English translation of the original paper is found in

Kaczmarz, S., 1993. "Approximate solution of systems of linear equations." *Int. J. Control* **57**: 1269–1271.

Estimation of continuous-time models is treated in, for instance:

Young, P. C., 1981. "Parameter estimation for continuous-time models: A survey." *Automatica* **17**: 23–29.

Unbehauen, H., and G. P. Rao, 1987. *Identification of Continuous-Time Systems.* Amsterdam: North-Holland.

Sastry, S., and M. Bodson, 1989. *Adaptive Control: Stability Convergence and Robustness.* Englewood Cliffs, N.J.: Prentice-Hall.

The LMS method is extensively treated in:

Widrow, B., and S. D. Stearns, 1985. *Adaptive Signal Processing.* Englewood Cliffs, N.J.: Prentice-Hall.

Haykins, S., 1991. *Adaptive Filter Theory*, 2nd edition. Englewood Cliffs, N.J.: Prentice-Hall.

A tutorial survey of algorithms for tracking time-varying systems is found in:

Ljung, L., and S. Gunnarsson, 1990. "Adaptation and tracking in system identification–A survey." *Automatica* **26**: 7–21.

CHAPTER 3

DETERMINISTIC SELF-TUNING REGULATORS

3.1 INTRODUCTION

Development of a control system involves many tasks such as modeling, design of a control law, implementation, and validation. The *self-tuning regulator* (STR) attempts to automate several of these tasks. This is illustrated in Fig. 3.1, which shows a block diagram of a process with a self-tuning regulator. It is assumed that the structure of a process model is specified. Parameters of the model are estimated on-line, and the block labeled "Estimation" in Fig. 3.1 gives an estimate of the process parameters. This block is a recursive estimator of the type discussed in Chapter 2. The block labeled "Controller design" contains computations that are required to perform a design of a controller with a specified method and a few design parameters that can be chosen externally. The design problem is called the *underlying design problem* for systems with known parameters. The block labeled "Controller" is an implementation of the controller whose parameters are obtained from the control design.

The name "self-tuning regulator" comes from one of the early papers. The main reason for using an adaptive controller is that the process or its environment is changing continuously. It is difficult to analyze such systems. To simplify the problem, it can be assumed that the process has constant but unknown parameters. The term *self-tuning* was used to express the property that the controller parameters converge to the controller that was designed if the process was known. An interesting result was that this could happen even if the model structure was incorrect.

The tasks shown in the block diagram can be performed in many different ways. There are many possible choices of model and controller structures.

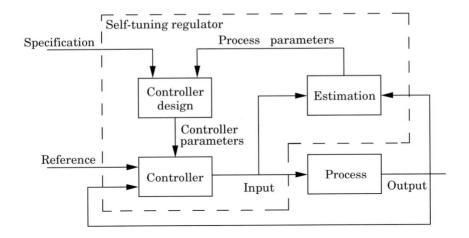

Figure 3.1 Block diagram of a self-tuning regulator.

Estimation can be performed continuously or in batches. In digital implementations, which are most common, different sampling rates can be used for the controller and the estimator. It is also possible to use hybrid schemes in which control is performed continuously and the parameters are updated discretely. Parameter estimation can be done in many ways, as was discussed in Chapter 2. There is also a large variety of techniques that can be used for control system design. It is also possible to consider nonlinear models and nonlinear design techniques. Although many estimation methods will provide estimates of parameter uncertainties, these are typically not used in the control design. The estimated parameters are treated as if they are true in designing the controller. This is called the *certainty equivalence principle*.

The controller shown in Fig. 3.1 is thus a very rich structure. Only a few possibilities have been investigated. The choice of model structure and its parameterization are important issues for self-tuning regulators. A straightforward approach is to estimate the parameters of the transfer function of the process. This gives an *indirect adaptive algorithm*. The controller parameters are not updated directly, but rather indirectly via the estimation of the process model.

Often, the model can be reparameterized such that the controller parameters can be estimated directly. That is, a *direct adaptive algorithm* is obtained (compare with the discussion of the direct MRAS in Section 1.4). There has been some confusion in the nomenclature. In the self-tuning context, indirect methods have often been called *explicit* self-tuning control, since the process parameters have been estimated. Direct updating of the controller parameters has been called *implicit* self-tuning control. In the early papers on adaptive control a direct adaptive controller was often referred to as an adaptive controller without identification. It is convenient to divide the algorithms into indirect and direct self-tuners, but the distinction should not be overemphasized. The basic idea in both types of algorithms is to identify some parameters that are related to the process and/or the specifications of the closed-loop system.

The purpose of this chapter is to present the basic ideas and to illustrate some properties of self-tuning regulators. It is assumed that the process model and the controller are linear systems. The discussion will also be restricted to single-input, single-output (SISO) systems. In most cases we will assume that the controller is sampled and that estimation and control are performed with the same sampling rates. Recursive least squares will be used for parameter estimation, and the design method is a deterministic pole placement. The reasons for these choices are mostly didactic; we would like to present simple methods that can be used in practice. Least-squares estimation was discussed in Chapter 2. In Section 3.2 we present the design method used in a simple setting. A straightforward combination of least-squares estimation and pole placement design gives an indirect self-tuning regulator. The sampled version is described in Section 3.3, and the continuous-time version is described in Section 3.4. In Section 3.5 we show how a direct self-tuning regulator is obtained. In this section we also discuss hybrid algorithms that combine features of direct and indirect algorithms. In Section 3.6 we discuss how to modify the adaptive controllers so that they can deal with disturbances.

3.2   POLE PLACEMENT DESIGN

A simple method for control design will now be presented. The idea is to determine a controller that gives desired closed-loop poles. In addition it is required that the system follows command signals in a specified manner. This is a simple method that, properly applied, can give practically useful controllers as well as useful understanding of adaptive control. It is also the key to understand the similarities between the self-tuning regulator and the model reference adaptive controller.

Process Model

It is assumed that the process is described by the single-input, single-output (SISO) system

$$A(q)y(t) = B(q)\,(u(t) + v(t))$$

where y is the output, u is the input of the process, and v is a disturbance. The disturbances can enter the system in many ways. Here it has been assumed that they enter at the process input. For linear systems in which the superposition principle holds, an equivalent input disturbance can always be found. Furthermore, A and B are polynomials in the forward shift operator q. The polynomials have the degrees $\deg A = n$ and $\deg B = \deg A - d_0$. Pa-

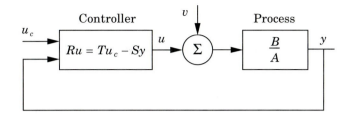

Figure 3.2   A general linear controller with two degrees of freedom.

rameter d_0, which is called the *pole excess*, represents the integer part of the ratio of time delay and sampling period. It is sometimes convenient to write the process model in the delay operator q^{-1}. This can be done by introducing the *reciprocal polynomial*

$$A^*(q^{-1}) = q^{-n}A(q)$$

where $n = \deg A$. The model can then be written as

$$A^*(q^{-1})y(t) = B^*(q^{-1})\left(u(t - d_0) + v(t - d_0)\right)$$

where

$$A^*(q^{-1}) = 1 + a_1q^{-1} + \ldots + a_nq^{-n}$$
$$B^*(q^{-1}) = b_0 + b_1q^{-1} + \ldots + b_mq^{-m}$$

with $m = n - d_0$. Notice that since n was defined as the degree of the system, we have $n \geq m + d_0$, and trailing coefficients of A^* may thus be zero.

We will mostly deal with discrete time systems. Since the design method is purely algebraic, we can handle continuous systems simultaneously by writing the model

$$Ay(t) = B\left(u(t) + v(t)\right) \tag{3.1}$$

where A and B denote polynomials in either the differential operator $p = d/dt$ or the forward shift operator q. It is assumed that A and B are *relatively prime*, that is, that they do not have any common factors. Further, it is assumed that A is *monic*, that is, that the coefficient of the highest power in A is unity.

A general linear controller can be described by

$$Ru(t) = Tu_c(t) - Sy(t) \tag{3.2}$$

where R, S, and T are polynomials. This control law represents a negative feedback with the transfer operator $-S/R$ and a feedforward with the transfer operator T/R. It thus has two degrees of freedom. A block diagram of the closed-loop system is shown in Fig. 3.2. Elimination of u between Eqs. (3.1) and (3.2) gives the following equations for the closed-loop system:

$$y(t) = \frac{BT}{AR + BS} u_c(t) + \frac{BR}{AR + BS} v(t)$$
$$u(t) = \frac{AT}{AR + BS} u_c(t) - \frac{BS}{AR + BS} v(t)$$

(3.3)

The closed-loop characteristic polynomial is thus

$$AR + BS = A_c \qquad (3.4)$$

The key idea of the design method is to specify the desired closed-loop characteristic polynomial A_c. The polynomials R and S can then be solved from Eq. (3.4). Notice that in the design procedure we consider polynomial A_c to be a design parameter that is chosen to give desired properties to the closed-loop system. Equation (3.4), which plays a fundamental role in algebra, is called the *Diophantine equation*. It is also called the *Bezout identity* or the *Aryabhatta equation*. The equation always has solutions if the polynomials A and B do not have common factors. The solution may be poorly conditioned if the polynomials have factors that are close. The solution can be obtained by introducing polynomials with unknown coefficients and solving the linear equations obtained. The solution of the equation is discussed in detail in Chapter 11.

Model-Following

The Diophantine equation (3.4) determines only the polynomials R and S. Other conditions must be introduced to also determine the polynomial T in the controller (3.2). To do this, we will require that the response from the command signal u_c to the output be described by the dynamics

$$A_m y_m(t) = B_m u_c(t) \qquad (3.5)$$

It then follows from Eqs. (3.3) that the following condition must hold:

$$\frac{BT}{AR + BS} = \frac{BT}{A_c} = \frac{B_m}{A_m} \qquad (3.6)$$

This model-following condition says that the response of the closed-loop system to command signals is as specified by the model (3.5). Whether model-following can be achieved depends on the model, the system, and the command signal. If it is possible to make the error equal to zero for all command signals, then *perfect model-following* is achieved.

The consequences of the model-following condition will now be explored. Equation (3.6) implies that there are cancellations of factors of BT and A_c. Factor the B polynomial as

$$B = B^+ B^- \qquad (3.7)$$

where B^+ is a monic polynomial whose zeros are stable and so well damped that they can be canceled by the controller and B^- corresponds to unstable or poorly damped factors that cannot be canceled. It thus follows that B^- must be a factor of B_m. Hence

$$B_m = B^- B'_m \tag{3.8}$$

Since B^+ is canceled, it must be a factor of A_c. Furthermore, it follows from Eq. (3.6) that A_m must also be a factor of A_c. The closed-loop characteristic polynomial thus has the form

$$A_c = A_o A_m B^+ \tag{3.9}$$

Since B^+ is a factor of B and A_c, it follows from Eq. (3.4) that it also divides R. Hence

$$R = R' B^+ \tag{3.10}$$

and the Diophantine equation (3.4) reduces to

$$AR' + B^- S = A_o A_m = A'_c \tag{3.11}$$

Introducing Eqs. (3.7), (3.8), and (3.9) into Eq. (3.6) gives

$$T = A_o B'_m \tag{3.12}$$

Causality Conditions

To obtain a controller that is causal in the discrete-time case or proper in the continuous-time case, we must impose the conditions

$$\begin{aligned} \deg S &\le \deg R \\ \deg T &\le \deg R \end{aligned} \tag{3.13}$$

The Diophantine equation (3.4) has many solutions because if R^0 and S^0 are solutions, then so are

$$\begin{aligned} R &= R^0 + QB \\ S &= S^0 - QA \end{aligned} \tag{3.14}$$

where Q is an arbitrary polynomial. Since there are many solutions, we may select the solution that gives a controller of lowest degree. We call this the minimum-degree solution. Since $\deg A > \deg B$, the term of highest order on the left-hand side of Eq. (3.4) is AR. Hence

$$\deg R = \deg A_c - \deg A$$

Because of Eqs. (3.14) there is always a solution such that $\deg S < \deg A = n$. We can thus always find a solution in which the degree of S is at most $\deg A - 1$. This is called the *minimum-degree solution* to the Diophantine equation. The condition $\deg S \le \deg R$ thus implies that

$$\deg A_c \ge 2 \deg A - 1$$

It follows from Eq. (3.12) that the condition $\deg T \leq \deg R$ implies that

$$\deg A_m - \deg B'_m \geq \deg A - \deg B^+$$

Adding $\deg B^-$ to both sides, we find that this is equivalent to $\deg A_m - \deg B_m \geq d_0$. This means that in the discrete-time case the time delay of the model must be at least as large as the time delay of the process, which is a very natural condition. Summarizing, we find that the causality conditions (3.13) can be written as

$$\deg A_c \geq 2 \deg A - 1$$
$$\deg A_m - \deg B_m \geq \deg A - \deg B = d_0 \tag{3.15}$$

It is natural to choose a solution in which the controller has the lowest possible degree. In the discrete-time case it is also reasonable to require that there be no extra delay in the controller. This implies that polynomials R, S, and T should have the same degrees. The following design procedure is then obtained.

A L G O R I T H M 3.1   Minimum-degree pole placement (MDPP)

Data: Polynomials A, B.

Specifications: Polynomials A_m, B_m, and A_o.

Compatibility Conditions:

$$\deg A_m = \deg A$$
$$\deg B_m = \deg B$$
$$\deg A_o = \deg A - \deg B^+ - 1$$
$$B_m = B^- B'_m$$

Step 1:   Factor B as $B = B^+ B^-$, where B^+ is monic.

Step 2:   Find the solution R' and S with $\deg S < \deg A$ from

$$AR' + B^- S = A_o A_m$$

Step 3:   Form $R = R'B^+$ and $T = A_o B'_m$, and compute the control signal from the control law

$$Ru = Tu_c - Sy \qquad \qquad \Box$$

There are special cases of the design procedure that are of interest.

All zeros are canceled   The design procedure simplifies significantly in the special case in which all process zeros are canceled; then $\deg A_o = \deg A - \deg B - 1$. It is natural to choose $B_m = A_m(1)q^{d_0}$. Then the factorization in Step 1 is very simple, and we get $B^- = b_0$, $B^+ = B/b_0$. Furthermore,

$T = A_m(1)q^{d_0}/b_0$, and the closed-loop characteristic polynomial becomes $A_c = B^+ A_c'$. The Diophantine equation in Step 2 reduces to

$$AR' + b_0 S = A_c' = A_o A_m$$

This equation is easy to solve because R' is the quotient and $b_0 S$ is the remainder when $A_o A_m$ is divided by A. However, all process zeros must be stable and well damped to allow cancellation.

No zeros are canceled   The factorization in Step 2 also becomes very simple if no zeros are canceled. We have $B^+ = 1$, $B^- = B$, and $B_m = \beta B$, where $\beta = A_m(1)/B(1)$. Furthermore, $\deg A_o = \deg A - \deg B - 1$ and $T = \beta A_o$. The closed-loop characteristic polynomial is $A_c = A_o A_m$, and the Diophantine equation in Step 2 becomes

$$AR + BS = A_c = A_o A_m$$

Examples

The model-following design is illustrated by three examples

EXAMPLE 3.1   Model-following with zero cancellation

Consider a continuous-time process described by the transfer function

$$G(s) = \frac{1}{s(s + 1)} \tag{3.16}$$

This can be regarded as a normalized model for a motor. The pulse transfer operator for the sampling period $h = 0.5$ s is

$$H(q) = \frac{B(q)}{A(q)} = \frac{b_0 q + b_1}{q^2 + a_1 q + a_2} = \frac{0.1065q + 0.0902}{q^2 - 1.6065q + 0.6065} \tag{3.17}$$

We have $\deg A = 2$ and $\deg B = 1$. The design procedure thus gives a first-order controller, and the closed-loop system will be of third order. The sampled data system has a zero in -0.84 and poles in 1 and 0.61. Let the desired closed-loop system be

$$\frac{B_m(q)}{A_m(q)} = \frac{b_{m0} q}{q^2 + a_{m1} q + a_{m2}} = \frac{0.1761q}{q^2 - 1.3205q + 0.4966} \tag{3.18}$$

This corresponds to a natural frequency of 1 rad/s and a relative damping of 0.7. Parameter b_{m0} is chosen so that the static gain is unity. This model satisfies the compatibility conditions because it has the same pole excess as the process and the process zero is stable although poorly damped. To apply the design procedure in Algorithm 3.1, we first factor the polynomial B, and we obtain

$$B^+(q) = q + b_1/b_0$$
$$B^-(q) = b_0$$
$$B_m'(q) = b_{m0}q/b_0$$

Since the process is of second order, the polynomials R, S, and T will all be of first order. Polynomial R' is thus of degree zero. Since the polynomial is monic, we have $R' = 1$. Since $\deg B^+ = 1$, it follows from the compatibility conditions that $\deg A_o = 0$. Choose

$$A_o(q) = 1$$

The Diophantine equation (3.11) then becomes

$$(q^2 + a_1 q + a_2) \cdot 1 + b_0(s_0 q + s_1) = q^2 + a_{m1} q + a_{m2}$$

Equating coefficients of equal power of q gives

$$a_1 + b_0 s_0 = a_{m1}$$
$$a_2 + b_0 s_1 = a_{m2}$$

These equations can be solved if $b_0 \neq 0$. The solution is

$$s_0 = \frac{a_{m1} - a_1}{b_0}$$
$$s_1 = \frac{a_{m2} - a_2}{b_0}$$

The controller is thus characterized by the polynomials

$$R(q) = B^+ = q + \frac{b_1}{b_0}$$
$$S(q) = s_0 q + s_1$$
$$T(q) = A_o B'_m = \frac{b_{m0} q}{b_0} \qquad \qquad \Box$$

The process in Example 3.1 has a zero that is stable but poorly damped. The continuous-time equivalent corresponds to a zero with relative damping $\zeta = 0.06$. We will therefore also determine a controller that does not cancel the zero. This is done in the next example.

EXAMPLE 3.2   **Model-following without zero cancellation**

Consider the same process as in Example 3.1, but use a control design in which there is no cancellation of the process zero. Since the process is of second order, the minimum-degree solution has polynomials R, S, and T of first order and the closed-loop system will be of third order. Since no zero is canceled, it follows from the compatibility condition in Algorithm 3.1 that $\deg A_0 = 1$. Since no process zeros are canceled, we have

$$B^+ = 1$$
$$B^- = B = b_0 q + b_1$$

It also follows from the compatibility conditions that the model must have the same zero as the process. The desired closed-loop transfer operator is thus

$$H_m(q) = \beta \frac{b_0 q + b_1}{q^2 + a_{m1} q + a_{m2}} = \frac{b_{m0} q + b_{m1}}{q^2 + a_{m1} q + a_{m2}}$$

where $b_{m0} = \beta b_0$ and

$$\beta = \frac{1 + a_{m1} + a_{m2}}{b_0 + b_1}$$

which gives unit steady state gain. The Diophantine equation (3.4) becomes

$$(q^2 + a_1 q + a_2)(q + r_1) + (b_0 q + b_1)(s_0 q + s_1) = (q^2 + a_{m1} q + a_{m2})(q + a_o) \quad (3.19)$$

Putting $q = -b_1/b_0$ and solving for r_1, we get

$$r_1 = \frac{b_1}{b_0} + \frac{(b_1^2 - a_{m1} b_0 b_1 + a_{m2} b_0^2)(-b_1 + a_o b_0)}{b_0(b_1^2 - a_1 b_0 b_1 + a_2 b_0^2)}$$

$$= \frac{a_o a_{m2} b_0^2 + (a_2 - a_{m2} - a_o a_{m1}) b_0 b_1 + (a_o + a_{m1} - a_1) b_1^2}{b_1^2 - a_1 b_0 b_1 + a_2 b_0^2} \quad (3.20)$$

Notice that the denominator is zero if polynomials $A(q)$ and $B(q)$ have a common factor. Equating coefficients of terms q^2 and q^0 in Eq. (3.19) gives

$$s_0 = \frac{b_1(a_o a_{m1} - a_2 - a_{m1} a_1 + a_1^2 + a_{m2} - a_1 a_o)}{b_1^2 - a_1 b_0 b_1 + a_2 b_0^2}$$

$$+ \frac{b_0(a_{m1} a_2 - a_1 a_2 - a_o a_{m2} + a_o a_2)}{b_1^2 - a_1 b_0 b_1 + a_2 b_0^2}$$

$$s_1 = \frac{b_1(a_1 a_2 - a_{m1} a_2 + a_o a_{m2} - a_o a_2)}{b_1^2 - a_1 b_0 b_1 + a_2 b_0^2} \quad (3.21)$$

$$+ \frac{b_0(a_2 a_{m2} - a_2^2 - a_o a_{m2} a_1 + a_o a_2 a_{m1})}{b_1^2 - a_1 b_0 b_1 + a_2 b_0^2}$$

Furthermore, it follows from Eq. (3.12) that

$$T(q) = \beta A_o(q) = \beta(q + a_o) \qquad \Box$$

Since the design method is purely algebraic, there is no difference between discrete-time systems and continuous-time systems. We illustrate this by an example.

EXAMPLE 3.3   **Continuous-time system**

The process discussed in Examples 3.1 and 3.2 has the transfer function

$$G(s) = \frac{b}{s(s + a)}$$

with $a = 1$ and $b = 1$. The design procedure given by Algorithm 3.1 will now be used to find a continuous-time controller. Since the process is of second order, the closed-loop system will be of third order and the minimum-degree controller is of first order. Polynomial A_m has degree two, B_m is a constant, and A_o has degree one. We choose

$$A_o(s) = s + a_o$$

and let the desired response be specified by the transfer function.

$$\frac{B_m(s)}{A_m(s)} = \frac{\omega^2}{s^2 + 2\zeta\omega s + \omega^2}$$

The Diophantine equation (3.4) becomes

$$s(s + a)(s + r_1) + b(s_0 s + s_1) = (s^2 + 2\zeta\omega s + \omega^2)(s + a_o)$$

Equating coefficients of equal powers of s gives the equations

$$a + r_1 = 2\zeta\omega + a_o$$
$$ar_1 + bs_0 = \omega^2 + 2\zeta\omega a_o$$
$$bs_1 = \omega^2 a_o$$

If $b \neq 0$, these equations can be solved, and we get

$$r_1 = 2\zeta\omega + a_o - a$$
$$s_0 = \frac{a_o 2\zeta\omega + \omega^2 - ar_1}{b}$$
$$s_1 = \frac{\omega^2 a_o}{b}$$

Furthermore, we have $B^+ = 1$, $B^- = b$, and $B'_m = \omega^2/b$. It then follows from Eq. (3.12) that

$$T(s) = B'_m(s)A_o(s) = \frac{\omega^2}{b}(s + a_o) \qquad \square$$

An Interpretation of Polynomial A_o

It is possible to give an interpretation of the polynomial A_o that appears in the minimum-degree pole placement solution in the case in which no process zeros are canceled. To do this, we observe that the pole placement problem can also be solved with state feedback and an observer. The closed-loop dynamics are then composed of two parts: one that corresponds to the state feedback and another that corresponds to the observer dynamics. For a system of degree n it is also known that it is sufficient to use an observer of degree $n - 1$. When no process zeros are canceled, the closed-loop characteristic polynomial in our case is $A_m A_o$, where A_m is of degree n and A_o is of degree $n - 1$. By this analogy we can interpret the polynomial A_m as being associated with the state feedback and A_o as being associated with the observer. We will therefore call A_o the *observer polynomial*. In a system with state feedback it is also natural to introduce the command signals in such a way that they do not generate observer errors. This means that the observer polynomial is canceled in the transfer function from command signal to process output.

Relations to Model-Following

Many other design methods can be related to pole placement. We will now show that pole placement can be interpreted as a model-following design. This is of interest because much work on MRAS is formulated in terms of model-following. Model-following generally means that the response of a closed-loop system to command signals is specified by a given model. This means that both poles and zeros of the model are specified by the user. Pole placement, on the other hand, specifies only the closed-loop poles. In the minimum-degree pole placement procedure we did, however, introduce some auxiliary conditions that included the process zeros. We will now show that the control law given by Eq. (3.2) can be interpreted as model-following. It follows from Eqs. (3.11) and (3.12) that

$$\frac{T}{R} = \frac{A_o B'_m}{R} = \frac{(AR' + B^- S)B'_m}{A_m R} = \frac{AB_m}{BA_m} + \frac{SB_m}{RA_m}$$

The control law of Eq. (3.2) can be written as

$$u = \frac{T}{R}u_c - \frac{S}{R}y = \frac{AB_m}{BA_m}u_c + \frac{SB_m}{RA_m}u_c - \frac{S}{R}y$$

$$= \frac{AB_m}{BA_m}u_c - \frac{S}{R}(y - y_m)$$

A block diagram representation of this controller is given in Fig. 3.3. The figure shows that the controller can be interpreted as a combination of a feedforward controller and a feedback controller. The feedforward controller attempts to cancel the plant dynamics and replace it with the response of the model B_m/A_m. Also the feedback attempts to make the output follow this model. It is thus clear that the control law (3.2) can indeed be interpreted as a model-following algorithm.

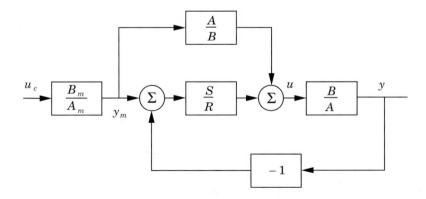

Figure 3.3   Alternative representation of model-following based on output feedback.

Notice that Fig. 3.3 is useful for the purpose of giving insight but that the controller cannot be implemented as shown in the figure because the inverse process model A/B is generally not realizable. Furthermore, A/B will be unstable if the system is non-minimum phase. However, the cascade combination of the reference model and the inverse process model is realizable if the model-following problem is well posed, that is, if Eqs. (3.13) are satisfied. Notice that the reference model and the inverse process model can be nonlinear without causing any stability problems because they appear only as part of a feedforward compensator.

Summary

In this section we have presented a straightforward design procedure that is relatively easy to use. The key problem in applying pole placement is to choose the desired closed-loop poles and the desired response to command signals. The choice is easy for low-order systems, but it may be difficult for systems of high order when many poles must be specified. Bad choices may result in a closed-loop system with poor sensitivity. In later chapters we will discuss this problem in more detail.

In the sampled-data case the sampling interval is a crucial design parameter. It is important to choose the sampling interval in relation to the desired closed-loop poles.

3.3   INDIRECT SELF-TUNING REGULATORS

Methods for estimating parameters of the model given by Eq. (3.1) were presented in Chapter 2. These methods will now be combined with the design method of Section 3.2 to obtain a simple self-tuning regulator. For simplicity it will be assumed that the disturbance v in Eq. (3.1) is zero.

Estimation

Several of the recursive estimation methods outlined in Chapter 2 can be used to estimate the coefficients of the A and B polynomials. The equations for recursive least-squares estimation will be used. The process model (3.1) can be written explicitly as

$$y(t) = -a_1 y(t-1) - a_2 y(t-2) - \ldots - a_n y(t-n)$$
$$+ b_0 u(t-d_0) + \ldots + b_m u(t-d_0-m)$$

Notice that the degree of the system is $\max(n, d_0 + m)$. The model is linear in the parameters and can written as

$$y(t) = \varphi^T(t-1)\theta$$

where

$$\theta^T = \left(\begin{array}{ccccccc} a_1 & a_2 & \ldots & a_n & b_0 & \ldots & b_m \end{array} \right)$$

$$\varphi^T(t-1) = \left(\begin{array}{cccccc} -y(t-1) & \ldots & -y(t-n) & u(t-d_0) & \ldots & u(t-d_0-m) \end{array} \right)$$

The least-squares estimator with exponential forgetting is given by

$$
\begin{aligned}
\hat{\theta}(t) &= \hat{\theta}(t-1) + K(t)\varepsilon(t) \\
\varepsilon(t) &= y(t) - \varphi^T(t-1)\hat{\theta}(t-1) \\
K(t) &= P(t-1)\varphi(t-1)\left(\lambda + \varphi^T(t-1)P(t-1)\varphi(t-1)\right)^{-1} \\
P(t) &= \left(I - K(t)\varphi^T(t-1)\right)P(t-1)/\lambda
\end{aligned}
\tag{3.22}
$$

(Compare with Eq. (2.21).) If the input signal to the process is sufficiently exciting and the structure of the estimated model is compatible with the process, the estimates will converge to their true values. It takes $\max(n, m + d_0)$ sampling periods before the regression vector is defined. In the deterministic case it takes at least $n + m + 1$ additional sampling periods to determine the $n + m + 1$ parameters of the model, assuming that the process input is persistently exciting. It thus takes at least

$$N = n + m + 1 + \max(n, m + d_0) \tag{3.23}$$

sampling periods for the algorithm to converge. With recursive least squares initialized with a large P-matrix it may take a few more steps. Since the process input is generated by feedback, it may be difficult to assert that it is persistently exciting. Presence of process noise may also make convergence much slower. Convergence issues will be discussed further in Chapter 6.

An Indirect Self-Tuner

Combining the recursive least squares (RLS) estimator given by Eqs. (3.22) with the minimum-degree pole placement method (MDPP) for controller design given by Algorithm 3.1, we obtain the following self-tuning regulator.

ALGORITHM 3.2   Indirect self-tuning regulator using RLS and MDPP

Data: Given specifications in the form of a desired closed-loop pulse transfer operator B_m/A_m and a desired observer polynomial A_o.

Step 1:   Estimate the coefficients of the polynomials A and B in Eq. (3.1) using the recursive least-squares method given by Eqs. (3.22).

Step 2:   Apply the minimum-degree pole placement method given by Algorithm 3.1 where polynomials A and B are the estimates obtained in Step 1. The polynomials R, S, and T of the control law are then obtained.

Step 3: Calculate the control variable from Eq. (3.2), that is,

$$Ru(t) = Tu_c(t) - Sy(t)$$

Repeat Steps 1, 2, and 3 at each sampling period. Notice that there are some variations in the algorithm depending on the cancellations of the process zeros. Also notice that it is not necessary to perform Steps 1 and 2 at each sampling interval. ☐

Examples

The properties of indirect self-tuning regulators are illustrated by the following two examples.

EXAMPLE 3.4   Indirect self-tuner with cancellation of process zero

Let the process be the same as in Example 3.1 and assume that the process zero is canceled. The specifications are the same as in Example 3.1, that is, to obtain a closed-loop characteristic polynomial A_m. The parameters of the model

$$y(t) + a_1 y(t-1) + a_2 y(t-2) = b_0 u(t-1) + b_1 u(t-2)$$

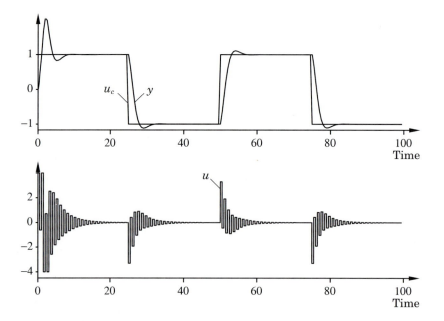

Figure 3.4 Output and input in using an indirect self-tuning regulator to control the system in Example 3.1. Notice the "ringing" in the control signal due to cancellation of the zero at -0.84.

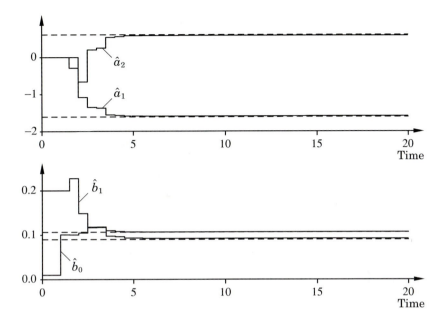

Figure 3.5   Parameter estimates corresponding to the simulation in Fig. 3.4. The true parameters are shown by dashed lines.

which has the same structure as Eq. (3.17), are estimated by using the least-squares algorithm. Algorithm 3.2 is used for the self-tuning regulator. The calculations, which were done in Example 3.1, give the control law

$$u(t) + r_1 u(t-1) = t_0 u_c(t) - s_0 y(t) - s_1 y(t-1)$$

The controller parameters were expressed as functions of the model parameters and the specifications. Figure 3.4 shows the process output and the control signal in a simulation of the process with the self-tuner when the command signal is a square wave. The output converges to the model output after an initial transient. The control signal has a severe oscillation ("ringing") with a period of two sampling periods. This is due to the cancellation of the process zero at $z = -b_1/b_0 = -0.84$. This oscillation is a consequence of a bad choice of the underlying design methodology. The initial transient depends critically on the initial values of the estimator. In this particular case these values were $\hat{a}_1(0) = \hat{a}_2(0) = 0$, $\hat{b}_0(0) = 0.01$, and $\hat{b}_1(0) = 0.2$. Notice that it is necessary that $\hat{b}_0 \neq 0$. (Compare with Example 3.1.) The initial covariance matrix was diagonal with $P(1,1) = P(2,2) = 100$ and $P(3,3) = P(4,4) = 1$. The reason for using different values for parameters $\hat{a}_i$ and $\hat{b}_i$ is that these parameters differ by an order of magnitude.

The parameter estimates are shown in Fig. 3.5. The behavior of the estimates depends critically on the initial values of the estimator. Notice that the estimates converge quickly. They are close to their correct values already at time $t = 5$. The estimates obtained at time $t = 100$ are

$$\hat{a}_1(100) = -1.60 \quad (-1.6065) \qquad \hat{b}_0(100) = 0.107 \quad (0.1065)$$
$$\hat{a}_2(100) = 0.60 \quad (0.6065) \qquad \hat{b}_1(100) = 0.092 \quad (0.0902)$$

These values are quite close to the true values, which are given in parentheses. The controller parameters obtained at time $t = 100$ are

$$r_1(100) = 0.85 \quad (0.8467) \qquad t_0(100) = 1.65 \quad (1.6531)$$
$$s_0(100) = 2.64 \quad (2.6852) \qquad s_1(100) = -0.99 \quad (-1.0321)$$

□

The system in Example 3.4 behaves quite well, apart from the "ringing" control signal. This can be avoided by using a design in which the process zero is not canceled. The consequences of this are illustrated in the next example.

EXAMPLE 3.5   **Indirect self-tuner without cancellation of process zero**

Consider the same process as in Example 3.4, but use a control design in which there is no cancellation of the process zero. The parameters are estimated in the same way as in Example 3.4, but the control law is now computed as in Example 3.2. Polynomial A_o is of first order. As in the previous examples the initial transient depends critically on the initial state of the

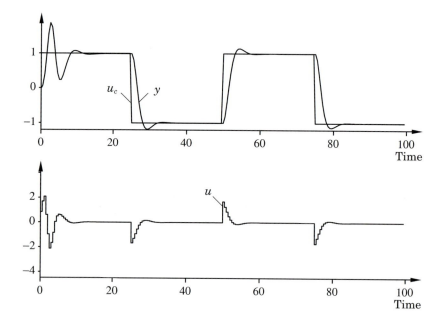

Figure 3.6   Same as in Fig. 3.4 but without cancellation of the process zero.

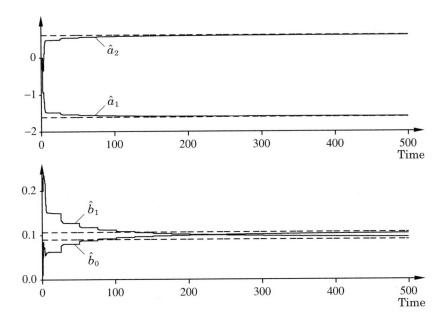

Figure 3.7 Parameter estimates corresponding to the simulation in Fig. 3.6. The true parameter values are indicated by dashed lines.

recursive estimator. For the design calculation it must be required that initial values are chosen so that polynomials A and B do not have a common factor. In this case the initial estimates were chosen to be $\hat{a}_1(0) = \hat{a}_2(0) = 0$, $\hat{b}_0(0) = 0.01$, and $\hat{b}_1(0) = 0.2$. The P-matrix was initialized as a diagonal matrix with $P(1,1) = P(2,2) = 100$ and $P(3,3) = P(4,4) = 1$ as in Example 3.4. Figure 3.6 shows results of a simulation of the direct algorithm with $a_o = 0$. Notice that the behavior of the process output is quite similar to that in Fig. 3.4 but that there is no "ringing" in the control signal. The parameter estimates are shown in Fig. 3.7. The values obtained at time $t = 100$ are

$$\hat{a}_1(100) = -1.57 \quad (-1.6065) \qquad \hat{b}_0(100) = 0.092 \quad (0.1065)$$
$$\hat{a}_2(100) = 0.57 \quad (0.6065) \qquad \hat{b}_1(100) = 0.112 \quad (0.0902)$$

The true values are given in parentheses. The controller parameters at time $t = 100$ are

$$r_1(100) = 0.114 \quad (0.1111) \qquad t_0(100) = 0.86 \quad (0.8951)$$
$$s_0(100) = 1.44 \quad (1.6422) \qquad s_1(100) = -0.58 \quad (-0.7471)$$

A comparison of Fig. 3.5 and Fig. 3.7 shows that it takes significantly longer for the estimates to converge when no zero is canceled. The reason for this is that the excitation is not as good as when there was "ringing" in the control signal.

There is very little excitation of the system in the periods when the output and the control signals are constant. This explains the steplike behavior of the estimates.

It may seem surprising that the controller already gives the correct steady-state value at time $t = 20$ when the parameter estimates differ so much from their correct values. The controller parameters are

$$r_1(20) = 0.090 \quad (0.1111) \qquad t_0(20) = 0.83 \quad (0.8951)$$

$$s_0(20) = 1.13 \quad (1.6422) \qquad s_1(20) = -0.29 \quad (-0.7471)$$

Since the process has integral action, we have $A(1) = 0$. It then follows from Eq. (3.3) that the static gain from command signal to output is

$$\frac{B(1)T(1)}{A(1)R(1) + B(1)S(1)} = \frac{T(1)}{S(1)}$$

To obtain the correct steady-state value, it is thus sufficient that the controller parameters are such that $S(1) = T(1)$, which in the special case is the same as $t_0 = s_0 + s_1$. When no poles are canceled, it follows from Eq. (3.12) that

$$T(1) = A_o(1)B'_m(1) = A_o(1)\frac{A_m(1)}{\hat{B}(1)}$$

where $\hat{B}$ is the estimated B polynomial. Hence

$$\frac{T(1)}{S(1)} = \frac{A_o(1)A_m(1)}{\hat{B}(1)S(1)} = 1$$

where the last equality follows from Eq. (3.11). Notice that we have $A(1) = 0$. We thus obtain the rather surprising conclusion that the adaptive controller in this case will automatically have parameters such that there will be no steady-state error. $\qquad\qquad\qquad\qquad\qquad\qquad\qquad\qquad\qquad\qquad\qquad\qquad\square$

These examples indicate that the indirect self-tuning algorithm behaves as can be expected and that the estimate of convergence time given by Eq. (3.23) is reasonable. The examples also show the importance of using a good underlying control design. With model-following design it is recommended that cancellation of process zeros is avoided.

Summary

The indirect self-tuning regulator based on model-following given by Algorithm 3.1 is a straightforward application of the idea of self-tuning. The adaptive controller has states that correspond to the parameter estimate φ, the covariance matrix P, the regression vector φ, and the states required for the implementation of the control law. The controller in Example 3.4 has 20 state variables; updating of the covariance matrix P alone requires ten states. The complete codes for the controllers in the examples are listed in the problems at the end of this chapter.

The algorithm can be generalized in many different ways by choosing other recursive estimation methods and other control design techniques. The idea is easy to apply. A detailed discussion of practical implementation is given in Chapter 11.

3.4 CONTINUOUS-TIME SELF-TUNERS

Continuous-time self-tuners can be derived in the same way as discrete-time self-tuners. To show this, consider a system that can be described by the model (3.1) with $v = 0$, that is,

$$A(p)y(t) = B(p)u(t)$$

where $A(p)$ and $B(p)$ are polynomials in the differential operator, $p = d/dt$:

$$A(p) = p^n + a_1 p^{n-1} + \cdots + a_n$$
$$B(p) = b_1 p^{n-1} + \cdots + b_n$$

A self-tuning regulator can be obtained by applying Algorithm 3.1. The only complication is that we now must apply recursive least-squares estimation to the continuous-time model. This was discussed in Section 2.3. Let us recall the key idea. Since it is undesirable to take derivatives, a stable filtering transfer function H_f with a pole excess of n or more is introduced.

If we introduce the filtered signals

$$y_f(t) = H_f y(t) \qquad u_f(t) = H_f u(t)$$

the model (3.1) can be written as

$$p^n y_f(t) = \varphi^T(t)\theta$$

where

$$\varphi(t) = \left(-p^{n-1}y_f \quad \cdots \quad -y_f \quad p^{n-1}u_f \quad \cdots \quad u_f \right)^T$$
$$\theta = \left(a_1 \quad \cdots \quad a_n \quad b_1 \quad \cdots \quad b_n \right)^T$$

By using least squares with exponential forgetting the parameter estimate is then obtained from Theorem 2.5:

$$\frac{d\hat{\theta}(t)}{dt} = P(t)\varphi(t)\left(p^n y_f(t) - \varphi^T(t)\hat{\theta}(t) \right)$$
$$\frac{dP(t)}{dt} = \alpha P(t) - P(t)\varphi(t)\varphi^T(t)P(t)$$

We illustrate the procedure given by Algorithm 3.1 by an example.

EXAMPLE 3.6   **Continuous-time self-tuner**

Consider the system in Example 3.3, in which the process has the transfer function

$$G(s) = \frac{b}{s(s+a)}$$

with $a = 1$ and $b = 1$. Notice that the process has only two unknown parameters, a and b. The regressor filters in the estimator are chosen to be

$$H_f(s) = \frac{1}{A_m(s)}$$

Furthermore, we use an estimator without forgetting, that is, $\alpha = 0$. Assume that it is desired to obtain a closed-loop system with the transfer function

$$G_m(s) = \frac{\omega^2}{s^2 + 2\zeta\omega s + \omega^2}$$

The observer polynomial is chosen to be $A_o(s) = s + a_o$ with $a_o = 2$. The specifications are the same as in Example 3.4, that is, $\zeta = 0.7$ and $\omega = 1$. In Example 3.3 we solved the design problem when the parameters a and b are known. We found that the controller has the form

$$u(t) = -\frac{s_0 p + s_1}{p + r_1} y(t) + \frac{t_0(p + a_o)}{p + r_1} u_c(t)$$

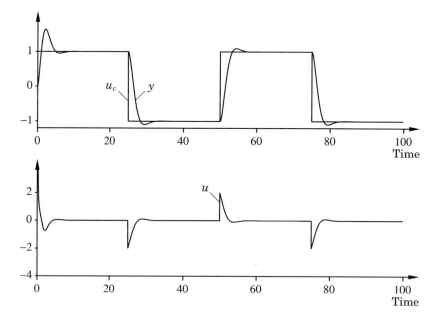

Figure 3.8   Output and input when using a continuous-time indirect self-tuning regulator to control the process in Example 3.6.

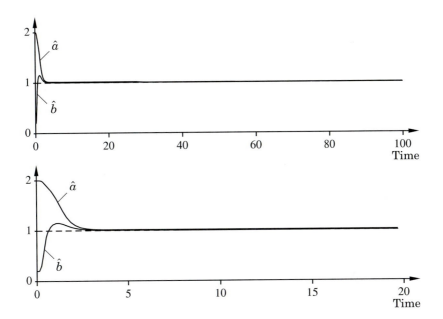

Figure 3.9   Continuous-time parameter estimates corresponding to the simulation in Fig. 3.8. The lower part shows the estimates in an extended time scale.

where the controller parameters are given by

$$r_1 = 2\zeta\omega + a_o - a$$

$$s_0 = \frac{a_o 2\zeta\omega + \omega^2 - a r_1}{b}$$

$$s_1 = \frac{\omega^2 a_o}{b}$$

$$t_0 = \frac{\omega^2}{b}$$

Figure 3.8 shows the process output and the control signal in a simulation. The initial transient depends critically on the initial values of the estimator. In this case we have chosen $\hat{a}(0) = 2$ and $\hat{b}(0) = 0.2$. The initial covariance is diagonal with $P(1,1) = P(2,2) = 100$. The parameter estimates are shown in Fig. 3.9. The estimates obtained at $t = 100$ are

$$\hat{a}(100) = 1.004 \quad (1.0000) \qquad \hat{b}(100) = 1.001 \quad (1.0000)$$

where the true values are given in parentheses. Notice that only two parameters are estimated in this case, whereas four parameters were estimated in Examples 3.4 and 3.5.   □

3.5   DIRECT SELF-TUNING REGULATORS

The design calculations in the indirect self-tuners may be time-consuming and poorly conditioned for some parameter values. It is possible to derive other algorithms in which the design calculations are simplified or even eliminated. The idea is to use the design equations to reparameterize the model in terms of the parameters of the controller. This reparameterization is also the key to understanding the relations between model-reference adaptive systems and self-tuning regulators.

Consider a process described by Eq. (3.1) with $v = 0$, that is,

$$Ay(t) = Bu(t)$$

and let the desired response be given by Eq. (3.5):

$$A_m y_m(t) = B_m u_c(t)$$

The process model will now be reparameterized in terms of the controller parameters. To do this, consider the Diophantine equation (3.11),

$$A_o A_m = AR' + B^- S$$

as an operator identity, and let it operate on $y(t)$. This gives

$$A_o A_m y(t) = R'Ay(t) + B^- Sy(t) = R'Bu(t) + B^- Sy(t)$$

It follows from Eq. (3.10) that

$$R'B = R'B^+ B^- = RB^-$$

Hence

$$A_o A_m y(t) = B^- (Ru(t) + Sy(t)) \qquad (3.24)$$

Notice that this equation can be considered a process model that is parameterized in the coefficients of the polynomials B^-, R, and S. If the parameters in the model given by Eq. (3.24) are estimated, the control law is thus obtained directly without any design calculations. Notice that the model Eq. (3.24) is nonlinear in the parameters because the right-hand side is multiplied by B^-. The difficulties caused by this can be avoided in the special case of minimum-phase systems in which $B^- = b_0$, which is a constant.

Minimum-Phase Systems

If the process dynamics is minimum phase, we have $\deg A_o = \deg A - \deg B - 1$, B^- is simply a constant, and Eq. (3.24) becomes

$$A_m A_o y(t) = b_0 (Ru(t) + Sy(t)) = \tilde{R}u(t) + \tilde{S}y(t) \qquad (3.25)$$

where R is monic, $\tilde{R} = b_0R$, and $\tilde{S} = b_0S$. Since R and $\tilde{R}$ differ only by R being monic, we will *not* use a separate notation in the following discussion. When it is necessary, we will simply note whether or not R is monic.

When all process zeros are canceled, it is also natural to choose specifications so that

$$B_m = q^{d_0}A_m(1)$$

where $d_0 = \deg A - \deg B$. This gives response with minimal delay and unit static gain.

By introducing the parameter vector

$$\theta = \begin{pmatrix} r_0 & \cdots & r_\ell & s_0 & \cdots & s_\ell \end{pmatrix}$$

and the regression vector $\qquad b_o$

$$\varphi(t) = \begin{pmatrix} u(t) & \cdots & u(t-\ell) & y(t) & \cdots & y(t-\ell) \end{pmatrix}$$

the model given by Eq. (3.25) can be written as

$$\eta(t) = A_o^*\left(q^{-1}\right)A_m^*\left(q^{-1}\right)y(t) = \varphi^T(t-d_0)\theta \tag{3.26}$$

Since $\eta(t)$ can be computed from $y(t)$, it can be regarded as an auxiliary output, and a recursive estimate of the parameters can now be obtained as described in Chapter 2.

This estimation method works very well if there is little noise, but the operation $A_o^*(q^{-1})A_m^*(q^{-1})y(t)$ may amplify noise significantly. The following method can be used to overcome this. Rewrite Eq. (3.25) as

$$y(t) = \frac{1}{A_oA_m}\left(Ru(t) + Sy(t)\right) = R^*u_f(t-d_0) + S^*y_f(t-d_0) \tag{3.27}$$

where

$$u_f(t) = \frac{1}{A_o^*(q^{-1})A_m^*(q^{-1})}u(t) \qquad\qquad f \; = \; \text{FILTER}$$

$$y_f(t) = \frac{1}{A_o^*(q^{-1})A_m^*(q^{-1})}y(t) \tag{3.28}$$

and $d_0 = \deg A - \deg B$. We have further assumed that $\deg R = \deg S = \deg(A_oA_m) - d_0 = \ell$. Equation (3.27) can be used for least-squares estimation. If we introduce

$$\theta = \begin{pmatrix} r_0 & \cdots & r_\ell & s_0 & \cdots & s_\ell \end{pmatrix}$$

and

$$\varphi(t) = \begin{pmatrix} u_f(t) & \cdots & u_f(t-\ell) & y_f(t) & \cdots & y_f(t-\ell) \end{pmatrix}$$

it can be written as

$$y(t) = \varphi^T(t-d_0)\theta$$

The estimates are then obtained recursively from Eqs. (3.22). The following adaptive control algorithm is then obtained.

ALGORITHM 3.3   Simple direct self-tuner

Data: Given specifications in terms of A_m, B_m, and A_o and the relative degree d_0 of the system.

Step 1:   Estimate the coefficients of the polynomials R and S in the model (3.27), that is,

$$y(t) = R^* u_f(t - d_0) + S^* y_f(t - d_0)$$

by recursive least squares, Eqs. (3.22).

Step 2:   Compute the control signal from

$$R^* u(t) = T^* u_c(t) - S^* y(t)$$

where R and S are obtained from the estimates in Step 1 and

$$T^* = A_o^* A_m(1) \tag{3.29}$$

with $\deg A_o = d_0 - 1$. Repeat Steps 1 and 2 at each sampling period.   □

Equation (3.29) is obtained from the observation that the closed-loop transfer operator from command signal u_c to process output is

$$\frac{TB}{AR + BS} = \frac{Tb_0 B^+}{b_0 A_o A_m B^+} = \frac{T}{A_o A_m}$$

Requiring that this be equal to $q^{d_0} A_m(1)/A_m$ gives Eq. (3.29).

Remark 1.   A comparison with Algorithm 3.2 shows that the step corresponding to control design is missing in Algorithm 3.3. This motivates the name "direct algorithm."

Remark 2.   Notice that it is necessary to know the relative degree d_0 of the plant *a priori*.

Remark 3.   The polynomials R and S contain the factor b_0. Notice that the polynomial R is not monic and that the parameter r_0 must be different from zero. Otherwise, the control law given by Eq. (3.2) is not causal. Since d_0 is the relative degree of the plant, the true value of $r_0 = b_0$ is different from zero. Any consistent estimate of the parameter will thus be different from zero. The estimate obtained for finite time may, however, be zero. In practice it is therefore essential to take some precautions.

Remark 4.   Notice that the assumption $B^- = b_0$ implies that all process zeros are canceled. This is the reason why the algorithm requires the plant to be minimum phase.

Examples

Properties of direct self-tuners will now be illustrated by some examples.

EXAMPLE 3.7   **Direct self-tuner with $d_0 = 1$**

Consider the system in Example 3.1. Since $\deg A = 2$ and $\deg B = 1$, we have $\deg A_m = 2$ and $\deg A_o = 0$. Hence $A_o = 1$, and we will choose $B_m = qA_m(1)$. Equation (3.29) in Algorithm 3.3 then gives $T = qA_m(1)$. The controller structure is given by $\deg R = \deg S = \deg T = \deg A - 1 = 1$. The model given by Eq. (3.27) therefore becomes

$$y(t) = r_0 u_f(t-1) + r_1 u_f(t-2) + s_0 y_f(t-1) + s_1 y_f(t-2) \tag{3.30}$$

where

$$u_f(t) + a_{m1}u_f(t-1) + a_{m1}u_f(t-2) = u(t)$$
$$y_f(t) + a_{m1}y_f(t-1) + a_{m1}y_f(t-2) = y(t)$$

It is now straightforward to obtain a direct self-tuner by applying Algorithm 3.3. The parameters of the model given by Eq. (3.30) are thus estimated, and the control signal is then computed from

$$\hat{r}_0 u(t) + \hat{r}_1 u(t-1) = \hat{t}_0 u_c(t) - \hat{s}_0 y(t) - \hat{s}_1 y(t-1)$$

where $\hat{r}_0$, $\hat{r}_1$, $\hat{s}_0$, and $\hat{s}_1$ are the estimates obtained and $\hat{t}_0$ is given by Eq. (3.29), that is,

$$\hat{t}_0 = 1 + a_{m1} + a_{m2}$$

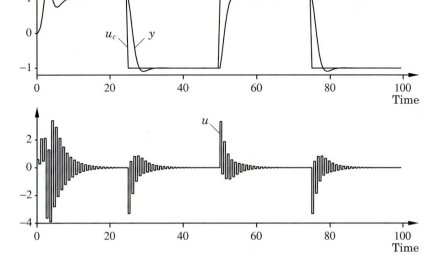

Figure 3.10   Command signal u_c, process output y, and control signal u when the process given by Eq. (3.16) is controlled by using a direct self-tuner with $d_0 = 1$. Compare with Fig. 3.4.

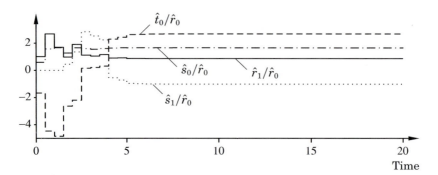

Figure 3.11   Parameter estimates corresponding to the simulation shown in Fig. 3.10: $\hat{r}_1/\hat{r}_0$ (solid line), $\hat{t}_0/\hat{r}_0$ (dashed line), $\hat{s}_0/\hat{r}_0$ (dash-dot line), $\hat{s}_1/\hat{r}_0$ (dotted line).

Notice that the estimate of r_0 must be different from zero for the controller to be causal.

Figure 3.10 shows the process inputs and outputs in a simulation of the direct algorithm, and Fig. 3.11 shows the parameter estimates. The initial transient depends strongly on the initial conditions. At $t = 100$ the controller parameters are

$$\frac{\hat{r}_1(100)}{\hat{r}_0(100)} = 0.850 \quad (0.8467) \qquad \frac{\hat{t}_0(100)}{\hat{r}_0(100)} = 1.65 \quad (1.6531)$$

$$\frac{\hat{s}_0(100)}{\hat{r}_0(100)} = 2.68 \quad (2.6852) \qquad \frac{\hat{s}_1(100)}{\hat{r}_0(100)} = -1.03 \quad (-1.0321)$$

The controller parameters are divided by $\hat{r}_0$ to make a direct comparison with Examples 3.1 and 3.3. The correct values are given in parentheses. A comparison of Fig. 3.4 and Fig. 3.10 shows that the direct and indirect algorithms have very similar behavior. The limiting control law is the same in both cases. There is "ringing" in the control signal because of the cancellation of the process zero. □

In a practical case the time delay and the order of the process that we would like to control are not known. It is therefore natural to consider these variables as design parameters that are chosen by the user. The parameter d_0 is of particular importance for a direct algorithm. In the next example we show that "ringing" can be avoided simply by increasing the value of d_0.

EXAMPLE 3.8   **Direct self-tuner with $d_0 = 2$**

In the derivation of the direct algorithm the parameter d_0 was the pole excess of the plant. Assume for a moment that we do not know the value of d_0 and that we treat it as a design parameter instead. Figure 3.12 shows a simulation of the direct algorithm used in Example 3.7 but with $d_0 = 2$ instead of $d_0 = 1$. All

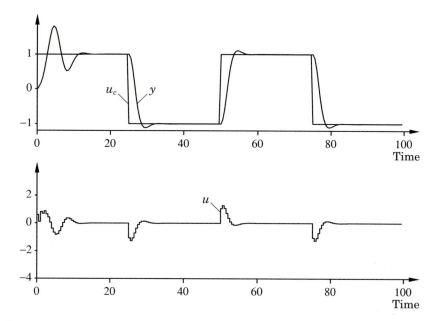

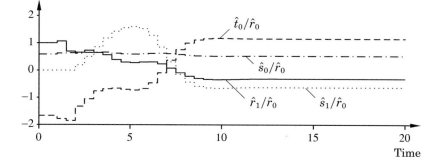

Figure 3.12   Command signal u_c, process output y, and control signal u when the process described by Eq. (3.16) is controlled with a direct self-tuner with $d_0 = 2$.

the other parameters are the same. Notice that the behavior of the system is quite reasonable without any "ringing" in the control signal. Figure 3.13 shows the parameter estimates. The estimates obtained at time $t = 100$ correspond to the controller parameters

$$\frac{\hat{r}_1(100)}{\hat{r}_0(100)} = -0.337 \qquad \frac{\hat{s}_0(100)}{\hat{r}_0(100)} = 1.20 \qquad \frac{\hat{s}_1(100)}{\hat{r}_0(100)} = -0.67 \qquad \frac{\hat{t}_0(100)}{\hat{r}_0(100)} = 0.52$$

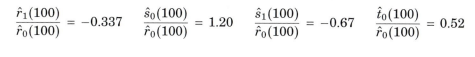

Figure 3.13   Parameter estimates corresponding to Fig. 3.12: $\hat{r}_1/\hat{r}_0$ (solid line), $\hat{t}_0/\hat{r}_0$ (dashed line), $\hat{s}_0/\hat{r}_0$ (dash-dot line), $\hat{s}_1/\hat{r}_0$ (dotted line).

We thus find the interesting and surprising result that cancellation of the process zero can be avoided by increasing the parameter d_0. This observation will be explained later when we will be analyzing the algorithms.   □

Feedforward Control

A nice feature of the direct self-tuner is that it is easy to include feedforward. Let v be a disturbance that can be measured. By estimating parameters in the model

$$y(t) = \frac{1}{A_o A_m} \left(Ru(t) + Sy(t) - Uv(t) \right) \tag{3.31}$$

and using the control law

$$Ru(t) = Tu_c(t) - Sy(t) - Uv(t)$$

we obtain a self-tuning controller that combines feedback and feedforward. The term Tu_c in the control law can also be viewed as a feedforward term.

In Algorithm 3.3, polynomials R and S are estimated and the polynomial T is computed. This means that the different terms of the control law are treated differently. It is possible to obtain an algorithm in which all coefficients of the control law are estimated by treating Tu_c as a feedforward term that is adapted. To do this, we first notice that the desired response is given by

$$y_m(t) = \frac{B_m}{A_m} u_c(t) = \frac{T}{A_o A_m} u_c(t)$$

It follows from Eq. (3.27) that error $e(t) = y(t) - y_m(t)$ is given by

$$e(t) = \frac{1}{A_o A_m} \left(Ru(t) + Sy(t) - Tu_c(t) \right)$$

$$= R^* u_f(t - d_0) + S^* y_f(t - d_0) - T^* u_{cf}(t - d_0) \tag{3.32}$$

where u_f, y_f, and u_{cf} are the filtered signals defined by Eqs. (3.28) and

$$u_{cf}(t) = \frac{1}{A_o^*(q^{-1}) A_m^*(q^{-1})} u_c(t)$$

Furthermore, $\deg T = \deg R = \deg S = \deg(A_o A_m) - d_0$ and $\deg A_m - \deg B_m = d_0$. An algorithm that is analogous to Algorithm 3.3, in which the parameters of the feedforward polynomial T are also estimated is now easily obtained by estimating the parameters in Eq. (3.32).

Non-minimum-Phase (NMP) Systems

The case in which process zeros cannot be canceled will now be discussed. Consider the transformed process model Eq. (3.24), that is,

$$A_o A_m y(t) = B^- \left(Ru(t) + Sy(t) \right)$$

where $\deg R = \deg S = \deg(A_o A_m) - \deg B^-$. If we introduce

$$\mathcal{R} = B^- R \qquad \text{and} \qquad \mathcal{S} = B^- S$$

the equation can be written as

$$y(t) = \frac{1}{A_o A_m} (\mathcal{R}u(t) + \mathcal{S}y(t)) = \mathcal{R}^* u_f(t - d_0) + \mathcal{S}^* y_f(t - d_0) \qquad (3.33)$$

where u_f and y_f are the filtered inputs and outputs given by Eqs. (3.28). Notice that the polynomial $\mathcal{R}$ is not monic. The polynomials $\mathcal{R}$ and $\mathcal{S}$ have a common factor, which represents poorly damped zeros. This factor should be canceled before the control law is calculated. The following direct adaptive control algorithm is then obtained.

ALGORITHM 3.4   Direct self-tuning regulator for NMP systems

Data: Given specifications in terms of A_m, B_m, and A_o and the relative degree d_0 of the system.

Step 1:   Estimate the coefficients of the polynomials $\mathcal{R}$ and $\mathcal{S}$ in the model of Eq. (3.33) by recursive least squares.

Step 2:   Cancel possible common factors in $\mathcal{R}$ and $\mathcal{S}$ to obtain R and S.

Step 3:   Calculate the control signal from Eq. (3.2) where R and S are those obtained in Step 2 and T is given by Eq. (3.12).

Repeat Steps 1, 2, and 3 at each sampling period.   □

This algorithm avoids the nonlinear estimation problem, but more parameters have to be estimated than when Eq. (3.24) is used because the parameters of the polynomial B^- are estimated twice. The estimation is straightforward, however, because the model is linear in the parameters. The Euclidean algorithm in Chapter 11 can be used in Step 2 to eliminate common factors of polynomials $\mathcal{R}$ and $\mathcal{S}$. This step is crucial because an unstable common factor may cause instabilities.

Calculation of polynomial T should be avoided. To do this, notice that

$$y_m = \frac{B^- B'_m}{A_m} u_c$$

The error $e = y - y_m$ can then be written as

$$\begin{aligned} e(t) &= \frac{B^-}{A_o A_m} (Ru(t) + Sy(t) - Tu_c(t)) \\ &= \mathcal{R}^* u_f(t - d_0) + \mathcal{S}^* y_f(t - d_0) - \mathcal{T}^* u_{cf}(t - d_0) \end{aligned} \qquad (3.34)$$

By basing parameter estimation on this equation, estimates of polynomials $\mathcal{R}$, $\mathcal{S}$, and $\mathcal{T}$ can be determined. Notice that to estimate coefficients of $\mathcal{T}$, it is necessary that the command signal be persistently exciting.

Mixed Direct and Indirect Algorithms

Another direct algorithm can be derived in the particular case in which no process zeros are canceled. In this case we have $B^- = B$, and the model Eq. (3.24) becomes

$$A_o A_m y(t) = B\Big(Ru(t) + Sy(t)\Big)$$

which can also be written as

$$y(t) = \frac{B}{A_o A_m}\big(Ru(t) + Sy(t)\big) = B^*\big(R^* u_f(t - d_0) + S^* y_f(t - d_0)\big) \qquad (3.35)$$

The following algorithm is a hybrid algorithm that combines features of direct and indirect schemes.

ALGORITHM 3.5   A hybrid self-tuner

Data: Given polynomials A_o and A_m.

Step 1:   Estimate parameters of polynomials A and B in the model

$$Ay = Bu$$

Step 2:   Estimate parameters of polynomials R and S in Eq. (3.35) where B is the estimate obtained in Step 1.

Step 3:   Use the control law

$$Ru = Tu_c - Sy$$

where R and S are obtained from Step 2 and $T = t_0 A_o$ where

$$t_0 = \frac{A_m(1)}{B(1)}$$

$\square$

Remark 1.   Instead of being computed, polynomial T can also be estimated by replacing Step 2 by the following step:

Step 2':   Estimate parameters of polynomials R and S and t_0 from the model

$$e(t) = y(t) - y_m(t) = \frac{B}{A_o A_m}\big(Ru(t) + Sy(t) - t_0 A_o u_c(t)\big)$$

$$= B^*\big(R^* u_f(t - d_0) + S^* y_f(t - d_0) - t_0 A_o^* u_{cf}(t - d_0)\big) \qquad (3.36)$$

where B is the polynomial obtained in Step 1. It is then assumed that $\deg A_m = \deg A$.

Remark 2.   Instead of the Diophantine equation being solved at each step, two process models are estimated. This implies that an additional iteration of the least-squares estimator has to be done at each sampling time.

3.6   DISTURBANCES WITH KNOWN CHARACTERISTICS

So far, we have concentrated on servo problems that are common in aerospace and mechatronics. In process control, regulation problems are more common. It is then important to consider attenuation of disturbances that act on the process. The disturbance may enter the process in many different ways. For simplicity we will assume that it enters at the process input as shown in Figure 3.2. This assumption is not very restrictive. If the disturbance is denoted by v, the system is then described by Eq. (3.1). We will first use an example to illustrate that load disturbances will cause problems.

EXAMPLE 3.9   **Effect of Load Disturbances**

Consider the system in Example 3.5, that is, an indirect self-tuning regulator with no zero cancellation. We will now make a simulation that is identical to the one shown in Fig. 3.6 except that the load disturbance will be $v(t) = 0.5$ for $t \geq 40$. A forgetting factor $\lambda = 0.98$ has also been introduced; otherwise, the conditions are identical to those in Example 3.5. The behavior of the system is shown in Fig. 3.14. Compare Fig. 3.14 with Fig. 3.6. Figure 3.14 shows that a load disturbance may be disastrous. It follows from the discussion in Example 3.5 that the correct steady-state value will always be reached provided that

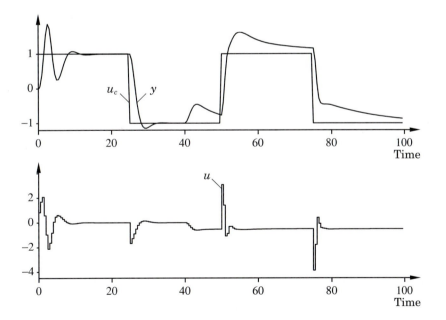

Figure 3.14   Output and control signal when for a system with an indirect self-tuner without zero canceling when there is a load disturbance in the form of a step at the process input at time $t = 40$.

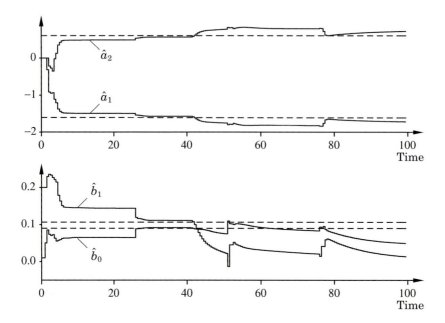

Figure 3.15   Parameter estimates corresponding to Fig. 3.14.

the steps are sufficiently long. Notice that the response is strongly asymmetric. The reason for this is that the controller parameters change rapidly when the control signal changes; see Fig. 3.15, which shows the parameter estimates. Rapid changes of the estimates in response to command signals indicates that the model structure is not correct. The parameter estimates also change significantly at the step in the load disturbance. When the command signal is constant, the parameters appear to settle at constant values that are far from the true parameters.   □

There are many ways to deal with disturbances. The *internal model principle* is used in this section. An alternative is to estimate the disturbance and compensate for it in a feedforward fashion. An in-depth discussion of different methods and their advantages and disadvantages is found in Chapter 11.

A Modified Design Procedure

The pole placement procedure can be modified to take disturbances into account. In many cases the important disturbances have known characteristics. This can be captured by assuming that the disturbance v in the model (3.1) is generated by the dynamical system

$$A_d v = e \tag{3.37}$$

where e is a pulse, a set of widely spread pulses, white noise, or the equivalent continuous-time concepts. For example, a step disturbance is generated in discrete-time systems by

$$A_d(q) = q - 1$$

and in continuous-time systems by

$$A_d(p) = p$$

With the controller of Eq. (3.2) we find

$$y = \frac{BT}{AR + BS} u_c + \frac{BR}{A_d(AR + BS)} e$$

$$u = \frac{AT}{AR + BS} u_c - \frac{BS}{A_d(AR + BS)} e$$

(3.38)

The closed-loop characteristic polynomial thus contains the disturbance dynamics as a factor. This polynomial typically has roots on the stability boundary or in the unstable region. It follows from Eqs. (3.38) that to maintain a finite output in case of these disturbances, A_d must be a factor of R. This would make y finite, but the controlled input u may be infinite. This is, of course, necessary to compensate for an infinite disturbance.

It has already been mentioned that the Diophantine equation has many solutions. Compare with Eqs. (3.14). If R^0 and S^0 are solutions to the Diophantine equation

$$AR^0 + BS^0 = A_c^0$$

it follows that

$$R = XR^0 + YB$$
$$S = XS^0 - YA$$

(3.39)

satisfies the equation

$$AR + BS = XA_c^0$$

If a controller R^0 S^0 that gives the characteristic polynomial A_c^0 has been obtained, we can thus obtain a controller with characteristic polynomial XA_c^0 by using the controller (3.39). Suppose that we have designed a controller R^0 and S^0 and that we would like to have a new controller in which $R = R'A_d$. We then choose a stable polynomial X that represents the additional closed-loop poles, and we determine R' and Y such that

$$R = A_d R' = XR^0 + YB$$

(3.40)

The new controller is then given by Eqs. (3.39).

Integral Action

In the special case in which the disturbance is a constant, that is, $A_d = q - 1$, we have to add an additional closed-loop pole. Hence

$$X = q + x_0$$

and Eq. (3.40) becomes

$$(q - 1)R' = (q + x_0)R^0 + y_0B$$

Putting $q = 1$ gives one equation to solve for y_0. Hence

$$y_0 = -\frac{(1 + x_0)R^0(1)}{B(1)} \tag{3.41}$$

Inserting X and $Y = y_0$ into Eqs. (3.39) gives the new controller.

Modifications of the Estimator

Disturbances will change the relations between the inputs and the outputs in the model. Load disturbances such as steps will have a particularly bad effect on the low-frequency properties of the model. Several ways to deal with this problem are discussed in Section 11.5. One possibility is to include the disturbance in the model and estimate it; another, which we will use here, is to filter the signal so that the effect of the disturbance is not so large. In the model given by Eq. (3.1) the equation error is $B(q)v$. This could be a very large quantity if $B(1) \neq 0$ and v is a large step. If the disturbance v in Eq. (3.1) can be described by Eq. (3.37) we find that Eq. (3.1) can be written as

$$A_dAy(t) = A_dB(u(t) + v(t)) = A_dBy(t) + e(t)$$

Hence

$$Ay_f(t) = Bu_f(t) + e(t) \tag{3.42}$$

By introducing the filtered signals $y_f = A_dy$ and $u_f = A_du$ we thus obtain a model in which the equation error is e instead of v, where e is significantly smaller than v. For example, if v is a step and $A_d = q - 1$ as in Example 3.9, we find that e is zero except at the time where the step in v occurs.

   The next example shows that the difficulties encountered in Example 3.9 can be avoided by using a self-tuner with a modified estimator and a modified control design.

EXAMPLE 3.10   **Load disturbances: Modified estimator and controller**

We now show that the difficulties found in Example 3.9 can be avoided by modifying the estimator and the controller. We first introduce a controller that has integral action by applying the design procedure that we have just

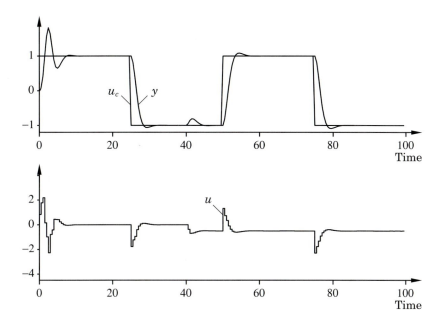

Figure 3.16 Output and control signal with an indirect self-tuner with integral action and a modified estimator.

described. To do this, we consider the same system as in Example 3.5 where the controller was defined by

$$R^0 = q + r_1 \qquad S^0 = s_0 q + s_1$$

The closed-loop characteristic polynomial A_c has degree three. To obtain a controller with integral action, the order of the closed-loop system is increased by introducing an extra closed-loop pole at $q = -x_0 = 0$. It then follows from Eq. (3.41) that

$$y_0 = -\frac{1 + r_1}{b_0 + b_1}$$

Hence $X = q$ and $Y = y_0$, and Eqs. (3.39) now give

$$R = q(q + r_1) + y_0(b_0 q + b_1) = (q - 1)(q - b_1 y_0)$$
$$S = q(s_0 q + s_1) - y_0(q^2 + a_1 q + a_2) = (s_0 - y_0)q^2 + (s_1 - a_1 y_0)q - a_2 y_0$$

The estimates are based on the model (3.42) with $A_d = q - 1$ to reduce the effects of the disturbances. Figure 3.16 shows a simulation corresponding to Fig. 3.14 with the modified self-tuning regulator. A comparison with Fig. 3.14 shows a significant improvement. The load disturbance is reduced quickly. Because of the integral action the control will decrease with a magnitude corresponding to the load disturbance shortly after $t = 40$. The parameter

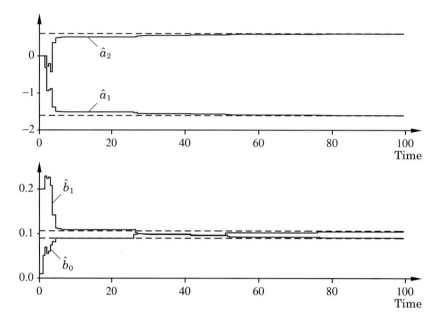

Figure 3.17   Parameter estimates corresponding to Fig. 3.16.

estimates are shown in Fig. 3.17, which indicates the advantages in using the modified estimator. Notice in particular that there is a very small change in the estimates when the load disturbance occurs.   □

A Direct Self-tuner with Integral Action

It is also straightforward to introduce integrators in the direct self-tuners. Consider a process model given by

$$A(q)y(t) = B(q)(u(t) + v(t)) \tag{3.43}$$

where $d = \deg A(q) - \deg B(q)$. It is assumed that v is constant or changes infrequently. Let the desired response to command signals be given by

$$A_m(q)y(t) = A_m(1)u_c(t - d) \tag{3.44}$$

where $\deg A_m \geq d$. Let the observer polynomial be $A_o(q)$. The design equation is

$$AR + BS = B^+A_oA_m \tag{3.45}$$

where $B = b_0B^+$. If we require that the regulator has integral action, we find that the polynomial R has the form

$$R = R'B^+ = R'_1B^+(q - 1) = R'_1B^+\Delta \tag{3.46}$$

Equation (3.45) then becomes

$$A\Delta R_1' + b_0 S = A_o A_m \tag{3.47}$$

Hence

$$
\begin{aligned}
A_o A_m y &= A R_1' \Delta y + b_0 S y \\
&= B R_1' \Delta u + b_0 R' \Delta v + b_0 S y \\
&= b_0 (R' \Delta u + S y) + b_0 R' \Delta v
\end{aligned} \tag{3.48}
$$

where Eq. (3.43) was used to obtain the second equality. Notice that the last term will vanish after a transient if v is constant. If we rewrite Eq. (3.48) in the backwards operator, ignoring v, we get

$$A_o^*(q^{-1})A_m^*(q^{-1})y(t+d) = b_0\left(R'^*(q^{-1})\Delta^*(q^{-1})u(t) + S^*(q^{-1})y(t)\right) \tag{3.49}$$

This equation can be used as a basis for parameter estimation, but there are several drawbacks in doing so. First, the operation $A_o^* A_m^*$ is a high-pass filter that is very sensitive to noise. Furthermore, it follows from Eq. (3.47) that

$$b_0 S^*(1) = A_o^*(1)A_m^*(1) = A_o(1)A_m(1) \tag{3.50}$$

All the parameters in the S polynomial are thus not free. If all parameters are estimated, there is, of course, no guarantee that Eq. (3.50) holds. However, it is easy to find a remedy. A polynomial S^* with the property given by Eq. (3.50) can be written as

$$
\begin{aligned}
b_0 S^* &= A_o(1)A_m(1) + (1 - q^{-1})S'^*(q^{-1}) \\
&= A_o(1)A_m(1) + S'^*(q^{-1})\Delta^*
\end{aligned}
$$

Equation (3.49) then becomes

$$
\begin{aligned}
A_o^*(q^{-1})A_m^*(q^{-1})y(t+d) &- A_o(1)A_m(1)y(t) \\
&= b_0\left(R'^*(q^{-1})\Delta^* u(t) + S'^*(q^{-1})\Delta^* y(t)\right) \\
&= \mathcal{R}^*(q^{-1})\Delta^* u(t) + \mathcal{S}^*(q^{-1})\Delta^* y(t)
\end{aligned} \tag{3.51}
$$

Division by $A_o^* A_m^*$ now gives

$$y(t+d) - \frac{A_o(1)A_m(1)}{A_o^*(q^{-1})A_m^*(q^{-1})}y(t) = \mathcal{R}^*(q^{-1})u_f(t) + \mathcal{S}^*(q^{-1})y_f(t) \tag{3.52}$$

where

$$u_f(t) = \frac{1 - q^{-1}}{A_o^*(q^{-1})A_m^*(q^{-1})}u(t)$$

$$y_f(t) = \frac{1 - q^{-1}}{A_o^*(q^{-1})A_m^*(q^{-1})}y(t)$$

Notice that the difference operation eliminates levels and that division by $A_o^* A_m^*$ corresponds to low-pass filtering. Thus the net effect is that the signals are band-pass filtered with filters that are matched to the desired closed-loop dynamics and the specified observer polynomial.

To complete the algorithm, it now remains to specify how the control law is obtained from the estimated parameters. To obtain the response to command signals given by Eq. (3.44), it follows from Eq. (3.51) that

$$\mathcal{R}^*(q^{-1})\Delta^* u(t) + \mathcal{S}^*(q^{-1})\Delta^* y(t) + A_o(1)A_m(1)y(t) = A_o^*(q^{-1})A_m(1)u_c(t)$$

A controller with integral action may perform poorly if there are actuators that saturate. The feedback loop is broken during saturation, and the integrator may drift to undesirable values. This phenomenon, which is called windup, can be avoided if the control algorithm is modified to

$$
\begin{aligned}
A_o^*(q^{-1})&\Big(\bar{u}(t) - A_m(1)u_c(t)\Big) \\
&= -A_o(1)A_m(1)y(t) - \mathcal{S}^*(q^{-1})\Delta^* y(t) \\
&\quad - \Big(\mathcal{R}^*(q^{-1})\Delta^* - A_o^*(q^{-1})\Big)u(t)
\end{aligned}
\tag{3.53}
$$

$$u(t) = \mathrm{sat}\,\bar{u}(t)$$

The windup phenomenon is discussed in detail in Section 11.2. In summary, Algorithm 3.6 is obtained.

ALGORITHM 3.6   A direct self-tuning algorithm

Step 1:   Estimate the parameters in Eq. (3.52) by recursive least squares.

Step 2:   Compute the control signal from Eqs. (3.53) by using the estimates from Step 1.   □

This algorithm may be viewed as a practical version of Algorithm 3.3.

3.7   CONCLUSIONS

Deterministic self-tuning regulators have been developed in this chapter. The controllers may be viewed as an attempt to automate the steps of modeling and control design that are normally done by a control system designer. By specifying a model structure, modeling reduces to recursive parameter estimation. Control design results in a map from process parameters to controller parameters. Simple estimation methods (least squares) and simple control design techniques (pole placement) have been used in this chapter. The control design was based on the certainty equivalence principle, which means that the uncertainties in the estimates are neglected in computing the control law. Two classes of algorithms have been discussed: indirect and direct algorithms. The indirect algorithms are a straightforward implementation in which process parameters are estimated and the controller parameters are computed by using some design equations. In the direct algorithms the controller parameters are estimated directly. To do this, design equations are used to reparameterize the process model in the controller parameters. This makes it possible to establish relations between MRAS and STR, as is discussed in Chapter 5.

PROBLEMS

3.1 In sampling a continuous-time process model with $h = 1$ the following pulse transfer function is obtained:

$$H(z) = \frac{z + 1.2}{z^2 - z + 0.25}$$

The design specification states that the discrete-time closed-loop poles should correspond to the continuous-time characteristic polynomial

$$s^2 + 2s + 1$$

(a) Design a minimal-order discrete-time indirect self-tuning regulator. The controller should have integral action and give a closed-loop system having unit gain in stationary. Determine the Diophantine equation that solves the design problem.

(b) Suggest a design that includes direct estimation of the controller parameters. Discuss why a well-working direct self-tuning regulator is more difficult to design for this process than is an indirect self-tuning regulator.

3.2 Consider the process

$$G(s) = \frac{1}{s(s + a)}$$

where a is an unknown parameter. Assume that the desired closed-loop system is

$$G_m(s) = \frac{\omega^2}{s^2 + 2\zeta\omega s + \omega^2}$$

Construct continuous- and discrete-time indirect self-tuning algorithms for the system.

3.3 Consider the system

$$G(s) = G_1(s)G_2(s)$$

where

$$G_1(s) = \frac{b}{s + a}$$

$$G_2(s) = \frac{c}{s + d}$$

where a and b are unknown parameters and c and d are known. Construct discrete-time direct and indirect self-tuning algorithms for the partially known system.

3.4   A process has the transfer function

$$G(s) = \frac{b}{s(s + 1)}$$

where b is a time-varying parameter. The system is controlled by a proportional controller

$$u(t) = k\left(u_c(t) - y(t)\right)$$

It is desirable to choose the feedback gain so that the closed-loop system has the transfer function

$$G(s) = \frac{1}{s^2 + s + 1}$$

Construct a continuous-time indirect self-tuning algorithm for the system.

3.5   The code for simulating Examples 3.4 and 3.5 is listed below. Study the code and try to understand the details.

```
DISCRETE SYSTEM reg

"Indirect Self-Tuning Regulator based on the model
"    H(q)=(b0*q+b1)/(q^2+a1*q+a2)
"using standard RLS estimation and pole placement design
"Polynomial B is canceled if cancel>0.5

INPUT ysp y                "set point and process output
OUTPUT u                   "control variable
STATE ysp1 y1 u1 v1        "controller states
STATE th1 th2 th3 th4      "parameter estimates
STATE f1   f2  f3  f4      "regression variables
STATE p11 p12 p13 p14      "covariance matrix
STATE     p22 p23 p24
STATE         p33 p34
STATE             p44
NEW   nysp1 ny1 nu1 nv1
NEW nth1 nth2 nth3 nth4
NEW nf1 nf2 nf3 nf4
NEW n11 n12 n13 n14 n22 n23 n24 n33 n34 n44
TIME t
TSAMP ts

INITIAL
"Compute sampled Am and Ao
a=exp(-z*w*h)
am1=-2*a*cos(w*h*sqrt(1-z*z))
```

```
am2=a*a
aop=IF w*To>100 THEN 0 ELSE -exp(-h/To)
ao=IF cancel>0.5 THEN 0 ELSE -aop

SORT
"1.0 Parameter Estimation
"1.1 Computation of P*f and estimator gain k
pf1=p11*f1+p12*f2+p13*f3+p14*f4
pf2=p12*f1+p22*f2+p23*f3+p24*f4
pf3=p13*f1+p23*f2+p33*f3+p34*f4
pf4=p14*f1+p24*f2+p34*f3+p44*f4
denom=lambda+f1*pf1+f2*pf2+f3*pf3+f4*pf4
k1=pf1/denom
k2=pf2/denom
k3=pf3/denom
k4=pf4/denom

"1.2 Update estimates and covariances
eps=y-f1*th1-f2*th2-f3*th3-f4*th4
nth1=th1+k1*eps
nth2=th2+k2*eps
nth3=th3+k3*eps
nth4=th4+k4*eps
n11=(p11-pf1*k1)/lambda
n12=(p12-pf1*k2)/lambda
n13=(p13-pf1*k3)/lambda
n14=(p14-pf1*k4)/lambda
n22=(p22-pf2*k2)/lambda
n23=(p23-pf2*k3)/lambda
n24=(p24-pf2*k4)/lambda
n33=(p33-pf3*k3)/lambda
n34=(p34-pf3*k4)/lambda
n44=(p44-pf4*k4)/lambda

"1.3 Update and filter regression vector
nf1=-y
nf2=f1
nf3=u
nf4=f3


"2.0 Control design
"2.1 Rename parameters
a1=nth1
a2=nth2
```

```
b0=nth3
b1=nth4

"2.2 Solve the polynomial identity AR+BS=AoAm
n=b1*b1-a1*b0*b1+a2*b0*b0
r10=(ao*am2*b0^2+(a2-am2-ao*am1)*b0*b1+(ao+am1-a1)*b1^2)/n
w1=(a2*am1+a2*ao-a1*a2-am2*ao)*b0
s00=(w1+(-a1*am1-a1*ao-a2+a1^2+am2+am1*ao)*b1)/n
w2=(-a1*am2*ao+a2*am2+a2*am1*ao-a2^2)*b0
s10=(w2+(-a2*am1-a2*ao+a1*a2+am2*ao)*b1)/n

"2.3 Compute polynomial T=Ao*Am(1)/B(1)
bs=b0+b1
as=1+am1+am2
bm0=as/bs

"2.4 Choose control algorithm
r1=IF cancel>0.5 THEN b1/b0 ELSE r10
s0=IF cancel>0.5 THEN (am1-a1)/b0 ELSE s00
s1=IF cancel>0.5 THEN (am2-a2)/b0 ELSE s10
t0=IF cancel>0.5 THEN as/b0 ELSE bm0
t1=IF cancel>0.5 THEN 0 ELSE bm0*ao

"3.0 Control law with anti-windup
v=-ao*v1+t0*ysp+t1*ysp1-s0*y-s1*y1+(ao-r1)*u1
u=IF v<-ulim THEN -ulim ELSE IF v<ulim THEN v ELSE ulim

"3.1 Update controller state
ny1=y
nu1=u
nv1=v
nysp1=ysp

"4.0  Update sampling time
ts=t+h

"Parameters
lambda:1          "forgetting factor
To:200            "observer time constant
z:0.7             "desired closed loop damping
w:1               "desired closed loop natural frequency
h:1               "sampling period
ulim:1            "limit of control signal
cancel:1          "switch for cancellation
th1:-2            "initial estimates
```

```
th2:1
th3:0.01
th4:0.01
p11:100          "initial covariances
p22:100
p33:100
p44:100

END
```

3.6 Consider the simulation of the indirect self-tuning regulator in Example 3.5. Investigate how the transient behavior of the algorithm depends on the initial values of θ and P and the forgetting factor.

3.7 Consider the indirect self-tuning regulator in Example 3.5. Make a simulation over longer time periods, and investigate how the parameters approach their true values. Also explore how the convergence rate depends on the forgetting factor λ.

3.8 Consider the indirect self-tuning regulator in Example 3.5. Show that no steady-state error is obtained if

$$\hat{a}_1 + \hat{a}_2 = 1$$

Modify the simulation used to generate Figs. 3.6 and 3.7, plot the parameter combination $\hat{a}_1 + \hat{a}_2$, and check how well the above condition is satisfied.

3.9 Consider the indirect self-tuning regulator in Example 3.5. Change the specifications on the closed-loop system, and investigate how the behavior of the system changes.

3.10 Consider the indirect self-tuning regulator in Example 3.5. Modify the simulation program so that the parameters of the process can be changed. Investigate experimentally how well the adaptive system can follow reasonable parameter variations.

3.11 Apply the indirect self-tuning regulator in Example 3.5 to a process with the transfer function

$$G(s) = \frac{1}{(s+1)^2}$$

Study and explain the behavior of the error when the reference signal is a square wave.

3.12 The code for simulating Example 3.6 is listed below. Study the code and try to understand all the details.

```
CONTINUOUS SYSTEM reg
"Continuous time STR for the system b/[s(s+a)]
```

```
"Desired response given by am2/(s^2+am1*s+am2)
"Observer polynomial s+ao

INPUT y ysp
OUTPUT u
STATE yf yf1 uf uf1 xu
STATE th1 th2
STATE p11 p12 p22
DER dyf dyf1 duf duf1 dxu
DER dth1 dth2
DER dp11 dp12 dp22

"Filter input and output
dyf=yf1
dyf1=-am1*yf1+am2*(y-yf)
duf=uf1
duf1=-am1*uf1+am2*(u-uf)

"Update parameter estimate
f1=-yf1
f2=uf
e=dyf1-f1*th1-f2*th2
pf1=p11*f1+p12*f2
pf2=p12*f1+p22*f2
dth1=pf1*e
dth2=pf2*e

"Update covariance matrix
dp11=alpha*p11-pf1*pf1
dp12=alpha*p12-pf1*pf2
dp22=alpha*p22-pf2*pf2
det=p11*p22-p12*p12


"Control design
a=th1
b=th2
r1=am1+ao-a
s0=(am2+am1*ao-a*r1)/b
s1=am2*ao/b
t0=am2/b

"Control signal computation
dxu=-ao*xu-(s1-ao*s0)*y+(ao-r1)*u
v=t0*ysp-s0*y+xu
```

```
u=if v<-ulim then -ulim else if v>ulim then ulim else v

"Parameters
am1:1.4
am2:1
alpha:0
ao:2
ulim:4
END
```

3.13 Consider the simulation of the continuous-time indirect self-tuning regulator in Example 3.6. Investigate how the transient behavior of the algorithm depends on the initial values of θ and P.

3.14 Consider the indirect self-tuning regulator in Example 3.6. Make a simulation, and investigate how the convergence rate depends on the forgetting factor α.

3.15 Consider the system in Problem 1.9.

(a) Sample the system, and determine a discrete-time controller for the known nominal system such that the specifications are satisfied.

(b) Use a direct self-tuning controller, and study the transient for different initial conditions and different values of the variable parameters of the system.

(c) Assume that $e = 0$ and that u_c is a square wave. Simulate a self-tuning controller for different prediction horizons.

(d) Investigate the behavior when the disturbance d is a step. What happens when the controller does not have an integrator?

REFERENCES

The pole placement design is extensively discussed in:

Åström, K. J., and B. Wittenmark, 1990. *Computer Controlled Systems—Theory and Design*, 2nd edition. Englewood Cliffs, N.J.: Prentice-Hall.

It is possible to solve the Diophantine equation using polynomial calculations. Solution of the Diophantine equation is discussed in:

Blankinship, W. A., 1963. "A new version of the Euclidean algorithm." *American Mathematics Monthly* **70**: 742–745.

Kučera, V., 1979. *Discrete Linear Control—The Polynomial Equation Approach*. New York: John Wiley.

Ježek, J., 1982. "New algorithm for minimal solution of linear polynomial equations." *Kybernetica* **18**: 505–516.

There are many papers, reports, and books about self-tuning algorithms. Some fundamental references are given in this section. The first publication of the self-tuning idea is probably:

Kalman, R. E., 1958. "Design of a self-optimizing control system." *Trans. ASME* **80**: 468–478.

In this paper, least-squares estimation combined with deadbeat control is discussed. No analysis is given of the properties of the closed-loop system. A prototype special-purpose computer was built to implement the controller, but the development was hampered by hardware problems. The main development of the theory for self-tuning controllers was first done for discrete-time systems with stochastic noise. This type of self-tuning controllers is discussed in the next chapter. Two similar algorithms based on least-squares estimation and minimum-variance control were presented at an IFAC symposium in Prague 1970:

Peterka, V., 1970. "Adaptive digital regulation of noisy systems." *Preprints 2nd IFAC Symposium on Identification and Process Parameter Estimation.* Prague.

Wieslander, J., and B. Wittenmark, 1971. "An approach to adaptive control using real time identification." *Automatica* **7**: 211–217.

The first thorough presentation and analysis of a self-tuning regulator was given in:

Åström, K. J., and B. Wittenmark, 1972. "On the control of constant but unknown systems." *5th IFAC World Congress.* Paris.

A revised version of this paper, in which the phrase "self-tuning regulator" was coined, is:

Åström, K. J., and B. Wittenmark, 1973. "On self-tuning regulators." *Automatica* **9**: 185–199.

The preceding papers inspired intensive research activity in adaptive control based on the self-tuning idea. A comprehensive treatment of the fundamental theory of adaptive control, especially self-tuning algorithms, is given in:

Goodwin, G. C., and K. S. Sin, 1984. *Adaptive Filtering Prediction and Control,* Information and Systems Science Series. Englewood Cliffs, N.J.: Prentice-Hall.

Pole placement and model-reference-type self-tuners are treated in:

Wellstead, P. E., J. M. Edmunds, D. Prager, and P. Zanker, 1979. "Self-tuning pole/zero assignment regulators." *Int. J. Control* **30**: 1–26.

Åström, K. J., and B. Wittenmark, 1980. "Self-tuning regulators based on pole-zero placement." *IEE Proceedings Part D* **127**: 120–130.

Continuous-time self-tuning regulators are discussed in:

Egardt, B., 1979. *Stability of Adaptive Controllers,* Lecture Notes in Control and Information Sciences. Berlin: Springer-Verlag.

Gawthrop, P. J., 1987. *Continuous-Time Self-Tuning Control I.* Letchworth, U.K.: Research Studies Press.

The book by Egardt also gives a unification of MRAS and STR.

CHAPTER 4

STOCHASTIC AND PREDICTIVE SELF-TUNING REGULATORS

———

4.1 INTRODUCTION

In Chapter 3 the key issue was to find self-tuning controllers that give desired responses to command signals. In this chapter we discuss self-tuners for the regulation problem. The key issue is now to design a controller that reduces disturbances as well as possible. Stochastic models are useful to describe disturbances. For this reason we start in Section 4.2 by describing a simple stochastic control problem. This leads to a minimum-variance controller and its generalization, the moving-average controller. In Section 4.3 we present a direct adaptive controller that has the surprising property that the moving-average controller is an equilibrium solution. This surprising property was one of the motivating factors in the original work on the self-tuning regulator. The minimum-variance controller has the drawback that its properties are critically dependent on the sampling period. In Section 4.4 some extensions are therefore presented. Linear quadratic Gaussian self-tuners are discussed in Section 4.5, and adaptive predictive control is discussed in Section 4.6.

4.2 DESIGN OF MINIMUM-VARIANCE AND MOVING-AVERAGE CONTROLLERS

In this section we derive controllers for linear stochastic systems. It is assumed that the process can be described by a pulse transfer function and that the disturbances acting on the system are filtered white noise. A steady-state regulation problem is considered. The criterion is based on the mean square deviations of the output and the control signal.

Process Model

Assume that the process dynamics are characterized by

$$x(t) = \frac{B_1(q)}{A_1(q)} u(t)$$

where $A_1(q)$ and $B_1(q)$ are polynomials in the forward shift operator without any common factors.

It is assumed that the action of the disturbances on the system can be described as filtered white noise. Since the system is linear, we can reduce all disturbances to an equivalent disturbance v at the system output. The output is thus given by

$$y(t) = x(t) + v(t)$$

where

$$v(t) = \frac{C_1(q)}{A_2(q)} e(t)$$

$C_1(q)$ and $A_2(q)$ are polynomials in the forward shift operator without any common factors, and $\{e(t)\}$ is a sequence of independent random variables (white noise) with zero mean and standard deviation σ.

The process can now be reduced to the standard form

$$A(q)y(t) = B(q)u(t) + C(q)e(t) \tag{4.1}$$

where

$$\begin{aligned} A &= A_1 A_2 \\ B &= B_1 A_2 \\ C &= C_1 A_1 \end{aligned} \tag{4.2}$$

Because of the assumptions, the three polynomials have no common factor. The model (4.1) is thus a minimal representation. The polynomials are normalized such that both the A and C polynomials are monic, that is, the leading coefficients are unity. Finally, the C polynomial can be multiplied by an arbitrary power of q without changing the correlation structure of $C(q)e(t)$. This is used to normalize C such that

$$\deg C = \deg A = n$$

The A and B polynomials may have zeros inside or outside the unit disc. It is assumed that the zeros of the C polynomial are inside the unit disc. By spectral factorization the polynomial $C(q)$ can be changed so that all its zeros are inside the unit disc or on the unit circle. An example shows how this is done.

EXAMPLE 4.1   **Modification of the polynomial** C

Consider the polynomial

$$C(z) = z + 2$$

which has the zero $z = -2$ outside the unit disc. Consider the signal

$$v(t) = C(q)e(t)$$

where $\{e(t)\}$ is a sequence of uncorrelated random variables with zero mean and unit variance. The spectral density of v is given by

$$\Phi(e^{i\omega h}) = \frac{1}{2\pi} C(e^{i\omega h})C(e^{-i\omega h})$$

Because

$$
\begin{aligned}
C(z)C(z^{-1}) &= (z + 2)(z^{-1} + 2) = (1 + 2z^{-1})(1 + 2z) \\
&= (2z + 1)(2z^{-1} + 1) \\
&= 4(z + 0.5)(z^{-1} + 0.5)
\end{aligned}
$$

the signal v may also be represented as

$$v(t) = C^*(q)e(t)$$

where

$$C^*(z) = 2z + 1$$

is the reciprocal of the polynomial $C(z)$ (see Section 3.2).   ☐

If the calculations (4.2) give a polynomial C that has zeros outside the unit disc, the polynomial is factored as

$$C = C^+C^-$$

where C^- contains all factors with zeros outside the unit disc. The C polynomial is then replaced by C^+C^{-*}. The model (4.1) is an *innovations representation*. It will be shown later that $e(t)$ is the innovation or the error in predicting the signal $y(t)$ over one sampling period. The C polynomial can be interpreted as the characteristic polynomial of the estimator or predictor.

Criteria

In steady-state regulation it makes sense to express the criteria in terms of the steady-state variances of the output and the control signals. This leads to the performance criterion

$$J = E\{y^2(t) + \rho u^2(t)\} \tag{4.3}$$

where E denotes mathematical expectation with respect to the noise process acting on the system. The control law minimizing (4.3) is the *linear quadratic*

Gaussian (LQG) controller. If $\rho = 0$, then the resulting controller is called the *minimum-variance (MV) controller.*

The properties of the control signal when the minimum-variance controller is used depend critically on the sampling interval. A short sampling interval gives large variance in the control signal, and a long sampling interval gives a low variance. Notice that the loss function (4.3) is defined in discrete time, that is, only the behavior at the sampling instances is considered.

To define the design problem, it is also necessary to define the admissible controllers. It will be assumed that $u(t)$ is allowed to be a function of $y(t)$, $y(t-1), \ldots, u(t-1), u(t-2), \ldots$.

Minimum-Variance Control

It is now assumed that $\rho = 0$ and that the process is minimum-phase, that is, that the B polynomial has all zeros inside the unit disc. Before we solve the general problem, we consider a simple example.

EXAMPLE 4.2   Minimum-variance control of a first-order system

Consider the first-order system

$$y(t+1) + ay(t) = bu(t) + e(t+1) + ce(t) \tag{4.4}$$

where $|c| < 1$ and $\{e(t)\}$ is a sequence of independent random variables with unit variance.

Consider the output at time $t+1$. From (4.4) it follows that by using $u(t)$ it is possible to change $y(t+1)$ arbitrarily. Further, $e(t+1)$ is independent of $y(t)$ and $u(t)$; thus

$$\text{var } y(t+1) \geq \text{var } e(t+1) = 1$$

Given measurements up to time t, we can use Eq. (4.4) to compute $e(t)$. The controller

$$u(t) = \frac{ay(t) - ce(t)}{b} \tag{4.5}$$

gives

$$y(t+1) = e(t+1) \tag{4.6}$$

which gives the lower bound of the variance of y. If Eq. (4.5) is used all the time, then from Eq. (4.6), it follows that $y(t) = e(t)$, and we get the controller

$$u(t) = \frac{a-c}{b} y(t) \tag{4.7}$$

□

The minimum-variance controller can in the general case be derived by using similar ideas as Example 4.2. Define

$$d_0 = \deg A - \deg B$$

as the pole excess of the system. This is the same as the time delay in the system. The input at time t will influence the output first at time $t + d_0$. Now consider

$$y(t + d_0) = \frac{B}{A} u(t + d_0) + \frac{C}{A} e(t + d_0) \qquad (4.8)$$

Let the polynomial F of degree $d_0 - 1$ be the quotient, and let the polynomial G of degree $n - 1$ be the remainder when $q^{d_0-1}C$ is divided by A. Hence

$$\frac{q^{d_0-1}C(q)}{A(q)} = F(q) + \frac{G(q)}{A(q)}$$

This can be interpreted as a Diophantine equation,

$$q^{d_0-1}C(q) = A(q)F(q) + G(q) \qquad (4.9)$$

Hence the output at $t + d_0$ can be written as

$$y(t + d_0) = \frac{B}{A}u(t + d_0) + Fe(t + 1) + \frac{qG}{A} e(t)$$

where

$$F(q) = q^{d_0-1} + f_1 q^{d_0-2} + \ldots + f_{d_0-1} \qquad (4.10)$$
$$G(q) = g_0 q^{n-1} + g_1 q^{n-2} + \ldots + g_{n-1} \qquad (4.11)$$

From Eq. (4.1) we can determine $e(t)$:

$$e(t) = \frac{A}{C} y(t) - \frac{B}{C} u(t)$$

From the measurement of $y(t)$ and $u(t)$ it is thus possible to compute the noise sequence, the innovations. This equation is an observer in which the dynamics are given by the C polynomial. It now follows that

$$y(t + d_0) = Fe(t + 1) + \left(\frac{B}{A}q^{d_0} - \frac{qGB}{AC} \right) u(t) + \frac{qGA}{AC} y(t)$$

$$= Fe(t + 1) + \frac{Bq}{AC} \left(q^{d_0-1}C - G \right) u(t) + \frac{qG}{C} y(t)$$

$$= Fe(t + 1) + \frac{qBF}{C} u(t) + \frac{qG}{C} y(t) \qquad (4.12)$$

where Eq. (4.9) has been used to obtain the last equality. The polynomials qG, qBF, and C are all of degree n. This implies that we have divided $y(t + d_0)$ in two parts. The first part, $F(q)e(t + 1)$, depends on the noise acting on the system from $t + 1, \ldots, t + d_0$. The second part,

$$\hat{y}(t + d_0|t) = \frac{qBF}{C} u(t) + \frac{qG}{C} y(t) \qquad (4.13)$$

depends on measured outputs and applied inputs, including the $u(t)$ that we want to determine. From Eqs. (4.12) it follows that $\hat{y}(t + d_0|t)$ is the mean square prediction of $y(t + d_0)$ given data up to and including time t. The prediction error is given by

$$\tilde{y}(t + d_0|t) = y(t + d_0) - \hat{y}(t + d_0|t) = F(q)e(t + 1)$$

and the variance of the prediction error is

$$\text{var } \tilde{y}(t + d_0|t) = \sigma^2(1 + f_1^2 + f_2^2 + \ldots + f_{d_0-1}^2)$$

Minimum variance of the output is now obtained by the control law

$$u(t) = -\frac{G(q)}{B(q)F(q)} y(t) \tag{4.14}$$

Using this controller gives

$$
\begin{aligned}
y(t + d_0) &= F(q)e(t + 1) \\
&= e(t + d_0) + f_1 e(t + d_0 - 1) + \ldots + f_{d_0-1}e(t + 1) \tag{4.15}
\end{aligned}
$$

and the minimum output variance is

$$\text{var } y(t) = \sigma^2(1 + f_1^2 + f_2^2 + \ldots + f_{d_0-1}^2)$$

which is the same as the variance of the prediction error. Using the controller (4.14) gives the closed-loop characteristic equation

$$q^{d_0-1}C(q)B(q) = 0$$

This implies that there are $d_0 - 1$ poles at the origin, n poles at the zeros of the C polynomial, which are inside the unit disc, and $n - d_0$ poles at the zeros of the B polynomial. Since the system was assumed to be minimum-phase, these poles are also inside the unit disc. Observe that minimum-variance control is the same as predicting the output d_0 steps ahead and then choosing the control signal such that the predicted value is equal to the desired reference value. See Fig. 4.1.

The minimum-variance controller can be interpreted as a pole placement controller, which was discussed in Section 3.2. This is seen by multiplying Eq. (4.9) by B, that is,

$$q^{d_0-1}CB = AR + BS \tag{4.16}$$

where

$$R = BF$$
$$S = G$$

The pole placement design leads to the controller

$$u(t) = -\frac{S}{R} y(t) = -\frac{G}{BF} y(t)$$

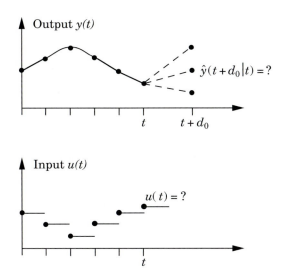

Figure 4.1   Minimum variance control is based on prediction d_0 steps ahead.

Nonminimum-Phase Systems

When the system is nonminimum phase, it is not possible to place some of the closed-loop poles at the zeros of the B polynomial. It can be shown that the optimal controller minimizing Eq. (4.3) with $\rho = 0$ gives the following closed-loop characteristic equation:

$$q^{d_0-1}B^+(q)B^{-*}(q)C(q) = 0$$

that is, the process zeros outside the unit disc, $B^-(q)$, are replaced by the zeros defined by the reciprocal polynomial, $B^{-*}(q)$. See Åström and Wittenmark (1990) in the references at the end of the chapter. The controller

$$u(t) = -\frac{S}{R}y(t)$$

is now obtained from the Diophantine equation

$$q^{d_0-1}B^+B^{-*}C = AR + BS \qquad (4.17)$$

Compare Eq. (4.16).

Moving-Average Controller

The minimum-variance controller leads to a closed-loop system in which the output is a moving average of order $d_0 - 1$ (see Eq. (4.15)). It is possible to design controllers such that the output is a moving average of higher order.

Instead of placing $d_0 - 1$ closed-loop poles at the origin, we may place $d - 1$ poles, where $d \geq d_0$.

The moving-average controller can be derived as follows. Factor the B polynomial as

$$B(q) = B^+(q)B^-(q)$$

where B^+ corresponds to well-damped zeros. To obtain a unique factorization, it is assumed that B^+ is monic. Determine R and S from

$$q^{d-1}B^+C = AR + BS \qquad (4.18)$$

It follows that B^+ must be a factor of R, that is, $R = R_1 B^+$. With the feedback law

$$u(t) = -\frac{S}{R}\, y(t)$$

we get

$$Ay(t) = B\left(-\frac{S}{R}\right) y(t) + Ce(t)$$

or

$$y(t) = \frac{CR}{AR + BS}\, e(t) = \frac{CB^+R_1}{q^{d-1}B^+C}\, e(t)$$

$$= \frac{R_1}{q^{d-1}}\, e(t) = \left(1 + r_1 q^{-1} + \ldots + r_{d-1}q^{-d+1}\right) e(t)$$

where $\deg R_1 = d - 1$ with

$$d = \deg A - \deg B^+$$

Since the controlled output is a moving-average process of order $d - 1$, we call the strategy *moving-average (MA) control*. Notice that no zeros are canceled if

$$B^+ = 1$$

which means that

$$d = \deg A = n$$

The minimum-variance controller and the moving-average controller are similar. The only difference is the value of the integer d, which controls the number of process zeros that are canceled. With $d = d_0$, all process zeros are canceled; with $d = \deg A = n$, no process zeros are canceled.

EXAMPLE 4.3   Moving-average controller

Consider the system (4.1) with

$$A(q) = q^2 + a_1 q + a_2$$
$$B(q) = b_0 q + b_1$$
$$C(q) = q^2 + c_1 q + c_2$$

In this case, $d_0 = 1$. The minimum-variance controller is obtained from Eq. (4.9), giving the controller

$$u(t) = -\frac{(c_1 - a_1) + (c_2 - a_2)q^{-1}}{b_0 + b_1 q^{-1}} y(t)$$

and the closed-loop system is

$$y(t) = e(t)$$

The minimum-variance controller can be used only if $|b_1/b_0| < 1$, that is, for the minimum-phase case.

The moving-average controller is obtained by solving Eq. (4.18). In this case, $d = 2$ and $B^+(q) = 1$. This gives the Diophantine equation

$$q(q^2 + c_1 q + c_2) = (q^2 + a_1 q + a_2)(q + r_1) + (b_0 q + b_1)(s_0 q + s_1)$$

Notice that this is the same as Eq. (3.19) with $A_o(q) = q$ and $A_m(q) = C(q)$. The solution is thus given by Eqs. (3.20) and (3.21):

$$r_1 = \frac{(a_2 - c_2)b_0 b_1 + (c_1 - a_1)b_1^2}{b_1^2 + a_1 b_0 b_1 + a_2 b_0^2}$$

$$s_0 = \frac{b_1(a_1^2 - a_2 - c_1 a_1 + c_2) + b_0(c_1 a_2 - a_1 a_2)}{b_1^2 + a_1 b_0 b_1 + a_2 b_0^2}$$

$$s_1 = \frac{b_1(a_1 a_2 - c_1 a_2) + b_0(a_2 c_2 - a_2^2)}{b_1^2 + a_1 b_0 b_1 + a_2 b_0^2}$$

The closed-loop system is

$$y(t) = (1 + r_1 q^{-1})e(t) \qquad \qquad \square$$

LQG Control

The pole placement and LQG problems are closely related. In the LQG formulation a loss function is specified. Minimization of the loss function leads to a fixed-gain controller that can be interpreted in terms of pole placement. The details are given in Section 4.5. To obtain the LQG solution, it is first necessary to solve the spectral factorization problem, that is, to find the nth-order monic, stable polynomial $P(q)$ that satisfies

$$rP(q)P(q^{-1}) = \rho A(q)A(q^{-1}) + B(q)B(q^{-1}) \qquad (4.19)$$

The LQG-controller is then obtained as the solution to the Diophantine equation

$$C(q)P(q) = A(q)R(q) + B(q)S(q) \qquad (4.20)$$

To get a unique solution with $\deg R = \deg S = n$, it is necessary to make some further restrictions to the solution given by Eq. (4.20). See Theorem 4.3 in Section 4.5. The interpretation of Eq. (4.20) is that the LQG-controller places the closed-loop poles in $P(q)$, given by the spectral factorization, and in $C(q)$, which characterizes the disturbances.

Summary

The minimum-variance controller, the moving-average controller, and the LQG-controller can all be interpreted as pole placement design as discussed in Section 3.2. The minimum-variance controller is obtained by solving the Diophantine equation (4.16) for the minimum-variance case or Eq. (4.17) for the nonminimum-variance case. The moving-average controller is given by Eq. (4.18) and the LQG-controller by Eq. (4.20). The closed-loop characteristic polynomial is chosen differently for each of the design methods.

4.3   STOCHASTIC SELF-TUNING REGULATORS

Indirect Self-tuning Regulator

A straightforward way to make a self-tuning regulator for the process (4.1) is to estimate the parameters in the A, B, and C polynomials by using, for instance, the extended least squares (ELS) algorithm or the recursive maximum-likelihood (RML) algorithm. (See Section 2.2.) The estimated parameters are then used in the design equation (4.9) if minimum-variance control is desired or in Eq. (4.20) if LQG control is desired.

> **EXAMPLE 4.4   Stochastic indirect self-tuning regulator**
>
> Consider the process (4.4) in Example 4.2 with $a = -0.9, b = 3$, and $c = -0.3$. The minimum-variance controller is given by the proportional controller

$$u(t) = \frac{a - c}{b} y(t) = -s_0 y(t) = -0.2y(t)$$

This gives the closed-loop system

$$y(t) = e(t)$$

The ELS method is used to estimate the unknown parameters a, b, and c. The estimates are obtained from Eq. (2.21) with

$$\theta^T = \begin{pmatrix} a & b & c \end{pmatrix}$$
$$\varphi^T(t-1) = \begin{pmatrix} -y(t-1) & u(t-1) & \varepsilon(t-1) \end{pmatrix}$$
$$\varepsilon(t) = y(t) - \varphi^T(t-1)\hat{\theta}(t-1)$$

The controller is

$$u(t) = \frac{\hat{a}(t) - \hat{c}(t)}{\hat{b}(t)} y(t)$$

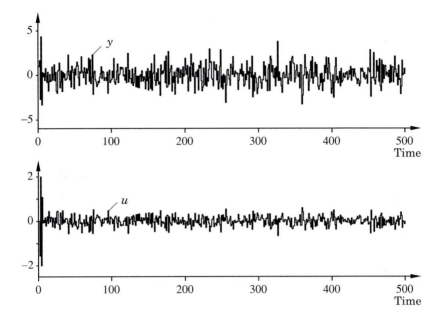

Figure 4.2   Output and input when an indirect self-tuning regulator based on minimum-variance control is used to control the system in Example 4.4.

Figure 4.2 shows the result of a simulation of the algorithm. The initial values in the simulation are

$$\hat{a}(0) = 0$$
$$\hat{b}(0) = 1$$
$$\hat{c}(0) = 0$$
$$P(0) = 100I$$

Figure 4.3 shows the accumulated loss

$$V(t) = \sum_{i=1}^{t} y^2(i)$$

when the optimal minimum-variance controller and the indirect self-tuning regulator are used. The curve of the accumulated loss of the STR is almost parallel to the optimal curve. This means that the performance of the self-tuning regulator is almost optimal except for a short startup transient. Figure 4.4 shows the estimated process parameters. The parameter estimates have not converged to the true values during the simulated period. However, the controller parameter $\hat{s}_0(t) = (\hat{a}(t) - \hat{c}(t))/\hat{b}(t)$ converges faster, as can be seen in Fig. 4.5. For a fixed controller the closed-loop system is stable when $-0.03 < s_0 < 0.63$. Notice that during some of the first steps the controller

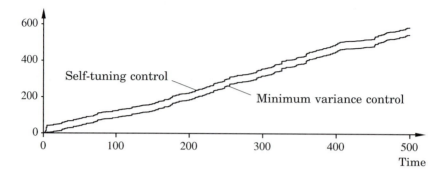

Figure 4.3 The accumulated loss when a self-tuning regulator and the optimal minimum-variance controller are used on the system in Example 4.4.

parameter $\hat{s}_0(t)$ is such that the closed-loop system would be unstable if the controller were frozen to those values.

The reason for the poor convergence of the three estimated process parameters is that the controller converges rapidly to a minimum-variance controller. After that there is poor excitation of the process. The example shows that the self-tuning controller compares well with the optimal controller for the known system. From the control law it can be seen that there may be numerical problems when $\hat{b}(t)$ is small.   □

Direct Minimum-Variance and Moving-Average STR

The design calculations for the indirect self-tuning regulators include the solution of a system of equations such as the Diophantine equation (4.18) or (4.20). The time to solve the Diophantine equation may be long in comparison with the sampling period. A self-tuning regulator that directly estimates the controller parameters eliminates the design calculations. It is thus desirable to

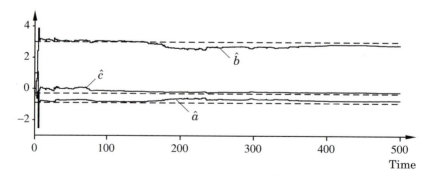

Figure 4.4 The estimated parameters $\hat{a}(t)$, $\hat{b}(t)$, and $\hat{c}(t)$ when the system in Example 4.4 is controlled. The dashed lines correspond to the true parameter values.

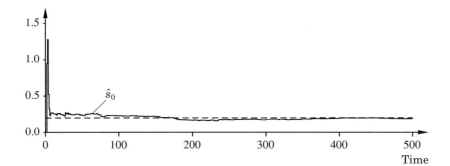

Figure 4.5 The controller parameter $\hat{s}_0(t)$ when the system in Example 4.4 is controlled. The dashed line is the optimal parameter for the minimum-variance controller.

construct direct self-tuning algorithms. Deterministic direct self-tuners were discussed in Section 3.5. The idea is to use the specification and the process model to make a reparameterization of the system. The same idea will now be used for stochastic systems of the form (4.1). In Section 4.2 it was shown that minimum-variance control is the same as predicting the output d_0 steps ahead and then determining the control signal $u(t)$ such that the predicted value is equal to the desired output. Consider the reparameterization (4.12), and rewrite the model in the backward shift operator. This gives

$$y(t + d_0) = \frac{1}{C^*}\left(R^*u(t) + S^*y(t)\right) + R_1^*e(t + d_0) \tag{4.21}$$

where $R_1^* = F^*$ and $\deg R_1 = d_0 - 1$. Using Eq. (4.18), we get, in the same way,

$$y(t + d) = \frac{B^{-*}}{C^*}\left(R^*u(t) + S^*y(t)\right) + R_1^*e(t + d) \tag{4.22}$$

where $\deg R_1 = d - 1$.

The factors $1/C^*$ and B^{-*}/C^* in Eqs. (4.21) and (4.22), respectively, can be interpreted as filters for the regressors. (Compare Section 3.5.) Both equations are now written in predictor form, where the controller polynomials R and S appear directly in the model. These equations can be used as a motivation for the following algorithm.

ALGORITHM 4.1   Basic direct self-tuning algorithm

Data: Given the prediction horizon d, let k and l be the degrees of the R^* and S^* polynomials, respectively. Let Q^*/P^* be a stable filter.

Step 1:   Estimate the coefficients of the polynomials R^* and S^* of the model

$$y(t + d) = R^*(q^{-1})u_f(t) + S^*(q^{-1})y_f(t) + \varepsilon(t + d) \tag{4.23}$$

where

$$R^*(q^{-1}) = r_0 + r_1 q^{-1} + \cdots + r_k q^{-k}$$
$$S^*(q^{-1}) = s_0 + s_1 q^{-1} + \cdots + s_l q^{-l}$$

and

$$u_f(t) = \frac{Q^*(q^{-1})}{P^*(q^{-1})} u(t)$$

$$y_f(t) = \frac{Q^*(q^{-1})}{P^*(q^{-1})} y(t)$$

using Eq. (2.21) with

$$\varepsilon(t) = y(t) - R^* u_f(t - d) - S^* y_f(t - d) = y(t) - \varphi^T(t - d)\hat{\theta}(t - 1)$$

$$\varphi^T(t) = \frac{Q^*(q^{-1})}{P^*(q^{-1})} \left(u(t) \quad \cdots \quad u(t-k) \quad y(t) \quad \cdots \quad y(t-l) \right)$$

$$\theta^T = \left(r_0 \quad \cdots \quad r_k \quad s_0 \quad \cdots \quad s_l \right)$$

Step 2:   Calculate the control signal from

$$R^*(q^{-1})u(t) = -S^*(q^{-1})y(t) \tag{4.24}$$

with R^* and S^* given by the estimates obtained in Step 1.

Repeat Steps 1 and 2 at each sampling period.   □

Remark 1.   Notice that this algorithm is the same as Algorithm 3.3 when $u_c = 0$, but with different filters.

Remark 2.   The parameter r_0 can either be estimated or be assumed to be known. In the latter case it is convenient to write R^* as

$$R^*(q^{-1}) = r_0 \left(1 + r_1' q^{-1} + \cdots + r_k' q^{-k} \right)$$

and use

$$\varepsilon(t) = y(t) - r_0 u_f(t - d) - \varphi^T(t - d)\hat{\theta}(t - 1)$$

$$\varphi^T(t) = \frac{Q^*(q^{-1})}{P^*(q^{-1})} \left(r_0 u(t - 1) \quad \cdots \quad r_0 u(t - k) \quad y(t) \quad \cdots \quad y(t-l) \right)$$

$$\theta^T = \left(r_1' \quad \cdots \quad r_k' \quad s_0 \quad \cdots \quad s_l \right)$$

Asymptotic Properties

The models of Eqs. (4.21) and (4.22) can be interpreted as reparameterizations of the process model of Eq. (4.1) in terms of the controller parameters. They are identical to the model of Eq. (4.23) in Algorithm 4.1 if the filter Q^*/P^* is chosen to be $1/C^*$ and B^{*-}/C^*, respectively. The regression vector is then uncorrelated with the errors, and the least-squares estimate can be expected to converge to the true parameters. The C^* and B^{-*} polynomials are not known, however. The surprising result is that the algorithm also self-tunes to the correct controller even when the filter is not correct. This property inspired the authors of this book to introduce the term "self-tuning." The following result shows that the correct controller parameters are equilibrium values for Algorithm 4.1 for an incorrect choice of Q^*/P^* also. A more detailed analysis of stability and convergence is found in Chapter 6.

THEOREM 4.1 Asymptotic properties 1

Let Algorithm 4.1 with $Q^*/P^* = 1$ be used with a least-squares estimator. The parameter $r_0 = b_0$ can be either fixed or estimated. Assume that the regression vectors are bounded, and assume that the parameter estimates converge. The closed-loop system obtained in the limit is then characterized by

$$\overline{y(t+\tau)y(t)} = 0 \qquad \tau = d, d+1, \ldots, d+l$$
$$\overline{y(t+\tau)u(t)} = 0 \qquad \tau = d, d+1, \ldots, d+k \tag{4.25}$$

where the overbar indicates a time average. Also, k and l are the degrees of the polynomials R^* and S^*, respectively.

Proof: The model of Eq. (4.23) can be written as

$$y(t+d) = \varphi^T(t)\theta + \varepsilon(t+d)$$

and the control law becomes

$$\varphi^T(t)\hat{\theta}(t+d) = 0 \tag{4.26}$$

At an equilibrium the estimated parameters $\hat{\theta}$ are constant. Furthermore, they satisfy the normal equations (2.5), which in this case are written as

$$\frac{1}{t} \sum_{k=1}^{t} \varphi(k)y(k+d) = \frac{1}{t} \sum_{k=1}^{t} \varphi(k)\varphi^T(k)\hat{\theta}(t+d)$$

By using the control law it follows from Eq. (4.26) that

$$\lim_{t\to\infty} \frac{1}{t} \sum_{k=1}^{t} \varphi(k)y(k+d) = \lim_{t\to\infty} \frac{1}{t} \sum_{k=1}^{t} \varphi(k)\varphi^T(k) \left(\hat{\theta}(t+d) - \hat{\theta}(k+d) \right)$$

If the estimate $\hat{\theta}(t)$ converges as $t \to \infty$ and the regression vector $\varphi(k)$ is bounded, the right-hand side goes to zero. Equation (4.25) now follows from $Q^*/P^* = 1$ and the definition of the regression vector in Algorithm 4.1. $\qquad\square$

Stronger statements can be made if more is assumed about the system to be controlled.

THEOREM 4.2   Asymptotic properties 2

Assume that Algorithm 4.1 with least-squares estimation is applied to Eq. (4.1) and that

$$\min(k, l) \geq n - 1 \tag{4.27}$$

If the asymptotic estimates of R^* and S^* are relatively prime, the equilibrium solution is such that

$$\overline{y(t + \tau)y(t)} = 0 \qquad \tau = d, d + 1, \ldots \tag{4.28}$$

that is, the output is a moving-average process of order $d - 1$.

Proof: The closed-loop system is described by

$$R^* u(t) = -S^* y(t)$$
$$A^* y(t) = B^* u(t - d_0) + C^* e(t)$$

Hence

$$(A^* R^* + q^{-d_0} B^* S^*)y = R^* C^* e$$
$$(A^* R^* + q^{-d_0} B^* S^*)u = -S^* C^* e$$

Introduce the signal w defined by

$$(A^* R^* + q^{-d_0} B^* S^*)w = C^* e \tag{4.29}$$

Hence

$$y = R^* w \qquad \text{and} \qquad u = -S^* w \tag{4.30}$$

The condition of Eq. (4.25) then implies that

$$\overline{R^* w(t) y(t + \tau)} = 0 \qquad \tau = d, d + 1, \ldots, d + l$$
$$\overline{S^* w(t) y(t + \tau)} = 0 \qquad \tau = d, d + 1, \ldots, d + k$$

If we introduce

$$C_{wy}(\tau) = \overline{w(t) y(t + \tau)}$$

the preceding equations can be written as

$$
\begin{pmatrix}
r_0 & r_1 & r_2 & \cdots & r_k & 0 & \cdots & 0 \\
0 & r_0 & r_1 & r_2 & \cdots & r_k & & \\
\vdots & \ddots & \ddots & \ddots & \ddots & & \ddots & \\
0 & \cdots & 0 & r_0 & r_1 & r_2 & \cdots & r_k \\
s_0 & s_1 & s_2 & \cdots & s_l & 0 & \cdots & 0 \\
0 & s_0 & s_1 & s_2 & \cdots & s_l & & \\
\vdots & \ddots & \ddots & \ddots & \ddots & & \ddots & \\
0 & \cdots & 0 & s_0 & s_1 & s_2 & \cdots & s_l
\end{pmatrix}
\begin{pmatrix}
C_{wy}(d) \\
\vdots \\
C_{wy}(d + k + l)
\end{pmatrix}
= 0
$$

Since the Sylvester matrix on the left is nonsingular when R^* and S^* are relatively prime (compare Section 11.4), it follows that

$$C_{wy}(\tau) = 0 \qquad \tau = d, d+1, \ldots, d+k+l$$

The covariance function satisfies the equation

$$F^*(q^{-1})C_{wy}(\tau) = 0 \qquad \tau \geq 0$$

The system of Eq. (4.29) has the order

$$n + k = n + \max(k, l)$$

If

$$k + l + 1 \geq n + \max(k, l)$$

or, equivalently,

$$\min(k, l) \geq n - 1$$

it follows that

$$C_{wy}(\tau) = 0 \qquad \tau = d, d+1, \ldots$$

It also follows from Eq. (4.30) that

$$C_y(\tau) = 0 \qquad \tau = d, d+1, \ldots$$

which completes the proof. □

Remark 1. The algorithm thus drives the correlation of the output to zero starting at lag $\tau = d$. It follows from Theorem 4.1 that the correlations at lags $d, d+1, \ldots, d+l$ will always be zero at equilibrium. If there are enough parameters in the controller, the covariance of the output will be zero for all higher lags. Notice that the condition of Eq. (4.28) is easily checked by monitoring the covariances of the output.

Remark 2. It is possible to influence cancellation of the process zeros simply by choosing the integer d. With $d = d_0$ a controller that cancels all zeros is obtained. With $d = n$ the controller will not cancel any process zeros. □

Theorems 4.1 and 4.2 imply that if the estimates converge, and if there are sufficiently many parameters in the controller, then Algorithm 4.1 will converge to the moving-average controller.

Examples

The properties of the minimum-variance and moving-average self-tuners are illustrated with two examples.

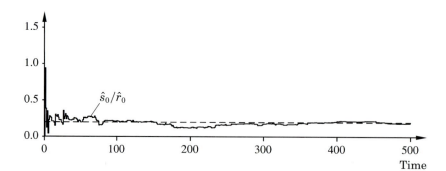

Figure 4.6   The parameter $\hat{s}_0/\hat{r}_0$ in the controller, when the process in Example 4.5 is controlled by using the direct minimum-variance self-tuning controller.

EXAMPLE 4.5   **Direct minimum-variance self-tuning regulator**

Consider the same process as in Example 4.4. The process model of Eq. (4.23) is now

$$y(t + 1) = r_0 u(t) + s_0 y(t) + \varepsilon(t + 1)$$

It is assumed that r_0 is fixed to the value $\hat{r}_0 = 1$. Notice that this is different from the true value, which is 3. The parameter s_0 is estimated by using the least-squares method. The control law becomes

$$u(t) = -\frac{\hat{s}_0}{\hat{r}_0} y(t)$$

Figure 4.6 shows $\hat{s}_0/\hat{r}_0$, which is seen to converge rapidly to a value corresponding to the value of the optimal minimum-variance controller, even if $\hat{r}_0$ is not equal to its true value. This is also seen in Fig. 4.7, which shows the loss function when the self-tuner and the optimal minimum-variance controller are used. Compare Figs. 4.3 and 4.5.   □

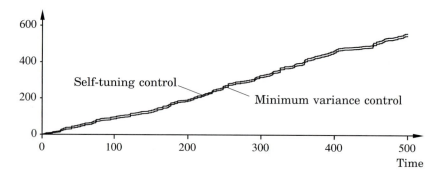

Figure 4.7   The loss function when the direct self-tuning regulator and the optimal minimum-variance controller are used on the system in Example 4.5.

EXAMPLE 4.6   **MA control of a nonminimum-phase system**

Consider an integrator with a time delay τ. For the sampling period $h > \tau$ the system is described by

$$A(q) = q(q - 1)$$
$$B(q) = (h - \tau)q + \tau = (h - \tau)(q + b)$$

where

$$b = \frac{\tau}{h - \tau} \quad \text{and} \quad d_0 = 1$$

The noise is assumed to be characterized by

$$C(q) = q(q + c) \qquad |c| < 1$$

The sampled-data system is nonminimum-phase if $\tau > h/2$. This implies that the basic minimum-variance self-tuner can be used only if $\tau < h/2$. Let the

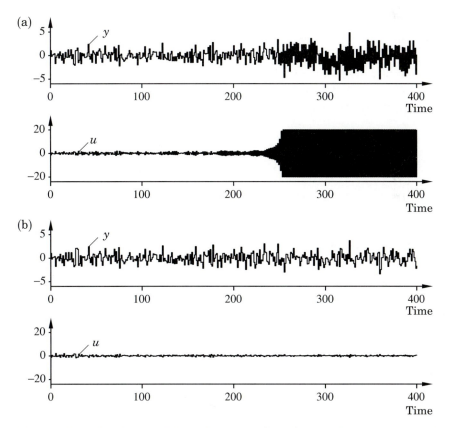

Figure 4.8   Simulation of the self-tuning algorithm on the integrator with time delay in Example 4.6. At $t = 100$ the delay is changed from 0.4 to 0.6. (a) $d = 1$; (b) $d = 2$.

controller have the structure

$$u(t) = -\hat{s}_0(t)y(t) - \hat{r}_1(t)u(t-1)$$

Simulations of the system are shown in Fig. 4.8 for $h = 1$ and $c = -0.8$. The time delay is initially 0.4 and is increased to 0.6 at time $t = 100$, at which time the sampled-data system gets a zero outside the unit circle. Figure 4.8(a) shows the results obtained with $d = 1$, the minimum-variance structure. The parameters first converge toward the minimum-variance controller. At $t = 100$ the sampled-data system gets a zero outside the unit circle. The self-tuning regulator then tries to cancel the zero, and the closed-loop system becomes unstable after some time. It does not become unstable exactly at $t = 100$ because it takes a while for the controller parameters to change. The control signal is limited to ±20, which explains why the signals do not grow exponentially. The forgetting factor is $\lambda = 0.99$. Figure 4.8(b) shows the results for the algorithm with $d = 2$. The moving-average controller is a stable equilibrium for both $\tau = 0.4$ and $\tau = 0.6$. There will be a shift in the parameter values when the delay is changed, but the closed-loop system is stable.

The controller that gives the smallest attainable variance of the output gives the standard deviations 1.000 and 1.004 when $\tau = 0.4$ and 0.6, respectively, while the moving-average controller gives the standard deviations 1.003 and 1.007 when $\tau = 0.4$ and 0.6, respectively. Degradation in the performance when the moving-average controller is used in this example is thus minor. $\square$

4.4   UNIFICATION OF DIRECT SELF-TUNING REGULATORS

The moving-average self-tuner is attractive because of its simplicity. It is easy to explain intuitively how the algorithm works, and the algorithm is easy to implement. This has led to great interest in the algorithm. The algorithm can be explained as follows: Determine the structure of a predictor that can be used to predict the output d steps ahead. The parameters of the predictor are estimated in real time. On the basis of the estimated parameters the control signal is determined such that the predicted output of the process is equal to the reference value. The algorithm has been analyzed extensively. The closed-loop bandwidth depends critically on the sampling period h and the prediction horizon d, so both must be chosen with care. The algorithm may result in a controller in which process zeros are canceled; the cancellations depend on the choice of prediction horizon. Many variants of the algorithm have been suggested. A number of these can be described in a unified framework, as we will demonstrate.

Consider the model of Eq. (4.1), and introduce the filtered output

$$y_f(t) = \frac{Q^*(q^{-1})}{P^*(q^{-1})} y(t)$$

where Q^* and P^* are stable polynomials. The filtered output satisfies the equation

$$A^*(q^{-1})P^*(q^{-1})y_f(t) = B^*(q^{-1})Q^*(q^{-1})u(t - d_0) + C^*(q^{-1})Q^*(q^{-1})e(t)$$

Introduce the identity

$$C^*(q^{-1})Q^*(q^{-1}) = A^*(q^{-1})P^*(q^{-1})R_1^*(q^{-1}) + q^{-d_0}S^*(q^{-1})$$

Then

$$y_f(t + d_0) = \frac{1}{C^*Q^*}\left(S^*y_f(t) + B^*Q^*R_1^*u(t)\right) + R_1^*e(t + d_0)$$

Introducing

$$y_f'(t) = \frac{1}{Q^*(q^{-1})}\, y_f(t) = \frac{1}{P^*(q^{-1})}\, y(t)$$

gives the model

$$y_f(t + d_0) = \frac{1}{C^*}\left(S^*y_f'(t) + B^*R_1^*u(t)\right) + R_1^*e(t + d_0) \qquad (4.31)$$

By analogy with Eq. (4.21) this model structure could be used with Algorithm 4.1 to derive a self-tuning regulator for minimization of the variance of y_f. This reparameterized model now suggests the following generalized self-tuning algorithm.

ALGORITHM 4.2   Generalized direct self-tuning algorithm

Data: Given the prediction horizon, d, the order of the controller, $\deg R^*$ and $\deg S^*$, the stable observer polynomial, A_o^*, and the stable polynomials Q^* and P^*, define the filtered signals

$$y_f(t) = \frac{Q^*}{P^*}\, y(t) \qquad y_f'(t) = \frac{1}{P^*}\, y(t)$$

Step 1:   Estimate the coefficients of the polynomials R^* and S^* of the model

$$y_f(t + d) = \frac{R^*}{A_o^*}\, u(t) + \frac{S^*}{A_o^*}\, y_f'(t) + \varepsilon(t + d) \qquad (4.32)$$

using the least-squares method.

Step 2:   Calculate the control signal from

$$u(t) = -\frac{S^*}{R^*}\, y_f'(t)$$

with R^* and S^* given by the estimates obtained in Step 1.

Repeat Steps 1 and 2 at each sampling period. $\qquad\qquad\qquad\qquad\qquad\square$

From Eq. (4.31) and Theorems 4.1 and 4.2 it follows that if the estimates converge, then the closed-loop system will be

$$y_f(t) = R_1^* e(t)$$

or

$$y(t) = \frac{P^* R_1^*}{Q^*} e(t) \tag{4.33}$$

where R_1^* is given by the identity

$$C^* Q^* = A^* P^* R_1^* + q^{-d} B^{-*} S^* \tag{4.34}$$

and the control signal is given by

$$u(t) = -\frac{S^*}{R^*} y_f'(t) = -\frac{S^*}{R^* P^*} y(t) \tag{4.35}$$

where $R^* = B^{+*} R_1^*$. The closed-loop poles will thus be influenced by Q^*, and additional zeros can be introduced through P^*. The introduction of the filter Q^*/P^* gives what is sometimes called a *detuned minimum-variance* algorithm.

Algorithm 4.2 is essentially the same as Algorithm 4.1 applied to filtered signals. The filter Q^*/P^* and the prediction horizon will determine the pulse transfer operator of the closed-loop system. The optimal observer polynomial is C^*, which is unknown. Instead, an approximation A_o^* is used. The observer polynomial A_o^* will determine the convergence properties. This will not influence the asymptotic properties as long as the filter Q^*/P^* and its inverse are stable.

Minimum-variance control may result in large control signals. One way to decrease the variation of the control signal is to generalize the loss function such that it also contains a penalty on the control signal. Linear quadratic controllers are of this type; a minor drawback with linear quadratic self-tuning regulators is the computational burden. One way to simplify the problems is to use a loss function of the form

$$E\left\{ \left(P^*(q^{-1}) y(t + d_0) \right)^2 + \left(Q^*(q^{-1}) u(t) \right)^2 \Big| \mathcal{Y}_t \right\}$$

where

$$\mathcal{Y}_t = \{ y(t), y(t-1), \ldots, y(0), u(t), u(t-1), \ldots, u(0) \}$$

that is, the data available at time t. The resulting controller is sometimes called a *generalized minimum-variance controller*. This controller can be interpreted in the same framework as above. To illustrate this, assume that $P^* = 1$ and that $Q^* = \sqrt{\rho}$. This gives the loss function

$$E\left\{ y^2(t + d_0) + \rho u^2(t) \Big| \mathcal{Y}_t \right\} \tag{4.36}$$

Notice that the loss function depends only on the output y at time $t + d_0$, that is, at only one time instant. Loss functions of the form (4.36) are sometimes called one-stage loss functions.

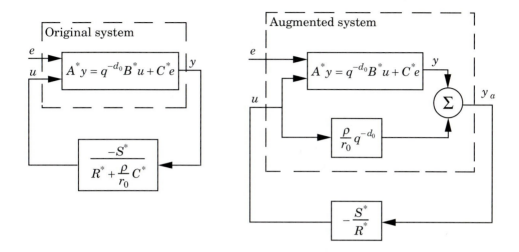

Figure 4.9   Equivalent systems.

Assume that the process is governed by Eq. (4.1). By using the representation of the process dynamics given by Eq. (4.21) it can be shown that the control law that minimizes Eq. (4.36) is

$$\left(R^* + \frac{\rho}{r_0} C^* \right) u(t) = -S^* y(t) \tag{4.37}$$

where

$$R^* = R_1^* B^*$$

and R^* and S^* are given by Eq. (4.16).

By using the same idea it is possible to construct a new system, which has Eq. (4.37) as its minimum-variance controller. Augment the original system with a parallel connection with the pulse transfer operator $\rho q^{-d_0}/r_0$ (see Fig. 4.9). This is in fact a standard technique to obtain an equivalent controller with a bounded gain. The input-output relation of the augmented system is

$$A^* y_a(t) = \left(B^* + \frac{\rho}{r_0} A^* \right) u(t - d_0) + C^* e(t)$$

The minimum-variance control law for this system is given by

$$R_1^* \left(B^* + \frac{\rho}{r_0} A^* \right) u(t) = -S^* y_a(t) \tag{4.38}$$

where R_1^* and S^* satisfy Eq. (4.16). It follows from Fig. 4.9 that

$$y_a(t) = y(t) + \frac{\rho}{r_0} q^{-d_0} u(t)$$

Then Eq. (4.38) can be written as

$$\left(R_1^* B^* + \frac{\rho}{r_0} A^* R_1^* \right) u(t) = -S^* \left(y(t) + \frac{\rho}{r_0} q^{-d_0} u(t) \right)$$

or

$$\left(R_1^* B^* + \frac{\rho}{r_0} (A^* R_1^* + q^{-d_0} S^*) \right) u(t) = -S^* y(t)$$

Equation (4.16) gives

$$\left(R_1^* B^* + \frac{\rho}{r_0} C^* \right) u(t) = -S^* y(t)$$

which is identical to Eq. (4.37). Notice that with the control law of Eq. (4.38) the canceled factor is not B^* but $B^* + \rho A^*/r_0$. This implies that problems can be expected when the system is nonminimum-phase and close to the stability boundary.

In the generalized minimum-variance control algorithm it is assumed that $C^*(q^{-1}) = 1$. The algorithm can thus be obtained simply by adding a parallel path to the original system and applying an ordinary self-tuning regulator based on minimum-variance control to the augmented system. The control gain is adjusted simply by changing the parameter ρ of the parallel path.

The preceding analysis shows that Algorithm 4.2 is very flexible. It can be used for many different types of specifications, not only for minimum-variance control. This is very important for the implementation of self-tuning regulators.

Self-tuning Feedforward Control

Feedforward control is a very useful way to reduce the influence of known disturbances. Examples of measurable disturbances can be temperatures and concentrations in incoming product streams in chemical processes, outdoor temperature in climate control systems, and thickness of the paper in paper machines. Command signals can also be interpreted as a measurable distur-bance. The controller in Eq. (3.2) can be interpreted as feedforward from the command signal. To use feedforward, it is necessary to know the dynamics of the process. It is, however, also possible to establish self-tuning feedforward compensation. One way to do this is to postulate a model structure of the form

$$y(t + d) = R^* u(t) + S^* y(t) + T^* v(t) + \varepsilon(t + d)$$

where $v(t)$ is the measurable disturbance acting on the system. The signal v could also be the reference value. The polynomials R^*, S^*, and T^* are estimated in the usual way, and the control law is chosen to be

$$u(t) = -\frac{S^*}{R^*} y(t) - \frac{T^*}{R^*} v(t)$$

Self-tuning feedforward control has been used successfully in many industrial applications.

Examples

The behavior of Algorithm 4.2 is illustrated through two examples.

EXAMPLE 4.7   **Effect of filtering**

Consider the process

$$y(t) + ay(t-1) = bu(t-1) + e(t) + ce(t-1)$$

where $a = -0.9$, $b = 3$, and $c = -0.3$, which is the same process as in Examples 4.4 and 4.5. Let the filter be

$$\frac{Q^*}{P^*} = \frac{1 + q_1 q^{-1}}{1 + p_1 q^{-1}}$$

The identity of Eq. (4.34) gives the solution

$$s_0 = c + q_1 - a - p_1$$
$$s_1 = cq_1 - ap_1$$

The control law is given by Eq. (4.35), with

$$R_1^* P^* B^{+*} = b(1 + p_1 q^{-1})$$

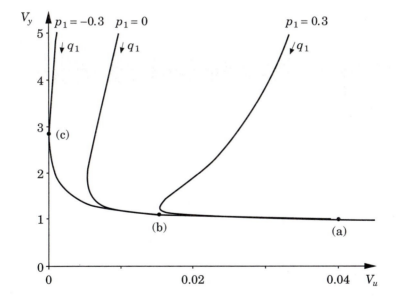

Figure 4.10   The output variance, V_y, and input variance, V_u, as functions of q_1 of the system in Example 4.7 when $p_1 = -0.3$, 0, and 0.3. Three different cases are indicated by dots: (a) $p_1 = q_1 = 0$; (b) $p_1 = 0$, $q_1 = -0.3$; (c) $p_1 = -0.3$, $q_1 = -0.9$.

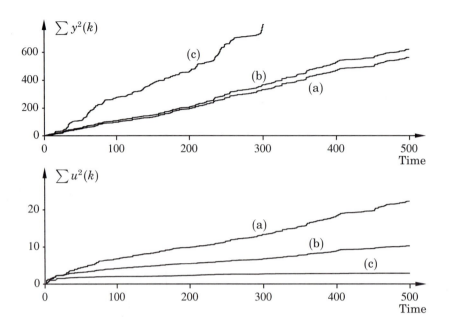

Figure 4.11 Simulation of the generalized self-tuning algorithm on the system in Example 4.7 when (a) $p_1 = q_1 = 0$ (minimum-variance control); (b) $p_1 = 0$, $q_1 = -0.3$; (c) $p_1 = -0.3$, $q_1 = -0.9$ (open-loop system).

The closed-loop system becomes

$$y(t) = \frac{1 + p_1 q^{-1}}{1 + q_1 q^{-1}} e(t)$$

$$u(t) = -\frac{s_0 + s_1 q^{-1}}{b(1 + q_1 q^{-1})} e(t)$$

There are many different ways to choose the filter Q^*/P^*. In principle it should be a phase-advance network. This implies that the closed-loop system given by Eq. (4.33) will be low-pass filtered. Figure 4.10 shows how the output and input variances change with q_1 for some values of p_1. Case (a) in Fig. 4.10 corresponds to minimum-variance control. In case (b) the output variance is increased by 10%, and the input variance is reduced by about 60% compared with the minimum-variance case. In case (c) the input variance is zero; that is, the system is open-loop. Figure 4.11 shows the accumulated losses for the input and the output when the generalized self-tuning algorithm is used. Cases (a), (b), and (c) are the same as in Fig. 4.10.   □

EXAMPLE 4.8   Generalized minimum variance self-tuning controller

The self-tuning controller that minimizes Eq. (4.36) will now be used to control the same system as in the previous example. The controller in Eq. (4.37), with

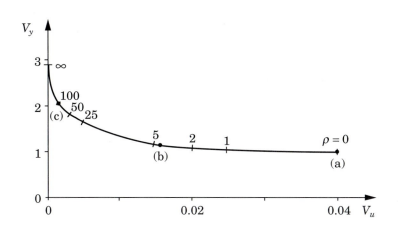

Figure 4.12   The output variance V_y as a function of the input variance V_u in Example 4.8 for different values of ρ: (a) $\rho = 0$; (b) $\rho = 4$; (c) $\rho = 100$.

R^* and S^* given by Eq. (4.16), is

$$u(t) = -\frac{c - a}{b + \rho(1 + aq^{-1})} \, y_a(t)$$

Figure 4.12 shows the output variance as a function of the input variance for different values of ρ. The curve has the same gross behavior as shown in Fig. 4.10. However, the parameter ρ may be easier to choose than the filter in Example 4.7. Figure 4.13 shows the accumulated losses of the output and the input for different values of ρ when the self-tuner in Algorithm 4.1 is used on the augmented system shown in Fig. 4.9. Compare Fig. 4.11.   □

Summary

There are many ways to make direct self-tuning regulators with good properties. The amount of computation is moderate, since the design calculations are eliminated. It has been shown that the generalized direct self-tuning algorithm, Algorithm 4.2, is very flexible. By using the filter Q^*/P^* and the prediction horizon, it is possible to determine the behavior of the closed-loop system. It is possible to choose, for instance, moving-average control, generalized minimum-variance control, or pole-zero placement control.

The observer polynomial does not influence the asymptotic properties. It will instead influence the transient properties and can be used to improve the convergence properties of the algorithm. The robustness and sensitivity of the algorithm are also influenced by the filter Q^*/P^*.

For simplicity, Algorithm 4.2 has been derived for the regulator case, in which the reference value is equal to zero. It is easy to modify the algorithm

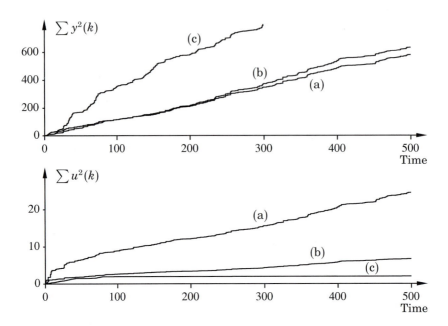

Figure 4.13 Simulation of the generalized minimum variance self-tuning algorithm on the system in Example 4.8 when (a) $\rho = 0$ (minimum-variance control); (b) $\rho = 4$; (c) $\rho = 100$ ("almost" open-loop control).

such that the output follows a reference trajectory; some ideas are suggested in the problems at the end of this chapter and in Section 11.3.

4.5   LINEAR QUADRATIC STR

The linear quadratic design procedure can also be used as the design method in a self-tuning regulator. Consider the process model

$$A(q)y(t) = B(q)u(t) + C(q)e(t) \tag{4.39}$$

and the steady-state loss function

$$J_{yu} = E\left\{(y(t) - y_m(t))^2 + \rho u^2(t)\right\} \tag{4.40}$$

The optimal feedback law that minimizes Eq. (4.40) for the system of Eq. (4.39) is given by the following theorem.

THEOREM 4.3   LQG control

Consider the system in Eq. (4.39). Let the monic polynomials $A(q)$ and $C(q)$ have degree n. Assume that $C(q)$ has all its zeros inside the unit disc, and assume that there is no nontrivial polynomial that divides $A(q)$, $B(q)$, and $C(q)$. Let $A_2(q)$ be the greatest common divisor of $A(q)$ and $B(q)$, let $A_2^+(q)$ of degree l be the factor of $A_2(q)$ with all its zeros inside the unit disc, and let $A_2^-(q)$ of degree m be the factor of $A(q)$ that has all its zeros outside the unit disc or on the unit circle.

The admissible control law that minimizes Eq. (4.40) with $\rho > 0$ is then given by

$$R(q)u(t) = -S(q)y(t) + T(q)y_m(t) \tag{4.41}$$

where R and S are of degree $n + m$

$$\begin{aligned} R(q) &= A_2^-(q)\tilde{R}(q) \\ S(q) &= z^m \tilde{S}(q) \end{aligned} \tag{4.42}$$

and $\tilde{R}(q)$ and $\tilde{S}(q)$ satisfy the Diophantine equation

$$A_1(q)A_2^-(q)\tilde{R}(q) + q^m B_1(q)\tilde{S}(q) = P_1(q)C(q) \tag{4.43}$$

with $\deg \tilde{R}(q) = \deg \tilde{S}(q) = n$ and $\tilde{S}(0) = 0$. Furthermore,

$$\begin{aligned} A(q) &= A_1(q)A_2(q) \\ B(q) &= B_1(q)A_2(q) \\ \tilde{B}(q) &= B_1(q)A_2^+(q) \end{aligned}$$

The polynomial $P(q)$ is given by

$$P(q) = A_2^+(q)P_1(q) \tag{4.44}$$

where $P_1(q)$ is the solution of the spectral factorization problem

$$rP_1(q)P_1(q^{-1}) = \rho A_1(q)A_2^-(q)A_1(q^{-1})A_2^-(q^{-1}) + B_1(q)B_1(q^{-1}) \tag{4.45}$$

with $\deg P_1(q) = \deg A_1(q) + \deg A_2^-(q)$. The polynomial $T(q)$ is given by

$$T(q) = t_0 q^m C(q)$$

where

$$t_0 = P_1(1)/B_1(1) \qquad\qquad \square$$

A proof of the theorem is found in Åström and Wittenmark (1990).

Remark.   By using Eqs. (4.42) the identity (4.43) can be written as

$$A(q)R(q) + B(q)S(q) = A_2(q)P_1(q)C(q)$$

The LQG solution can thus be interpreted as a pole-placement controller, where the poles are positioned at the zeros of A_2, P_1, and C. The controller also has the property that A_2^- divides R. This is an example of the *internal model principle*. Using the internal model principle implies that a model of the disturbance is included in the controller. $\square$

To solve the design problem, it is necessary to solve the spectral factoriza-tion problem of Eq. (4.45) and to solve the Diophantine equation Eq. (4.43). The solution to the LQG problem given by Theorem 4.3 is closely related to the pole placement design problem. The solution to the spectral factorization problem gives the desired closed-loop poles. The second part of the algorithm can be interpreted as a pole placement problem.

An alternative solution to the design problem is to use a state space formulation. The process model of Eq. (4.39) can be written in state space form as

$$x(t+1) = \bar{A}x(t) + \bar{B}u(t) + \bar{K}e(t)$$
$$y(t) = \bar{C}x(t) + e(t) \tag{4.46}$$

where the matrices $\bar{A}$, $\bar{B}$, $\bar{C}$, and $\bar{K}$ are given in the canonical form

$$\bar{A} = \begin{pmatrix} -a_1 & 1 & 0 & \cdots & 0 \\ \vdots & & \ddots & & \\ -a_{n-1} & 0 & & \cdots & 1 \\ -a_n & 0 & & \cdots & 0 \end{pmatrix}$$

$$\bar{B} = \begin{pmatrix} 0 & \cdots & 0 & b_0 & \cdots & b_m \end{pmatrix}^T$$

$$\bar{C} = \begin{pmatrix} 1 & 0 & \cdots & 0 \end{pmatrix}$$

$$\bar{K} = \begin{pmatrix} c_1 - a_1 & \cdots & c_n - a_n \end{pmatrix}^T$$

where $m = n - d_0$. The model in Eq. (4.46) is called the *innovation model*, and $\bar{K}$ is the optimal steady-state gain in the Kalman filter, that is, $\hat{x}(t+1|t) = x(t+1)$. It is also possible to derive the filter for $\hat{x}(t|t)$, which is given by

$$\hat{x}(t|t) = (qI - \bar{A} + \bar{K}\bar{C})^{-1} (\bar{B}u(t) + \bar{K}y(t))$$

By using the definitions of $\bar{A}$, $\bar{K}$, and $\bar{C}$ it is easily seen that $\det(qI - \bar{A} + \bar{K}\bar{C}) = C(q)$. That is, the optimal observer polynomial is equal to $C(q)$.

Introduce the loss function

$$J_x = E \left\{ \sum_{t=1}^{N} x^T(t)Q_1 x(t) + \rho u^2(t) + x^T(N)Q_0 x(N) \right\} \tag{4.47}$$

The optimal controller is given by

$$u(t) = -L(t)\hat{x}(t|t) \tag{4.48}$$

where $L(t)$ is a time-varying feedback gain given through a Riccati equation

$$S(t) = \left(\bar{A} - \bar{B}L(t) \right)^T S(t+1) \left(\bar{A} - \bar{B}L(t) \right) + Q_1 + \rho L^T(t)L(t)$$
$$L(t) = \left(\rho + \bar{B}^T S(t+1)\bar{B} \right)^{-1} \bar{B}^T S(t+1)\bar{A} \tag{4.49}$$

with $S(N) = Q_0$. The limiting controller

$$\bar{L} = \lim_{t\to\infty} L(t)$$

is such that the closed-loop characteristic equation is

$$P(q) = \det(q - \bar{A} + \bar{B}\bar{L}) = 0$$

where $P(q)$ is the same as in Eq. (4.44).

The two solutions to the LQG control problem suggest two ways to construct indirect linear quadratic self-tuning regulators. In both algorithms it is first necessary to estimate the A, B, and C polynomials in the process model of Eq. (4.39). This can be done by using the recursive maximum-likelihood method or the extended least-squares method. This leads to the following algorithm.

ALGORITHM 4.3   Indirect LQG-STR based on spectral factorization

Data: Given specifications in the form of the parameter ρ in the loss function of Eq. (4.40) and the order of the system.

Step 1:   Estimate the coefficients of the polynomials A, B, and C in Eq. (4.39).

Step 2:   Replace A, B, and C with the estimates obtained in Step 1 and solve the spectral factorization problem of Eq. (4.45) to obtain $P(q)$.

Step 3:   Solve the Diophantine equation of Eq. (4.43).

Step 4:   Calculate the control signal from Eq. (4.41).

Repeat Steps 1, 2, 3, and 4 at each sampling period.   □

The state space formulation gives the following algorithm.

ALGORITHM 4.4   Indirect LQG-STR based on the Riccati equation

Data: Given specifications in the form of the parameters Q_0, Q_1, and ρ in the loss function of Eq. (4.47) and the order of the system.

Step 1:   Estimate the coefficients of the polynomials A, B, and C in Eq. (4.39).

Step 2:   Replace A, B, and C with the estimates obtained in Step 1 and solve the algebraic Riccati equation or iterate Eqs. (4.49) to obtain $\bar{L}$.

Step 3:   Calculate the control signal from Eq. (4.48).

Repeat Steps 1, 2, and 3 at each sampling period.   □

Notice that if $Q_1 = \bar{C}^T\bar{C}$, the steady-state solution to Eqs. (4.49) will give the same result as the minimization of Eq. (4.40). Algorithms 4.3 and 4.4 are indirect algorithms that are able to handle nonminimum-phase systems and varying time delays. The computations are more extensive for these algorithms than for the simple self-tuning regulators discussed above.

Solution of the spectral factorization or the Riccati equation is the major computation in an LQG self-tuner. These calculations can be made in many different ways. The Riccati equation can be solved by using an eigenvalue method or by some iterative method. The iterative methods will in general lead to shorter code. In general the Riccati equation is iterated several steps. To guarantee that the calculations can be done in a prescribed sampling interval, it is necessary to truncate the iterations; it is important that a reasonable result be obtained when the iteration is truncated. For instance, the polynomial P in the spectral factorization must be stable. This is guaranteed for some algorithms. In some algorithms it is suggested that the Riccati equation be iterated only one step at each sampling.

4.6   ADAPTIVE PREDICTIVE CONTROL

Algorithm 4.1 is one way to make a controller with a variable prediction horizon. The underlying control problem is the moving-average controller. The moving-average controller may also be used for nonminimum-phase systems, as was illustrated in Section 4.3.

In using the minimum-variance controller or the moving-average controller the output is predicted only at *one* future time. The prediction horizon d is then a design parameter. The predicted output can also be computed for different prediction horizons and then used in a loss function. Several ways to achieve predictive control have been suggested in the literature; we now discuss and analyze some of these. The case with known parameters is first analyzed before the adaptive versions are discussed.

Predictive control algorithms are based on an assumed model of the process and on an assumed scenario for the future control signals. This gives a sequence of control signals. Only the first one is applied to the process, and a new sequence of control signals is calculated when a new measurement is obtained. This is called a *receding-horizon controller*. There are many variants of predictive control, for instance, *model predictive control*, *dynamic matrix control*, *generalized predictive control*, and *extended horizon control*. The methodology has been used extensively in chemical process control.

Output Prediction

One basic idea in the predictive control algorithms is to rewrite the process model to get an explicit expression for the output at a future time. Compare Eq. (4.22). Consider the deterministic process

$$A^*(q^{-1})y(t) = B^*(q^{-1})u(t - d_0) \tag{4.50}$$

and introduce the identity

$$1 = A^*(q^{-1})F_d^*(q^{-1}) + q^{-d}G_d^*(q^{-1}) \tag{4.51}$$

where

$$\deg F_d^* = d - 1$$
$$\deg G_d^* = n - 1$$

The subscript d is used to indicate that the prediction horizon is d steps. It is assumed that $d \geq d_0$. The polynomial identity of Eq. (4.51) can be used to predict the output d steps ahead. Hence

$$y(t + d) = A^*F_d^*y(t + d) + G_d^*y(t) = B^*F_d^*u(t + d - d_0) + G_d^*y(t) \tag{4.52}$$

Compare Eq. (4.12). Introduce

$$B^*(q^{-1})F_d^*(q^{-1}) = R_d^*(q^{-1}) + q^{-(d-d_0+1)}\bar{R}_d^*(q^{-1})$$

where

$$\deg R_d^* = d - d_0$$
$$\deg \bar{R}_d^* = n - 2$$

The coefficients of R_d^* are the first $d - d_0 + 1$ terms of the pulse response of the open-loop system. This can be seen as follows:

$$\begin{aligned}
\frac{q^{-d_0}B^*}{A^*} &= q^{-d_0}B^* \left(F_d^* + q^{-d}\frac{G_d^*}{A^*} \right) \\
&= q^{-d_0}R_d^*(q^{-1}) + q^{-(d+1)}\bar{R}_d^*(q^{-1}) + \frac{B^*(q^{-1})G_d^*(q^{-1})}{A^*(q^{-1})}q^{-(d+d_0)}
\end{aligned} \tag{4.53}$$

The powers of the last two terms are at least $-(d + 1)$. It then follows that R_d^* is the first part of the pulse response, since $\deg R_d^* = d - d_0$.

Equation (4.52) can be written as

$$\begin{aligned}
y(t + d) &= R_d^*(q^{-1})u(t + d - d_0) + \bar{R}_d^*(q^{-1})u(t - 1) + G_d^*(q^{-1})y(t) \\
&= R_d^*(q^{-1})u(t + d - d_0) + \bar{y}_d(t)
\end{aligned} \tag{4.54}$$

$R_d^*(q^{-1})u(t + d - d_0)$ depends on $u(t), \ldots, u(t + d - d_0)$, and $\bar{y}_d(t)$ is a function of $u(t - 1), u(t - 2), \ldots$, and $y(t), y(t - 1), \ldots$. The variable $\bar{y}_d(t)$ can be interpreted as the constrained prediction of $y(t + d)$ under the assumption that $u(t)$ and future control signals are zero. The output at time $t + d$ thus depends on future control signals (if $d > d_0$), the control signal to be chosen, and old inputs and outputs. If $d > d_0$, it is necessary to make some assumptions about the future control signals. One possibility is to assume that the control signal will remain constant, that is, that

$$u(t) = u(t + 1) = \cdots = u(t + d - d_0) \tag{4.55}$$

Another way is to determine the control law that brings $y(t + d)$ to a desired value while minimizing the control effort over the prediction horizon, that is, to minimize

$$\sum_{k=t}^{t+d} u(k)^2 \tag{4.56}$$

A third way is to assume that the increment of the control signal will be zero after some time. This is used, for instance, in generalized predictive control (GPC), which is discussed below.

Constant Future Control

Consider Eq. (4.54) and assume that the predicted output is equal to the desired output, that is, $y(t + d) = y_m(t + d)$. If we assume that Eq. (4.55) holds, then $u(t)$ should be chosen such that

$$y_m(t + d) = \left(R_d^*(1) + q^{-1}\bar{R}_d^*(q^{-1}) \right) u(t) + G_d^*(q^{-1})y(t)$$

This gives the control law

$$u(t) = \frac{y_m(t + d) - G_d^*(q^{-1})y(t)}{R_d^*(1) + \bar{R}_d^*(q^{-1})q^{-1}} \tag{4.57}$$

This control signal is then applied to the process. At the next sampling instant a new measurement is obtained, and the control law of Eq. (4.57) is used again. Note that the value of the control signal is changed rather than kept constant, as was assumed when Eq. (4.57) was derived. The receding-horizon control principle is thus used. Note that the control law is time-invariant, in contrast to a fixed-horizon linear quadratic controller.

We now analyze the closed-loop system when Eq. (4.57) is used to control the process of Eq. (4.50). It is now necessary to make the calculations in the forward shift operator, since poles at the origin may otherwise be overlooked. The identity of Eq. (4.51) can be written in the forward shift operator as

$$q^{n+d-1} = A(q)F_d(q) + G_d(q) \tag{4.58}$$

The characteristic polynomial of the closed-loop system is

$$P(q) = A(q)\left(q^{n-1}R_d(1) + \bar{R}_d(q) \right) + G_d(q)B(q) \tag{4.59}$$

where

$$\deg P = \deg A + n - 1 = 2n - 1$$

The design equation (Eq. 4.58) can now be used to rewrite $P(q)$:

$$\begin{aligned} B(q)q^{n+d-1} &= A(q)B(q)F_d(q) + G_d(q)B(q) \\ &= A(q)\left(q^{n-1}R_d(q) + \bar{R}_d(q) \right) + G_d(q)B(q) \end{aligned}$$

Hence

$$A(q)\bar{R}_d(q) + G_d(q)B(q) = B(q)q^{n+d-1} - A(q)q^{n-1}R_d(q)$$

which gives

$$P(q) = q^{n-1}A(q)R_d(1) + q^{n-1}\left(q^d B(q) - A(q)R_d(q)\right)$$

If the process is stable, it follows from Eq. (4.53) that the last term vanishes as $d \to \infty$. Thus

$$\lim_{d\to\infty} P(q) = q^{n-1}A(q)R_d(1) \qquad \text{if } A(z) \text{ is a stable polynomial}$$

The properties of the predictive control law are illustrated by an example.

EXAMPLE 4.9   Predictive control

Consider the process model

$$y(t + 1) + ay(t) = bu(t)$$

The identity of Eq. (4.58) gives

$$q^d = (q + a)(q^{d-1} + f_1 q^{d-2} + \cdots + f_{d-1}) + g_0$$

Hence

$$F(q) = q^{d-1} - aq^{d-2} + a^2 q^{d-3} + \cdots + (-a)^{d-1}$$
$$G(q) = (-a)^d$$
$$R_d(q) = bF(q)$$
$$\bar{R}_d(q) = 0$$

and the control law becomes, when $y_m = 0$,

$$u(t) = -\frac{(-a)^d}{b(1 - a + \ldots + (-a)^{d-1})}\, y(t) = -\frac{(-a)^d(1 + a)}{b(1 - (-a)^d)}\, y(t)$$

The characteristic polynomial of the closed-loop system is

$$P(q) = q + a + \frac{(-a)^d(1 + a)}{1 - (-a)^d}$$

which has the pole

$$p_d = -\frac{a + (-a)^d}{1 - (-a)^d}$$

If $a \leq 0$ the location of the pole is given by

$$0 \leq p_d < -a \qquad |a| \leq 1 \quad \text{(stable open-loop system)}$$
$$0 \leq p_d < 1 \qquad |a| > 1 \quad \text{(unstable open-loop system)}$$

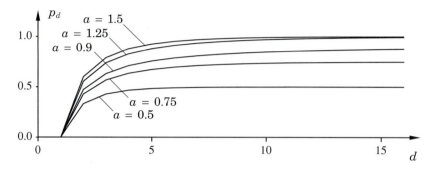

Figure 4.14 The closed-loop pole $p_d = (a^d - a)/(a^d - 1)$ as function of d for different values of a.

The closed-loop pole for different values of a and d is shown in Fig. 4.14. The example indicates that it can be sufficient to use a prediction horizon of five to ten samples.   □

It is possible to generalize the result of Example 4.9 to higher-order systems. The conclusion is that the closed-loop response will be slow for slow or unstable systems when the prediction horizon increases. The restriction of Eq. (4.55) is then not very useful.

Minimum Control Effort

The control strategy that brings $y(t+d)$ to $y_m(t+d)$ while minimizing Eq. (4.56) will now be derived. Equation (4.54) is

$$y(t + d) = R_d^*(q^{-1})u(t + d - d_0) + \bar{y}_d(t)$$
$$= r_{d0}u(t + v) + \cdots + r_{dv}u(t) + \bar{y}_d(t)$$

where $v = d - d_0$. The condition

$$y(t + d) = y_m(t + d) = \bar{y}_d(t) + R_d^*(q^{-1})u(t + d - d_0)$$

can be regarded as a constraint while minimizing Eq. (4.56). Introducing the Lagrangian multiplier λ gives the loss function

$$2J = u(t)^2 + \cdots + u(t + v)^2 + 2\lambda \left(y_m(t + d) - \bar{y}_d(t) - R_d^*(q^{-1})u(t + v) \right)$$

Equating the partial derivatives with respect to $u(t), \cdots, u(t + v)$ and λ to zero gives

$$u(t) = \lambda r_{dv}$$

$$\vdots$$

$$u(t + v) = \lambda r_{d0}$$
$$y_m(t + d) - \bar{y}_d(t) = r_{d0}u(t + v) + \cdots + r_{dv}u(t)$$

This set of equations gives

$$u(t) = \frac{y_m(t+d) - \bar{y}_d(t)}{\mu}$$

where

$$\mu = \frac{\displaystyle\sum_{i=0}^{v} r_{di}^2}{r_{dv}}$$

Using the definition of $\bar{y}_d(t)$ gives

$$\mu u(t) = y_m(t+d) - \bar{R}_d^* u(t-1) - G_d^* y(t)$$

or

$$u(t) = \frac{y_m(t+d) - G_d^* y(t)}{\mu + q^{-1}\bar{R}_d^*} = \frac{y_m(t+d+n-1) - G_d(q)y(t)}{\mu q^{n-1} + \bar{R}_d(q)} \tag{4.60}$$

Using Eq. (4.60) and the model of Eq. (4.50) gives the closed-loop characteristic polynomial

$$P(q) = A(q)\left(q^{n-1}\mu + \bar{R}_d(q)\right) + G_d(q)B(q)$$

This is of the same form as Eq. (4.56), with $R_d(1)$ replaced by μ. This implies that the closed-loop poles approach the zeros of $q^{n-1}A(q)$ when $A(q)$ is stable and when $d \to \infty$. What will happen when the open-loop system is unstable? Consider the following example.

EXAMPLE 4.10 **Minimum-effort control**

Consider the same system as in Example 4.9. The minimum-effort controller is in this case given by

$$\mu = b\frac{1 + a^2 + \cdots + a^{2(d-1)}}{(-a)^{d-1}} = \frac{b(a^{2d} - 1)}{(-a)^{d-1}(a^2 - 1)}$$

which gives (when $y_m = 0$)

$$u(t) = -\frac{(-a)^d}{\mu} y(t) = \frac{a^{2d-1}(a^2 - 1)}{b(a^{2d} - 1)} y(t)$$

The pole of the closed-loop system is

$$p_d = -a + \frac{a^{2d-1}(a^2 - 1)}{a^{2d} - 1} = -\frac{a^{2d-1} + a}{a^{2d} - 1}$$

which gives

$$\lim_{d\to\infty} p_d = -a \qquad |a| \le 1 \quad \text{(stable open-loop system)}$$

$$\lim_{d\to\infty} p_d = -1/a \qquad |a| > 1 \quad \text{(unstable open-loop system)}$$

For this example the minimum-effort controller gives a better closed-loop system than if the future control is assumed to be constant. □

Generalized Predictive Control (GPC)

The predictive controllers discussed so far have considered the output at only one future instant of time. Different generalizations of predictive control have been suggested, in which different loss functions are minimized. One possibility is to use

$$
J(N_1, N_2, N_u) = E \left\{ \sum_{k=N_1}^{N_2} \Big(y(t+k) - y_m(t+k) \Big)^2 + \sum_{k=1}^{N_u} \rho \Delta u(t+k-1)^2 \right\}
$$

$$(4.61)$$

where

$$
\Delta = 1 - q^{-1}
$$

is the difference operator. Different choices of N_1, N_2, and N_u give rise to the different schemes suggested in the literature.

The methodology of generalized predictive control is illustrated by using the loss function of Eq. (4.61) and the process model

$$
A^*(q^{-1})y(t) = B^*(q^{-1})u(t-d_0) + \frac{e(t)}{\Delta}
$$

$$(4.62)$$

This model is sometimes called the *CARIMA* (controlled auto-regressive integrating moving-average) model. It has the advantage that the controller will automatically contain an integrator. (Compare Section 3.6.) As with Eq. (4.50), the following identity is introduced:

$$
1 = A^*(q^{-1})F_d^*(q^{-1})(1 - q^{-1}) + q^{-d}G_d^*(q^{-1})
$$

$$(4.63)$$

This can be used to determine the output d steps ahead:

$$
y(t+d) = F_d^* B^* \Delta u(t+d-d_0) + G_d^* y(t) + F_d^* e(t+d)
$$

F_d^* is of degree $d-1$. The optimal mean squared error predictor, given measured output up to time t and any given input sequence, is

$$
\hat{y}(t+d) = F_d^* B^* \Delta u(t+d-d_0) + G_d^* y(t)
$$

$$(4.64)$$

Suppose that the future desired outputs, $y_m(t+k)$, $k = 1, 2, \ldots$ are available. The loss function of Eq. (4.61) can now be minimized, giving a sequence of future control signals. Notice that the expectation in Eq. (4.61) is made with respect to data obtained up to time t, assuming that no future measurements are available. That is, it is assumed that the computed control sequence is applied to the system. However, only the first element of the control sequence is used. The calculations are repeated when a new measurement is obtained. The resulting controller belongs to the confusingly named class called *open-loop-optimal-feedback* control. As the name suggests, it is assumed that feedback is used, but it is computed only on the basis of the information available at the present time.

In analogy with Eq. (4.54) we get

$$y(t + 1) = R_1^*(q^{-1})\Delta u(t + 1 - d_0) + \bar{y}_1(t) + F_1^* e(t + 1)$$
$$y(t + 2) = R_2^*(q^{-1})\Delta u(t + 2 - d_0) + \bar{y}_2(t) + F_2^* e(t + 2)$$
$$\vdots$$
$$y(t + N) = R_N^*(q^{-1})\Delta u(t + N - d_0) + \bar{y}_N(t) + F_N^* e(t + N)$$

Each output value depends on future control signals (if $d > d_0$), measured inputs, and future noise signals. The equations above can be written as

$$\mathbf{y} = \mathbf{R}\Delta\mathbf{u} + \bar{\mathbf{y}} + \mathbf{e}$$

where

$$\mathbf{y} = \left(\begin{array}{ccc} y(t + 1) & \cdots & y(t + N) \end{array} \right)^T$$
$$\Delta\mathbf{u} = \left(\begin{array}{ccc} \Delta u(t + 1 - d_0) & \cdots & \Delta u(t + N - d_0) \end{array} \right)^T$$
$$\bar{\mathbf{y}} = \left(\begin{array}{ccc} \bar{y}_1(t) & \cdots & \bar{y}_N(t) \end{array} \right)^T$$
$$\mathbf{e} = \left(\begin{array}{ccc} F_1^* e(t + 1) & \cdots & F_N^* e(t + N) \end{array} \right)^T$$

From Eq. (4.53) it follows that the coefficients of R_d^* are the first $d - d_0 + 1$ terms of the pulse response of $q^{-d_0} B^*/(A^*\Delta)$, which are the same as the first $d - d_0 + 1$ terms of the step response of $q^{-d_0} B^*/A^*$. The matrix $\mathbf{R}$ is thus a lower triangular matrix:

$$\mathbf{R} = \left(\begin{array}{cccc} r_0 & 0 & \cdots & 0 \\ r_1 & r_0 & \cdots & 0 \\ \vdots & & \ddots & \vdots \\ r_{N-1} & r_{N-2} & \cdots & r_0 \end{array} \right)$$

If there is a dead time in the system, $d_0 > 1$, then the first $d_0 - 1$ rows of $\mathbf{R}$ will be zero. Also introduce

$$\mathbf{y_m} = \left(\begin{array}{ccc} y_m(t + 1) & \cdots & y_m(t + N) \end{array} \right)^T$$

The expected value of the loss function can be written as

$$J(1, N, N) = E\left\{ (\mathbf{y} - \mathbf{y_m})^T(\mathbf{y} - \mathbf{y_m}) + \rho\Delta\mathbf{u}^T\Delta\mathbf{u} \right\}$$
$$= (\mathbf{R}\Delta\mathbf{u} + \bar{\mathbf{y}} - \mathbf{y_m})^T(\mathbf{R}\Delta\mathbf{u} + \bar{\mathbf{y}} - \mathbf{y_m}) + \rho\Delta\mathbf{u}^T\Delta\mathbf{u} \quad (4.65)$$

Minimization of this expression with respect to $\Delta\mathbf{u}$ gives

$$\Delta\mathbf{u} = (\mathbf{R}^T\mathbf{R} + \rho I)^{-1}\mathbf{R}^T(\mathbf{y_m} - \bar{\mathbf{y}}) \quad (4.66)$$

The first component in $\Delta\mathbf{u}$ is $\Delta u(t)$, which is the control signal applied to the system. Notice that the controller automatically has an integrator. This is necessary to compensate for the drifting noise term in Eq. (4.62).

Notice that $\mathbf{R}$ is independent of the measurements and the old control signals. Only $\mathbf{y_m}$ and $\bar{\mathbf{y}}$ depend on the measurements. The controller (4.66) is thus a time-invariant controller if the process is time-invariant. The predictive controller can thus be interpreted in terms of a pole placement controller. For instance, $N_u = N_1 = n + 1$, $N_2 \geq 2(n + 1) - 1$, and $\rho = 0$ leads to a deadbeat controller.

The calculation of Eq. (4.66) involves the inversion of an $N \times N$ matrix, where N is the prediction horizon in the loss function. To decrease the computations, it is possible to introduce constraints on the future control signals. For instance, it can be assumed that the control increments are zero after $N_u < N$ steps:

$$\Delta u(t + k - 1) = 0 \qquad k > N_u$$

This implies that the control signal is assumed to be constant after N_u steps. Compare the constraint of Eq. (4.55). The control law (Eq. 4.66) then changes to

$$\Delta\mathbf{u} = (\mathbf{R_1}^T\mathbf{R_1} + \rho I)^{-1}\mathbf{R_1}^T(\mathbf{y_m} - \bar{\mathbf{y}}) \tag{4.67}$$

where $\mathbf{R_1}$ is the $N \times N_u$ matrix

$$\mathbf{R_1} = \begin{pmatrix} r_0 & 0 & \cdots & 0 \\ r_1 & r_0 & \cdots & 0 \\ \vdots & & \ddots & \vdots \\ & & \cdots & r_0 \\ \vdots & & & \vdots \\ r_{N-1} & r_{N-2} & \cdots & r_{N-N_u} \end{pmatrix}$$

The matrix to be inverted is now of order $N_u \times N_u$.

One advantage of the receding horizon controllers is that it is possible to include constraints in the states and the control signal. References to this are given at the end of the chapter. One disadvantage with the GPC is that there are many parameters to determine, and it is not obvious how to choose the parameters to get a stable closed-loop system.

The output and control horizons can be chosen as follows: The lower limit N_1 in Eq. (4.61) indicates the first output that will be used in the loss function. The first output that is influenced by $u(t)$ is $y(t+d_0)$. If the time delay is known, then $N_1 = d_0$ is the obvious choice. When the time delay is unknown, $N_1 = 1$ or $N_1 = d_{0min}$ could be used, where d_{0min} is an estimate of the lower limit of the delay. For unknown delays the order of the B polynomial should be increased to make it possible to include all possible values of d_0. This will make the adaptive GPC quite insensitive to variations in the time delay.

The maximum output horizon N_2 can be chosen such that $N_2 h$ is of the same magnitude as the rise time of the plant, where h is the sampling time of the controller. If the system is nonminimum phase, then N_2 should be chosen such that N_2 exceeds the degree of the B polynomial. This will imply that the maximum output horizon is longer than a possible negative-going nonminimum-phase transient.

The control horizon N_u is an important design parameter. As a rule, N_u should be longer the more complex the process is. For processes that are unstable or close to the stability boundary it is necessary to use a N_u that is at least equal to the number of unstable or poorly damped poles. For simpler processes, $N_u = 1$ often gives good results.

To make the generalized predictive controller adaptive, it is necessary at each step of time to estimate the A^* and B^* polynomials. The predicted values for different prediction horizons are computed, and the control signal is calculated from Eq. (4.67). The adaptive generalized predictive controller is thus an indirect control algorithm. The predictions of Eq. (4.64) can be computed recursively, which will simplify the computations. Finally, N_u is usually chosen to be small, which implies that only a low-order matrix needs to be inverted. The adaptive version of GPC has shown good performance and a certain degree of robustness with respect to the choice of model order and poorly known time delays.

To investigate the closed-loop properties of the system in using GPC, we first determine the control signal $\Delta u(t)$ from Eq. (4.67):

$$\Delta u(t) = \begin{pmatrix} 1 & 0 & \cdots & 0 \end{pmatrix} \left(\mathbf{R_1}^T \mathbf{R_1} + \rho I\right)^{-1} \mathbf{R_1}^T \left(\mathbf{y_m} - \bar{\mathbf{y}}\right)$$

$$= \begin{pmatrix} \alpha_1 & \cdots & \alpha_N \end{pmatrix} \left(\mathbf{y_m} - \bar{\mathbf{y}}\right)$$

Further, from Eq. (4.62), using Eq. (4.54),

$$\bar{\mathbf{y}} = \begin{pmatrix} \bar{R}_1^* \Delta u(t-1) + G_1^* y(t) \\ \vdots \\ \bar{R}_N^* \Delta u(t-1) + G_N^* y(t) \end{pmatrix} = \begin{pmatrix} \dfrac{\bar{R}_1^* A^* \Delta}{B^*} q^{d_0-1} + G_1^* \\ \vdots \\ \dfrac{\bar{R}_N^* A^* \Delta}{B^*} q^{d_0-1} + G_N^* \end{pmatrix} y(t)$$

The closed-loop system has the characteristic equation

$$A^* \Delta + \begin{pmatrix} \alpha_1 & \cdots & \alpha_N \end{pmatrix} \begin{pmatrix} \bar{R}_1^* A^* \Delta q^{d_0-1} + B^* G_1^* \\ \vdots \\ \bar{R}_N^* A^* \Delta q^{d_0-1} + B^* G_N^* \end{pmatrix}$$

The identity of Eq. (4.63) gives

$$B^* = A^* \Delta B^* F_d^* + q^{-d} G_d^* B^*$$

$$= A^* \Delta (R_d^* + q^{-(d-d_0+1)} \bar{R}_d^*) + q^{-d} G_d^* B^*$$

This gives the characteristic equation

$$A^*\Delta + \begin{pmatrix} \alpha_1 & \cdots & \alpha_N \end{pmatrix} \begin{pmatrix} (B^* - A^*\Delta R_1^*)q \\ \vdots \\ (B^* - A^*\Delta R_N^*)q^N \end{pmatrix}$$

$$= A^*\Delta + \sum_{i=1}^{N} \alpha_i q^i (B^* - A^*\Delta R_i^*) \qquad (4.68)$$

Equation (4.68) gives an expression for the closed-loop characteristic equation, but it is still difficult to draw any general conclusions about the properties of the closed-loop system even when the process is known.

If $N_u = 1$, then

$$\alpha_i = \frac{r_i}{\rho + \sum_{j=1}^{N} r_j^2}$$

If ρ is sufficiently large, the closed-loop system becomes unstable if the open-loop process is unstable. However, if both the control and output horizons are increased, the problem is the same as a finite-horizon linear quadratic control problem and should thus have better stability properties.

The model predictive controllers such as GPC have the drawback that there are many parameters to choose. Even if there are rules of thumb for choosing the parameters, it is sometimes difficult to determine the parameters such that the closed-loop system is stable (see Problem 4.12). This difficulty exists both when the process is known and when an adaptive GPC algorithm is used.

It is easily seen that the GPC control problem can be interpreted as a stationary LQG control problem but with time-varying weighting matrices or as a finite horizon LQG problem. Compare the loss functions (4.47) and (4.61). The stationary LQG problem and the associated Riccati equation have been extensively studied, and there is much knowledge about the properties of the closed-loop system. The drawback of the infinite-horizon LQG formulation is that it cannot handle constraints in the states or the control signal. Different ways to formulate and solve the constrained receding horizon problem are given in the references at the end of the chapter.

4.7   CONCLUSIONS

This chapter has reviewed different self-tuning regulators. The basic idea is to make a separation between the estimation of the unknown parameters of the process and the design of the controller. The estimated parameters are assumed to be equal to the true parameters in making the design of the controller. It is sometimes of interest to include the uncertainties of the parameter estimates in the design. Such controllers are discussed in Chapter 7. By combining different estimation schemes and design methods, it is possible to derive self-tuners with different properties. In this chapter, only the basic ideas and the asymptotic properties are discussed. The convergence of the estimates and the stability of the closed-loop system are discussed in Chapter 6.

The most important aspect of self-tuning regulators is the issue of parameterization. A reparameterization can be achieved by using the process model *and* the desired closed-loop response. The goal of the reparameterization is to make a direct estimation of the controller parameters, which usually implies that the new model should be linear in the controller parameters.

Only a few of the proposed self-tuning algorithms have been treated in this chapter. Different combinations of estimation methods and underlying control problems give algorithms with different properties. One goal of the chapter has been to give a feel for how self-tuning algorithms can be developed and analyzed. It is important that the desired closed-loop specifications are carefully chosen in applying a self-tuner. A design method that is unsuitable when the process is known will not become better when the process is unknown.

It is also possible to derive self-tuning regulators for multi-input, multi-output (MIMO) systems. The MIMO case is more difficult to analyze. One main difficulty is to define what the necessary *a priori* knowledge is in the MIMO case. It is quite straightforward to derive a self-tuning algorithm corresponding to the generalized direct self-tuning regulator for the restricted case when the delays between the different inputs and outputs are known.

PROBLEMS

4.1   Consider the process and controller in Example 4.4. The controller parameter $\hat{s}_0$ may be very large if $\hat{b}$ is small. Discuss alternatives to ensure that the controller parameter stays bounded.

4.2   Consider the basic direct self-tuning controller in Algorithm 4.1. Discuss different ways to incorporate reference values in the controller. What are the properties of the following three ways for taking care of the reference value?

  (a) Use the difference $y - u_c$ instead of y in the algorithm, and introduce an integrator in the controller.

  (b) Estimate the parameters using the model

$$y(t + d) = R^*u + S^*y - T^*u_c + \varepsilon$$

  and let the controller be

$$R^*u = -S^*y + T^*u_c$$

  (c) Use the difference $u_c - y$ instead of y in the algorithm, and introduce an integrator in the controller.

4.3   Show that the control equation (4.37) minimizes the loss function (4.36).

4.4 Consider the system in Example 4.6. Assume that the process is known. Compute the optimal minimum-variance controller and the least attainable output variance when (a) $\tau = 0.4$ (the minimum-phase case) and (b) $\tau = 0.6$ (the nonminimum-phase case). (*Hint*: Use Theorem 4.3 for the nonminimum-phase case.)

4.5 Make the same calculations as in Problem 4.4 but for the moving-average controller with $d = 2$.

4.6 Consider the generalized minimum-variance controller of Eq. (4.37). Compute the closed-loop characteristic equation. Discuss when the design method may give an unstable closed-loop system. For instance, is it useful for the process in Example 4.6 when $\tau = 0.6$?

4.7 Consider the process in Example 4.6 when $\tau = 0.6$ and $C = 0$. Use Eq. (4.67) to compute the closed-loop poles for different values of N when $N_u = 1$.

4.8 Show that the moving-average controller with $B^{+*} = 1$ and $d = n$ corresponds to a state deadbeat controller.

4.9 Consider the process in Example 4.3. Assume that

$$A(q) = q^2 - 1.5q + 0.7$$
$$B(q) = q + b_1$$
$$C(q) = q^2 - q + 0.2$$

Determine the variance of the closed-loop system as a function of b_1 when the moving-average controller is used. Compare with the lowest achievable variance.

4.10 Show that the control law (4.66) minimizes the loss function (4.65).

4.11 Consider the process in Example 4.5. Investigate through simulation what values of $\hat{r}_0$ can be used. Make the simulations with and without bounds on the control signal. How sensitive is the choice of initial values in the algorithm?

4.12 Consider the system (4.62) with $e(t) = 0$ and

$$A^*(q^{-1}) = 1 - 4q^{-1} + 4q^{-2} = (1 - 2q^{-1})^2$$
$$B^*(q^{-1}) = q^{-1} - 1.999q^{-2}$$

The open-loop process is unstable, and there is a near pole-zero cancellation. Assume that $\rho = 0.1$ and compute the generalized predictive controller that minimizes Eq. (4.65) for different values of N. How large must N be to get a stable closed-loop system? (The problem is adopted from Bitmead *et al.* (1990).) (*Hint*: Don't give up until $N > 25$.)

4.13 Consider the system in Problem 1.9.

    (a) Sample the system and assume that e is discrete-time measurement noise. Determine the minimum-variance controller for the system.

    (b) Simulate a self-tuning moving-average controller for different prediction horizons.

4.14 Make the same investigation as in Problem 4.12 but for the process in Problem 1.10.

REFERENCES

There are many papers, reports, and books about self-tuning algorithms. Some fundamental references are given in this section. The first publication of the self-tuning idea is probably:

Kalman, R. E., 1958. "Design of a self-optimizing control system." *Trans. ASME* **80**: 468–478.

In this paper, least-squares estimation combined with deadbeat control is discussed. Two similar algorithms based on least-squares estimation and minimum-variance control were presented at an IFAC symposium in Prague 1970:

Peterka, V., 1970. "Adaptive digital regulation of noisy systems." *Preprints 2nd IFAC Symposium on Identification and Process Parameter Estimation*. Prague.

Wieslander, J., and B. Wittenmark, 1971. "An approach to adaptive control using real time identification." *Automatica* **7**: 211–217.

The first thorough presentation and analysis of a self-tuning regulator were given in:

Åström, K. J., and B. Wittenmark, 1972. "On the control of constant but unknown systems." *Proceedings of the 5th IFAC World Congress*, Pt 3, Paper 37.5. Paris.

A revised version of this paper, in which the phrase "self-tuning regulator" was coined, is:

Åström, K. J., and B. Wittenmark, 1973. "On self-tuning regulators." *Automatica* **9**: 185–199.

Different aspects of the basic self-tuning regulator described in Algorithm 4.1 are given in the thesis:

Wittenmark, B., 1973. "A self-tuning regulator." Ph.D. thesis TFRT-1003, Department of Automatic Control, Lund Institute of Technology, Lund, Sweden.

The generalized minimum-variance self-tuner was presented in:

Clarke, D. W., and P. J. Gawthrop, 1975. "A self-tuning controller." *IEE Proc.* **122**: 929–934.

The papers above inspired intensive research activity in adaptive control based on the self-tuning idea. A comprehensive treatment of the fundamental theory of adaptive control, especially self-tuning algorithms, is given in:

Goodwin, G. C., and K. S. Sin, 1984. *Adaptive Filtering Prediction and Control*, Information and Systems Science Series. Englewood Cliffs, N.J.: Prentice-Hall.

A more recent state-of-the-art article is:

Ren, W., and P. R. Kumar, 1994. "Stochastic adaptive prediction and model reference control." *IEEE Trans. Automat. Contr.* **AC-39**(10).

The problem of controlling nonminimum-phase plants is discussed in:

Åström, K. J., 1980. "Direct methods for nonminimum phase systems." *Proceedings of the 19th Conference on Decision and Control*, pp. 611–615. Albuquerque, N.M.

Clarke, D. W., 1984. "Self-tuning control of nonminimum-phase systems." *Automatica* **20**(5, Special Issue on Adaptive Control): 501–517.

Åström, K. J., and B. Wittenmark, 1985. "The self-tuning regulator revisited." *Preprints 7th IFAC Symposium on Identification and System Parameter Estimation.* York, U.K.

In the latter, the moving-average controller is presented. Algorithm 4.2 can be used to explain the pole-zero assignment controller in:

Wellstead, P. E., J. M. Edmunds, D. Prager, and P. Zanker, 1979. "Self-tuning pole/zero assignment regulators." *Int. J. Control* **30**: 1–26.

In Åström and Wittenmark (1985), the moving-average controller is presented. It also gives a motivation for the more heuristically introduced model-reference self-tuner in Clarke (1984), where a prediction model of the form of Eq. (4.31) is used but with different filtering.

Multivariable self-tuning regulators are treated in:

Borisson, U., 1979. "Self-tuning regulators for a class of multivariable systems." *Automatica* **15**: 209–215.

Goodwin, G. C., and R. S. Long, 1980. "Generalization of results on multivariable adaptive control." *IEEE Trans. Automat. Contr.* **AC-25**: 1241–1245.

Koivo, H., 1980. "A multivariable self-tuning controller." *Automatica* **16**: 351–356.

Johansson, R., 1983. "Multivariable adaptive control." Ph.D. thesis TFRT-1024, Department of Automatic Control, Lund Institute of Technology, Lund, Sweden.

Dugard, L., G. C. Goodwin, and X. Xianya, 1984. "The role of the interactor matrix in multivariable stochastic adaptive control." *Automatica* **20**(5, Special Issue on Adaptive Control): 701–709.

Elliott, H., and W. A. Wolovich, 1984. "Parameterization issues in multivariable adaptive control." *Automatica* **20**(5, Special Issue on Adaptive Control): 533–545.

Johansson, R., 1986. "Parametric models of linear multivariable systems for adaptive control." *IEEE Trans. Automat. Contr.* **AC-32**: 303–313.

Wittenmark, B., R. H. Middleton, and G. C. Goodwin, 1987. "Adaptive decoupling of multivariable systems." *Int. J. Control* **46**: 1993–2009.

Model predictive control is discussed in:

Richalet, J. A., A. Rault, J. L. Testud, and J. Papon, 1978. "Model predictive heuristic control: Applications to industrial processes." *Automatica* **14**: 413–428.

Cutler, C. R., and B. C. Ramaker, 1980. "Dynamic matrix control—A computer control algorithm." Paper WP5-B, *Preprints Joint Automatic Control Conference*. San Francisco, Calif.

Ydstie, B. E., 1982. "Robust adaptive control of chemical processes." Ph.D. thesis, Imperial College, University of London.

Ydstie, B. E., 1984. "Extended horizon adaptive control." Paper 14.4/E-4, *Preprints 9th IFAC World Congress*. Budapest.

De Keyser, R. M. C., and A. R. Van Cauwenberghe, 1985. "Extended prediction self-adaptive control." *Preprints 7th IFAC Symposium on Identification and System Parameter Estimation*, pp. 1255–1260. York, UK.

Clarke, D. W., C. Mohtadi, and P. S. Tuffs, 1987a. "Generalized predictive control. Part I: The basic algorithm." *Automatica* **23**: 137–148.

Clarke, D. W., C. Mohtadi, and P. S. Tuffs, 1987b. "Generalized predictive control. Part II: Extensions and interpretations." *Automatica* **23**: 149–160.

Clarke, D. W., and C. Mohtadi, 1989. "Properties of generalized predictive control." *Automatica* **25**: 859–875.

Garcia, C. E., and M. Morari, 1989. "Model predictive control: theory and practice–A survey." *Automatica* **25**: 335–348.

Bitmead, R. R., M. Gevers, and V. Wertz, 1990. *Adaptive Optimal Control: The Thinking Man's GPC*. Englewood Cliffs, N.J.: Prentice-Hall.

Clarke, D. W., ed., 1994. *Advances in Model-Based Predictive Control*. Oxford, U.K.: Oxford University Press.

Stability of receding horizon controllers with and without constraints are discussed in Bitmead et al. (1990) and in:

Kwon, W. H., and A. E. Pearson, 1977. "A modified quadratic cost problem and feedback stabilization of a linear system." *IEEE Trans. Automat. Contr.* **AC-22**: 838–842.

Clarke, D. W., and R. Scattolini, 1991. "Constrained receding-horizon predictive control." *Proc. IEE Pt. D* **138**: 347–354.

Mosca, E., and J. Zhang, 1992. "Stable redesign of predictive control." *Automatica* **28**: 1229–1233.

Rawlings, J. B., and K. R. Muske, 1993. "The stability of constrained receding horizon control." *IEEE Trans. Automat. Contr.* **AC-38**: 1512–1516.

Michalska, H., and D. Q. Mayne, 1993. "Robust receding horizon control of constrained nonlinear systems." *IEEE Trans. Automat. Contr.* **AC-38**: 1623-1633.

Linear quadratic Gaussian self-tuning regulators are treated in:

Peterka, V., and K. J. Åström, 1973. "Control of multivariable systems with unknown but constant parameters." *Preprints 3rd IFAC Symposium on Identification and System Parameter Estimation*, pp. 535–544. The Hague, Netherlands.

Åström, K. J., and Z. Zhou-Ying, 1982. "A linear quadratic Gaussian self-tuner." *Recerche di Automatica* **13**: 106–122.

Mosca, E., G. Zappa, and C. Manfredi, 1982. "Progress in multistep horizon self-tuners: The MUSMAR approach." *Ricerche di Automatica* **13**(1): 85–105.

Åström, K. J., 1984. "LQG self-tuners." *Proceedings of the IFAC Workshop on Adaptive Systems in Control and Signal Processing, San Francisco 1983.* New York: Pergamon Press.

Greco, C., G. Menga, E. Mosca, and G. Zappa, 1984. "Performance improvements of self-tuning controllers by multistep horizons: The MUSMAR approach." *Automatica* **20**: 681–699.

Grimble, M. J., 1984. "Implicit and explicit LQG self-tuning controllers." *Automatica* **20**: 661–669.

Peterka, V., 1984. "Predictor-based self-tuning control." *Automatica* **20**: 39–50.

Clarke, D. W., P. P. Kanjilal, and C. Mohtadi, 1985a. "A generalized LQG approach to self-tuning control. Part I: Aspects of design." *Int. J. Control* **41**: 1509–1523.

Clarke, D. W., P. P. Kanjilal, and C. Mohtadi, 1985b. "A generalized LQG approach to self-tuning control. Part II: Implementation and simulation." *Int. J. Control* **41**: 1525–1544.

A detailed treatment of LQG self-tuners is given in:

Kárný, M., A. Halousková, J. Böhm, R. Kulhavý, and P. Nedoma, 1985. "Design of linear quadratic adaptive control: Theory and algorithms for practice." Supplement to *Kybernetica* **21**: 3–97.

It contains much information and many useful hints for practical applications.

Design methods for stochastic systems that are useful in self-tuning regulators are given in:

Åström, K. J., 1970. *Introduction to Stochastic Control Theory.* New York: Academic Press.

Åström, K. J., and B. Wittenmark, 1990. *Computer Controlled Systems: Theory and Design,* 2nd edition. Englewood Cliffs, N.J.: Prentice-Hall.

More about the Sylvester matrix can be found in:

Barnett, S., 1971. *Matrices in Control Theory.* New York: Van Nostrand Reinhold.

Barnett, S., 1983. *Polynomials and Linear Control Systems.* New York: Marcel Dekker.

CHAPTER 5

MODEL-REFERENCE ADAPTIVE SYSTEMS

5.1   INTRODUCTION

The model-reference adaptive system (MRAS) is an important adaptive controller. It may be regarded as an adaptive servo system in which the desired performance is expressed in terms of a reference model, which gives the desired response to a command signal. This is a convenient way to give specifications for a servo problem. A block diagram of the system is shown in Fig. 5.1. The system has an ordinary feedback loop composed of the process and the controller and another feedback loop that changes the controller parameters. The parameters are changed on the basis of feedback from the error, which is the

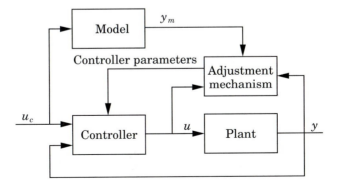

Figure 5.1   Block diagram of a model-reference adaptive system (MRAS).

difference between the output of the system and the output of the reference model. The ordinary feedback loop is called the inner loop, and the parameter adjustment loop is called the outer loop. The mechanism for adjusting the parameters in a model-reference adaptive system can be obtained in two ways: by using a gradient method or by applying stability theory.

In the MRAS the desired behavior of the system is specified by a model, and the parameters of the controller are adjusted based on the error, which is the difference between the outputs of the closed-loop system and the model. Model-reference adaptive systems were originally derived for deterministic continuous-time systems. Extensions to discrete-time systems and systems with stochastic disturbances were given later.

The presentation in this chapter follows the historical development. The MIT rule is derived in Section 5.2. This rule has one parameter, the adaptation gain, that must be chosen by the user. In Section 5.3 we discuss methods to determine the adaptation gain. Section 5.4 presents Lyapunov's stability theory, and Section 5.5 shows how this theory can be used to derive stable adaptation laws. These laws are similar to those obtained by the MIT rule. In Section 5.6 we introduce the theory for input-output stability. This gives another way of viewing adaptive control systems, which is presented in Section 5.7. In Section 5.8 we show how MRASs can be obtained for output feedback of general linear systems. Section 5.9 gives a comparison between self-tuning regulators and MRASs. Adaptive control of nonlinear systems is briefly discussed in Section 5.10. The chapter is summarized in Section 5.11. Further insight into model reference adaptive systems is given in Chapter 6.

5.2   THE MIT RULE

The MIT rule is the original approach to model-reference adaptive control. The name is derived from the fact that it was developed at the Instrumentation Laboratory (now the Draper Laboratory) at MIT.

To present the MIT rule, we will consider a closed-loop system in which the controller has one adjustable parameter θ. The desired closed-loop response is specified by a model whose output is y_m. Let e be the error between the output y of the closed-loop system and the output y_m of the model. One possibility is to adjust parameters in such a way that the loss function

$$J(\theta) = \frac{1}{2} e^2 \tag{5.1}$$

is minimized. To make J small, it is reasonable to change the parameters in the direction of the negative gradient of J, that is,

$$\frac{d\theta}{dt} = -\gamma \frac{\partial J}{\partial \theta} = -\gamma e \frac{\partial e}{\partial \theta} \tag{5.2}$$

This is the celebrated *MIT rule*. The partial derivative $\partial e/\partial \theta$, which is called the *sensitivity derivative* of the system, tells how the error is influenced by the adjustable parameter. If it is assumed that the parameter changes are slower than the other variables in the system, then the derivative $\partial e/\partial \theta$ can be evaluated under the assumption that θ is constant.

There are many alternatives to the loss function given by Eq. (5.1). If it is chosen to be

$$J(\theta) = |e| \tag{5.3}$$

the gradient method gives

$$\frac{d\theta}{dt} = -\gamma \frac{\partial e}{\partial \theta} \operatorname{sign} e \tag{5.4}$$

The first MRAS that was implemented was based on this formula. There are, however, many other possibilities, for example,

$$\frac{d\theta}{dt} = -\gamma \operatorname{sign}\left(\frac{\partial e}{\partial \theta}\right) \operatorname{sign}(e)$$

This is called the *sign-sign algorithm*. A discrete-time version of this algorithm is used in telecommunications, in which simple implementation and fast computations are required. (See Section 13.2.)

Adjusting many parameters   Equation (5.2) also applies when there are many parameters to adjust. The symbol θ should then be interpreted as a vector and $\partial e/\partial \theta$ as the gradient of the error with respect to the parameters.

Examples

We now give two examples that illustrate how the MIT rule is used to obtain a simple adaptive controller, and we also show some properties of adaptive systems.

EXAMPLE 5.1   Adaptation of a feedforward gain

Consider the problem of adjusting a feedforward gain. In this problem it is assumed that the process is linear with the transfer function $kG(s)$, where $G(s)$ is known and k is an unknown parameter. The underlying design problem is to find a feedforward controller that gives a system with the transfer function $G_m(s) = k_0 G(s)$, where k_0 is a given constant. With the feedforward controller

$$u = \theta u_c$$

where u is the control signal and u_c the command signal, the transfer function from command signal to the output becomes $\theta k G(s)$. This transfer function is equal to $G_m(s)$ if the parameter θ is chosen to be

$$\theta = \frac{k_0}{k}$$

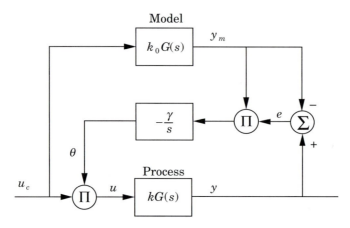

Figure 5.2   Block diagram of an MRAS for adjustment of a feedforward gain based on the MIT rule.

We will now use the MIT rule to obtain a method for adjusting the parameter θ when k is not known. The error is

$$e = y - y_m = kG(p)\theta u_c - k_0 G(p)u_c$$

where u_c is the command signal, y_m is the model output, y is the process output, θ is the adjustable parameter, and $p = d/dt$ is the differential operator. The sensitivity derivative is given by

$$\frac{\partial e}{\partial \theta} = kG(p)u_c = \frac{k}{k_0} y_m$$

The MIT rule then gives the following adaptation law:

$$\frac{d\theta}{dt} = -\gamma' \frac{k}{k_0} y_m e = -\gamma y_m e \tag{5.5}$$

where $\gamma = \gamma' k/k_0$ has been introduced instead of γ'. Notice that to have the correct sign of γ, it is necessary to know the sign of k. Equation (5.5) gives the law for adjusting the parameter. A block diagram of the system is shown in Fig. 5.2.

     The properties of the system can be illustrated by simulation. Figure 5.3 shows a simulation when the system has the transfer function

$$G(s) = \frac{1}{s+1}$$

The input u_c is a sinusoid with frequency 1 rad/s, and the parameter values are $k = 1$ and $k_0 = 2$. Figure 5.3 shows that the parameter converges toward the correct value reasonably fast when the adaptation gain is $\gamma = 1$ and that the process output approaches the model output. Figure 5.3 also shows that

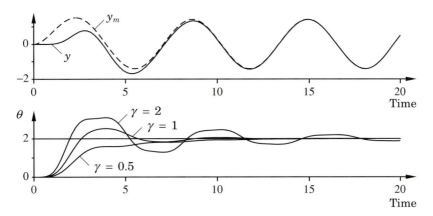

Figure 5.3 Simulation of an MRAS for adjusting a feedforward gain. The process (solid line) and the model (dashed line) outputs are shown in the upper graph for $\gamma = 1$. The controller parameter is shown in the lower graph when the adaptation gain γ has the values 0.5, 1, and 2.

the convergence rate depends on the adaptation gain. It is thus important to know a reasonable value of this parameter. Intuitively, we may expect that parameters converge slowly for small γ and that the convergence rate increases with γ. Simulation experiments indicate that this is true for small values of γ but also that the behavior is quite unpredictable for large γ. □

An example of a practical problem that fits this formulation is control of robots with unknown load, in which the process transfer function from the motor current to the angular velocity is

$$G(s) = \frac{k_I}{Js}$$

where k_I is the current to torque constant and J is the unknown moment of inertia. Another example is the dynamics of a CD player, in which the sensitivity of the laser diode is the unknown process parameter.

A remark on notation In analyzing the MRAS with time-varying parameters it is important to consider the fact that the parameter θ is time-varying. The expression

$$G(p)(\theta u)$$

where $p = d/dt$ is the differential operator should be interpreted as the differential operator $G(p)$ acting on the signal θu. When θ is time-varying, this is different from $\theta G(p) u$. For example, if $G(p) = p$, we have

$$G(p)(\theta u) = p(\theta u) = \theta \, \frac{du}{dt} + \frac{d\theta}{dt} \, u = \theta(pu) + u(p\theta)$$

Care must thus be taken in manipulating expressions and block diagrams.

Notice that no approximations were needed in Example 5.1. When the MIT rule is applied to more complicated problems, however, it is necessary to use approximations to obtain the sensitivity derivatives. This is illustrated by another example.

EXAMPLE 5.2   MRAS for a first-order system

Consider a system described by the model

$$\frac{dy}{dt} = -ay + bu \tag{5.6}$$

where u is the control variable and y is the measured output. Assume that we want to obtain a closed-loop system described by

$$\frac{dy_m}{dt} = -a_m y_m + b_m u_c$$

Let the controller be given by

$$u(t) = \theta_1 u_c(t) - \theta_2 y(t) \tag{5.7}$$

The controller has two parameters. If they are chosen to be

$$\theta_1 = \theta_1^0 = \frac{b_m}{b}$$
$$\theta_2 = \theta_2^0 = \frac{a_m - a}{b} \tag{5.8}$$

the input-output relations of the system and the model are the same. This is called perfect model-following.

To apply the MIT rule, introduce the error

$$e = y - y_m$$

where y denotes the output of the closed-loop system. It follows from Eqs. (5.6) and (5.7) that

$$y = \frac{b\theta_1}{p + a + b\theta_2} u_c$$

where $p = d/dt$ is the differential operator. The notation used is discussed in Section 1.5. The sensitivity derivatives are obtained by taking partial derivatives with respect to the controller parameters θ_1 and θ_2:

$$\frac{\partial e}{\partial \theta_1} = \frac{b}{p + a + b\theta_2} u_c$$

$$\frac{\partial e}{\partial \theta_2} = -\frac{b^2 \theta_1}{(p + a + b\theta_2)^2} u_c = -\frac{b}{p + a + b\theta_2} y$$

These formulas cannot be used directly because the process parameters a and b are not known. Approximations are therefore required. One possible approximation is based on the observation that $p + a + b\theta_2^0 = p + a_m$ when the

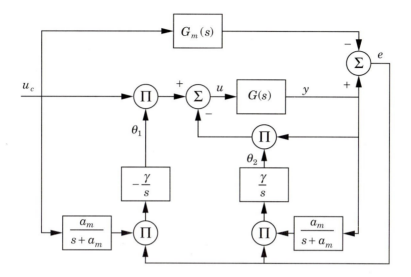

Figure 5.4   Block diagram of a model-reference controller for a first-order process.

parameters give perfect model-following. We will therefore use the approximation

$$p + a + b\theta_2 \approx p + a_m$$

which will be reasonable when parameters are close to their correct values. With this approximation we get the following equations for updating the controller parameters:

$$\frac{d\theta_1}{dt} = -\gamma \left(\frac{a_m}{p + a_m} u_c \right) e$$

$$\frac{d\theta_2}{dt} = \gamma \left(\frac{a_m}{p + a_m} y \right) e$$

(5.9)

In these equations we have combined parameters b and a_m with the adaptation gain γ', since they appear as the product $\gamma' b / a_m$. The sign of parameter b must be known to have the correct sign of γ. Notice that the filter has also been normalized so that its steady-state gain is unity.

The adaptive controller is a dynamical system with five state variables that can be chosen to be the model output, the parameters, and the sensitivity derivatives. A block diagram of the system is shown in Fig. 5.4. The behavior of the system is now illustrated by a simulation. The parameters are chosen to be $a = 1$, $b = 0.5$, and $a_m = b_m = 2$, the input signal is a square wave with amplitude 1, and $\gamma = 1$. Figure 5.5 shows the results of a simulation. Figure 5.6 shows the parameter estimates for different values of the adaptation gain γ. Notice that the parameters change most when the command signal changes

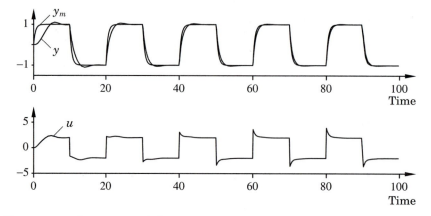

Figure 5.5 Simulation of the system in Example 5.2 using an MRAS. The parameter values are $a = 1$, $b = 0.5$, $a_m = b_m = 2$, and $\gamma = 1$.

and that the parameters converge very slowly. For $\gamma = 1$, the value used in Fig. 5.5, the parameters have the values $\theta_1 = 3.2$ and $\theta_2 = 1.2$ at time $t = 100$. These values are far from the correct values $\theta_1^0 = 4$ and $\theta_2^0 = 2$. However, the parameters will converge to the true values with increasing time. The convergence rate increases with increasing γ, as is shown in Fig. 5.6.

 The fact that the control is quite good even at time $t = 10$ is a reflection of the fact that the parameter estimates are related to each other in a very special way, although they are quite far from their true values. This is illustrated in Fig. 5.7, where parameter θ_2 is plotted as a function of θ_1 for a simulation with a duration of 500 time units. Figure 5.7 shows that parameters do indeed

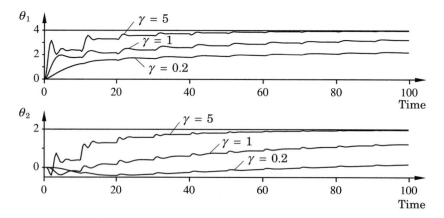

Figure 5.6 Controller parameters θ_1 and θ_2 for the system in Example 5.2 when $\gamma = 0.2$, 1 and 5.

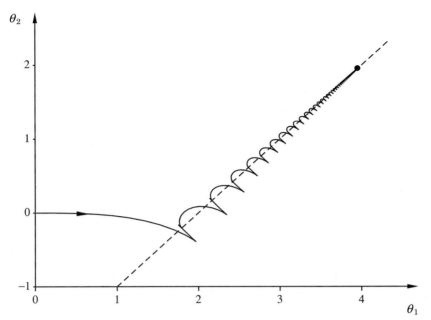

Figure 5.7   Relation between controller parameters θ_1 and θ_2 when the system in Example 5.2 is simulated for 500 time units. The dashed line shows the line $\theta_2 = \theta_1 - a/b$. The dot indicates the convergence point.

approach their correct values as time increases. The parameter estimates quickly approach the line $\theta_2 = \theta_1 - a/b$. This line represents parameter values such that the closed-loop system has correct steady-state gain.   □

Error and Parameter Convergence

The goal in model-reference adaptive systems is to drive the error $e = y - y_m$ to zero. This does not necessarily imply that the controller parameters approach their correct values, as is illustrated in the following example.

EXAMPLE 5.3   Lack of parameter convergence

Consider the simple system for updating a feedforward gain, discussed in Example 5.1. Assume that $G(s) = 1$. The process model is $y = ku$, the control law is $u = \theta u_c$, and the desired response is given by $y_m = k_0 u_c$. The error is

$$e = (k\theta - k_0)u_c = k(\theta - \theta^0)u_c$$

where $\theta^0 = k_0/k$. The MIT rule gives the following differential equation for the parameter:

$$\frac{d\theta}{dt} = -\gamma k^2 u_c^2 (\theta - \theta^0)$$

This equation has the solution

$$\theta(t) = \theta^0 + (\theta(0) - \theta^0)e^{-\gamma k^2 I_t} \tag{5.10}$$

where

$$I_t = \int_0^t u_c^2(\tau)\, d\tau$$

and $\theta(0)$ is the initial value of the parameter θ. The estimate converges toward its correct value only if the integral I_t diverges as $t \to \infty$. The convergence is exponential if the input signal is persistently exciting. (Compare with Section 2.4.) The error is given by

$$e(t) = ku_c(t)(\theta(0) - \theta^0)e^{-\gamma k^2 I_t}$$

Notice that the error will always go to zero as $t \to \infty$ because either the integral I_t diverges or else $u_c(t) \to 0$. However, the limiting value of the parameter θ will depend on the properties of the input signal. $\qquad\square$

Example 5.3 illustrates the fact that the error e goes to zero but that the parameters do not necessarily converge to their correct values. This is a characteristic feature of all adaptive systems. The input signal must have certain properties for the parameters to converge. The conditions required were discussed in Chapter 2; compare with the notion of persistent excitation, which was introduced in Section 2.4.

5.3   DETERMINATION OF THE ADAPTATION GAIN

In Section 5.2 we found that it was straightforward to obtain an adaptive system by using the MIT rule. The adaptive control laws had one parameter, the adaptation gain γ, which had to be chosen by the user. The simulation experiments indicated that the choice of the adaptation gain could be crucial. In this section we will discuss methods for determining the adaptation gain.

Consider the MRAS for adaptation of a feedforward gain in Example 5.1. We thus have a system with the transfer function $kG(s)$, where $G(s)$ is known and k is an unknown constant. It is assumed that $G(s)$ is stable. We would like to find a feedforward control that gives the transfer function $k_0G(s)$. The system is described by the following equations:

$$y = kG(p)u$$
$$y_m = k_0G(p)u_c$$
$$u = \theta u_c$$
$$e = y - y_m$$
$$\frac{d\theta}{dt} = -\gamma y_m e$$

where u_c is the command signal, y_m is the model output, y is the process output, θ is the adjustable parameter, and $p = d/dt$ is the differential operator. Elimination of the variables u and y in these equations gives

$$\frac{d\theta}{dt} + \gamma y_m \left(kG(p)\theta u_c\right) = \gamma y_m^2 \tag{5.11}$$

This equation is a compact description of the behavior of the parameters that we call the *parameter equation*. Notice that y_m may be considered a known function of time. If $G(s)$ is a rational transfer function, Eq. (5.11) is a linear time-varying ordinary differential equation. Such equations may exhibit very complicated behavior. It is not possible to give a simple analytical characterization of the properties of the system, particularly how they are influenced by the parameter γ.

A Thought Experiment

To get some insight into the behavior of the system given by Eq. (5.11), we consider an experiment with the adaptive system such that the equation simplifies considerably. An understanding of the behavior of the system under such circumstances will give us some insight, but it will of course not give the full picture.

Consider the following experiment: Assume that the value of parameter θ is fixed, that the adaptation mechanism is disconnected, and that a constant input signal u_c is applied. The adaptation mechanism is then connected when all signals have settled to steady-state values. The behavior of the parameter is then given by

$$\frac{d\theta}{dt} + \gamma y_m^o u_c^o \left(kG(p)\theta\right) = \gamma(y_m^o)^2 \tag{5.12}$$

which is a linear time-invariant system. This equation is linear with constant coefficients. Its stability is governed by the algebraic equation

$$s + \gamma y_m^o u_c^o kG(s) = 0 \tag{5.13}$$

We can immediately conclude that the behavior of the parameter is determined by the quantity

$$\mu = \gamma y_m^o u_c^o k \tag{5.14}$$

A picture of how the zeros of Eq. (5.13) vary with parameter μ is easily obtained by plotting the root locus with respect to the parameter. We can conclude that if Eq. (5.13) has zeros in the right half-plane, then the parameters will diverge even in the very special conditions of the experiment. Intuitively, we may also expect the analysis to approximately describe the case in which the command signal is changing slowly with respect to the dynamics of $G(s)$.

Equation (5.13) can also be used to determine a suitable value of the adaptation gain, as is illustrated in Example 5.4.

EXAMPLE 5.4   Choice of adaptation gain

Consider the system in Example 5.1 with $G(s) = 1/(s + 1)$, $k = 1$, and $k_0 = 2$. Assume that the reference signal has unit amplitude. Equation (5.13) then becomes

$$s^2 + s + \mu = s^2 + s + \gamma y_m^o u_c^o k = 0$$

A reasonable choice is to make $\gamma y_m^o u_c^o k = 1$. If we disregard the transients, the average value of $y_m u_c$ is 2. This gives $\gamma = 0.5$, which is the value used in one of the simulations in Fig. 5.3.   □

Normalized Algorithms

It follows from Eq. (5.13) that the adaptive system will be unstable if the transfer function $G(s)$ has pole excess larger than 1 and parameter μ in Eq. (5.14) is sufficiently large. The parameter μ is large if the signals are large or if the adaptation gain is large. The behavior of the system depends strongly on the signal levels. This will be illustrated by a numerical experiment.

EXAMPLE 5.5   Stability depends on the signal amplitudes

Consider the system in Example 5.1. Let the transfer function G be given by

$$G(s) = \frac{1}{s^2 + a_1 s + a_2}$$

Equation (5.13) then becomes

$$s^3 + a_1 s^2 + a_2 s + \mu = 0$$

where $\mu = \gamma y_m^o u_c^o k$. The equation has all its roots in the left half-plane if

$$\gamma y_m^o u_c^o k < a_1 a_2 \tag{5.15}$$

Since this inequality involves the magnitude of the command signal, it may happen that the equilibrium solution corresponding to one command signal is stable and the solution corresponding to another command signal is unstable. This is illustrated by the simulation results shown in Fig. 5.8, where parameters are chosen so that $k = a_1 = a_2 = 1$. In the simulation the adaptation rate γ was adjusted to give a good response when u_c is a square wave with unit amplitude. In this case we have $u_c^o = y_m^o = 1$, and inequality (5.15) gives the stability condition $\gamma < 1$. A reasonable value of γ is $\gamma = 0.1$, which was used in the simulation. Figure 5.8 shows clearly that the convergence rate depends on the magnitude of the command signal. Notice that the solution is unstable when the amplitude of u_c is 3.5. The approximate model predicts instability for u_c larger than 3.16. Also notice that the response is intolerably slow for low amplitudes of u_c.   □

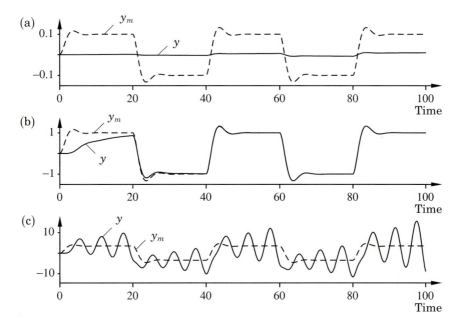

Figure 5.8   Simulation of the MRAS in Example 5.5. The command signal is a square wave with the amplitude (a) 0.1, (b) 1, and (c) 3.5. The model output y_m is a dashed line; the process output is a solid line. The following parameters are used: $k = a_1 = a_2 = \theta^0 = 1$, and $\gamma = 0.1$.

The example indicates clearly that the choice of adaptation gain is crucial and that the value chosen depends on the signal levels. Because of this it seems natural to modify the algorithm so that it does not depend on the signal levels. To do this, we will write the MIT rule as

$$\frac{d\theta}{dt} = \gamma \varphi e$$

where we have introduced $\varphi = -\partial e / \partial \theta$. Introduce the following modified adjustment rule:

$$\frac{d\theta}{dt} = \frac{\gamma \varphi e}{\alpha + \varphi^T \varphi} \tag{5.16}$$

where parameter $\alpha > 0$ is introduced to avoid difficulties when φ is small. Notice that we have written the equation in such a way that it also holds when θ is a vector; in that case, φ is also a vector of the same dimension.

If we repeat the analysis of the thought experiment, we find that Eq. (5.13) is replaced by

$$s + \gamma \frac{\varphi^o u_c^o}{\alpha + \varphi^{oT} \varphi^o} kG(s) = 0$$

Since φ^o is proportional to u_c^o, the roots of this equation will not change much with the signal levels. The adaptation rule given by Eq. (5.16) is called

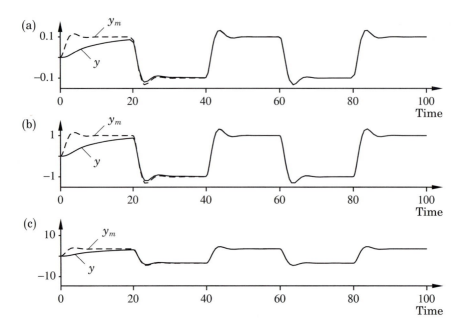

Figure 5.9 Simulation of the MRAS in Example 5.5 with the normalized MIT rule. The command signal is a square wave with the amplitude (a) 0.1, (b) 1, and (c) 3.5. Compare with Fig. 5.8. The model output y_m is a dashed line; the process output is a solid line. The parameters used are $k = a_1 = a_2 = \theta^0 = 1$, $\alpha = 0.001$, and $\gamma = 0.1$.

the *normalized MIT rule*. The improved performance with this algorithm is illustrated in Fig. 5.9. A comparison with Fig. 5.8 shows that normalization is useful.

Notice that the normalized adjustment rule performs very well even in the cases in which difficulties were encountered with the MIT rule. It is in fact possible to make the modified adjustment rule work very well over a wide range of command signal amplitudes. Notice that the normalization is obtained automatically with algorithms based on parameter estimation. (Compare with Example 2.16.)

Summary

Having derived the MIT rule and investigated some of its properties, we can now summarize some of the key issues. The model-reference control problem can be described as follows: Let the desired performance be specified by a reference model having the transfer function $G_m(s)$, and let the closed-loop transfer function of the plant be $G(s, \theta)$, where θ are the adjustable parameters. Furthermore, let u_c be the command signal. The model-reference adaptive system tries to change the controller parameters so that the error

$$e(t) = (G(p, \theta) - G_m(p))u_c(t)$$

goes to zero. The MIT rule given by

$$\frac{d\theta}{dt} = \gamma \varphi e$$

where $\varphi = -\partial e/\partial \theta$ and γ is the adaptation gain, can be interpreted as a gradient method for minimizing the error. The MIT rule can be applied in many different cases; a few examples have been given in this section. The choice of the adaptation gain is critical and depends on the signal levels. The normalized algorithm

$$\frac{d\theta}{dt} = \gamma \frac{\varphi e}{\alpha + \varphi^T \varphi}$$

is less sensitive to signal levels. Notice that a normalization of a similar type is obtained automatically in the self-tuning regulator. Compare with Eq. (3.22).

Preliminary numerical experiments indicate that the systems obtained with the MIT rule work as expected for small adaptation gains. Very complex behavior may be obtained for high adaptation gains. To proceed to develop our understanding of adaptive systems, we will investigate the stability problem.

5.4 LYAPUNOV THEORY

There is no guarantee that an adaptive controller based on the MIT rule will give a stable closed-loop system. It is clearly desirable to see whether there are other methods for designing adaptive controllers that can guarantee the stability of the system. As a first step in this direction we now present the Lyapunov stability theory. For the benefit of students who are encountering Lyapunov theory for the first time, we first prove a stability theory for time-invariant systems. We then state a more powerful theorem for time-varying systems, which can be used to design adaptive controllers.

Lyapunov's Theory for Time-invariant Systems

Fundamental contributions to the stability theory for nonlinear systems were made by the Russian mathematician Lyapunov in the end of the nineteenth century. Lyapunov investigated the nonlinear differential equation

$$\frac{dx}{dt} = f(x) \qquad f(0) = 0 \tag{5.17}$$

Since $f(0) = 0$, the equation has the solution $x(t) = 0$. To guarantee that a solution exists and is unique, it is necessary to make some assumptions about

$f(x)$. A sufficient assumption is that $f(x)$ is locally Lipschitz, that is,

$$\|f(x) - f(y)\| \leq L\|x - y\| \qquad L > 0$$

in the neighborhood of the origin. Lyapunov was interested in investigating whether the solution of Eq. (5.17) is stable with respect to perturbations. For this purpose he introduced the following stability concept.

DEFINITION 5.1   Lyapunov stability

The solution $x(t) = 0$ to the differential equation (5.17) is called *stable* if for given $\varepsilon > 0$ there exists a number $\delta(\varepsilon) > 0$ such that all solutions with initial conditions

$$\|x(0)\| < \delta$$

have the property

$$\|x(t)\| < \varepsilon \quad \text{for} \quad 0 \leq t < \infty \tag{5.18}$$

The solution is *unstable* if it is not stable. The solution is *asymptotically stable* if it is stable and δ can be found such that all solutions with $\|x(0)\| < \delta$ have the property that $\|x(t)\| \to 0$ as $t \to \infty$.   □

Remark 1.   If the solution is asymptotically stable for any initial value, then it is said to be *globally asymptotically stable*.

Remark 2.   Notice that Lyapunov stability refers to stability of a particular solution and not to the differential equation.   □

Lyapunov developed a method for investigating stability that is based on the idea of finding a function with special properties. To describe these, we first introduce the notion of positive definite functions.

DEFINITION 5.2   Positive definite and semidefinite functions

A continuously differentiable function $V : R^n \to R$ is called *positive definite* in a region $U \subset R^n$ containing the origin if

1. $V(0) = 0$
2. $V(x) > 0, \quad x \in U$ and $x \neq 0$

A function is called *positive semidefinite* if Condition 2 is replaced by $V(x) \geq 0$.   □

A positive definite function has level curves that enclose the origin. Curves corresponding to larger values of the function enclose curves that correspond to smaller values. The situation in the two-dimensional case is illustrated in Fig. 5.10. If we can find a function so that the velocity vector, $dx/dt = f(x)$, always points toward the interior of the level curves, then it seems intuitively clear that a solution that starts inside a given level curve can never pass to the outside of the same level curve. We have the following theorem.

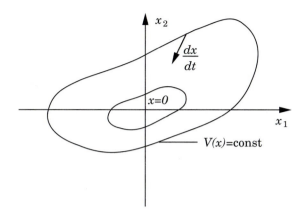

Figure 5.10   Illustration of Lyapunov's method for investigating stability.

THEOREM 5.1   Lyapunov's stability theorem: time-invariant systems

If there exists a function $V : R^n \to R$ that is positive definite such that its derivative along the solution of Eq. (5.17),

$$\frac{dV}{dt} = \frac{\partial V^T}{\partial x} \frac{dx}{dt} = \frac{\partial V^T}{\partial x} f(x) = -W(x) \tag{5.19}$$

is negative semidefinite, then the solution $x(t) = 0$ to Eq. (5.17) is stable. If dV/dt is negative definite, then the solution is also asymptotically stable. The function V is called a *Lyapunov function* for the system (5.17).

Moreover if

$$\frac{dV}{dt} < 0 \quad \text{and} \quad V(x) \to \infty \quad \text{when} \quad \|x\| \to \infty$$

then the solution is globally asymptotically stable.

Proof: Given $\varepsilon > 0$ such that $\{x \mid \|x\| \le \varepsilon\} \in U$, determine ℓ and δ such that

$$\ell = \min_{\|x\|=\varepsilon} V(x) = \max_{\|x\|\le\delta} V(x) \tag{5.20}$$

Consider initial conditions such that

$$\|x(0)\| < \delta$$

Since V is positive definite, it then follows from Definition 5.2 that

$$V(x(0)) < \ell$$

To prove that inequality (5.18) holds, we proceed by contradiction. Assume that t_1 is the smallest value such that $\|x(t_1)\| = \varepsilon$. It follows from Eq. (5.20) that

$$V(x(t_1)) \ge \ell$$

Furthermore,

$$V\left(x(t_1)\right) = V\left(x(0)\right) + \int_0^{t_1} \frac{dV}{dt} \, dt = V\left(x(0)\right) - \int_0^{t_1} W\left(x(s)\right) \, ds \qquad (5.21)$$

Since $W(x)$ is positive semidefinite, it follows that

$$V\left(x(t_1)\right) \leq V\left(x(0)\right) < \ell$$

and we have thus obtained a contradiction and it can be concluded that $\|x(t)\| < \varepsilon$ for all t, which by Definition 5.1 implies that the solution $x(t) = 0$ is stable. To prove asymptotic stability, we notice that it follows from Eq. (5.21) that

$$0 \leq \int_0^t W\left(x(s)\right) \, ds = V\left(x(0)\right) - V\left(x(t)\right) \leq \ell$$

Since $W(x)$ and $x(t)$ are continuous, it then follows that

$$\lim_{t \to \infty} W\left(x(t)\right) = 0$$

If $W(x)$ is positive definite, this implies that $x(t) \to 0$ as $t \to \infty$.   □

Remark.   Notice that it follows from the proof that if the derivative of the Lyapunov function is negative semidefinite, the solution converges to the set $\{x \mid W(x) = 0\}$.   □

Finding Lyapunov Functions

Lyapunov's theorem is very elegant. However, it is necessary to have methods for constructing Lyapunov functions. There is no universal method for constructing Lyapunov functions for a stable system. To apply the method, we therefore have to resort to trial and error. A good first attempt is to test quadratic functions. However, for linear systems we have the following important result.

THEOREM 5.2   Lyapunov functions for linear systems

Assume that the linear system

$$\frac{dx}{dt} = Ax \qquad (5.22)$$

is asymptotically stable. Then for each symmetric positive definite matrix Q there exists a unique symmetric positive definite matrix P such that

$$A^T P + PA = -Q \qquad (5.23)$$

Furthermore, the function
$$V(x) = x^T P x \tag{5.24}$$
is a Lyapunov function for Eq. (5.22).

Proof: Let Q be a symmetric positive definite matrix. Define

$$P(t) = \int_0^t e^{A^T(t-s)} Q e^{A(t-s)} \, ds$$

The matrix P is symmetric and positive definite because an integral of positive definite matrices is positive definite. The matrix P also satisfies

$$\frac{dP}{dt} = A^T P + P A + Q$$

Since the matrix A is stable, the limit

$$P_o = \lim_{t \to \infty} P(t)$$

exists. This matrix satisfies Eq. (5.23). It can also be shown that the solution to Eq. (5.23) is unique, which completes the argument.   □

For a stable linear system we can thus always find a quadratic Lyapunov function. To use Theorem 5.2 to construct a Lyapunov function, we simply choose a positive matrix Q and solve the linear equation (5.23) for P. The following example shows how it can be done.

EXAMPLE 5.6   **Lyapunov functions for a linear system**

Consider the linear system (5.22) with

$$A = \begin{pmatrix} a_1 & a_2 \\ a_3 & a_4 \end{pmatrix}$$

where it is assumed that all eigenvalues of A are in the left half-plane. Let the matrix Q be

$$Q = \begin{pmatrix} q_1 & 0 \\ 0 & q_2 \end{pmatrix}$$

where q_1 and q_2 are positive. Assume that the matrix P has the form

$$P = \begin{pmatrix} p_1 & p_2 \\ p_2 & p_3 \end{pmatrix}$$

Equation (5.23) then becomes

$$\begin{pmatrix} 2a_1 & 2a_3 & 0 \\ a_2 & a_1 + a_4 & a_3 \\ 0 & 2a_2 & 2a_4 \end{pmatrix} \begin{pmatrix} p_1 \\ p_2 \\ p_3 \end{pmatrix} = \begin{pmatrix} -q_1 \\ 0 \\ -q_2 \end{pmatrix}$$

This is a linear equation. Theorem 5.2 implies that it always has a solution when A is stable and that the solution is a positive definite matrix P.   □

Lyapunov Theory for Time-variable Systems

We now consider time-variable differential equations of the type

$$\frac{dx}{dt} = f(x,t) \tag{5.25}$$

The origin is an equilibrium point for Eq. (5.25) if $f(0,t) = 0$ $\forall t \geq 0$. It is assumed that f is such that solutions exist for all $t \geq t_0$. To guarantee this, it is assumed that f is piecewise continuous in t and locally Lipschitz in x in a neighborhood of $x(t) = 0$. We now investigate the stability of the solution $x(t) = 0$.

In the time-varying case the solution will depend on t as well as on the starting time t_0. This implies that the bound δ in Definition 5.1 will depend on ε and t_0. The definition on stability can be refined to give uniform stability properties with respect to the initial time. We have the following definition.

DEFINITION 5.3   Uniform Lyapunov stability

The solution $x(t) = 0$ of Eq. (5.25) is *uniformly stable* if for $\varepsilon > 0$ there exists a number $\delta(\varepsilon) > 0$, independent of t_0, such that

$$\|x(t_0)\| < \delta \Rightarrow \|x(t)\| < \varepsilon \qquad \forall t \geq t_0 \geq 0$$

The solution is *uniformly asymptotically stable* if it is uniformly stable and there is $c > 0$, independent of t_0, such that $x(t) \to 0$ as $t \to \infty$, uniformly in t_0, for all $\|x(t_0)\| < c$.   □

To state a stability theorem for solutions to Eq. (5.25), we first have to introduce the so-called *class K functions*.

DEFINITION 5.4   Class K function

A continuous function $\alpha\colon [0,a) \to [0,\infty)$ is said to belong to *class K* if it is strictly increasing and $\alpha(0) = 0$. It is said to belong to class K_∞ if $a = \infty$ and $\alpha(r) \to \infty$ as $r \to \infty$.   □

For time-varying systems the following stability theorem can now be stated.

THEOREM 5.3   Lyapunov's stability theorem: Time-varying systems

Let $x = 0$ be an equilibrium point for Eq. (5.25) and $D = \{x \in R^n \mid \|x\| < r\}$. Let V be a continuously differentiable function such that

$$\alpha_1(\|x\|) \leq V(x,t) \leq \alpha_2(\|x\|) \tag{5.26}$$

$$\frac{dV}{dt} = \frac{\partial V}{\partial t} + \frac{\partial V}{\partial x} f(x,t) \leq -\alpha_3(\|x\|)$$

for $\forall t \geq 0$, where α_1, α_2, and α_3 are class K functions. Then $x = 0$ is uniformly asymptotically stable.

Proof: A proof can be found in Khalil (1992).   □

Remark 1.   The derivative of V along the trajectories of Eq. (5.25) is now given by

$$\frac{dV}{dt} = \frac{\partial V}{\partial t} + \frac{\partial V}{\partial x} f(x, t)$$

Remark 2.   A function $V(x, t)$ satisfying the left inequality of (5.26) is said to be positive definite. A function satisfying the right inequality of (5.26) is said to be *decrescent*.

Remark 3.   To show stability for time-variable systems, it is necessary to bound the function $V(x, t)$ by a function that doesn't depend on t.   □

When using Lyapunov theory on adaptive control problems, we often find that dV/dt only is negative semidefinite. This implies that additional conditions must be imposed on the system. The following lemma gives a useful result.

LEMMA 5.1   Barbalat's lemma

If g is a real function of a real variable t, defined and uniformly continuous for $t \geq 0$, and if the limit of the integral

$$\int_0^t g(s)\, ds$$

as t tends to infinity exists and is a finite number, then

$$\lim_{t \to \infty} g(t) = 0 \qquad\qquad □$$

Remark.   A consequence of Barbalat's lemma is that if $g \in L_2$ and dg/dt is bounded, then

$$\lim_{t \to \infty} g(t) = 0 \qquad\qquad □$$

When applying Lyapunov theory to an adaptive control problem, we get a time derivative of the Lyapunov function V, which depends on the control signal and other signals in the system. If these signals are bounded, Lemma 5.1 and the remark that follows can be used on dV/dt to prove stability. We have the following theorem.

THEOREM 5.4   Boundedness and convergence set

Let $D = \{x \in R^n \,|\, \|x\| < r\}$ and suppose that $f(x, t)$ is locally Lipschitz on $D \times [0, \infty)$. Let V be a continuously differentiable function such that

$$\alpha_1(\|x\|) \leq V(x, t) \leq \alpha_2(\|x\|)$$

and

$$\frac{dV}{dt} = \frac{\partial V}{\partial t} + \frac{\partial V}{\partial x} f(x,t) \le -W(x) \le 0$$

$\forall t \ge 0$, $\forall x \in D$, where α_1 and α_2 are class K functions defined on $[0,r)$ and $W(x)$ is continuous on D. Further, it is assumed that dV/dt is uniformly continuous in t.

Then all solutions to Eq. (5.25) with $\|x(t_0)\| < \alpha_2^{-1}(\alpha_1(r))$ are bounded and satisfy

$$W(x(t)) \to 0 \quad \text{as} \quad t \to \infty$$

Moreover, if all the assumptions hold globally and α_1 belongs to class K_∞, the statement is true for all $x(t_0) \in R^n$. □

A proof of a slight modification of this theorem can be found in Khalil (1992). The theorem states that the states of the system are bounded and that they approach the set $\{x \in D \mid W(x) = 0\}$. In the theorem it is assumed that dV/dt is uniformly continuous, that is, that the continuity is independent of t. A sufficient condition for this is that $\ddot{V}$ is bounded.

5.5   DESIGN OF MRAS USING LYAPUNOV THEORY

We will now show how Lyapunov's stability theory can be used to construct algorithms for adjusting parameters in adaptive systems. To do this, we first derive a differential equation for the error, $e = y - y_m$. This differential equation contains the adjustable parameters. We then attempt to find a Lyapunov function and an adaptation mechanism such that the error will go to zero. When using the Lyaponov theory for adaptive systems, we find that dV/dt is usually only negative semidefinite. The procedure is to determine the error equation and a Lyapunov function with a bounded second derivative. Theorem 5.4 is then used to show boundedness and that the error goes to zero. To show parameter convergence, it is necessary to impose further conditions, such as persistently excitation and uniform observability, on the reference signal and the system. (See the references in the end of the chapter.) We start with a simple example.

EXAMPLE 5.7   **First-order MRAS based on stability theory**

Consider the problem in Example 5.2. The desired response is given by

$$\frac{dy_m}{dt} = -a_m y_m + b_m u_c$$

where $a_m > 0$ and the reference signal is bounded. The process is described by

$$\frac{dy}{dt} = -ay + bu$$

The controller is

$$u = \theta_1 u_c - \theta_2 y$$

Introduce the error

$$e = y - y_m$$

Since we are trying to make the error small, it is natural to derive a differential equation for the error. We get

$$\frac{de}{dt} = -a_m e - (b\theta_2 + a - a_m)y + (b\theta_1 - b_m)u_c$$

Notice that the error goes to zero if the parameters are equal to the values given by Eqs. (5.8). We will now attempt to construct a parameter adjustment mechanism that will drive the parameters θ_1 and θ_2 to their desired values. For this purpose, assume that $b\gamma > 0$ and introduce the following quadratic function:

$$V(e, \theta_1, \theta_2) = \frac{1}{2}\left(e^2 + \frac{1}{b\gamma}(b\theta_2 + a - a_m)^2 + \frac{1}{b\gamma}(b\theta_1 - b_m)^2\right)$$

This function is zero when e is zero and the controller parameters are equal to the correct values. For the function to qualify as a Lyapunov function the derivative dV/dt must be negative. The derivative is

$$\frac{dV}{dt} = e\frac{de}{dt} + \frac{1}{\gamma}(b\theta_2 + a - a_m)\frac{d\theta_2}{dt} + \frac{1}{\gamma}(b\theta_1 - b_m)\frac{d\theta_1}{dt}$$

$$= -a_m e^2 + \frac{1}{\gamma}(b\theta_2 + a - a_m)\left(\frac{d\theta_2}{dt} - \gamma y e\right)$$

$$+ \frac{1}{\gamma}(b\theta_1 - b_m)\left(\frac{d\theta_1}{dt} + \gamma u_c e\right)$$

If the parameters are updated as

$$\frac{d\theta_1}{dt} = -\gamma u_c e$$

$$\frac{d\theta_2}{dt} = \gamma y e$$

(5.27)

we get

$$\frac{dV}{dt} = -a_m e^2$$

The derivative of V with respect to time is thus negative semidefinite but not negative definite. This implies that $V(t) \le V(0)$ and thus that e, θ_1, and θ_2 must be bounded. This implies that $y = e + y_m$ also is bounded. To use Theorem 5.4, we determine

$$\frac{d^2 V}{dt^2} = -2a_m e\frac{de}{dt} = -2a_m e\left(-a_m e - (b\theta_2 + a - a_m)y + (b\theta_1 - b_m)u_c\right)$$

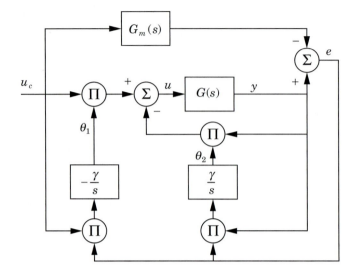

Figure 5.11   Block diagram of an MRAS based on Lyapunov theory for a first-order system. Compare with the controller based on the MIT rule for the same system in Fig. 5.4.

Since u_c, e, and y are bounded, it follows that $\ddot{V}$ is bounded; hence dV/dt is uniformly continuous. From Theorem 5.4 it now follows that the error e will go to zero. However, the parameters will not necessarily converge to their correct values; it is shown only that they are bounded. To have parameter convergence, it is necessary to impose conditions on the excitation of the system. (Compare with Example 5.3.)

The adaptation rule given by Eqs. (5.27) is similar to the MIT rule given by Eqs. (5.9), but the sensitivity derivatives are replaced by other signals. A block diagram of the system is shown in Fig. 5.11. Compare with the corresponding block diagram for the system with the MIT rule in Fig. 5.4. The only difference is that there is no filtering of the signals u_c and y with the Lyapunov rule. In both cases the adjustment law can be written as

$$\frac{d\theta}{dt} = \gamma \varphi e \qquad (5.28)$$

where θ is a vector of parameters and

$$\varphi = \left(\begin{array}{cc} -u_c & y \end{array} \right)^T$$

for the Lyapunov rule and

$$\varphi = \frac{a_m}{p + a_m} \left(\begin{array}{cc} -u_c & y \end{array} \right)^T$$

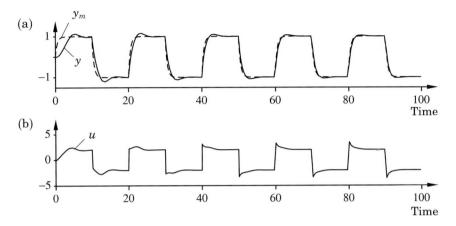

Figure 5.12   Simulation of the system in Example 5.7 using an adaptive controller based on Lyapunov theory. The parameter values are $a = 1, b = 0.5$, $a_m = b_m = 2$, and $\gamma = 1$. (a) Process (solid line) and model (dashed line) outputs. (b) Control signal.

for the MIT rule. The adjustment rule obtained from Lyapunov theory is simpler because it does not require filtering of the signals. Figure 5.12 shows a simulation of the system for the case $G(s) = 0.5/(s+1)$ and $G_m(s) = 2/(s+2)$. The behavior is quite similar to that obtained with the MIT rule in Fig. 5.5. Notice, however, that arbitrary large values of the adaptation gain γ can be used with the Lyapunov approach.

Figure 5.13 shows the parameter estimates in the simulation for different values of adaptation gain γ. For comparison we have also shown the parameters obtained with the MIT rule.   □

State Space Systems

We will now show how Lyapunov's theory can be used to derive stable MRASs for general linear systems. The idea is the same as used previously. It can be described as follows:

1.   Find a controller structure.
2.   Derive the error equation.
3.   Find a Lyapunov function and use it to derive a parameter updating law such that the error will go to zero.

Consider a linear system described by

$$\frac{dx}{dt} = Ax + Bu \qquad (5.29)$$

Assume that it is desired to find a control law so that the response to command

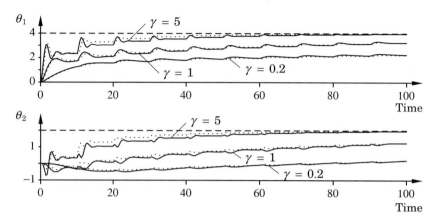

Figure 5.13   Controller parameters θ_1 and θ_2 for the system in Example 5.7 when $\gamma = 0.2$, 1, and 5. The dotted lines are the parameters obtained with the MIT rule. Compare Fig. 5.6.

signals is given by

$$\frac{dx_m}{dt} = A_m x_m + B_m u_c \tag{5.30}$$

A general linear control law for the system given by Eq. (5.29) is

$$u = M u_c - Lx \tag{5.31}$$

The closed-loop system then becomes

$$\frac{dx}{dt} = (A - BL)x + BM u_c = A_c(\theta)x + B_c(\theta)u_c \tag{5.32}$$

The control law can be parameterized in different ways. All parameters in the matrices L and M may be chosen freely. There may also be constraints among the parameters. The general case can be captured by assuming that the closed-loop system is described by Eq. (5.32), where matrices A_c and B_c depend on a parameter θ.

Compatibility conditions   It is not always possible to find parameters θ such that Eq. (5.32) is equivalent to Eq. (5.30). A sufficient condition is that there exists a parameter value θ^0 such that

$$\begin{aligned} A_c(\theta^0) &= A_m \\ B_c(\theta^0) &= B_m \end{aligned} \tag{5.33}$$

This condition for perfect model-following is fairly stringent. When all parameters in the control law can be chosen freely, it implies that

$$\begin{aligned} A - A_m &= BL \\ B_m &= BM \end{aligned}$$

This means that the columns of matrices $A - A_m$ and B_m are linear combinations of the columns of matrix B. If these conditions are satisfied and the columns of B and B_m are linearly independent, then the matrices L and M are given by

$$L = (B^T B)^{-1} B^T (A - A_m) = (B_m^T B)^{-1} B_m^T (A - A_m)$$

$$M = (B^T B)^{-1} B^T B_m = (B_m^T B)^{-1} B_m^T B_m$$

The error equation   Introduce the error defined as

$$e = x - x_m$$

Subtracting Eq. (5.30) from Eq. (5.29) gives

$$\frac{de}{dt} = \frac{dx}{dt} - \frac{dx_m}{dt} = Ax + Bu - A_m x_m - B_m u_c$$

Adding and subtracting $A_m x$ from the right-hand side give

$$\frac{de}{dt} = A_m e + (A - A_m - BL) x + (BM - B_m) u_c$$

$$= A_m e + (A_c(\theta) - A_m) x + (B_c(\theta) - B_m) u_c$$

$$= A_m e + \Psi (\theta - \theta^0) \tag{5.34}$$

To obtain the last equality, it has been assumed that the conditions for exact model-following are satisfied. This is required for θ^0 to exist. To derive a parameter adjustment law, we introduce the Lyapunov function

$$V(e, \theta) = \frac{1}{2} \left(\gamma e^T P e + (\theta - \theta^0)^T (\theta - \theta^0) \right)$$

where P is a positive definite matrix. The function V is positive definite. To find out whether it can be a Lyapunov function, we calculate its total time derivative

$$\frac{dV}{dt} = -\frac{\gamma}{2} e^T Q e + \gamma(\theta - \theta^0) \Psi^T P e + (\theta - \theta^0)^T \frac{d\theta}{dt}$$

$$= -\frac{\gamma}{2} e^T Q e + (\theta - \theta^0)^T \left(\frac{d\theta}{dt} + \gamma \Psi^T P e \right)$$

where Q is positive definite and such that

$$A_m^T P + P A_m = -Q$$

Notice that it follows from Theorem 5.2 that a pair of positive definite matrices P and Q with this property always exist if A_m is stable.

If the parameter adjustment law is chosen to be

$$\frac{d\theta}{dt} = -\gamma \Psi^T P e \tag{5.35}$$

we get

$$\frac{dV}{dt} = -\frac{\gamma}{2} e^T Q e$$

The time derivative of the Lyapunov function is negative semidefinite. By using Lemma 5.1 in the same way as in Example 5.7 it can be shown that the error goes to zero. Notice that we have assumed that all states x are measurable.

Adaptation of a Feedforward Gain

We now attempt to use Lyapunov theory to derive parameter adjustment laws for the problem of adjusting a feedforward gain. We consider the case in which the plant has transfer function $kG(s)$, where $G(s)$ is known and k is unknown. The desired response is given by the transfer function $k_0 G(s)$. This problem was discussed previously in Examples 5.1 and 5.3. The error is given by

$$e = (kG(p)\theta - k_0 G(p))u_c = kG(p)(\theta - \theta^0)u_c$$

where $\theta^0 = k_0/k$. To use Lyapunov theory, we first introduce a state space representation of the transfer function G. The relation between the parameter θ and the error e can then be written as

$$\frac{dx}{dt} = Ax + B(\theta - \theta^0)u_c \qquad (5.36)$$

$$e = Cx$$

If the homogeneous system $\dot{x} = Ax$ is asymptotically stable, there exist positive definite matrices P and Q such that

$$A^T P + PA = -Q \qquad (5.37)$$

Choose the following function as a candidate for a Lyapunov function:

$$V = \frac{1}{2}\left(\gamma x^T P x + (\theta - \theta^0)^2\right)$$

The time derivative of V along the differential equation (Eqs. 5.36) is given by

$$\frac{dV}{dt} = \frac{\gamma}{2}\left(\frac{dx^T}{dt} P x + x^T P \frac{dx}{dt}\right) + (\theta - \theta^0)\frac{d\theta}{dt}$$

Using Eqs. (5.36), we get

$$\frac{dV}{dt} = \frac{\gamma}{2}\left(\left(Ax + Bu_c(\theta - \theta^0)\right)^T P x + x^T P \left(Ax + Bu_c(\theta - \theta^0)\right)\right)$$
$$+ (\theta - \theta^0)\frac{d\theta}{dt}$$
$$= -\frac{\gamma}{2} x^T Q x + (\theta - \theta^0)\left(\frac{d\theta}{dt} + \gamma u_c B^T P x\right)$$

If the parameter adjustment law is chosen to be

$$\frac{d\theta}{dt} = -\gamma u_c \, B^T P x \tag{5.38}$$

we find that the derivative of the Lyapunov function will be negative as long as $x \neq 0$. The state vector x and the error $e = Cx$ will go to zero as t goes to infinity. Notice, however, that the parameter error $\theta - \theta^0$ will not necessarily go to zero.

Output feedback   The result obtained is quite restrictive because it requires that all state variables are known. A parameter adjustment law that uses output feedback can be obtained if the Lyapunov function can be chosen so that

$$B^T P = C$$

where C is the output matrix of the system in Eq. (5.34). With this choice of P it follows that

$$B^T P x = C x = e$$

and the adjustment rule becomes

$$\frac{d\theta}{dt} = -\gamma u_c e$$

The appropriate condition is given by the celebrated Kalman-Yakubovich lemma. The following definition is needed to state this lemma.

DEFINITION 5.5   Positive real transfer function

A rational transfer function G with real coefficients is *positive real* (PR) if

$$\text{Re } G(s) \geq 0 \quad \text{for} \quad \text{Re } s \geq 0 \tag{5.39}$$

A transfer function G is *strictly positive real* (SPR) if $G(s - \varepsilon)$ is positive real for some real $\varepsilon > 0$.   □

The concept of SPR is discussed further in Section 5.6. Let it suffice to mention that $G(s) = 1/(s + 1)$ is SPR and $G(s) = 1/s$ is PR but not SPR. The following result gives a state space interpretation of SPR.

LEMMA 5.2   Kalman-Yakubovich lemma

Let the time-invariant linear system

$$\frac{dx}{dt} = Ax + Bu$$
$$y = Cx$$

be completely controllable and completely observable. The transfer function

$$G(s) = C(sI - A)^{-1}B$$

is strictly positive real if and only if there exist positive definite matrices P and Q such that

$$A^T P + PA = -Q$$

and

$$B^T P = C \qquad\qquad \square$$

A proof of this result is given in Section 5.6. There is a more general version of the theorem that applies to systems with a direct term from input to output. The simpler version is sufficient for our purposes.

THEOREM 5.5   MRAS using the Lyapunov rule

Consider the problem of adapting a feedforward gain. Assume that the transfer function G is strictly positive real. Then the parameter adjustment rule

$$\frac{d\theta}{dt} = -\gamma u_c e \qquad\qquad (5.40)$$

where γ is a positive constant, makes the output error e in Eqs. (5.36) go to zero. $\qquad\qquad \square$

The control law of Eq. (5.40) is very similar to the control law obtained by the MIT rule, Eq. (5.5). This is illustrated in Fig. 5.14, which shows block diagrams of both systems. The only difference between the systems is that the connection to the first multiplier comes from the model output for the MIT rule and from the command signal for the Lyapunov rule. This seemingly small difference has major consequences, however.

A remark on the assumptions   It may seem strange that such drastically different behaviors can be obtained by minor modifications of the system. It also seems strange that it is possible to use arbitrarily high adaptation gains. This is because the assumption that a transfer function is positive real is very strong. It follows from Definition 5.5 that Re $G(i\omega) \geq 0$ if the transfer function $G(s)$ is positive real. This means that the Nyquist curve of G is in the right half-plane. Such a system is stable under proportional feedback with arbitrarily high gain. The closed-loop system can be made arbitrarily insensitive to the gain variations. The result is of limited practical value because of the strong assumptions that are made.

Summary

In this section we have shown that it is possible to construct parameter adjustment rules based on Lyapunov's stability theory. The adjustment rules obtained in this way guarantee that the error goes to zero, but it cannot be asserted that the parameters converge to their correct values. The adjustment rules obtained are similar to those obtained by the MIT rule. However, the rules

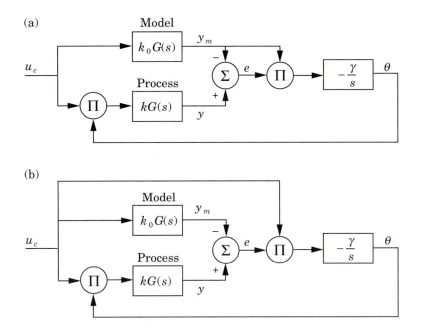

Figure 5.14 Block diagrams of the adaptive systems for feedforward gain compensation obtained by (a) the MIT rule and (b) the Lyapunov rule.

are not normalized. The adjustment rules have the remarkable property that arbitrarily high adaptation gains can be used. This property depends on the strong assumptions that are made. This is discussed further in Chapter 6.

5.6 BOUNDED-INPUT, BOUNDED-OUTPUT STABILITY

Systems can be described from two points of view: the internal or state space view or the external or input-output view. The state space approach is based on a detailed description of the inner structure of the system. In the input-output approach, a system is considered to be a black box that transforms inputs to outputs. In Section 5.5 we approached stability from the state space view. In this section we develop stability theory from the input-output view. In the next section the results are applied to design of adaptive controllers.

We start with a brief presentation of the operator view of dynamical systems. This leads naturally to the concept of bounded-input, bounded-output (BIBO) stability. The fundamental results like the small gain theorem and the passivity theorem are then presented. In Section 5.5 we found that the notion of positive real was essential. This notion, which is closely related to passivity, will also be discussed.

The Operator View of Dynamical Systems

Signals are elements of a normed space X, which we call the signal space. A system S is considered as an operator $S : X \rightarrow X$. For simplicity we consider systems with one input and one output and the signals are real functions from R to R. Several choices of norms are considered, for example, the L_2 norm

$$\|u\| = \left(\int_{-\infty}^{\infty} u^2(t)\, dt \right)^{\frac{1}{2}}$$

or the sup norm

$$\|u\| = \sup_{0 \leq t \leq \infty} |u(t)|$$

A drawback of using L_2 is that it must be assumed *a priori* that all signals go to zero as $t \rightarrow \infty$. The notion of extended space is introduced to avoid this assumption. This is introduced as follows.

Let Y be the space of real-valued functions on $[0, \infty)$. Let x be an element of Y. The *truncation* of x at $T > 0$ is defined as

$$x_T(t) = \begin{cases} x(t) & 0 \leq t \leq T \\ 0 & t > T \end{cases}$$

DEFINITION 5.6   Extended space

If X is a normed linear subspace of Y, then the extended space X_e is the set $\{x \in Y \mid x_T \in X$ for some fixed $T \geq 0\}$.   □

The extended L_2 space is denoted L_{2e}. There is now a simple way to introduce the notion of the gain of a system.

DEFINITION 5.7   Gain of a nonlinear system

Let the signal space be X_e. The *gain $\gamma(S)$ of a system S* is defined as

$$\gamma(S) = \sup_{u \in X_e} \frac{\|Su\|}{\|u\|}$$

where u is the input signal to the system.

Remark.   The gain is thus the smallest value such that

$$\|Su\| \leq \gamma(S)\|u\| \quad \text{for all } u \in X_e$$

We use supremum because the maximum of $\|Su\|/\|u\|$ may not be assumed for a signal in the class that we are considering.   □

We illustrate the definition with a few examples.

EXAMPLE 5.8 **Linear systems with signals in L_{2e}**

Let the signal space be L_{2e}. Consider a linear system with the transfer function $G(s)$. Assume that $G(s)$ has no poles in the closed right half-plane and that the system is initially at rest. Let u be the input and y the output, and let U and Y be the corresponding Laplace transforms. It follows from Parseval's theorem, Theorem 2.8, that

$$\|y\|^2 = \int_0^\infty y^2(t)\,dt = \frac{1}{2\pi}\int_{-\infty}^\infty Y(i\omega)Y(-i\omega)\,d\omega$$

$$= \frac{1}{2\pi}\int_{-\infty}^\infty G(i\omega)U(i\omega)G(-i\omega)U(-i\omega)\,d\omega$$

$$\leq \max_\omega |G(i\omega)|^2 \frac{1}{2\pi}\int_{-\infty}^\infty U(i\omega)U(-i\omega)\,d\omega$$

$$= \max_\omega |G(i\omega)|^2 \int_0^\infty u^2(t)\,dt = \max_\omega |G(i\omega)|^2 \cdot \|u\|^2$$

Hence

$$\|y\| \leq \max_\omega |G(i\omega)| \cdot \|u\|$$

The gain is thus less than $\max|G(i\omega)|$. We get equality in the above equation if u is a sinusoid with the frequency that maximizes $|G(i\omega)|$. However, such a signal is not in L_{2e}. The value of $\|y\|$ can be made arbitrarily close to $\max|G(i\omega)|$ with a truncated sinusoid in L_{2e} by making T sufficiently large. The gain of the system is thus

$$\gamma(G) = \max_\omega |G(i\omega)| \tag{5.41}$$

$\square$

EXAMPLE 5.9 **Linear system with sup norm**

Consider a stable linear system with impulse response $h(t)$. We have

$$y(t) = \int_0^\infty h(\tau)u(t-\tau)\,d\tau$$

Using the sup norm, we get

$$|y(t)| = \left|\int_0^\infty h(\tau)u(t-\tau)\,d\tau\right| \leq \sup_t |u(t)| \int_0^\infty |h(\tau)|\,d\tau$$

This gives

$$\sup_t |y(t)| \leq \gamma(G) \cdot \sup_t |u(t)|$$

where the gain of the system is given by

$$\gamma(G) = \int_0^\infty |h(\tau)|\,d\tau$$

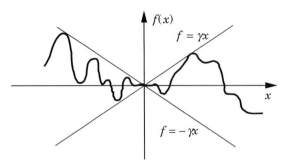

Figure 5.15   Illustration of the gain of a static nonlinearity.

If we let $u_0 = \max_t |u(t)|$, the maximum is assumed for the signal

$$u(s) = u_0 \, \text{sign} \, (h(t - s))$$

However, this signal is not in L_{2e}. Since the system is stable, we can get arbitrarily close with a signal in L_{2e} by making T sufficiently large.   □

EXAMPLE 5.10   **Static nonlinear system**

Consider a static system that is described by the nonlinear equation

$$y(t) = f(u(t))$$

For all norms we have

$$|y(t)| \leq \max_u |f(u(t))|$$

The gain of the system is thus given by

$$\gamma = \max_u \frac{|f(u)|}{|u|}$$

The gain of a static system has a simple interpretation. A function whose norm is γ can be bounded between the straight lines $y = \pm\gamma u$, as is illustrated in Fig. 5.15.   □

Having defined the gain of a system, we can now define stability.

DEFINITION 5.8   **BIBO stability**

A system is called *bounded-input, bounded-output (BIBO) stable* if the system has bounded gain.   □

Notice that this definition refers to stability of a system and not stability of a particular solution. Also notice that a system with bounded gain is BIBO stable but that the converse is not true. The static system $y = u^2$ does not have finite gain, but it is BIBO stable.

Stability Criteria

Having defined the notion of stability, we now give criteria for stability. For this purpose, consider the simple feedback system in Fig. 5.16. We are interested in determining when the gain from u to y is bounded. We have the following theorem.

THEOREM 5.6   The small gain theorem

Consider the system in Fig. 5.16. Let γ_1 and γ_2 be the gains of the systems H_1 and H_2. The closed-loop system is BIBO stable if

$$\gamma_1\gamma_2 < 1 \tag{5.42}$$

and its gain is less than

$$\gamma = \frac{\gamma_1}{1 - \gamma_1\gamma_2} \tag{5.43}$$

Outline of proof: For a rigorous proof it must first be established that y exists. If this is true, we have

$$y = H_1 e = H_1(u - H_2 y)$$

Hence

$$\|y\| \le \|H_1 u\| + \|H_1 H_2 y\| \le \gamma_1\|u\| + \gamma_1\gamma_2\|y\|$$

Because of Eq. (5.42) we can solve for $\|y\|$. Hence

$$\|y\| \le \frac{\gamma_1}{1 - \gamma_1\gamma_2}\|u\| = \gamma\|u\|$$

which proves BIBO stability and gives the expression (5.43) for the gain of the system.   □

Remark 1.   The result has a strong intuitive interpretation. It simply says that if the total gain around the loop is less than 1, then the closed-loop system is stable.

Remark 2.   For the special case of linear systems with L_2 norms it follows from Example 5.8 that the gain is the maximum magnitude of the transfer function.

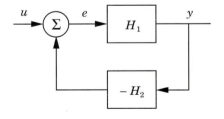

Figure 5.16   Block diagram of a simple feedback loop.

The theorem can be interpreted as an extension of the Nyquist theorem. The condition (5.42) implies that the loop gain is always less than 1. From this interpretation we can also conclude that the result is quite conservative.

Passivity

We now present another stability theorem that is also based on the input-output point of view. The starting point is the notion of passivity, which is an abstract formulation of the idea of energy dissipation. Passive systems are common in engineering. A system composed only of components like resistors, capacitors, and inductors is one example from electrical engineering. A system composed of masses, springs, and dashpots is an example from mechanical engineering. When dealing with electrical systems, we will consider two-port systems in which the current is the input and the voltage is the output. The same concepts apply to mechanical systems, in which the variables are position and force.

Passivity is naturally associated with power dissipation. Such a concept can be defined for linear as well as nonlinear systems. Roughly speaking, the passivity theorem says that a feedback connection of one passive system and one strictly passive system is stable. To state the result formally, we need an abstract notion of passivity. We start with the operator view of systems, in which a system is represented by an operator mapping signals to signals. The signal space is assumed to be L_{2e} with a scalar product defined by

$$\langle x \,|\, y \rangle = \int_0^\infty x(s)y(s)\,ds = \int_0^T x(s)y(s)\,ds$$

We have the following definition.

DEFINITION 5.9   Passive system
A system with input u and output y is *passive* if

$$\langle y \,|\, u \rangle \geq 0$$

The system is *input strictly passive* (ISP) if there exists $\varepsilon > 0$ such that

$$\langle y \,|\, u \rangle \geq \varepsilon \|u\|^2$$

and *output strictly passive* (OSP) if there exists $\varepsilon > 0$ such that

$$\langle y \,|\, u \rangle \geq \varepsilon \|y\|^2$$

$\square$

Notice that in electrical systems the power is proportional to the product of current and voltage. The definition is thus a very natural abstraction. The following example illustrates the definition of passivity.

EXAMPLE 5.11   **Static nonlinear systems**

Consider a static nonlinear system characterized by the function $f : R \to R$. We have

$$\langle y \,|\, u \rangle = \int_0^\infty f(u(t))u(t) \, dt$$

The right-hand side is thus nonnegative if

$$xf(x) \geq 0 \qquad\qquad (5.44)$$

which is the condition for passivity. This condition means that the graph of the curve f is entirely in the first and the third quadrants. The system is input strictly passive if

$$xf(x) \geq \delta |x|^2$$

It is output strictly passive if

$$xf(x) \geq \delta f^2(x)$$

A static system with $f(x) = x + x^3$ is thus input strictly passive, and a static system with $f(x) = x/(1 + |x|)$ is output strictly passive. $\square$

Positive Real Functions

For linear systems the concept of passivity is closely related to the properties positive real and strictly positive real introduced in Definition 5.5 in Section 5.5. The notion of positive real did actually originate from an effort to characterize driving point impedance functions for linear circuits composed of passive components. The driving point impedance function is the transfer function from current to voltage across two terminals in a circuit. The driving point admittance function is the transfer function from voltage to current. In circuit theory it was established that such impedance functions have certain properties that were taken as the definition of positive real. In this section we discuss some properties of positive real functions. It follows from Definition 5.5 that if the transfer function $G(s)$ is PR (SPR), then its inverse $1/G(s)$ is also PR (SPR). This is a direct consequence of the symmetry of admittance functions and impedance functions. It does not matter whether we consider current or voltage as the input to a circuit. Positive real functions can be characterized in many different ways. An alternative to Definition 5.5 that is easier to use is given by the following theorem.

THEOREM 5.7   **Conditions for positive realness**

A rational transfer function $G(s)$ with real coefficients is PR if and only if the following conditions hold.

(i)   The function has no poles in the right half-plane.

(ii) If the function has poles on the imaginary axis or at infinity, they are simple poles with positive residues.

(iii) The real part of G is nonnegative along the $i\omega$ axis, that is,

$$\text{Re}\,(G(i\omega)) \geq 0 \qquad (5.45)$$

A transfer function is SPR if conditions (i) and (iii) hold and if condition (ii) is replaced by the condition that $G(s)$ has no poles or zeros on the imaginary axis.

Proof: Assume that $G(s)$ is PR. Since it is rational, the only singularities are poles. A function assumes all values around a pole. According to Definition 5.5 the function has positive real part for $\text{Re}\,s \geq 0$. Hence it cannot have poles in this region. Equation (5.45) follows by setting $s = i\omega$ in Definition 5.5. Furthermore, $G(s)$ cannot have multiple poles at infinity because the condition $\text{Re}\,G(s) \geq 0$ for $\text{Re}\,s \geq 0$ would then be violated. For the same reason a pole at infinity must also have positive residue.

We have thus shown the necessity. To show sufficiency, we use the fact that a function that is analytic in a region assumes its largest values on the boundary. Consider the function

$$F(s) = e^{G(s)}$$

We have

$$|F(s)| = e^{\text{Re}\,G(s)} \qquad (5.46)$$

Let the region D be bounded by the imaginary axis and an infinite half-circle to the right with the imaginary axis as a diameter. Let Γ be the boundary of D. Assume that conditions (i), (ii), and (iii) hold. Because of condition (iii) we have $|F(s)| > 1$ on the imaginary axis. It now remains to investigate the value of F on the large half-circle. It follows from condition (ii) that G has at most one pole at infinity. We have three cases: $G(s)$ may go to zero; it may go to a constant, which must be positive because of condition (iii); or it may go to infinity as ks, where the constant k must be positive because of condition (ii). We can thus conclude that $|F(s)| > 1$ on Γ. Since F is analytic in D, the condition then also holds on D, and Eq. (5.45) then follows. Notice that it also follows that the function $G(s)$ does not have any zeros inside D.   □

We now illustrate the different passivity concepts on linear time-invariant systems.

EXAMPLE 5.12   Linear time-invariant systems

Consider a linear time-invariant system with the transfer function $G(s)$. Assume that $G(s)$ has no poles in the closed right half-plane. It follows from

Parseval's theorem that

$$\langle y\,|\,u \rangle = \int_0^\infty y(t)u(t)\,dt = \frac{1}{2\pi} \int_{-\infty}^\infty Y(i\omega)U(-i\omega)\,d\omega$$

$$= \frac{1}{2\pi} \int_{-\infty}^\infty G(i\omega)U(i\omega)U(-i\omega)\,d\omega$$

$$= \frac{1}{\pi} \int_0^\infty \mathrm{Re}\,\{G(i\omega)\}\,U(i\omega)U(-i\omega)\,d\omega \tag{5.47}$$

where Y and U are the Laplace transforms of y and u, respectively. If $G(i\omega)$ is positive real (see Definition 5.5), we have $\mathrm{Re}\,G(i\omega) \geq 0$, and we get

$$\langle y\,|\,u \rangle \geq 0$$

which shows that the system is passive. It follows from Definition 5.9 that a positive real transfer function is input strictly passive if

$$\mathrm{Re}\,G(i\omega) \geq \varepsilon > 0$$

and output strictly passive if

$$\mathrm{Re}\,G(i\omega) \geq \varepsilon|G(i\omega)|^2$$

The transfer function $G(s) = s + 1$ is thus SPR and ISP but not OSP. The transfer function $G(s) = 1/(s + 1)$ is SPR and OSP but not ISP. The transfer function

$$G(s) = \frac{s^2 + 1}{(s + 1)^2}$$

is OSP and ISP but not SPR. □

In control systems applications it is common for transfer functions to be proper or strictly proper. The output strict passivity is therefore the concept that is normally used in these applications.

Proof of the Kalman-Yakubovich Lemma

Having developed the notion of SPR, we can now give a proof of the Kalman-Yakubovic lemma, which was given as Lemma 5.2 in Section 5.5. Consider the linear system

$$\frac{dx}{dt} = Ax + Bu$$
$$y = Cx \tag{5.48}$$

which is assumed to be completely controllable and completely observable. The system has the transfer function

$$G(s) = C\,(sI - A)^{-1}B \tag{5.49}$$

We will prove that a necessary and sufficient condition for $G(s)$ to be SPR is that there exist positive definite matrices P and Q such that

$$A^T P + PA = -Q \tag{5.50}$$

and

$$B^T P = C \tag{5.51}$$

We will first prove necessity. If we use $V = x^T Px$ as a Lyapunov function, it follows from Theorem 5.1 that the system (5.48) is stable. This implies that the transfer function $G(s)$ is analytic in the closed right half-plane. To prove that $G(s)$ is SPR, it remains to verify condition (iii) in Theorem 5.7. It follows from Eq. (5.50) that

$$-sP - A^T P + sP - PA = (-sI - A)^T P + P(sI - A) = Q$$

To obtain this equation, we have added and subtracted sP. Multiplying the equation with $B^T(-sI - A)^{-T}$ from the left and $(sI - A)^{-1}B$ from the right gives

$$B^T P(sI - A)^{-1}B + B^T(-sI - A)^{-T} PB = B^T(-sI - A)^{-T} Q(sI - A)^{-1}B \tag{5.52}$$

Since $G^T(-s) = G(-s)$, Eq. (5.49) now implies that

$$2\,\mathrm{Re}\,G(i\omega) = G(i\omega) + G(-i\omega) = B^T(-i\omega I - A)^{-T} Q(i\omega I - A)^{-1}B \geq 0$$

It now follows from Theorem 5.7 that $G(s)$ is PR. Replacing s by $s - \varepsilon$ in the above calculations, we find in a similar way that

$$\mathrm{Re}\,G(i\omega - \varepsilon) \geq 0$$

Since the matrix A has all its eigenvalues in the open left half-plane, it follows that the matrix $A + \varepsilon I$ is also stable. It now follows from Theorem 5.7 that $G(s)$ is SPR.

To prove sufficiency, we start with the assumption that the system (5.48) has a transfer function $G(s)$ that is SPR. The proof is based on a direct construction of the matrices P and Q. Consider the expression

$$G(s) + G(-s) = \frac{B(s)}{A(s)} + \frac{B(-s)}{A(-s)} = \frac{A(-s)B(s) + A(s)B(-s)}{A(s)A(-s)} = \frac{Q(s)}{A(s)A(-s)}$$

where

$$Q(s) = q_1(-1)^{n-1}s^{2(n-1)} + q_2(-1)^{n-2}s^{2(n-2)} + \ldots + q_n$$

Notice that polynomial $Q(s)$ has only terms of even power and that all coefficients q_i are positive, since $G(s)$ is SPR. Let Q be a diagonal matrix with elements q_i. Introduce the following realization of the transfer function:

$$\frac{dx}{dt} = \begin{pmatrix} -a_1 & -a_2 & \cdots & -a_{n-1} & -a_n \\ 1 & 0 & & 0 & 0 \\ \vdots & & & \vdots & \\ 0 & 0 & \cdots & 1 & 0 \end{pmatrix} x + \begin{pmatrix} 1 \\ 0 \\ \vdots \\ 0 \end{pmatrix} u$$

With this choice we have

$$(sI - A)^{-1}B = \frac{1}{A(s)} \begin{pmatrix} s^{n-1} \\ s^{n-2} \\ \vdots \\ 1 \end{pmatrix} \tag{5.53}$$

and

$$B^T(-sI - A)^{-T}Q(sI - A)^{-1}B = \frac{Q(s)}{A(s)A(-s)} = G(s) + G(-s) \tag{5.54}$$

Since $G(s)$ is SPR, the matrix A has no eigenvalues in the right half-plane or on the imaginary axis. Let P be the solution to Eq. (5.50). This matrix is positive definite because Q is positive definite and A has all its eigenvalues in the left half-plane. Furthermore, let $\tilde{C} = B^T P$. We now show that $\tilde{C} = C$. Since P is the solution to Eq. (5.50), it follows that

$$\tilde{C}(sI - A)^{-1}B + B^T(-sI - A)^{-T}\tilde{C}^T = B^T(-sI - A)^{-T}Q(sI - A)^{-1}B$$

But according to Eq. (5.54) the right-hand side is equal to $G(s) + G(-s)$. Since a partial fraction expansion is unique, it follows from Eq. (5.52) that

$$G(s) = C(sI - A)^{-1}B = \tilde{C}(sI - A)^{-1}B$$

which implies that $\tilde{C} = C$, and the theorem is proven. $\qquad\square$

Test for Positive Realness

It is useful to have an algorithm to test whether a function is positive real. Theorem 5.7 can be used for this purpose. Condition (i) is easily tested by an ordinary Routh-Hurwitz test. Condition (ii) is a straightforward calculation. To test condition (iii), we proceed as follows:

$$G(s) = \frac{B(s)}{A(s)}$$

then

$$\operatorname{Re} G(i\omega) = \operatorname{Re} \frac{B(i\omega)}{A(i\omega)} = \operatorname{Re} \frac{B(i\omega)A(-i\omega)}{A(i\omega)A(-i\omega)}$$

Since the denominator is nonnegative and $G(i\omega)$ is symmetric with respect to the real axis, it suffices to investigate whether the function

$$f(\omega) = \operatorname{Re} (B(i\omega)A(-i\omega))$$

is nonnegative for $\omega \geq 0$. Notice that f is an even function of ω. It is thus sufficient to investigate whether $f(\omega)$ has any real zeros. This can be verified directly by solving the equation $f(\omega) = 0$. There is also an indirect procedure. To describe this, introduce the polynomial

$$g(x) = f\left(\sqrt{x}\right)$$

The problem is thus to find whether the polynomial $g(x)$ has any zeros on the interval $(0, \infty)$. This classical problem can be solved as follows:

1. Let $g_1(x) = g(x)$, $g_2(x) = g'(x)$. Form a sequence of functions $\{g_1(x), g_2(x), \ldots, g_n(x)\}$ by letting $-g_{k+2}(x)$ be the remainder when dividing $g_k(x)$ by $g_{k+1}(x)$. Proceed until g_n is a constant.
2. Let $V(x)$ be the number of sign changes in the sequence $\{g_1(x), g_2(x), \ldots, g_n(x)\}$.
3. The number of real zeros of the function $g(x)$ in the interval $[a, b]$ is then $V(a) - V(b)$.

The function sequence $\{g_1(x), g_2(x), \ldots, g_n(x)\}$ is called a *Sturm sequence*. The procedure is illustrated by an example.

EXAMPLE 5.13   Second-order system

Consider the transfer function

$$G(s) = \frac{s^2 + 6s + 8}{s^2 + 4s + 3}$$

First notice that G has no poles in the right half-plane. Furthermore,

$$f(\omega) = \mathrm{Re}\left((-\omega^2 + 6i\omega + 8)(-\omega^2 - 4i\omega + 3)\right) = \omega^4 + 13\omega^2 + 24$$

Hence

$$g(x) = x^2 + 13x + 24$$

We get

$$g_1(x) = x^2 + 13x + 24$$
$$g_2(x) = 2x + 13$$
$$g_3(x) = \frac{73}{4}$$

Since $V(0) = 0$, $V(\infty) = 0$, $g(x)$ has no zeros on the positive real axis. The transfer function $G(s)$ is then SPR. $\square$

An Alternative Test

An alternative test for SPR for a system with a proper transfer function can be obtained from the proof of the Kalman-Yakubovich lemma. Write the matrix A

in controllable canonical form. Solve the equations

$$A^T P + PA = \begin{pmatrix} q_1 & 0 & \cdots & 0 \\ 0 & q_2 & \cdots & 0 \\ \vdots & & & \\ 0 & 0 & \cdots & q_n \end{pmatrix}$$

$$B^T P = C$$

where P is a symmetric matrix. This gives $n + n(n + 1)/2$ equations for the unknown elements of P and Q. The transfer function is SPR if $q_i > 0$ for $i = 1, \ldots, n$.

The Passivity Theorem

Having established a notion of passivity, we can now state a key result.

THEOREM 5.8   The passivity theorem

Consider a system obtained by connecting two systems H_1 and H_2 in a feedback loop as in Fig. 5.16. Let H_1 be strictly output passive and H_2 be passive. The closed-loop system is then BIBO stable.

Proof: Since H_1 is strictly output passive, we have

$$\langle y \,|\, e \rangle > \delta \|y\|^2$$

Since $e = u - H_2 y$, we have

$$\delta \|y\|^2 \le \langle y \,|\, e \rangle = \langle y \,|\, u - H_2 y \rangle = \langle y \,|\, u \rangle - \langle y \,|\, H_2 y \rangle \qquad (5.55)$$

Since H_2 is passive, we have

$$\langle y \,|\, H_2 y \rangle \ge 0$$

and it then follows from Eq. (5.55) that

$$\delta \|y\|^2 \le \langle y \,|\, u \rangle \le \|y\| \, \|u\|$$

where the last inequality follows from Schwartz inequality. We now get

$$\|y\| < \frac{1}{\delta} \|u\|$$

which proves the result.   □

Remark.   The passivity theorem may also be regarded as an extension of Nyquist's stability theorem. Instability is avoided by having a loop transfer function with a phase lag less than 180°.   □

Relations between Passivity and Small Gain Theorems

The small gain theorem (Theorem 5.6) and the passivity theorem (Theorem 5.8) are closely related. To investigate this connection further, we consider signal spaces that are inner product spaces and we show that the small gain theorem can be derived from the passivity theorem. We start with Fig. 5.16 and make a sequence of transformations of the feedback loop that are shown in Fig. 5.17.

Consider the closed-loop system in Fig. 5.17(a). Assume that the system H_1 is strictly output passive and that H_2 is passive. In Fig. 5.17(b) we have introduced two loops that cancel each other. The input-output relations of the encircled loops are $(I + H_1)^{-1}H_1$ and $I - H_2$, respectively. These two systems are shown in Fig. 5.17(c), where we have also added two loops and two gains $(1/2$ and $2)$ that cancel each other. The transfer functions of the encircled loops

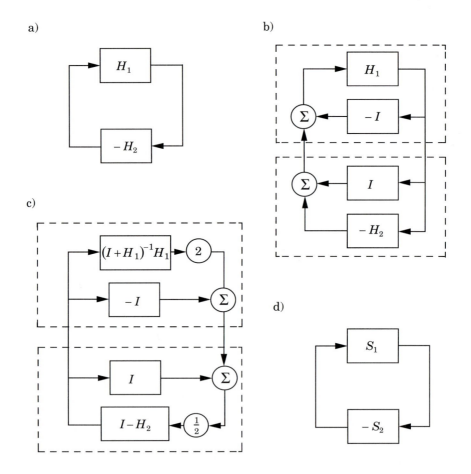

Figure 5.17   Four equivalent systems.

are

$$S_1 = 2(H_1 + I)^{-1}H_1 - (I + H_1)^{-1}(I + H_1) = (H_1 + I)^{-1}(H_1 - I)$$

and

$$S_2 = -\left(I - \frac{1}{2}(I - H_2)\right)^{-1}\frac{1}{2}(I - H_2) = (H_2 + I)^{-1}(H_2 - I)$$

The system obtained after the transformations is shown in Fig. 5.17(d).

The systems in Fig. 5.17(a) and Fig. 5.17(d) are equivalent. We use their equivalence to prove the result. First we observe that if the system $(H + I)^{-1}$ exists, it commutes with H. To prove this, use the identity

$$H + H^2 = H(H + I) = (H + I)H$$

and multiply from the left and the right by $(I + H)^{-1}$; then

$$(H + I)^{-1}H = H(H + I)^{-1}$$

Subtracting $(H + I)^{-1}$ from both sides gives

$$(H + I)^{-1}(H - I) = (H - I)(H + I)^{-1}$$

The systems S and H are related through

$$S = (H - I)(H + I)^{-1} = (H + I)^{-1}(H - I)$$

The input-output relation for the system S is

$$y = Su = (H - I)(H + I)^{-1}u$$

Introduce

$$x = (H + I)^{-1}u$$

We find that

$$y = (H - I)x$$
$$u = (H + I)x$$

Hence

$$\|y\|^2 = \langle y|y \rangle = \langle Hx - x|Hx - x \rangle = \langle Hx|Hx \rangle + \langle x|x \rangle - 2\langle Hx|x \rangle$$

Similarly, we find that

$$\|u\|^2 = \langle u|u \rangle = \langle Hx + x|Hx + x \rangle = \langle Hx|Hx \rangle + \langle x|x \rangle + 2\langle Hx|x \rangle$$

Hence

$$\|y\|^2 = \|u\|^2 - 4\langle Hx|x \rangle \tag{5.56}$$

If H is passive, we have $\langle Hx|x \rangle \geq 0$; hence $\|y\| \leq \|u\|$, which implies that $\gamma(S) \leq 1$. Similarly, we find that $\gamma(S) < 1$ if H is strictly output passive.

It follows from Eq. (5.56) that

$$\langle Hx|x \rangle = \frac{\|u\|^2 - \|y\|^2}{4} = \frac{\|u\|^2 - \|Su\|^2}{4} \geq (1 - \gamma(S)) \frac{\|u\|^2}{4}$$

This implies that H is passive if $\gamma(S) \leq 1$ and strictly output passive if $\gamma(S) < 1$.

Notice that the argument would be the same if S and H were complex numbers. The result is an example of the equivalence between complex numbers and operators on inner product spaces.

5.7   APPLICATIONS TO ADAPTIVE CONTROL

The results from input-output stability theory are now used to construct adjustment rules for adaptive systems. So that we can focus on the principles and avoid unnecessary details, only the problem of adjusting a feedforward gain is considered in this section.

Consider a system with transfer function $kG(s)$ where $G(s)$ is known and k is an unknown constant. We will determine an adaptive feedforward compensation so that the transfer function from command signal to output is $k_0 G(s)$. This problem was previously considered in Examples 5.1 and 5.3. A parameter adjustment law was also derived for the problem in Section 5.5 using Lyapunov theory. This control law can be represented by the block diagram in Fig. 5.14(b). According to Theorem 5.5 the adaptive system will be stable if the transfer function $G(s)$ is SPR. This condition indicates that the result is related to passivity theory. To establish this, we redraw the block diagram as in Fig. 5.18, which gives a configuration in which the passivity theorem can be applied. To use the passivity theorem, we must investigate the properties of the dashed block in Fig. 5.18. We have the following lemma.

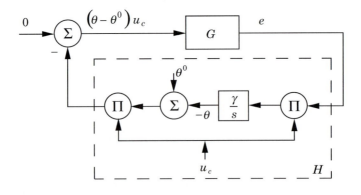

Figure 5.18   Representation of the system with adjustable feedforward gain when using the control law of Eq. 5.40. Compare with Fig. 5.14(b).

L E M M A 5.3   Property of positive real systems

Let r be a bounded square integrable function, and let $G(s)$ be a transfer function that is positive real. The system whose input-output relation is given by

$$y = r\,(G(p)ru)$$

is then passive.

Proof: It follows that

$$\langle y \,|\, u \rangle = \int\limits_0^\infty y(\tau)u(\tau)\,d\tau = \int\limits_0^\infty (u(\tau)r(\tau))\,(G(p)ru)(\tau)\,d\tau$$

$$= \int\limits_0^\infty w(\tau)\,(G(p)w)\,(\tau)\,d\tau = \langle w \,|\, Gw \rangle$$

where $w = ru$. Since $G(s)$ is positive real, it follows from Example 5.12 that $\langle w|Gw \rangle \geq 0$, which proves the result.   □

By invoking the passivity theorem (Theorem 5.8) we can now obtain an alternative proof of Theorem 5.5. Figure 5.18 shows that the model-reference system can be viewed as a feedback connection of two systems. One system is linear with the transfer function G. It has the signal $(\theta - \theta^0)u_c$ as the input and the model error as the output. The other system has the model error e as the input and the quantity $-(\theta - \theta^0)u_c$ as the output. Since an integrator is positive real, it follows from Lemma 5.3 that the system H is passive. If the transfer function G is proper and strictly positive real, it follows from Example 5.12 that $G(s)$ is output strictly proper. The passivity theorem (Theorem 5.8) then implies that the closed-loop system is BIBO stable. In Fig. 5.18 there are no external inputs, as in Fig. 5.16. The system in Fig. 5.18 may have initial conditions, however, because the process and the model may have different initial conditions. The integrator may also have an initial condition that can be thought of as being generated by an external input signal. Such an input signal can always be chosen to be zero for $t \geq 0$. We thus have a situation covered by Theorem 5.6, where the input signal u is bounded in L_2. The error $e(t)$ goes to zero as t goes to infinity. Notice that the MRAS is stable for all values of $\gamma > 0$ when the SPR condition is satisfied. This implies that the adaptation can be made arbitrarily fast.

Design of Stable Adjustment Mechanisms

The passivity theorem gives a convenient way to construct stable adjustment laws. We simply try to introduce some compensating network so that the transfer function relating the error to $(\theta - \theta^0)u_c$ is strictly positive real, as

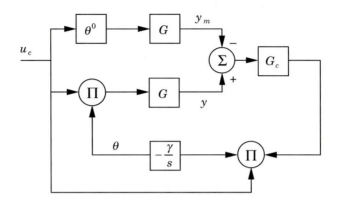

Figure 5.19   A stable parameter adjustment law is obtained if GG_c is SPR.

is illustrated in Fig. 5.19. For systems with output feedback, the problem is to find a compensator G_c such that the transfer function GG_c is strictly positive real. This can be done by using the Kalman-Yakubovich lemma (Lemma 5.2). With pure feedforward control it is natural to assume that G is stable. It can then be written as

$$G(s) = \frac{B(s)}{A(s)}$$

where $A(s)$ has all its zeros in the left half-plane. For a stable polynomial $A(s)$ a polynomial $C(s)$ such that $C(s)/A(s)$ is SPR can always be found. To do this, we introduce the following canonical realization of $1/A(s)$:

$$\frac{dx}{dt} = \begin{pmatrix} -a_1 & -a_2 & \cdots & -a_{n-1} & -a_n \\ 1 & 0 & & 0 & 0 \\ & \vdots & & & \\ 0 & 0 & & 1 & 0 \end{pmatrix} x + \begin{pmatrix} 1 \\ 0 \\ \vdots \\ 0 \end{pmatrix} u$$

Choose a symmetric positive definite matrix Q and solve the equation

$$A^T P + PA = -Q$$

The coefficients of a C polynomial such that $C(s)/A(s)$ is SPR are then the first row of the P matrix.

The polynomial $C(s)$ will have a degree that is at most equal to $\deg A - 1$. For systems with stable zeros and pole excess 1 it is thus possible to find a stable adjustment rule by choosing $G_c(s) = C(s)/B(s)$. However, for systems with higher pole excess than 1 the compensator required to make GG_c strictly positive real will contain derivatives. We will show how to deal with the case in which the pole excess is higher by introducing the augmented error.

PI Adjustments

All adjustment laws discussed so far have been integral controllers. That is, the parameter has always been obtained as the output of an integrator. There are, of course, many other possibilities for choosing the adaptation mechanism H in Fig. 5.18. For instance, it can be expected that quicker adaptation can be achieved by using a proportional and integral adjustment law. This means that the control law of Eq. (5.40) is replaced by

$$\theta(t) = -\gamma_1 u_c(t)e(t) - \gamma_2 \int^t u_c(\tau)e(\tau)\, d\tau \qquad (5.57)$$

Since a system with the transfer function

$$H(s) = \gamma_1 + \gamma_2/s$$

is output strictly passive for positive γ_1 and γ_2, it follows from Theorem 5.8 (the passivity theorem) that Eq. (5.57) gives a stable adjustment law if GG_c is positive real.

The Augmented Error

Some progress has now been made to construct stable parameter adjustment rules for the problem of adjusting a feedforward gain. Passivity theory gave good insight and led to the idea of filtering the model error so that GG_c is SPR. However, we have not solved the problem in which G has a pole excess larger than 1. To do this, we first factor the transfer function G as

$$G = G_1 G_2$$

where the transfer function G_1 is SPR. The error $e = y - y_m$ can then be written as

$$
\begin{aligned}
e &= G(\theta - \theta^0)u_c = (G_1 G_2)(\theta - \theta^0)u_c \\
&= G_1\left(G_2(\theta - \theta^0)u_c + (\theta - \theta^0)G_2 u_c - (\theta - \theta^0)G_2 u_c\right) \\
&= G_1\left((\theta - \theta^0)G_2 u_c\right) - G_1\left((\theta - \theta^0)G_2 u_c - G_2(\theta - \theta^0)u_c\right)
\end{aligned}
$$

Introduce the *augmented error* ε defined by

$$\varepsilon = e + \eta$$

where η is the *error augmentation* defined by

$$
\begin{aligned}
\eta &= G_1(\theta - \theta^0)G_2 u_c - G(\theta - \theta^0)u_c \\
&= G_1(\theta G_2 u_c) - G\theta u_c
\end{aligned}
$$

The second equality follows because $G\theta^0 u = \theta^0 G u$ when θ^0 is constant. The augmented error is thus obtained by adding a correction term η to the error. The correction term vanishes when the parameter θ is constant. It follows that

$$\varepsilon = G_1 \left((\theta - \theta^0) G_2 u_c \right) = G_1 (\theta - \theta^0) \bar{u}_c \tag{5.58}$$

where $\bar{u}_c$ is the reference signal filtered through G_2. Equation (5.58) is an error model similar to the ones used previously, and we have the following theorem.

THEOREM 5.9 Stability using augmented error

Consider a model-reference system for adaptation of a feedforward gain for a system with the transfer function G. Let $G_1 G_2$ be a factorization of G such that G_1 is SPR. The parameter adjustment law

$$\frac{d\theta}{dt} = -\gamma \varepsilon (G_2 u_c) \tag{5.59}$$

where

$$\varepsilon = e + G_1(\theta G_2 u_c) - G(\theta u_c) \tag{5.60}$$

gives a closed-loop system in which the error goes to zero as t goes to infinity.

Proof: Since G_1 is SPR, the discussion of the error model shows that $\varepsilon \in L_2$.

Remark 1. The trivial factorization with $G_1 = 1$ is one possibility.

Remark 2. If the input signal is persistently exciting, it can be shown that the parameters also converge.

Remark 3. Notice that G_2 must be minimum phase to establish that θ converges to θ^0. The reason is that we have to go "backwards" through G_2 to show that $\theta - \theta^0$ goes to zero if the output e goes to zero. That is, the inverse of G_2 must be stable. This is a condition that will be seen again in the general case in Section 5.8. □

A block diagram of the system with augmented error is shown in Fig. 5.20. To implement the augmented error, it is necessary to introduce realizations of the transfer functions G_1 and G_2. The augmented error was introduced by Monopoli. It was a key idea for adaptive control systems having pole excess larger than 1. Application of the idea to general linear systems is discussed in Section 5.8. In Section 5.9 we show that the augmented error appears naturally in the self-tuning regulator.

Summary

The problem of adjusting the gain in a known system has been used to introduce some ideas in the design of stable model-reference adaptive systems. It was first shown that adjustment rules could be obtained for systems in which the plant is strictly positive real. The parameter adjustment rules were similar to those obtained by the gradient method.

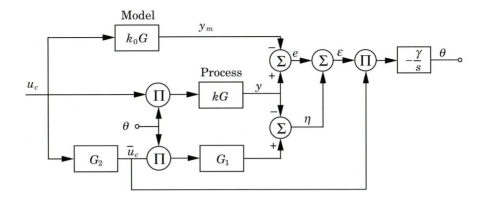

Figure 5.20   Block diagram of a model-reference adaptive system based on the augmented error.

The class of systems could then be extended by using adjustment rules in which the error is filtered. In this way the problem can be solved for stable minimum-phase systems that have pole excess less than 1. The idea of augmented error was introduced to solve the problem of higher pole excess.

5.8   OUTPUT FEEDBACK

We now derive an MRAS for adjusting the parameters of a controller based on output feedback in a fairly general case. A process with one input and one output is considered. It is assumed that the dynamics are linear and that the control problem is formulated as model-following. The key assumption is that the controller can be parameterized in such a way that the error is linear in the controller parameters. The derivation of the MRAS is described as follows:

Step 1:   Find a controller structure that admits perfect output tracking.

Step 2:   Derive an error model of the form

$$\varepsilon = G_1(p)\left\{\varphi^T(t)(\theta^0 - \theta)\right\} \tag{5.61}$$

where G_1 is a strictly positive real transfer function, θ^0 is the process parameters, and θ is the controller parameters. The right-hand side should be expressed in computable quantities.

Step 3:   Use the parameter adjustment law

$$\frac{d\theta}{dt} = \gamma\varphi\varepsilon \tag{5.62}$$

or the normalized law

$$\frac{d\theta}{dt} = \gamma\frac{\varphi\varepsilon}{\alpha + \varphi^T\varphi} \tag{5.63}$$

Notice that the error ε in Eq. (5.61) is linear in the parameters, a condition that imposes restrictions on the models and controllers that can be dealt with. A model of the form (5.61) is typically obtained by algebraic manipulations, filtering, and error augmentation.

We now show one way to apply the design procedure.

Finding a Controller Structure

The first step in the design procedure is to find a suitable controller structure. The tools for doing this were developed in Section 3.2. Let the process be described by the continuous-time model

$$Ay(t) = b_0 Bu(t) \tag{5.64}$$

where it is assumed that the polynomials A and B do not have common factors and the polynomial B is monic and assumed to have all its zeros in the left half-plane. Furthermore, the polynomial is normalized so that B is monic. The variable b_0 is called the *instantaneous gain* or the *high-frequency gain*. A general linear controller can be written as

$$Ru(t) = -Sy(t) + Tu_c(t) \tag{5.65}$$

where u_c is the command signal. Since the polynomial B is stable, the corresponding poles can be canceled by the controller. This corresponds to $R = R_1 B$. The closed-loop system obtained when the controller is applied to the process (5.64) is described by

$$(AR_1 + b_0 S)y = b_0 Tu_c \tag{5.66}$$

If polynomial T is chosen to be $T = t_0 A_0$, where A_0 is a stable monic polynomial and R_1 and S satisfy

$$AR_1 + b_0 S = A_0 A_m \tag{5.67}$$

it is possible to achieve perfect model-following with the model

$$A_m y_m(t) = b_0 t_0 u_c(t) \tag{5.68}$$

The Error Model

Having obtained a suitable controller structure, we now proceed to derive an error model. It follows from Eq. (5.67) that

$$A_o A_m y = AR_1 y + b_0 Sy = R_1 b_0 Bu + b_0 Sy \tag{5.69}$$

where the first equality follows from Eq. (5.67) and the second from Eq. (5.64). Introduce the error

$$e = y - y_m$$

It follows from Eqs. (5.69) and (5.68) that

$$A_o A_m e = A_o A_m (y - y_m) = b_0 (Ru + Sy - Tu_c)$$

or

$$e = \frac{b_0}{A_o A_m} (Ru + Sy - Tu_c)$$

This expression is not yet a suitable error model, because the transfer function $b_0/(A_o A_m)$ is not SPR. Therefore introduce the filtered error

$$e_f = \frac{Q}{P} e = \frac{Q}{P} (y - y_m)$$

where Q is a polynomial whose degree is not greater than $\deg A_o A_m$ such that

$$\frac{b_0 Q}{A_o A_m} \tag{5.70}$$

is SPR. The filtered error can be written as

$$e_f = \frac{b_0 Q}{A_o A_m} \left(\frac{R}{P} u + \frac{S}{P} y - \frac{T}{P} u_c \right)$$

Let $P = P_1 P_2$, where P_2 is a stable monic polynomial of the same degree as R. Rewrite R/P as

$$\frac{R}{P} = \frac{R - P_2 + P_2}{P_1 P_2} = \frac{1}{P_1} + \frac{R - P_2}{P}$$

The filtered error then becomes

$$e_f = \frac{b_0 Q}{A_o A_m} \left(\frac{1}{P_1} u + \frac{R - P_2}{P} u + \frac{S}{P} y - \frac{T}{P} u_c \right)$$

Let k, l, and m be the degrees of the polynomials R, S, and T, respectively. Introduce a vector of true controller parameters

$$\theta^0 = \left(r_1' \ldots r_k' \; s_0 \ldots s_l \; t_0 \ldots t_m \right)^T \tag{5.71}$$

where r_i' are the coefficients of the polynomial $R - P_2$. Also introduce a vector of filtered input, output, and command signals

$$\varphi^T = \left(\frac{p^{k-1}}{P(p)} u \; \ldots \; \frac{1}{P(p)} u \quad \frac{p^\ell}{P(p)} y \; \ldots \; \frac{1}{P(p)} y \quad -\frac{p^m}{P(p)} u_c \; \ldots \; -\frac{1}{P(p)} u_c \right) \tag{5.72}$$

The filtered error can then be written as

$$e_f = \frac{b_0 Q}{A_o A_m} \left(\frac{1}{P_1} u + \varphi^T \theta^0 \right) \tag{5.73}$$

To obtain an error model, we must introduce a parameterization of the controller. In the nominal case in which the parameters are known, the control law can be expressed as

$$u = -P_1(\varphi^T \theta^0) = -P_1\left((\theta^0)^T \varphi\right) = -(\theta^0)^T(P_1\varphi) \tag{5.74}$$

where P_1 is a polynomial in the differential operator. Let θ denote the adjustable controller parameters. The feedback law

$$u = -P_1(\varphi^T \theta)$$

would give the desired error model. However, this control law is not realizable if P_1 has a degree greater than 1 because the term $P_1(\varphi^T \theta)$ contains derivatives of the parameters. However, the control law

$$u = -\theta^T(P_1\varphi) \tag{5.75}$$

is realizable because of Eq. (5.69). If we use this control law, it follows from Eq. (5.70) that the filtered error can be written as

$$\begin{aligned}
e_f &= \frac{b_0 Q}{A_o A_m}\left(\varphi^T \theta^0 - \frac{1}{P_1}\theta^T(P_1\varphi)\right)\\
&= \frac{b_0 Q}{A_o A_m}\left(\varphi^T \theta^0 - \varphi^T \theta - \frac{1}{P_1}\theta^T(P_1\varphi) + \varphi^T \theta\right)
\end{aligned}$$

Introduce the signals η and ε, defined by

$$\begin{aligned}
\eta &= \frac{1}{P_1}\theta^T(P_1\varphi) - \varphi^T \theta = -\left(\frac{1}{P_1}u + \varphi^T \theta\right)\\
\varepsilon &= e_f + \frac{b_0 Q}{A_o A_m}\eta = \frac{b_0 Q}{A_o A_m}\varphi^T(\theta^0 - \theta)
\end{aligned} \tag{5.76}$$

The signal ε is called the *augmented error*, and η is called the *error augmentation*. The augmented error is computed as follows:

$$\varepsilon = \frac{Q}{P}(y - y_m) + \frac{b_0 Q}{A_o A_m}\eta \tag{5.77}$$

With the chosen degrees of P and Q it is straightforward to verify that the computation does not require taking derivatives of the signals y, u, u_c, and y_m. The error model of Eq. (5.76) is also linear in the parameters, and the transfer function $b_0 Q/(A_o A_m)$ is SPR. The error model thus satisfies the requirements of Step 2, and the parameters can then be updated by Eq. (5.62) or Eq. (5.63). So far, the derivation has been done along the lines developed in Sections 5.3 and 5.4. However, to show the stability of the closed-loop system, it is not sufficient that the system (5.70) is SPR. It is also necessary that the signals in φ are bounded. This condition can be difficult to show. Furthermore, Eqs. (5.76) are valid only if the control signal is generated from Eq. (5.75). This implies, for instance, that the control signal cannot be saturated. Notice that it is necessary to know the parameter b_0 to compute the augmented error ε.

The derived algorithm thus requires that the high-frequency gain b_0 be known. If the parameter is not known, it can be estimated as follows. The error model of Eq. (5.73) can be written as

$$e_f = b_0 \left(\varphi_f^T \theta^0 + u_f \right) \tag{5.78}$$

where

$$\varphi_f = \frac{Q}{A_o A_m} \varphi$$

$$u_f = \frac{Q}{A_o A_m P_1} u$$

A simple gradient estimator for b_0 and θ^0 is then given by

$$\frac{d\theta}{dt} = \gamma' \hat{b}_0 \varphi_f \varepsilon_p = \gamma \varphi_f \varepsilon_p$$

$$\frac{d\hat{b}_0}{dt} = \gamma \left(\varphi_f^T \theta + u_f \right) \varepsilon_p \tag{5.79}$$

where ε_p is the prediction error

$$\varepsilon_p = e_f - \hat{e}_f = e_f - \hat{b}_0 \left(\varphi_f^T \theta + u_f \right) \tag{5.80}$$

Notice that $\hat{b}_0$ can be absorbed in the adaptation gain if its sign is known.

Realization

The equations needed to implement the general MRAS can now be summarized:

$$y_m = \frac{B_m}{A_m} u_c$$

$$e_f = \frac{Q}{P} e = \frac{Q}{P} (y - y_m)$$

$$\eta = - \left(\frac{1}{P_1} u + \varphi^T \theta \right)$$

$$\varepsilon = e_f + \frac{b_0 Q}{A_o A_m} \eta$$

$$\frac{d\theta}{dt} = \gamma \varphi \varepsilon$$

$$u = -\theta^T (P_1 \varphi)$$

A block diagram of the model-reference adaptive system is shown in Fig. 5.21. The block labeled "Filter" in Fig. 5.21 is a linear system that generates $P_1 \varphi$ and φ from the signals u_c, u, and y. The vector φ is composed of three parts having the same structure. It therefore suffices to discuss one part. Consider, for example, how to generate φ_u and $P_1\varphi_u$ where

$$P_1\varphi_u = \left(\frac{p^{k-1}}{P_2}u \ldots \frac{1}{P_2}u\right)^T = (x_1 \ldots x_k)^T = x^T$$

and

$$\varphi_u = \left(\frac{p^{k-1}}{P}u \ldots \frac{1}{P}u\right)^T$$

where $P = P_1P_2$ and $k = \deg R = \deg P_2$.

Let the polynomials P_1 and P_2 be

$$P_1 = p^n + \alpha_1 p^{n-1} + \cdots + \alpha_n$$
$$P_2 = p^k + \beta_1 p^{k-1} + \cdots + \beta_k$$

We also assume that $\deg P_1 > \deg P_2$. The vectors x and φ_u can then be realized as follows:

$$\frac{dx}{dt} = \begin{pmatrix} -\beta_1 & -\beta_2 & \cdots & -\beta_{k-1} & -\beta_k \\ 1 & 0 & & 0 & 0 \\ \vdots & & \ddots & & \\ 0 & 0 & & 1 & 0 \end{pmatrix} x + \begin{pmatrix} 1 \\ 0 \\ \vdots \\ 0 \end{pmatrix} u$$

$$\frac{dz}{dt} = \begin{pmatrix} -\alpha_1 & -\alpha_2 & \cdots & -\alpha_{n-1} & -\alpha_n \\ 1 & 0 & & 0 & 0 \\ \vdots & & \ddots & & \\ 0 & 0 & & 1 & 0 \end{pmatrix} z + \begin{pmatrix} 1 \\ 0 \\ \vdots \\ 0 \end{pmatrix} x_k$$

where $x_k = 1/P_2 \cdot u$ is the last element of the x vector. The elements of φ_u are the k last elements of the state vector z. Furthermore, $1/P_1 \cdot u$ can also be obtained from the generation of φ_u and $P_1\varphi_u$. To generate the full vectors φ and $P_1\varphi$, we thus need three realizations of the transfer functions P_1 and P_2. The block labeled "Filter" in Fig. 5.21 represents these systems.

Design Parameters

Several parameters must be chosen in the design procedure:

- The model transfer function B_m/A_m,
- The observer polynomial A_o,
- The degrees of polynomials R, S, and T, and
- The polynomials P_1, P_2, and Q.

Many different model-reference adaptive systems can be obtained by different choices of the design parameters. A popular choice of the polynomials is $P_1 = A_m, P_2 = A_o$, and $Q = A_o A_m$.

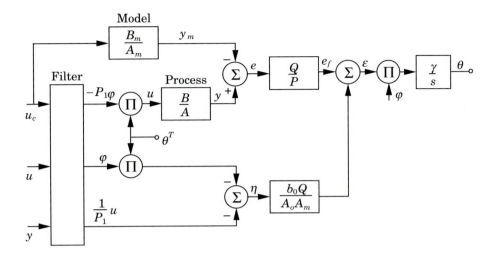

Figure 5.21 Block diagram of a model-reference adaptive system for a SISO system.

A Priori Knowledge

To apply the MRAS procedure, the plant must be minimum-phase and the following prior information must also be known:

- The sign of the instantaneous gain b_0,
- The pole excess of the plant, and
- The order of the plant or the controller complexity.

EXAMPLE 5.14   **Second-order MRAS**

The performance of the general MRAS is illustrated by a second-order example, given the system

$$G(s) = \frac{k}{s(s + a)}$$

and the model

$$G_m(s) = \frac{B_m}{A_m} = \frac{\omega^2}{s^2 + 2\zeta\omega s + \omega^2}$$

The polynomials A_o, R, S, and T can be chosen to be

$$A_o(s) = s + a_o$$
$$R(s) = s + r_1$$
$$S(s) = s_0 s + s_1$$
$$T(s) = t_0 s + t_1$$

The Diophantine equation (Eq. 5.67) gives the solution

$$r_1 = 2\zeta\omega + a_o - a$$
$$s_0 = (2\zeta\omega a_o + \omega^2 - ar_1)/k$$
$$s_1 = a_o\omega^2/k$$
$$t_0 = \omega^2/k$$
$$t_1 = a_o\omega^2/k$$

For simplicity we choose

$$Q(s) = A_o(s)A_m(s)$$
$$P_1(s) = A_m(s)$$
$$P_2(s) = A_o(s)$$

Figure 5.22 shows a simulation of the system with $\gamma = 1$, $\zeta = 0.7$, $\omega = 1$, $a_o = 2$, $a = 1$, and $k = 2$. In the simulation it is assumed that $\hat{b}_0 = b_0$. The used values of the filters P_1, P_2, Q, and A_o give a fairly rapid convergence of y to y_m. The parameter estimates at the end of the simulation are still far from the optimal values, but the error is small (see Fig. 5.22c). The controller parameters are shown in Fig. 5.23. The control law at $t = 150$ gives a closed-loop system with a pole in -1.95 and two complex poles corresponding to

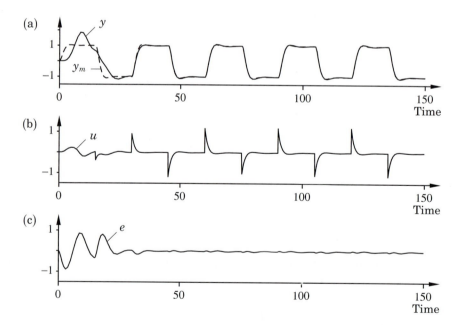

Figure 5.22 Simulation of the system in Example 5.14. (a) The process output (solid line) and the model output (dashed line). (b) The control signal. (c) The error $e = y - y_m$.

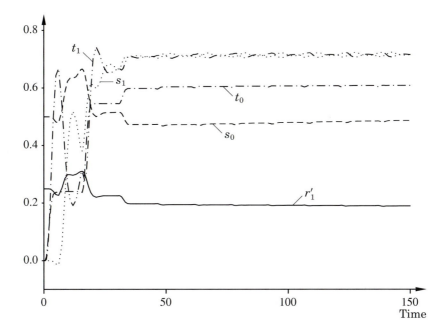

Figure 5.23 The controller parameters in the simulation of the system in Example 5.14.

$\omega = 0.84$ and $\zeta = 0.78$, which should be compared to the roots of A_oA_m, which are in -2, and complex poles corresponding to $\omega = 1$ and $\zeta = 0.7$.   □

5.9   RELATIONS BETWEEN MRAS AND STR

For a long time, model-reference adaptive systems and self-tuning regulators were regarded as two quite different approaches to adaptive control. In this section we will show that the methods are closely related. The key observation is that the direct self-tuner in which process zeros are canceled (Algorithm 3.3) can be interpreted as a MRAS.

An MRAS for a general continuous-time linear system was derived in Section 5.8. In the derivation it was assumed that the process was minimum-phase and that all its zeros were canceled in the design. We showed that the adjustment law for updating the parameters can be written as

$$\frac{d\theta}{dt} = \gamma \varphi_f \varepsilon \tag{5.81}$$

where φ_f is a filtered regression vector and ε is the augmented error given by Eq. (5.77), that is,

$$\varepsilon = \frac{Q}{P}(y - y_m) + \frac{b_o Q}{A_o A_m}\eta \tag{5.82}$$

Now consider a discrete-time direct self-tuner. When all process zeros are canceled, polynomial B^- is a constant and we get the process model

$$y(t) = \varphi_f^T (t - d_0) \theta$$

In the direct algorithm the estimated parameters are equal to the controller parameters. The least-squares method can be used for the estimation by using the residual

$$\varepsilon(t) = y(t) - \hat{y}(t) = y(t) - \varphi_f^T (t - d_0) \hat{\theta}(t - 1)$$

The parameter update can be written as

$$\hat{\theta}(t) = \hat{\theta}(t - 1) + P(t)\varphi_f^T (t - d_0) \varepsilon(t) \tag{5.83}$$

The residual is given by

$$\varepsilon(t) = y(t) - \hat{y}(t) = y(t) - y_m(t) + y_m(t) - \hat{y}(t) = e(t) + \eta(t) \tag{5.84}$$

A comparison of Eqs. (5.81) and (5.83) show that Eq. (5.83) can be interpreted as a discrete-time version of Eq. (5.81). Notice that the gain γ in the MRAS is replaced by the matrix $P(t)$. This matrix changes the gradient direction φ_f and gives an appropriate step length. Also notice that it follows from Eq. (5.84) that the error augmentation is simply $y - \hat{y}$. The augmented error that required a significant ingenuity to derive in the MRAS context is thus obtained directly from the least-squares equations in the STR. More filtering is required in the MRAS because of the continuous time formulation. Notice that it follows from Eq. (5.83) that

$$\varphi_f^T (t - d_0) = - \text{grad}_\theta \, \varepsilon(t)$$

The vector $\varphi_f^T (t - d_0)$ can be interpreted as the sensitivity derivative of the prediction error ε with respect to the parameter. The parameter update of Eq. (5.83) is thus a discrete-time version of the MIT rule. The main difference is that the model error $e(t) = y(t) - y_m(t)$ is replaced by the prediction error $\varepsilon(t)$.

Notice that in the identification-based schemes such as self-tuning controllers we normally attempt to obtain a form similar to

$$y(t) = \varphi_f^T \theta$$

With the model-reference approach, it is also possible to admit a model of the form

$$y(t) = G(p) (\varphi_f^T \theta)$$

where $G(p)$ is SPR. In summary we thus find that the MRAS-type algorithms can be obtained in a straightforward way as a direct self-tuning regulator based on a minimum-degree pole placement design with cancellation of the whole B polynomial.

5.10   NONLINEAR SYSTEMS

The Lyapunov method can also be used to find adaptive control laws for non-linear systems. This is a difficult problem because no general design methods are available. There is, however, much interest in adaptive control of nonlinear systems. For this reason we present some of the current ideas and illustrate them by a few examples.

Feedback Linearization

Before attempting to do adaptive control, we must first have a design method for the case in which the parameters are known. Feedback linearization is a design method that is similar in spirit to pole placement. It can be applied to certain classes of systems. We illustrate it through an example.

EXAMPLE 5.15   **Feedback linearization**

Consider the system

$$\frac{dx_1}{dt} = x_2 + f(x_1)$$

$$\frac{dx_2}{dt} = u$$

where f is a differentiable function. The first step is to introduce new coordinates

$$\xi_1 = x_1$$
$$\xi_2 = x_2 + f(x_1)$$

The equations then become

$$\frac{d\xi_1}{dt} = \xi_2$$

$$\frac{d\xi_2}{dt} = \xi_2 f'(\xi_1) + u$$

By introducing the control law

$$u = -a_2\xi_1 - a_1\xi_2 - \xi_2 f'(\xi_1) + v$$

we get a linear closed-loop system described by

$$\frac{d\xi}{dt} = \begin{pmatrix} 0 & 1 \\ -a_2 & -a_1 \end{pmatrix} \xi + \begin{pmatrix} 0 \\ 1 \end{pmatrix} v$$

This system is linear with the characteristic equation

$$s^2 + a_1 s + a_2 = 0$$

By transforming back to the original coordinates the control law can be written as

$$u = -a_2 x_1 - \left(a_1 + f'(x_1)\right)\left(x_2 + f(x_1)\right) + v \qquad \square$$

The closed-loop system obtained in the example will behave like a linear system. This is the reason why the method is called *feedback linearization*. The system in Example 5.15 is quite special. Applying the same procedure for a system described by

$$\frac{dx}{dt} = f(x) + u g(x)$$

we first pick

$$\xi_1 = h(x)$$

as a new state variable. The time derivative of ξ_1 is

$$\frac{d\xi_1}{dt} = h'(x)\left(f(x) + u g(x)\right)$$

If $h'(x)g(x) = 0$, we introduce the new state variable

$$\xi_2 = h'(x) f(x)$$

We proceed as long as the control variable u does not appear explicitly on the right-hand side. In this way we obtain the state variables $\xi_1 \ldots \xi_r$, which are combined to the vector $\xi \in R^r$, where $r \leq n$. We also introduce the new state variable $\eta_1 \ldots \eta_{n-r}$, which are combined into the vector $\eta \in R^{n-r}$. This can be done in many different ways. We obtain the following system of equations:

$$\frac{d\xi_1}{dt} = \xi_2$$

$$\frac{d\xi_2}{dt} = \xi_3$$

$$\vdots \qquad\qquad (5.85)$$

$$\frac{d\xi_r}{dt} = \alpha(\xi,\eta) + u\beta(\xi,\eta)$$

$$\frac{d\eta}{dt} = \gamma(\xi,\eta)$$

Notice that the state variables ξ represents a chain of r integrators, where the integer r is the nonlinear equivalence of pole excess. The variables η will not appear if $r = n$. This case corresponds to a system without zeros. This actually occurs in Example 5.15, where $r = n = 2$.

A design procedure, which is the nonlinear analog of pole placement, can be constructed if $\beta(\xi,\eta) \neq 0$. If this is the case, we can introduce the feedback law

$$u = \frac{1}{\beta(\xi,\eta)}\left(-a_r \xi_1 - a_{r-1}\xi_2 - \ldots - a_1 \xi_r - \alpha(\xi,\eta) + b_0 v\right)$$

The closed-loop system then becomes

$$\frac{d\xi}{dt} = \begin{pmatrix} 0 & 1 & 0 & \cdots & 0 \\ 0 & 0 & 1 & \cdots & \\ & \vdots & & & \\ -a_r & -a_{r-1} & -a_{r-2} & \cdots & -a_1 \end{pmatrix} \xi + \begin{pmatrix} 0 \\ 0 \\ \vdots \\ b_0 \end{pmatrix} v \tag{5.86}$$

$$\frac{d\eta}{dt} = \gamma(\xi, \eta)$$

The relation between v and ξ_1 is given by a *linear* dynamical system with the transfer function

$$\frac{\Xi_1(s)}{V(s)} = G(s) = \frac{b_0}{s^r + a_1 s^{r-1} + \ldots a_r}$$

This differential equation has a triangular structure. The part corresponding to the state vector ξ is a linear system that is decoupled from the variable η. If $\xi = 0$, the behavior of the system (5.86) is governed by

$$\frac{d\eta}{dt} = \gamma(0, \eta) \tag{5.87}$$

This equation represents the *zero dynamics*. It is necessary for this system to be stable if the proposed control design is going to work. For linear systems the zero dynamics are the dynamics associated with the zeros of the transfer function. Feedback linearization is the nonlinear analog of pole placement with cancellation of all process zeros.

Adaptive Feedback Linearization

We now show how feedback linearization can be extended to deal with the situation in which the process model has unknown parameters. The approach will be similar to the idea used to derive model-reference adaptive controllers. Let us start with an example that is an adaptive version of Example 5.15.

EXAMPLE 5.16   Adaptive feedback linearization

Consider the system

$$\frac{dx_1}{dt} = x_2 + \theta f(x_1)$$

$$\frac{dx_2}{dt} = u$$

where θ is an unknown parameter and f is a known differentiable function. Applying the certainty equivalence principle gives the following control law:

$$u = -a_2 x_1 - \left(a_1 + \hat{\theta} f'(x_1) \right) \left(x_2 + \hat{\theta} f(x_1) \right) + v \tag{5.88}$$

Introducing this into the system equations gives an error equation that is non-linear in the parameter error. This makes it very difficult to find a parameter adjustment law that gives a stable system. Therefore it is necessary to use another approach.

Proceeding as in Example 5.15 and introducing the new coordinates

$$\xi_1 = x_1$$
$$\xi_2 = x_2 + \hat{\theta} f(x_1)$$

where $\hat{\theta}$ is an estimate of θ, we have

$$\frac{d\xi_1}{dt} = \frac{dx_1}{dt} = x_2 + \theta f(x_1) = \xi_2 + (\theta - \hat{\theta}) f(\xi_1)$$

$$\frac{d\xi_2}{dt} = \frac{d\hat{\theta}}{dt} f(x_1) + \hat{\theta}\left(x_2 + \theta f(x_1)\right) f'(x_1) + u$$

Choosing the control law to be

$$u = -a_2\xi_1 - a_1\xi_2 - \hat{\theta}\left(x_2 + \hat{\theta}f(x_1)\right)f'(x_1) - f(x_1)\frac{d\hat{\theta}}{dt} + v \tag{5.89}$$

we get

$$\frac{d\xi}{dt} = \begin{pmatrix} 0 & 1 \\ -a_2 & -a_1 \end{pmatrix}\xi + \begin{pmatrix} f(\xi_1) \\ \hat{\theta}f(\xi_1)f'(\xi_1) \end{pmatrix}\tilde{\theta} + \begin{pmatrix} 0 \\ 1 \end{pmatrix}v$$

A comparison with the certainty equivalence control law given by Eq. (5.88) shows that the major difference is the presence of the term $d\hat{\theta}/dt$ in Eq. (5.89).

In analogy with the model-reference adaptive system, let us assume that it is desired to have a system in which the transfer function from command signal to output has the transfer function

$$G(s) = \frac{a_2}{s^2 + a_1 s + a_2}$$

Introduce the following realization of the transfer function:

$$\frac{dx_m}{dt} = \begin{pmatrix} 0 & 1 \\ -a_2 & -a_1 \end{pmatrix}x_m + \begin{pmatrix} 0 \\ a_2 \end{pmatrix}u_m$$

and let $e = \xi - x_m$ be the error vector. If we choose

$$v = a_2 u_m \tag{5.90}$$

we find that the error equation becomes

$$\frac{de}{dt} = \begin{pmatrix} 0 & 1 \\ -a_2 & -a_1 \end{pmatrix}e + \begin{pmatrix} f(\xi_1) \\ \hat{\theta}f(\xi_1)f'(\xi_1) \end{pmatrix}\tilde{\theta} = Ae + B\tilde{\theta}$$

where

$$A = \begin{pmatrix} 0 & 1 \\ -a_2 & -a_1 \end{pmatrix} \qquad B = \begin{pmatrix} f(\xi_1) \\ \hat{\theta}f(\xi_1)f'(\xi_1) \end{pmatrix}$$

The matrix A has all eigenvalues in the left half-plane if $a_1 > 0$ and $a_2 > 0$. It is then possible to find a matrix P such that

$$A^T P + PA = -I$$

Choosing the Lyapunov function

$$V = e^T Pe + \frac{1}{\gamma} \tilde{\theta}^2$$

we find that

$$\frac{dV}{dt} = e^T (A^T P + PA)e + 2\tilde{\theta} B^T Pe + \frac{2}{\gamma} \tilde{\theta} \frac{d\tilde{\theta}}{dt}$$

If the law for updating the parameters is chosen to be

$$\frac{d\hat{\theta}}{dt} = \gamma B^T Pe$$

we find that

$$\frac{d\tilde{\theta}}{dt} = \frac{d}{dt}(\theta - \hat{\theta}) = -\frac{d\hat{\theta}}{dt} = -\gamma B^T Pe$$

and the derivative of the Lyapunov function becomes

$$\frac{dV}{dt} = -e^T e$$

This function is negative as long as any component of the error vector is different from zero. With the control law given by (5.89) and (5.90) the tracking error will thus always go to zero.   □

Backstepping

Unfortunately, adaptive feedback linearization cannot be applied to all systems that can be linearized by feedback. The reason is that higher derivatives of the parameter estimate will appear in the control law for systems of higher order. There is, however, another nonlinear design technique called *backstepping* that can be used. We first introduce this method and later show how it can be used for adaptive control. In feedback linearization we introduced new state variables and a nonlinear feedback so that the equations describing the transformed variables had a particular structure. A similar idea is used in backstepping, but the transformed equations have a different form. To show the key ideas without too many technical complications, we consider a simple stabilization problem. To simplify the writing, we frequently drop the arguments of functions.

EXAMPLE 5.17   **Stabilization by backstepping**

Consider the system described by

$$\frac{dx_1}{dt} = x_2 + f(x_1)$$

$$\frac{dx_2}{dt} = x_3 \qquad\qquad (5.91)$$

$$\frac{dx_3}{dt} = u$$

Introduce $\xi_1 = x_1$. Then

$$\frac{d\xi_1}{dt} = x_2 + f(\xi_1) = -\xi_1 + x_2 + \xi_1 + f(\xi_1)$$

If we introduce the function

$$a_1(\xi_1) = \xi_1 + f(\xi_1)$$

and the state variable

$$\xi_2 = x_2 + a_1(\xi_1) \qquad\qquad (5.92)$$

the differential equation for ξ_1 can be written as

$$\frac{d\xi_1}{dt} = -\xi_1 + \xi_2$$

The derivative of the variable ξ_2 is given by

$$\frac{d\xi_2}{dt} = \frac{dx_2}{dt} + \frac{da_1}{d\xi_1}(-\xi_1 + \xi_2) = -\xi_2 + x_3 + \xi_2 + \frac{da_1}{d\xi_1}(-\xi_1 + \xi_2)$$

If we introduce the function

$$a_2(\xi_1, \xi_2) = \xi_2 + \frac{da_1}{d\xi_1}(-\xi_1 + \xi_2)$$

and the state variable

$$\xi_3 = x_3 + a_2(\xi_1, \xi_2)$$

the differential equation for ξ_2 can be written as

$$\frac{d\xi_2}{dt} = -\xi_2 + \xi_3$$

Taking derivatives of ξ_3 and using Eqs. (5.91), we find that

$$\frac{d\xi_3}{dt} = u + \frac{\partial a_2}{\partial \xi_1}(-\xi_1 + \xi_2) + \frac{\partial a_2}{\partial \xi_2}(-\xi_2 + \xi_3)$$

Introducing the function

$$a_3(\xi_1, \xi_2, \xi_3) = \xi_3 + \frac{\partial a_2}{\partial \xi_1}(-\xi_1 + \xi_2) + \frac{\partial a_2}{\partial \xi_2}(-\xi_2 + \xi_3)$$

we find that the differential equation for ξ_3 can be written as

$$\frac{d\xi_3}{dt} = -\xi_3 + a_3(\xi_1, \xi_2, \xi_3) + u$$

The feedback

$$u = -a_3(\xi_1, \xi_2, \xi_3)$$

gives the closed-loop system described by

$$\frac{d\xi}{dt} = \begin{pmatrix} -1 & 1 & 0 \\ 0 & -1 & 1 \\ 0 & 0 & -1 \end{pmatrix} \xi \tag{5.93}$$

This system is clearly stable, and its state ξ goes to zero exponentially. Notice that by a slight modification of the procedure we can have any number in the diagonal of the system matrix.

The transformation was obtained recursively. Notice that if the variable x_2 was a control variable that could be chosen freely, the "control law"

$$x_2 = -a_1(\xi_1)$$

would give

$$\frac{d\xi_1}{dt} = -\xi_1$$

The state variable ξ_2 defined by Eq. (5.92) can thus be interpreted as the difference between x_2 and the "stabilizing feedback" $-a_1(\xi_1)$.

Similarly, if x_3 was a control variable that could be chosen freely, the "control law"

$$x_3 = -a_2(\xi_1, \xi_2)$$

would give the closed-loop system

$$\frac{d\xi_1}{dt} = -\xi_1 + \xi_2$$

$$\frac{d\xi_2}{dt} = -\xi_2$$

The state variable ξ_3 can be interpreted as the difference between x_3 and the "stabilizing feedback" $-a_2(\xi_1, \xi_2)$.

The procedure was originally derived by applying this reasoning recursively, and the name "backstepping" derives from this.

In the example the system was transformed to a triangular form given by Eq. (5.93). There are many other possibilities. □

Adaptive Backstepping

The key idea of backstepping is to derive an error equation and to construct a control law and a parameter adjustment law such that the state of the error equation goes to zero. The idea is illustrated by a simple example.

EXAMPLE 5.18   **Adaptive stabilization by backstepping**

Consider the system

$$\frac{dx_1}{dt} = x_2 + \theta f(x_1)$$

$$\frac{dx_2}{dt} = x_3$$

$$\frac{dx_3}{dt} = u$$

where f is a known function and θ an unknown parameter. We derive a control law that stabilizes the system when the parameter θ is unknown. Introduce a new state variable $\xi_1 = x_1$. We write the derivative of ξ_1 as a sum of terms in which one of them depends on known quantities only. For this purpose we introduce the parameter estimate $\hat{\theta}$ and the error $\tilde{\theta} = \theta - \hat{\theta}$. The derivative of ξ_1 then becomes

$$\frac{d\xi_1}{dt} = -\xi_1 + \xi_1 + x_2 + \hat{\theta}f(\xi_1) + \tilde{\theta}f(\xi_1)$$

Introduce the next state variable ξ_2 as

$$\xi_2 = x_2 + a_1(\xi_1, \hat{\theta})$$

where

$$a_1(\xi_1, \hat{\theta}) = \xi_1 + \hat{\theta}f(\xi_1) \tag{5.94}$$

The differential equation for ξ_1 can then be written as

$$\frac{d\xi_1}{dt} = -\xi_1 + \xi_2 + \tilde{\theta}f \tag{5.95}$$

We now proceed to rewrite the derivative of ξ_2 as a sum of two terms in which the first depends only on ξ_1, ξ_2, and $\hat{\theta}$. Hence

$$\frac{d\xi_2}{dt} = \frac{dx_2}{dt} + \frac{\partial a_1}{\partial \xi_1} \cdot \frac{d\xi_1}{dt} + \frac{\partial a_1}{\partial \hat{\theta}} \cdot \frac{d\hat{\theta}}{dt}$$

Equation (5.95) gives the desired separation of terms in $d\xi_1/dt$. Some work is required to obtain a similar expression for $d\hat{\theta}/dt$. We have

$$\frac{d\xi_2}{dt} = x_3 + \frac{\partial a_1}{\partial \xi_1}\left(-\xi_1 + \xi_2 + \tilde{\theta}f\right) + \frac{\partial a_1}{\partial \hat{\theta}} \cdot \frac{d\hat{\theta}}{dt} \tag{5.96}$$

Following the idea of backstepping, we consider x_3 to be a control variable that can be chosen freely. The Lyapunov function

$$2V = \xi_1^2 + \xi_2^2 + \tilde{\theta}^2$$

can be used to find a control law and an adaptation law that stabilizes the error equation for variables ξ_1 and ξ_2. After some calculations we find that the derivative of V is given by

$$\frac{dV}{dt} = -\xi_1^2 + \xi_1\xi_2 + x_3\left(\xi_2 + \frac{\partial a_1}{\partial\hat{\theta}}\frac{d\hat{\theta}}{dt}\right) + \tilde{\theta}\left(\xi_1 f + \xi_2 \frac{\partial a_1}{\partial\xi_1}f(\xi_1) - \frac{d\hat{\theta}}{dt}\right)$$

The term containing $\tilde{\theta}$ can be eliminated by choosing

$$\frac{d\hat{\theta}}{dt} = b_2(\xi_1,\xi_2)$$

where

$$b_2 = \xi_1 f(\xi_1) + \xi_2 \frac{\partial a_2}{\partial\xi_1} f(\xi_1) \tag{5.97}$$

The function $b_2(\xi_1,\xi_2)$ can be interpreted as a good way to choose the parameter update rate $d\hat{\theta}/dt$ based on ξ_1 and ξ_2. The "control variable" x_3 can be chosen to give

$$\frac{dV}{dt} = -\xi_1^2 - \xi_2^2$$

Using b_2 as an estimate of $d\hat{\theta}/dt$, we now rewrite Eq. (5.96) as

$$\frac{d\xi_2}{dt} = -\xi_1 - \xi_2 + x_3 + \xi_1 + \xi_2 + \frac{\partial a_1}{\partial\xi_1}\left(-\xi_1 + \xi_2 + \tilde{\theta}f\right)$$

$$+ \frac{\partial a_1}{\partial\hat{\theta}}b_2 + \frac{\partial a_1}{\partial\hat{\theta}}\left(\frac{d\hat{\theta}}{dt} - b_2\right) \tag{5.98}$$

Now define

$$a_2\left(\xi_1,\xi_2,\hat{\theta}\right) = \xi_1 + \xi_2 + \frac{\partial a_1}{\partial\xi_1}\left(-\xi_1 + \xi_2\right) + \frac{\partial a_1}{\partial\hat{\theta}}b_2 \tag{5.99}$$

and introduce the state variable ξ_3 as

$$\xi_3 = x_3 + a_2(\xi_1,\xi_2,\hat{\theta})$$

The differential equation (5.98) can be written as

$$\frac{d\xi_2}{dt} = -\xi_1 - \xi_2 + \xi_3 + \frac{\partial a_1}{\partial\xi_1}\tilde{\theta}f + \frac{\partial a_1}{\partial\hat{\theta}}\left(\frac{d\hat{\theta}}{dt} - b_2\right) \tag{5.100}$$

The derivative of ξ_3 becomes

$$\frac{d\xi_3}{dt} = u + \frac{\partial a_2}{\partial\xi_1}\cdot\frac{d\xi_1}{dt} + \frac{\partial a_2}{\partial\xi_2}\cdot\frac{d\xi_2}{dt} + \frac{\partial a_2}{\partial\hat{\theta}}\cdot\frac{d\hat{\theta}}{dt} \tag{5.101}$$

Notice that the control variable u now appears explicitly on the right-hand side. In the stabilization problem the error is equal to the vector ξ and the error equation is obtained by combining Eqs. (5.95), (5.100), and (5.101). Following the general MRAS approach, we now attempt to find a feedback law and a parameter adjustment rule that stabilizes the error equation. Choosing

$$2V = \xi_1^2 + \xi_2^2 + \xi_3^2 + \tilde{\theta}^2$$

as a possible Lyapunov function, we get, after straightforward but tedious calculations,

$$
\frac{dV}{dt} = -\xi_1^2 - \xi_2^2 + \xi_2\xi_3 + \xi_2 \frac{\partial a_1}{\partial \hat{\theta}} \left(\frac{d\hat{\theta}}{dt} - b_2 \right)
$$
$$
+ \xi_3 \left(u + \frac{\partial a_1}{\partial \hat{\theta}} \left(\frac{d\hat{\theta}}{dt} - b_2 \right) + \frac{\partial a_2}{\partial \hat{\theta}} \frac{d\hat{\theta}}{dt} \right)
$$
$$
+ \tilde{\theta} \left(\xi_1 f + \xi_2 \frac{\partial a_1}{\partial \xi_1} f + \xi_3 \left(\frac{\partial a_2}{\partial \xi_1} + \frac{\partial a_1}{\partial \xi_1} \frac{\partial a_2}{\partial \xi_2} \right) f - \frac{d\hat{\theta}}{dt} \right)
$$

The term that contains $\tilde{\theta}$ can be eliminated by updating the parameters in the following way:

$$
\frac{d\hat{\theta}}{dt} = \xi_1 f + \xi_2 \frac{\partial a_1}{\partial \xi_1} f + c(\xi_1, \xi_2)\xi_3 \tag{5.102}
$$

where

$$
c(\xi_1, \xi_2) = \left(\frac{\partial a_2}{\partial \xi_1} + \frac{\partial a_1}{\partial \xi_1} \frac{\partial a_2}{\partial \xi_2} \right) f
$$

Furthermore, introducing

$$
b_3(\xi_1, \xi_2, \xi_3) = b_2 + c\xi_3
$$

and

$$
a_3 = \xi_2 + \xi_3 + \frac{\partial a_2}{\partial \xi_1}(-\xi_1 + \xi_2) + \frac{\partial a_2}{\partial \xi_2} \left(-\xi_1 - \xi_2 + \xi_3 - \xi_3^2 \frac{\partial a_1}{\partial \hat{\theta}} c \right) + \frac{\partial a_2}{\partial \hat{\theta}} b_3
$$

we find that

$$
\frac{d\hat{\theta}}{dt} - b_2 = c\xi_3
$$

The derivative of the Lyapunov function can then be written as

$$
\frac{dV}{dt} = -\xi_1^2 - \xi_2^2 - \xi_3^2 + \xi_3(u + a_3)
$$

The feedback law

$$
u = -a_3(\xi_1, \xi_2, \xi_3) \tag{5.103}
$$

gives

$$
\frac{dV}{dt} = -\xi_1^2 - \xi_2^2 - \xi_3^2
$$

and we find that dV/dt is negative as long as $|\xi| \neq 0$.   □

Summary

The examples given should give some of the flavor of nonlinear adaptive control. The results obtained depend on clever changes of coordinates. A reasonable characterization of the class of systems in which the methods apply is not available. Nevertheless, we can make some interesting observations from the examples. First, we can notice that the adaptive control laws that are obtained differ significantly from those obtained from the certainty equivalence principle. In the nonlinear approaches the control law and the rule for updating the parameters are obtained simultaneously. An estimate of the rate of change of the parameters appears in the feedback law. Many problems remain to be solved.

5.11 CONCLUSIONS

The fundamental ideas behind the MRAS have been covered in this chapter, including

- Gradient methods,
- Lyapunov and passivity design, and
- Augmented error.

In all cases the rule for updating the parameters is of the form

$$\frac{d\theta}{dt} = \gamma \varphi \varepsilon$$

or, in the normalized form,

$$\frac{d\theta}{dt} = \gamma \frac{\varphi \varepsilon}{\alpha + \varphi^T \varphi}$$

In the gradient method the vector φ is the negative gradient of the error with respect to the parameters. Estimation of parameters or approximations may be needed to obtain the gradient. In other cases, φ is a regression vector, which is found by filtering inputs, outputs, and command signals. The quantity ε is the augmented error, which also can be interpreted as the prediction error of the estimation problem. It is customary to use an augmented error that is linear in the parameters.

The gradient method is flexible and simple to apply to any system structure. The calculations required are the determination of the sensitivity derivative. Since the sensitivity derivative cannot be obtained for an unknown process, it is necessary to make several approximations. The initial values of the parameters must be such that the closed-loop system is stable. Empirical evidence indicates that the system is stable for small adaptation gains but that high gains lead to instability. It is difficult to find the bounds. In Chapter 6 we give more insight into the properties of the gradient method.

A general MRAS is derived in Section 5.8 on the basis of the model-following design in Chapter 3. This algorithm includes as special cases many of the MRAS designs given in the literature. The estimation of the parameters can be done in several ways other than those given in Eqs. (5.62) and (5.63). Various modifications are discussed in Chapter 6.

PROBLEMS

5.1 Consider the process

$$G(s) = \frac{1}{s(s+a)}$$

where a is an unknown parameter. Determine a controller that can give the closed-loop system

$$G_m(s) = \frac{\omega^2}{s^2 + 2\zeta\omega s + \omega^2}$$

Determine model-reference adaptive controllers based on gradient and stability theory, respectively. (Compare Problem 3.2.)

5.2 Consider the simple MRAS in Fig. 5.4 with $G = 1/s$. Let the parameter adjustment law be Eq. (5.57) (i.e., of PI type). Determine the differential equation for θ, and discuss how γ_1 and γ_2 influence the convergence rate.

5.3 Consider a position servo described by

$$\frac{dv}{dt} = -av + bu$$

$$\frac{dy}{dt} = v$$

where parameters a and b are unknown. Assume that the control law

$$u = \theta_1(u_c - y) - \theta_2 v$$

is used and that it is desired to control the system in such a way that the transfer function from command signal to process output is given by

$$G_m(s) = \frac{\omega^2}{s^2 + 2\zeta\omega s + \omega^2}$$

Determine an adaptive control law that adjusts parameter θ_1 and θ_2 so that the desired objective is obtained.

5.4 An integrator

$$G_p(s) = \frac{b}{s}$$

is to be controlled by a zero-order continuous-time controller

$$u(t) = -s_0 y(t) + t_0 u_c(t)$$

The desired response model is given by

$$G_m(s) = \frac{b_m}{s + a_m}$$

Derive, using the Lyapunov theory, a parameter update law of an MRAS guaranteeing that the error $e = y - y_m$ goes to zero. Try the Lyapunov function

$$V(x) = \frac{1}{2}\left(e^2 + \frac{1}{b}\left(bs_0 - a_m\right)^2 + \frac{1}{b}\left(bt_0 - b_m\right)^2\right)$$

where

$$e(t) = y(t) - y_m(t)$$

5.5 Consider the problem of adaptation of a feedforward gain in Example 5.1 when

$$G(s) = \frac{1}{(s + 1)(s + 2)}$$

(a) Introduce the augmented error, and determine an MRAS based on stability theory.

(b) Show that the derived adaptation law in part (a) gives a stable closed-loop system.

5.6 Determine conditions in which a second-order transfer function

$$G(s) = \frac{b_0 s^2 + b_1 s + b_2}{s^2 + a_1 s + a_2}$$

is strictly positive real.

5.7 Show that $B(s)/A(s)$ is SPR if $A(s)$ is a stable polynomial and the B polynomial is the first row of the P-matrix defined by the Lyapunov equation

$$A^T P + PA = -Q$$

where the matrix A is

$$A = \begin{pmatrix} -a_1 & -a_2 & \cdots & -a_{n-1} & -a_n \\ 1 & 0 & & 0 & 0 \\ \vdots & & \ddots & & \\ 0 & 0 & & 1 & 0 \end{pmatrix}$$

and Q is a symmetric positive definite matrix. Show that the system of equations for solving p_1, p_2, and p_3 in Example 5.6 has a unique solution only if all the eigenvalues of A are in the left half-plane.

5.8 Show that the transfer function

$$G(s) = 1 + s$$

is SPR and ISP but not OSP.

5.9 Show that the transfer function

$$G(s) = \frac{1}{s+1}$$

is SPR and OSP but not ISP.

5.10 Show that the transfer function

$$G(s) = \frac{s^2 + 1}{(s+1)^2}$$

is OSP and ISP but not SPR.

5.11 Consider the system

$$G(s) = G_1(s)G_2(s)$$

where

$$G_1(s) = \frac{b}{s+a}$$
$$G_2(s) = \frac{c}{s+d}$$

where a and b are unknown parameters and c and d are known. Discuss how to make an MRAS based on the gradient approach. (Compare Problem 3.3.) Let the desired model be described by

$$G_m(s) = \frac{\omega^2}{s^2 + 2\zeta\omega s + \omega^2}$$

5.12 A process has the transfer function

$$G(s) = \frac{b}{s(s+1)}$$

where b is a time-varying parameter. The system is controlled by a proportional controller

$$u(t) = k\left(u_c(t) - y(t)\right)$$

It is desirable to choose the feedback gain so that the closed-loop system has the transfer function

$$G(s) = \frac{1}{s^2 + s + 1}$$

Design an MRAS that gives the desired result, and investigate the system by simulation. (Compare Problem 3.4.)

5.13 The general MRAS procedure in Section 5.8 was derived for known instantaneous gain b_0. If b_0 is unknown, we may use the following augmented error:

$$\varepsilon = \frac{Q}{A_0 A_m}\left(\left(b_0 - \hat{b}_0\right)\left(\varphi^T \theta + \frac{u}{P_1}\right) + b_0 \varphi^T \left(\theta - \theta^0\right)\right)$$

where $\hat{b}_0$ is the estimate of b_0. Discuss how this augmented error can be
obtained and how it may be used to update the parameters b_0 and θ.

5.14 Study the parameter adjustment law in Example 5.2. Make a simulation
program that implements the adaptive system. Repeat the simulation in
Fig. 5.5. Investigate the behavior of the parameters and the error. Explore
how the behavior is influenced by the adaptation gain γ.

5.15 Repeat the simulation in Problem 5.4 for different types of input signals.
Change the amplitude and the nature of the signals. Can you find values
of the adaptation gain that work well for different inputs?

5.16 Consider the system in Example 5.5. Assume that u_c is a step that implies
that y_m will be time-varying. Investigate by analysis or simulate the
stability limit and compare with the limit obtained in the example, in
which u_c and y_m were constant.

5.17 Consider a first-order system with the transfer function

$$G(s) = \frac{b}{s + a}$$

where a and b are unknown parameters. Assume that the system is
controlled by the control law

$$u = \theta_1 u_c - \theta_2 y$$

Compare by simulation the properties of the systems obtained with the
MIT rule and the one derived from Lyapunov theory. Use the same pa-
rameter values as in Example 5.2. (*Hint*: The algorithms are given in
Examples 5.2 and 5.7.)

5.18 Investigate the properties of the system in Example 5.7 by simulation.

5.19 Investigate through simulation the convergence rate of the parameters
in Example 5.2 when the control law of Eqs. (5.9) is used. How will the
parameter adjustment change if an adaptation rule based on stability
theory is used? For instance, plot the phase plane for the parameters.

5.20 Consider the process

$$G(s) = \frac{50}{s(s + 4)}$$

and the criterion

$$\int_0^\infty \left((y - u_c)^2 + \rho u^2\right) dt$$

Let the control law have the form

$$u(t) = -s_o(y - u_c)$$

or

$$u(t) = -\frac{s_o p + s_1}{p + r_1}(y - u_c)$$

Determine the controller parameters through explicit minimization of the criterion, and let the gradients be obtained from an estimated model of the process. (*Hint*: See Trulsson and Ljung, 1985.)

5.21 Consider the system in Example 5.14. Figure 5.22(c) shows the rapid decrease in the error, while the parameters converge much more slowly. Explain the slow parameter convergence by analyzing the sensitivity of the closed-loop poles with respect to the estimated parameters.

5.22 Consider a system described by

$$G(s) = \frac{b}{s^2 + a}$$

where a and b are unknown parameters. Find a simple control law that can control the plant well, and derive an adaptive algorithm that gives good performance.

REFERENCES

The model-reference approach was developed by Whitaker and his colleagues around 1958. One early reference giving the basic ideas using the gradient method is:

Osburn, P. V., H. P. Whitaker, and A. Kezer, 1961. "New developments in the design of adaptive control systems." Paper No 61-39, February 1961, Institute of Aeronautical Sciences.

The problem with stability of the gradient method was analyzed by using Lyapunov stability theory in:

Butchart, R. L., and B. Shackcloth, 1965. "Synthesis of model reference adaptive control systems by Lyapunov's second method." *Proceedings of the 1965 IFAC Symposium on Adaptive Control.* Teddington, U.K.

and explored further in:

Parks, P. C., 1966. "Lyapunov redesign of model reference adaptive control systems." *IEEE Trans. Automat. Contr.* **AC-11**: 362–365.

The different approaches to MRAS are treated in:

Landau, Y. D., 1979. *Adaptive Control: The Model Reference Approach.* New York: Marcel Dekker.

Parks, P. C., 1981. "Stability and convergence of adaptive controllers: Continuous systems." In *Self-tuning and Adaptive Control: Theory and Applications*, eds. C. J. Harris and S. A. Billings. Stevenage, U.K.: Peter Peregrinus.

A comparison of the Lyapunov and the input-output stability approaches is given in:

Narendra, K. S., and L. S. Valavani, 1980. "A comparison of Lyapunov and hyperstability approaches to adaptive control of continuous systems." *IEEE Trans. Automat. Contr.* **AC-25**: 243–247.

The augmented error method was introduced in:

Monopoli, R. V., 1974. "Model reference adaptive control with an augmented error signal." *IEEE Trans. Automat. Contr.* **AC-19**: 474–484.

A unification of MRAS and self-tuning controllers is found in:

Egardt, B., 1979. "Unification of some continuous-time adaptive control schemes." *IEEE Trans. Automat. Contr.* **AC-24**: 588–592.

Stability of continuous-time MRAS is discussed in:

Morse, A. S., 1980. "Global stability of parameter-adaptive control systems." *IEEE Trans. Automat. Contr.* **AC-25**: 433–439.

Goodwin, G. C., and D. Q. Mayne, 1987. "A parameter estimation perspective of continuous time model reference adaptive control." *Automatica* **23**: 57–70.

The main problem in the stability analysis of adaptive controllers is the boundedness of the variables of the system. Proofs of boundedness and stability are found in:

Egardt, B., 1979. *Stability of Adaptive Controllers.* Lecture notes in Control and Information Sciences, vol. 20. Berlin: Springer-Verlag.

Narendra, K. S., A. M. Annaswamy, and R. P. Singh, 1985. "A general approach to the stability analysis of adaptive systems." *Int. J. Control* **41**: 193–216.

The error model plays an important role in the analysis of the MRAS. Different generic error models are discussed in:

Narendra, K. S., and Y.-H. Lin, 1980. "Design of stable model reference adaptive controllers." In *Applications of Adaptive Control*, eds. K. S. Narendra and R. V. Monopoli. New York: Academic Press.

PI adjustment of the parameters in the MRAS is discussed in Landau (1979) above and in:

Hang, C. C., and P. C. Parks, 1973. "Comparative studies of model reference adaptive control systems." *IEEE Trans. Automat. Contr.* **AC-18**: 419–428.

Textbooks on MRAS are, for instance, Landau (1979) and:

Narendra, K. S., and A. M. Annaswamy, 1989. *Stable Adaptive Systems.* Englewood Cliffs, N.J.: Prentice-Hall.

Sastry, S., and M. Bodson, 1989. *Adaptive Control: Stability, Convergence and Robustness.* Englewood Cliffs, N.J.: Prentice-Hall.

Lyapunov theory and passivity theory are basic tools for the stability analysis. Some general references are:

Hahn, W., 1967. *Stability of Motion*. Berlin: Springer-Verlag.

Popov, V. M., 1973. *Hyperstability of Control Systems*. Berlin: Springer-Verlag.

Vidyasagar, M., 1978. *Nonlinear Systems Analysis*. Englewood Cliffs, N.J.: Prentice-Hall.

Vidyasagar, M., 1986. "New directions of research in nonlinear system theory." *Proceedings IEEE* **74**: 1060–1091.

Slotine, J.-J. E., and W. Li, 1991. *Applied Nonlinear Control*. Englewood Cliffs, N.J.: Prentice-Hall.

Khalil, H. K., 1992. *Nonlinear Systems*. New York: Macmillan.

Early work on passivity and input-output stability was done by Popov and a little later by Zames and Sandberg. The theory is summarized in:

Desoer, C. A., and M. Vidyasagar, 1975. *Feedback Systems: Input-output Properties*. New York: Academic Press.

A proof of the Popov-Kalman-Yakubovich lemma is given in:

Lefschetz, S., 1965. *Stability of Nonlinear Control Systems*, pp. 114–118. New York: Academic Press.

Discrete-time MRAS is discussed in detail in Egardt (1979) and Landau (1979), above.

The explicit criterion minimization approach to adaptive control can be found in:

Tsypkin, Y. Z., 1971. *Adaptation and Learning in Automatic Systems*. New York: Academic Press.

Goodwin, G. C., and P. J. Ramadge, 1979. "Design of restricted complexity adaptive regulators." *IEEE Trans. Automat. Contr.* **AC-24**: 584–588.

Trulsson, E., and L. Ljung, 1985. "Adaptive control based on explicit criterion minimization." *Automatica* **21**: 385–399.

The backstepping method was invented by Kokotovic and his students. An overview of the method is given in the 1991 Bode Lecture; see:

Kokotovic, P. V., 1992. "The joy of feedback: nonlinear and adaptive control." *IEEE Control Systems Magazine* **12**(3): 7–17.

Additional details are found in:

Kokotovic, P. V., ed., 1991. *Foundations of Adaptive Control*. Berlin: Springer-Verlag.

Kokotovic, P. V., I. Kanellakopoulos, and M. Krstić, 1992. "On letting adaptive control be what it is: nonlinear feedback." *Proceedings of the IFAC Symposium on Adaptive Control and Adaptive Signal Processing*. Grenoble, France.

Kokotovic, P. V., M. Krstić, and I. Kanellakopoulos, 1992. "Backstepping to passivity: recursive design of adaptive systems." *Proceedings of the IEEE Conference on Decision and Control*, pp. 3276–3280. Tucson, Arizona.

CHAPTER 6

PROPERTIES OF ADAPTIVE SYSTEMS

6.1 INTRODUCTION

Some theoretical problems were discussed in earlier chapters in connection with description or derivation of specific algorithms. In particular we used equilibrium analysis to analyze the self-tuners and stability theory to derive some model-reference algorithms. In this chapter we attempt to bring together theory of a more general character. The theory has several different goals:

- To present some mathematical tools that are useful in analysis of adaptive systems.
- To analyze the behavior of adaptive systems in nonideal cases.
- To give ideas for new algorithms and for improvement of old algorithms.

In this chapter we focus on the first two issues. The behavior of specific algorithms can be understood through analysis of stability, convergence, and performance. Stability proofs require certain assumptions. It is also of considerable interest to understand what happens when the assumptions are violated. Analysis of performance may give useful insight into performance limits; it is helpful to know whether the performance of a particular algorithm is close to the theoretical limits. A good theory should also give clues to the construction of new algorithms.

Unfortunately, there is no collection of results that can be called a theory of adaptive control in the sense specified. There is instead a scattered body of results, which gives only partial results. One reason for this is that the behavior of adaptive systems is quite complex because of their time-varying and nonlinear character. Readers who are familiar only with linear systems theory, in which most problems can be answered in great detail, should thus be warned.

The closed-loop systems obtained with adaptive control are nonlinear and sometimes also stochastic. Such systems are also very difficult to analyze. To obtain some insight with a reasonable effort, it is therefore necessary to make some simplifications. It is often possible to analyze the equilibrium conditions. The local behavior in the neighborhood of the equilibria can also be explored by using linearization. The global behavior of the systems can, however, be very complex, particularly if the design parameters are chosen badly.

In Section 6.2 we show that the adaptive control problem has a special nonlinear structure that can be exploited in the analysis. We first show that very complex, even chaotic, behavior can be observed if the adaptation gain is chosen to be too high.

Section 6.3 presents an analysis of a system with adaptation of a feedforward gain. Such systems can be described by linear time-varying systems in which the time variation originates from the command signal. The particular case of periodic variations can be dealt with by so-called Floquet theory. The analysis reveals that very complex behavior can be obtained even in this simple case.

The properties of indirect discrete-time adaptive systems are investigated in Section 6.4. In this case it is natural to investigate parameter estimation and the control design separately. There is interaction between these problems because the identification is done in closed loop and the control design influences the signals generated by feedback. The analysis brings out the importance of persistency of excitation and the dangers with singularities in the control design. A consequence of this is that it is desirable to have as few parameters as possible and to have external excitation. In Section 6.5 we make a similar analysis of the direct algorithm. One of the conditions required for the proof is that the complexity of the model used must be at least as complex as the process to be controlled. A characteristic feature of direct adaptive algorithms is that the closed-loop behavior can converge to the desired behavior even if the parameters do not converge.

It is reasonable to assume that if the adaptation rate is small, the parameter estimates will change more slowly than the other variables in a system. The closed-loop system can then be viewed as having different time scales. This has been emphasized in the descriptions of both the self-tuning regulator and the model reference adaptive controller. The analysis can then be simplified by considering the slow and fast modes separately. Averaging analysis is a good analytical tool for this. A short presentation of this technique is given in Section 6.6. A significant advantage of the averaging technique is that it makes it possible to reduce the dimensionality of the problem to the number of parameters in the algorithm. The averaging method also makes it possible to explore the behavior in detail. A drawback of the averaging results is that they hold for small adaptation gains but the theory does not give quantitative results about smallness.

It is a characteristic feature of feedback that a controller can often be designed by using a simplified model of a real process. This is one of the reasons why automatic control has been so successful in applications. So far, we have analyzed the behavior of some adaptive algorithms under the simplifying assumption that the structure of the process is the same as the model used to design the adaptive controller. Having obtained the tool of averaging, we are in a position to investigate the consequences of the simplifying assumptions made in the earlier sections, and we can explore how adaptive systems behave in the presence of unmodeled dynamics, that is, when the order of the process is different from that of the model used to derive the adaptive controller. Analysis of several examples in Section 6.7 leads us to various modifications of the algorithms that will improve their robustness to unmodeled dynamics.

In Section 6.8 we show that averaging techniques can be used to analyze stochastic self-tuning regulators. The equilibrium points of the algorithms and their local behavior can often be obtained without too much effort. In Section 6.9, different ways are discussed to make the adaptive algorithms robust with respect to the assumptions made in the idealized cases.

6.2 NONLINEAR DYNAMICS

We have mentioned several times that adaptive systems are inherently non-linear. A natural approach to understand the behavior of adaptive systems is thus to use tools from the theory of nonlinear dynamical systems. We first investigate the structure of adaptive systems. This reveals that they have a very special structure. Some tools from dynamical systems are then reviewed briefly, and we apply them to a very simple system. This analysis reveals that adaptive systems behave in the expected way when the adaptation gain is small. However, the behavior can be very complex for large adaptation gains. The analysis also indicates the difficulties involved in the approach. We also investigate the special case of adaptation of a feedforward gain. In this case the problem is simplified significantly because it can be described as a linear time-varying system. A reasonably complete analysis can be performed when the command signal is periodic. This analysis reveals that the system is well behaved for small adaptation gains but that the behavior is quite complex for large adaptation gains.

Structure of Equations Describing Adaptive Systems

Consider a process controlled by an indirect adaptive controller as shown in Fig. 3.1. We will first consider the case in which parameters of a continuous-time model are estimated by using a gradient procedure. Assume that the system to be controlled is linear. Let ϑ denote the controller parameters and v the external driving signals. The signal v is typically composed of the command signal u_c and nonmeasurable disturbances acting on the process. With constant controller parameters the closed-loop system can be written as

$$\frac{d\xi}{dt} = A(\vartheta)\xi + B(\vartheta)v$$

$$\eta = \begin{pmatrix} e \\ \varphi \end{pmatrix} = C(\vartheta)\xi + D(\vartheta)v \tag{6.1}$$

The state vector ξ includes the states of the system, the reference model, the data filter, and the auxiliary state variables that may have to be introduced to calculate the error e and the regression vector φ used in the parameter adjustment mechanism. The vector η consists of the error and the regression vector that are used by the parameter estimator.

Furthermore, let θ denote the process parameters. A normalized gradient scheme for estimating the parameters can be described by

$$\frac{d\hat{\theta}}{dt} = \gamma \frac{\varphi(\vartheta,\xi)e(\vartheta,\xi)}{\alpha + \varphi(\vartheta,\xi)^T \varphi(\vartheta,\xi)} \tag{6.2}$$

The control design can be represented by a nonlinear function $\vartheta = \chi(\hat{\theta})$, which maps the estimated parameters into controller parameters. This map becomes the identity for direct algorithms.

For constant ϑ the system (6.1) is linear. The solution can then also be characterized by the operators G_{ev} and $G_{\varphi v}$, which relate e and φ to v. These operators depend on the controller parameters ϑ. Equation (6.2) can then be written as

$$\frac{d\hat{\theta}}{dt} = \gamma \frac{(G_{\varphi v}v)(G_{ev}v)}{\alpha + (G_{\varphi v}v)^T G_{\varphi v}v}$$

The adaptive system is thus described by Eqs. (6.1) and (6.2), which have a very special structure. Equation (6.1) is linear in the states and the external driving signals. The controller parameters appear in the coefficients of matrices A, B, C, and D. Nonlinearities appear in the product φe in Eq. (6.2), in the design map χ, and in the functions $A(\vartheta)$, $B(\vartheta)$, $C(\vartheta)$, and $D(\vartheta)$ in Eq. (6.1). The equations for an adaptive system have a similar form in the discrete-time case. For a system with recursive least-squares estimation the equations can be written as

$$\xi(t+1) = A(\vartheta)\xi(t) + B(\vartheta)v(t)$$

$$\eta(t) = \begin{pmatrix} e(t) \\ \varphi(t) \end{pmatrix} = C(\vartheta)\xi(t) + D(\vartheta)v(t)$$

$$\hat{\theta}(t+1) = \hat{\theta}(t) + P(t+1)\varphi(t)e(t) \tag{6.3}$$

$$P(t+1) = P(t) - P(t)\varphi(t)(\lambda + \varphi^T(t)P(t)\varphi(t))^{-1}\varphi^T(t)P(t)$$

It is useful to try to exploit the special structure of the equations to get a deeper understanding of adaptive systems. One special feature is that the state of the closed-loop system is naturally separated into two parts, ξ and $\hat{\theta}$. Moreover, it is reasonable to assume that $\hat{\theta}$ changes more slowly than ξ.

Analysis

Let us briefly summarize how a nonlinear system such as Eqs. (6.1) and (6.2) or Eqs. (6.3) can be analyzed. It is a comparatively simple task to find the equilibrium solutions by solving the algebraic equations

$$\frac{d\xi}{dt} = 0$$

$$\frac{d\hat{\theta}}{dt} = 0$$

for continuous-time systems. For discrete-time systems the equivalent equations become

$$\xi(t + 1) = \xi(t)$$

$$\hat{\theta}(t + 1) = \hat{\theta}(t)$$

It may happen that proper equilibria do not exist. In such cases there may be integral manifolds where the parameters $\hat{\theta}$ are constant although the state ξ varies with time. We are then led to averaging analysis, which is discussed in depth in Section 6.6. Equilibria having been found, it is natural to determine the local behavior by linearizing the equations around the equilibria and applying standard linear theory. A complication is that critical cases in which the eigenvalues are zero frequently occur. Having determined possible equilibria, we can proceed to investigate how the nature of the equilibria changes with important parameters of the system. It is of particular interest to investigate changes in which the nature of the local equilibria changes (bifurcation analysis). When the local properties are investigated, it is natural to proceed to find the global properties. There are no general tools for this, and we have to resort to simulations and approximations. Phase plane analysis is useful for two-dimensional systems.

Analysis of a Simple Discrete-Time System

To illustrate how the analysis can be done, we discuss a simple example. Consider a discrete-time adaptive controller that is based on estimation of the parameter θ in the model

$$y(t + 1) = \theta y(t) + u(t) \tag{6.4}$$

Let the controller be

$$u(t) = -\hat{\theta}(t)y(t) + y_0 \tag{6.5}$$

where $\hat{\theta}$ is an estimate of θ and y_0 the setpoint. If the process is indeed described by Eq. (6.4) and if the estimate $\hat{\theta}$ is correct, the controller gives a deadbeat response. The parameter is estimated by using a normalized gradient algorithm

$$\hat{\theta}(t + 1) = \hat{\theta}(t) + \gamma \frac{y(t)\left(y(t+1) - \hat{\theta}(t)y(t) - u(t)\right)}{\alpha + y^2(t)} \tag{6.6}$$

where γ and α are parameters. This is equal to Kaczmarz's projection algorithm if $\gamma = 1$ and $\alpha = 0$.

To analyze the closed-loop system, we must also have a description of the actual process. We assume that this is given by

$$y(t + 1) = \theta_0 y(t) + a + u(t) \tag{6.7}$$

Notice that, because of the presence of the parameter a on the right-hand side, this model is different from the model (6.4) used to design the adaptive controller. Equations (6.4), (6.5), (6.6), and (6.7) thus describe a very simple case of adaptive control of a process with a constant unmodeled disturbance. Using Eq. (6.5) to eliminate u in Eqs. (6.6) and (6.7), we find that the closed-loop system can be described by the equations

$$y(t + 1) = \left(\theta_0 - \hat{\theta}(t)\right) y(t) + a + y_0$$

$$\hat{\theta}(t + 1) = \hat{\theta}(t) + \gamma \frac{y(t)\left(\left(\theta_0 - \hat{\theta}(t)\right)y(t) + a\right)}{\alpha + y^2(t)} \tag{6.8}$$

This is a second-order nonlinear system. To explore the behavior of this system, we follow the procedure of nonlinear analysis.

Equilibrium Analysis   Equations (6.8) have the equilibrium solution

$$y = y_0$$

$$\hat{\theta} = \theta_0 + \frac{a}{y_0} \tag{6.9}$$

Notice that the equilibrium value of the output is always equal to the setpoint in spite of the disturbance. This is a phenomenon that we have observed before in adaptive systems. (Compare with Example 3.5 and Example 5.2.) Unmodeled dynamics, however, give a parameter error.

Linearizing Eqs. (6.8) around the equilibrium equations (6.9), we find that the system matrix is

$$A = \begin{pmatrix} -\dfrac{a}{y_0} & -y_0 \\[2mm] -\gamma\dfrac{a}{\alpha + y_0^2} & 1 - \gamma\dfrac{y_0^2}{\alpha + y_0^2} \end{pmatrix} \tag{6.10}$$

This matrix has the characteristic polynomial

$$z^2 + a_1 z + a_2$$

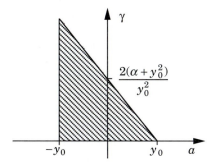

Figure 6.1 Stability region for the closed-loop system.

where

$$a_1 = \frac{a}{y_0} - 1 + \gamma \frac{y_0^2}{\alpha + y_0^2}$$

$$a_2 = -\frac{a}{y_0}$$

It follows from the stability criterion for discrete-time systems (Schur-Cohn) that the characteristic polynomial has all its roots inside the unit disc if

$$a_2 < 1$$
$$a_2 - a_1 + 1 > 0$$
$$a_2 + a_1 + 1 > 0$$

Inserting the expressions for a_1 and a_2 into these conditions gives

$$(i) \qquad \frac{a}{y_0} > -1$$

$$(ii) \qquad \gamma < 2\frac{(1 - a/y_0)(\alpha + y_0^2)}{y_0^2} \qquad (6.11)$$

$$(iii) \qquad \gamma > 0$$

The equilibrium is stable if parameters a and γ are inside the triangular region shown in Fig. 6.1. To have a stable equilibrium, it must thus be required that the magnitude of the disturbance a is less than the magnitude of the command signal y_0. In addition the adaptation gain γ should not be too large. It is interesting to see the consequences of unmodeled dynamics. If there are no unmodeled dynamics ($a = 0$), then the condition for local stability of the equilibrium becomes

$$0 < \gamma < 2\frac{\alpha + y_0^2}{y_0^2}$$

Stability is thus guaranteed simply by choosing a reasonable value of γ.

Global Properties   We now investigate the global properties when the parameters are chosen in such a way that the equilibrium is stable. To get some
guidelines for the analysis, we first simulate the system. In Fig. 6.2 we show
a phase portrait for the case in which $\alpha = 0.1$, $\gamma = 0.1$, $\theta_0 = 1.5$, $y_0 = 1$,
and $a = 0.9$. It follows from Eqs. (6.9) that the equations have an equilibrium
for $y = 1.0$ and $\hat{\theta} = 2.4$ and from condition (ii) in Eqs. (6.11) that the equilibrium is stable provided that $0 < \gamma < 0.22$. The equilibrium is thus stable
for the chosen value of the adaptation gain. Remember that the system is a
discrete-time system. The discrete solution points are connected with straight
lines to give a continuous graph. All trajectories shown in the simulation are
approaching the equilibrium. Solutions with initial values $\hat{\theta}(0) = 0$ appear to
have large excursions, and the trajectory with $\hat{\theta}(0) = 2.5$ seems to be oscillatory. To understand the behavior intuitively, we consider the equations for
y and $\hat{\theta}$ separately. It follows from Eqs. (6.8) that if $\hat{\theta}$ is constant, then the
motion of y is governed by

$$y(t + 1) = (\theta_0 - \hat{\theta})y(t) + a + y_0$$

This is a first-order difference equation with the equilibrium solution

$$y = f(\hat{\theta}) = \frac{a + y_0}{1 + \hat{\theta} - \theta_0} \tag{6.12}$$

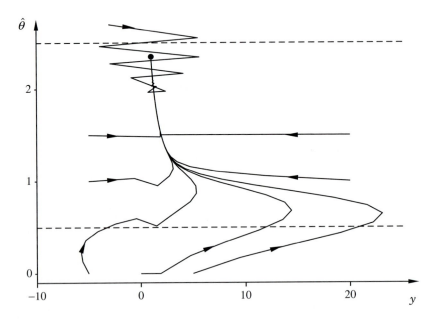

Figure 6.2 Phase portrait for the system in the stable case. Parameter
values are $\alpha = 0.1$, $\gamma = 0.1$, $\theta_0 = 1.5$, $y_0 = 1$, and $a = 0.9$. The dashed lines
indicate the interval $\theta_0 - 1 < \hat{\theta} < \theta_0 + 1$. The dot is the equilibrium point.

If parameter $\hat{\theta}$ is constant, the solution is stable if

$$\theta_0 - 1 < \hat{\theta} < \theta_0 + 1$$

and unstable otherwise. These bounds are shown as dashed lines in Fig. 6.2. If the parameter $\hat{\theta}$ is kept constant, y diverges monotonically at the lower bound and diverges in an oscillatory manner with period 2 at the upper bound. In reality, parameter $\hat{\theta}$ will of course change. The smaller the adaptation gain is, the smaller rate of change. With the numbers used in the simulation the bounds are 0.5 and 2.5. The behavior shown in Fig. 6.2 can thus be explained qualitatively. The solution approaches the curve (6.12) and then moves along this curve. The variable y appears to grow exponentially for $\hat{\theta} < 0.5$; it decays exponentially for $0.5 < \hat{\theta} < 1.5$ and decays in an oscillatory manner for $1.5 < \hat{\theta} < 2.5$. The variable grows in an oscillatory manner for $\hat{\theta} > 2.5$.

We now turn our attention to the equation for the parameter estimate. Introducing

$$\tilde{\theta} = \hat{\theta} - \theta_0$$

we find

$$\tilde{\theta}(t + 1) = \left(1 - \gamma \frac{y^2(t)}{\alpha + y^2(t)}\right) \tilde{\theta}(t) + \gamma \frac{ay(t)}{\alpha + y^2(t)} \tag{6.13}$$

This equation implies that the signals y and $\tilde{\theta}$ cannot be unbounded because Eq. (6.13) is always stable when γ is sufficiently small. For large values of $y(t)$ the added term is small, and the solution will decay. It thus appears as though the equilibrium solution that is locally stable may also be globally stable in this case. A more precise discussion of this is given in Section 6.5.

Unstable Local Equilibria   We now investigate what happens when parameters are such that the local equilibrium is unstable. We first observe that the instabilities may occur by violating any of the conditions given in Eqs. (6.11). Analyzing how the eigenvalues change with the parameters shows that the eigenvalue passes the unit circle with complex values if condition (i) is violated, through $z = -1$ if condition (ii) is violated and through $z = 1$ if condition (iii) is violated. We consider the situation in which the value of the adaptation gain is too large. Increasing the gain means that the solution will become unstable with period 2. Consider the case in which $\theta_0 = 1$, $\alpha = 0.1$, $y_0 = 1$, and $a = 0.9$. The equilibrium is $y = 1$ and $\theta = 1.9$. It follows from the stability criterion that the equilibrium is stable if $\gamma < 0.22$. With $\gamma = 0.5$ the linearized closed-loop system is unstable. Figure 6.3 shows a simulation of the system. The behavior of the system is typical for the case with unmodeled dynamics. The output y and the parameter estimate $\hat{\theta}$ appear to approach their equilibrium values. The equilibrium is unstable and a diverging oscillation with period 2 appears when y and $\hat{\theta}$ come sufficiently close to their equilibrium. The variables then oscillate with large excursions. When this happens, the modeling error becomes less significant, and the process output y and the parameter estimate $\hat{\theta}$ approach their equilibrium values. The process then repeats all over

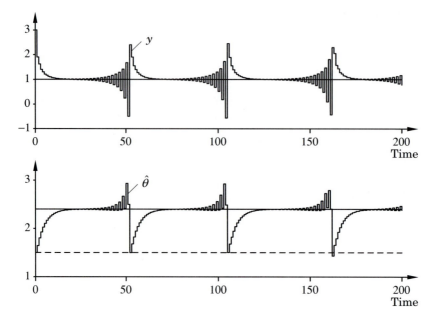

Figure 6.3   Simulation of a simple adaptive controller with unmodeled dynamics. The equilibrium values of y and $\hat\theta$ are indicated by solid straight lines. The true parameter value θ_0 is indicated by a dashed straight line.

again. The phenomenon that has been observed in many adaptive systems is called *bursting*.

The simulation shown in Fig. 6.3 represents a very complex behavior. Although essentially the same phenomenon repeats itself, the solution is *not* periodic. This is seen more clearly if the system is simulated for a longer time. Figure 6.4 shows a phase plane when the simulation time is extended to 10,000 time units. The solution is very irregular. There is, however, some pattern in the motion, as is indicated in the figure. For example, the state moves close to the curve given by Eq. (6.12) for part of the motion. The behavior shown is in fact an example of chaotic behavior. The pattern shown in Fig. 6.4 is called a *strange attractor*.

Structural Stability

Structural stability is an important concept in nonlinear dynamics. Intuitively, a system is structurally stable if small changes in the equations will not lead to drastic changes in the behavior of the system. A necessary condition for structural stability in the continuous-time case is that all equilibria are such that the linearized equations do not have eigenvalues whose real parts are zero. The equilibria are then said to be *hyperbolic*. Stability and structural

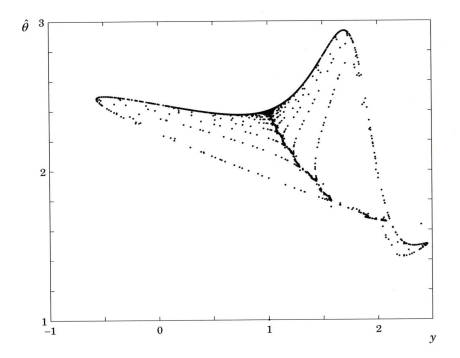

Figure 6.4   Phase plane plot corresponding to the case in Fig. 6.3 when over 10,000 time units are simulated.

stability in adaptive systems are closely related to persistency of excitation. We illustrate this by two examples.

EXAMPLE 6.1   **Lack of excitation leads to instability**

Consider the model-reference adaptive system shown in Fig. 5.14(b). Assume that the input signal is $u_c(t) = e^{-t}$. The system can then be described by the equations

$$\frac{de}{dt} = -e + k\tilde{\theta}u_c$$

$$\frac{d\tilde{\theta}}{dt} = -\gamma e u_c$$

$$\frac{du_c}{dt} = -u_c$$

where $\tilde{\theta} = \hat{\theta} - \theta_0$. The equilibrium is $e = \tilde{\theta} = u_c = 0$. Linearization around this point gives a linear system with the system matrix.

$$A = \begin{pmatrix} -1 & 0 & 0 \\ 0 & 0 & 0 \\ 0 & 0 & -1 \end{pmatrix}$$

This matrix has the eigenvalues -1, 0, and -1, and the system is clearly not stable. ☐

EXAMPLE 6.2   Persistency of excitation gives structural stability

Consider the same system as in Example 6.1, but assume now that the command signal is a step, that is, $u_c(t) = 1$. The system is then described by the equations

$$\frac{de}{dt} = -e + k\tilde{\theta}$$

$$\frac{d\tilde{\theta}}{dt} = -\gamma e$$

The equilibrium is $e = \tilde{\theta} = 0$. Linearization around this fixed point gives a linear system with the system matrix.

$$A = \begin{pmatrix} -1 & k \\ -\gamma & 0 \end{pmatrix}$$

This matrix has the characteristic polynomial

$$s^2 + s + \gamma k$$

and the equilibrium is thus stable if γk is positive. ☐

Figure 2.10 in Chapter 2, which illustrates a case of identification under closed-loop conditions, is a typical example of structural instability. Additional examples are given in Section 6.9.

6.3   ADAPTATION OF A FEEDFORWARD GAIN

The special case of adaptation of a feedforward gain has been discussed many times because of its simplicity. Let us therefore consider the structure of the equations in this case too. For the system in Fig. 5.14 we get

$$\frac{d\xi}{dt} = A\xi + B\theta u_c$$

$$e = C\xi - y_m \tag{6.14}$$

$$\varphi = \begin{cases} -y_m & \text{MIT rule} \\ -u_c & \text{Lyapunov rule} \end{cases}$$

where A, B, and C are matrices that give a realization of the transfer function $kG(s)$. Notice that in this case the matrices A, B, and C, the regression vector φ, and the error e do not depend on the controller parameters explicitly. Furthermore, the parameter is updated as

$$\frac{d\hat{\theta}}{dt} = \gamma \varphi e(\xi) \tag{6.15}$$

for a gradient scheme. If u_c is a function of time, then y_m is also a function of time, and Eqs. (6.14) and (6.15) are simply time-varying linear differential equations. Such equations can have a complex behavior. We illustrate this by an example before proceeding.

EXAMPLE 6.3   Adaptation of a feedforward gain

In Example 5.1 we derived an adaptation law for adjusting the feedforward gain by using the MIT rule. The behavior of the system was illustrated in Fig. 5.3. The system is described by

$$\frac{dy}{dt} = k\hat{\theta}(t)u_c(t) - y(t)$$

and the parameter adjustment rule is

$$\frac{d\hat{\theta}}{dt} = -\gamma y_m(t)e(t) = -\gamma y_m(t)\,(y(t) - y_m(t))$$

Since the signal y_m can be computed from the command signal u_c, both u_c and y_m can thus be regarded as known time-varying signals. The adaptive system is described by the equation

$$\frac{d}{dt}\begin{pmatrix} \hat{\theta} \\ y \end{pmatrix} = \begin{pmatrix} 0 & -\gamma y_m(t) \\ ku_c(t) & -1 \end{pmatrix}\begin{pmatrix} \hat{\theta} \\ y \end{pmatrix} + \begin{pmatrix} \gamma y_m^2(t) \\ 0 \end{pmatrix} \tag{6.16}$$

The system can thus be described by a time-varying linear differential equation of second order. In Fig. 6.5 we show three simulations for the case in which $G(s) = 1/(s + 1)$, $k = k_0 = 1$, and $\gamma = 11$. The reference signal is sinusoidal in all cases. The frequency is $\omega = 1$ in the first case, $\omega = 2$ in the second, and $\omega = 3$ in the third. The controller parameter converges to the correct value for $\omega = 1$ and $\omega = 3$, but it diverges for $\omega = 2$. We thus have a situation in which the system is stable for one input but unstable for another. The system is stable for low frequencies of the input signal. As the frequency increases, it becomes unstable. It becomes stable again as the frequency is increased further. This pattern repeats itself as the frequency is increased further.   □

Example 6.3 shows that the system has quite a complex behavior that cannot be explained by the intuitive argument of the previous section. To understand what is happening, we analyze the equations describing the system. Equation (6.16) can be written as

$$\frac{dx}{dt} = A(t)x + B(t) \tag{6.17}$$

This is a linear system with time-varying parameters. In the particular case in which the input u_c is periodic and we connect the adaptation when model output y_m has also become periodic, the system is also periodic. For such systems there is a well-developed theory that can be used to understand the behavior of the system.

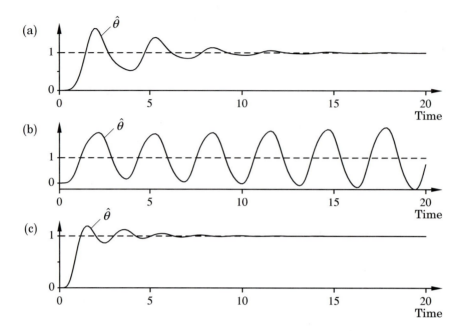

Figure 6.5 Behavior of the controller gain for an MRAS using the MIT rule. The input signal is a unit amplitude sinusoidal with frequency (a) 1; (b) 2; and (c) 3 rad/s. The system has the transfer function $G(s) = 1/(s+1)$, the parameters are $k = k_0 = 1$, and the adaptation gain is $\gamma = 11$. The dashed lines indicate the correct values of the gain.

Floquet Theory

To investigate the stability properties of (6.17), we consider the homogeneous part, when $A(t)$ is periodic with period τ and continuous for all t. The solution is given by

$$x(t) = \Phi(t, t_0)x(t_0)$$

where $\Phi(t, t_0)$ satisfies the linear matrix differential equation

$$\frac{d\Phi}{dt} = A(t)\Phi \qquad (6.18)$$

Since $A(t)$ is periodic with period τ, it follows that $A(t+\tau) = A(t)$. This implies that if $\Phi(t)$ is a solution, then $\Phi(t+\tau)$ is also a solution. Since the two solutions to Eq. (6.18) differ only in their initial conditions it follows that

$$\Phi(t + \tau) = \Phi(t)W \qquad (6.19)$$

where W is a nonsingular constant matrix. Since the matrix $\Phi(t)$ is nonsingular for all t, it follows that W is also nonsingular. By repeated use of this equation we find that

$$\Phi(t + n\tau) = \Phi(t)W^n$$

where $t < \tau$. We thus obtain the following result.

THEOREM 6.1   Stability of linear periodic system

The periodic differential equation (6.17) is stable if and only if all eigenvalues of the matrix W have magnitudes less than 1.   □

This result is actually all we need for stability analysis. We can, however, also obtain a slightly more general result. Notice that we can compute W simply by integrating the differential equation over one period with the initial condition equal to the identity matrix.

THEOREM 6.2   Solution of periodic systems

The solution to the matrix differential equation (6.18) has the form

$$\Phi(t) = D(t)e^{Ct}$$

where C is a constant matrix and D is periodic with period τ.

Proof: Since the matrix W in Eq. (6.19) is nonsingular, there exists a matrix C such that

$$W = e^{C\tau} \tag{6.20}$$

Introduce the matrix function $D(t)$ defined by

$$D(t) = \Phi(t)e^{-Ct}$$

Then

$$D(t + \tau) = \Phi(t + \tau)e^{-C(t+\tau)} = \Phi(t)We^{-C\tau}e^{-Ct} = D(t)$$

and the theorem is proven.   □

Remark.   From Eq. (6.20) we see that the differential equation (6.18) is stable if the matrix C has all its eigenvalues in the left half-plane, which means that the matrix W should have all its eigenvalues inside the unit disc. Stability can thus be determined by numerical integration over one period.   □

We now show how the results can be used to investigate the stability of the system in Example 6.3.

EXAMPLE 6.4   **Parametric excitation**

Consider the system in Example 6.3. Let the command signal be $u_c(t) = \sin \omega t$. After a transient the model output becomes

$$y_m(t) = \frac{1}{\sqrt{1 + \omega^2}} \sin (\omega t - \arctan (\omega))$$

To determine the stability of Eq. (6.16), we compute W by integrating Eq. (6.18) over one period, that is, $\tau = 2\pi/\omega$, with the initial condition $\Phi(0) = I$. Then from Eq. (6.19) we get $W = \Phi(\tau)$. Choosing $\omega = 2$ and integrating to $\tau = \pi$ give

$$\Phi(\tau) = \begin{pmatrix} 0.4373 & 0.7283 \\ 0.2389 & 0.4967 \end{pmatrix}$$

with eigenvalues 0.049 and 0.885 for $\gamma = 10$ and

$$\Phi(\tau) = \begin{pmatrix} 0.5609 & 0.9960 \\ 0.2642 & 0.5463 \end{pmatrix}$$

with eigenvalues 0.041 and 1.067 for $\gamma = 11$. It can thus be concluded that the adaptive system will be stable for $\gamma = 10$ but unstable for $\gamma = 11$.

This calculation can be repeated for many frequencies and many values of the adaptation gain to determine the values of ω and γ for which the system is stable. The result of such a calculation is shown in Fig. 6.6. Notice in particular that Fig. 6.6 explains the behavior observed in the numerical experiment in Example 6.3, in which the system goes through a region of instability as the frequency of the input signal increases. Notice, however, that the system is stable for low adaptation gains.   □

Example 6.3 indicates that even very simple adaptive systems can exhibit complex behavior. The mechanism of periodic excitation can also give rise to instabilities in more complex adaptive systems. The analysis can be made in the same way as for the simple example, but the details are much more complicated. The behavior is typically associated with periodic excitation and comparatively high adaptation gains. The phenomenon illustrated in Fig. 6.5

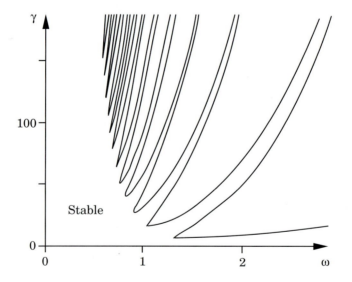

Figure 6.6 Stability region for adjustment of a feedforward gain with the MIT rule.

is an example of *parametric excitation*, that is, a system can be made unstable by changing its parameters periodically. A classical example is the Mathieu equation:

$$\frac{d^2 y}{dt^2} + \alpha \frac{dy}{dt} + (\beta + \gamma \cos \omega t)y = 0$$

For $\alpha = 0$ this equation describes a pendulum whose pivot point is oscillating vertically. It is well known that the normal equilibrium, with the pendulum hanging down, can be made unstable by a proper choice of the parameters.

EXAMPLE 6.5   **Lyapunov redesign**

In Example 6.3 we found that the MIT rule could give instabilities for large adaptation gains. Under the strong assumption that the transfer function of the process is strictly positive real, however, the control law derived from stability theory is stable for all values of the adaptation gain. We illustrate this in the simulation in Fig. 6.7, in which the Lyapunov rule is applied to the system in Example 6.3. Compare with Fig. 6.5.   □

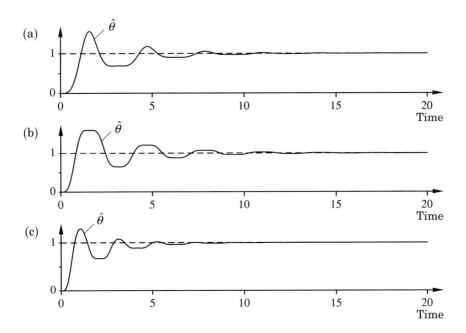

Figure 6.7   Behavior of the controller gain for an adaptive system based on Lyapunov stability theory when the input signal is a unit amplitude sinusoidal with frequency (a) 1; (b) 2; and (c) 3 rad/s. The system has the transfer function $G(s) = 1/(s + 1)$, the parameters are $k = k_0 = 1$, and the adaptation gain is $\gamma = 11$. The dashed lines indicate the correct values of the gain.

Summary

A discrete-time system and the feedforward gain example have been discussed in this section. The examples show that adaptive controllers can have rather strange properties. The phenomena could be explained by using simple mathematics, but there will be difficulties in the general cases. It is therefore appropriate to consider some simplified situations in the coming sections. First, indirect and direct self-tuning regulators are discussed under idealized assumptions. Second, the adaptive control problem is divided into two parts with different time scales, and averaging techniques are used to analyze properties of the closed-loop systems.

6.4   ANALYSIS OF INDIRECT DISCRETE-TIME SELF-TUNERS

In this section we analyze the properties of indirect discrete-time self-tuners of the type illustrated by the block diagram in Fig. 1.19. Since such controllers contain a recursive parameter estimator and a control design calculation, it is natural to investigate these subsystems separately. Since identification is performed in closed loop, there may also be undesirable effects due to interaction of control and identification. We start by investigating the properties of the recursive parameter estimator. Second, the design calculations must be considered. It is particularly important to understand when the design calculations are poorly conditioned so that small changes in process parameter estimates may cause large changes in the controller parameters.

It would be highly desirable to determine whether the adaptive system can track parameters of a time-varying system. This is a very difficult problem, and we therefore limit the analysis to the case in which the real system has constant parameters. This can be considered as a first test of an adaptive algorithm. To carry out the analysis, we also assume that the real system is described by models that are compatible with the models used for parameter estimation. In this case it makes sense to talk about the "true parameters." In reality, however, we also have to deal with the fact that the models that we use are approximations. This is called the nonideal case. This problem is discussed later in Section 6.9.

Properties of Recursive Estimators

To investigate recursive estimators, it is necessary to make some assumptions on how the data was generated. In this section we make the assumption that the data is generated by a model having the same structure as the model used in the estimation. It is also necessary to specify the nature of the disturbances—for example, whether they are deterministic or stochastic. We also find that it is important for there to be sufficient excitation and that identification under closed-loop conditions may cause difficulties.

The deterministic case, in which data is generated from a system that is compatible with the model used in the estimator, is particularly simple. In this case it is possible to derive general properties of the estimators.

Projection or Gradient Algorithms

The properties of the projection or gradient algorithm are now investigated in the ideal case in which data is generated by the model

$$y(t) = \varphi^T(t)\theta^0 \tag{6.21}$$

We have the following result.

THEOREM 6.3   Projection algorithm properties

Let the estimator

$$
\begin{aligned}
\hat{\theta}(t) &= \hat{\theta}(t-1) + \frac{\gamma\varphi(t)}{\alpha + \varphi^T(t)\varphi(t)}e(t) \\
e(t) &= y(t) - \varphi^T(t)\hat{\theta}(t-1) = \varphi^T(t)\left(\theta^0 - \hat{\theta}(t-1)\right)
\end{aligned}
\tag{6.22}
$$

with $\alpha \geq 0$ and $0 < \gamma < 2$, be applied to data generated by Eq. (6.21). It then follows that

(i) $$\|\hat{\theta}(t) - \theta^0\| \leq \|\hat{\theta}(t-1) - \theta^0\| \leq \|\hat{\theta}(0) - \theta^0\| \qquad t \geq 1$$

(ii) $$\lim_{t\to\infty} \frac{e(t)}{\sqrt{\alpha + \varphi^T(t)\varphi(t)}} = 0$$

(iii) $$\lim_{t\to\infty} \|\hat{\theta}(t) - \hat{\theta}(t-k)\| = 0 \qquad \text{for any finite } k$$

Proof: Introduce $\tilde{\theta}(t) = \hat{\theta}(t) - \theta^0$ and

$$V(t) = \tilde{\theta}^T(t)\tilde{\theta}(t) = \|\tilde{\theta}(t)\|^2$$

It follows that

$$e(t) = \varphi^T(t)\theta^0 - \varphi^T(t)\hat{\theta}(t-1) = -\varphi^T(t)\tilde{\theta}(t-1)$$

Subtracting θ^0 from both sides of the parameter equation in Eqs. (6.22) and taking the norm, we get

$$
\begin{aligned}
V(t) - V(t-1) &= 2\frac{\gamma\varphi^T(t)\tilde{\theta}(t-1)e(t)}{\alpha + \varphi^T(t)\varphi(t)} + \frac{\gamma^2\varphi^T(t)\varphi(t)e^2(t)}{(\alpha + \varphi^T(t)\varphi(t))^2} \\
&= \chi(t)\frac{\gamma e^2(t)}{\alpha + \varphi^T(t)\varphi(t)}
\end{aligned}
$$

where

$$\chi(t) = -2 + \frac{\gamma \varphi^T(t)\varphi(t)}{\alpha + \varphi^T(t)\varphi(t)} \leq -\delta < 0$$

and the inequality follows from $\alpha \geq 0$ and $0 < \gamma < 2$. Property (i) has thus been established. It follows from the preceding equation that

$$V(t) = V(0) + \sum_{k=1}^{t} \chi(k) \frac{\gamma e^2(k)}{\alpha + \varphi^T(k)\varphi(k)}$$

Hence

$$\sum_{k=1}^{t} \frac{\gamma e^2(k)}{\alpha + \varphi^T(k)\varphi(k)} \leq \frac{1}{\delta} (V(0) - V(t))$$

Since $0 \leq V(t) \leq V(0)$, it follows that the normalized error

$$\frac{e(t)}{\sqrt{\alpha + \varphi^T(t)\varphi(t)}}$$

is in l_2, that is, squared summable, and thus property (ii) follows. From Eqs. (6.22),

$$\|\hat{\theta}(t) - \hat{\theta}(t-1)\|^2 = \frac{\gamma^2 \varphi^T(t)\varphi(t)e^2(t)}{(\alpha + \varphi^T(t)\varphi(t))^2}$$

$$= \frac{\gamma^2 e^2(t)}{\alpha + \varphi^T(t)\varphi(t)} \left(1 - \frac{\alpha}{\alpha + \varphi^T(t)\varphi(t)}\right)$$

It follows from property (ii) that the right-hand side of the preceding equation goes to zero as $t \to \infty$ if $\alpha > 0$. Hence

$$\|\hat{\theta}(t) - \hat{\theta}(t-k)\|^2 = \left\|\sum_{i=1}^{k} \hat{\theta}(t-i+1) - \hat{\theta}(t-i)\right\|^2$$

$$\leq \sum_{i=1}^{k} \|\hat{\theta}(t-i+1) - \hat{\theta}(t-i)\|^2$$

where the right-hand side goes to zero as $t \to \infty$ for finite k. $\square$

Remark 1.   For $\gamma = 1$ and $\alpha = 0$ the algorithm reduces to Kaczmarz's projection algorithm.

Remark 2.   Notice that the result does *not* imply that the estimates $\hat{\theta}(t)$ converge.

Remark 3.   The function $V(t)$ can be interpreted as a discrete-time Lyapunov function. $\square$

Theorem 6.3 is useful because it gives some properties of the estimator that are valid no matter how the regressors $\varphi(t)$ are generated. Additional conditions are required to guarantee that the estimates converge. The theorem will also be useful to prove convergence of the indirect adaptive schemes.

Parameter Convergence of Gradient Algorithms

We now give conditions for the estimates to converge to the true parameter values. Notice that to pose such a problem, it is necessary to assume that data is generated by a model that is compatible with the model used to formulate the estimate. Parameter convergence is closely related to system identification. The properties of identifiability and persistency of excitation play an essential role. The convergence rate depends on the algorithm used and the amount of excitation. We first consider the gradient algorithm, which is simpler than the least-squares algorithm, although it converges at a considerably slower rate. A typical projection or gradient algorithm is given by Eqs. (6.22), where $\alpha \geq 0$ and $0 < \gamma < 2$. The estimation error is given by

$$\tilde{\theta}(t) = \hat{\theta}(t) - \theta^0 = A(t-1)\tilde{\theta}(t-1) \tag{6.23}$$

where

$$A(t-1) = I - \frac{\gamma \varphi(t)\varphi^T(t)}{\alpha + \varphi^T(t)\varphi(t)}$$

The problem of analyzing convergence rates is thus equivalent to analyzing the stability of Eq. (6.23). Notice that

$$A(t-1)\varphi(t) = \left(I - \frac{\gamma \varphi(t)\varphi^T(t)}{\alpha + \varphi^T(t)\varphi(t)}\right)\varphi(t) = \varphi(t)\left(1 - \frac{\gamma \varphi^T(t)\varphi(t)}{\alpha + \varphi^T(t)\varphi(t)}\right)$$

The second factor on the right-hand side is a scalar. This implies that the vector $\varphi(t)$ is an eigenvector to $A(t-1)$ with an eigenvalue that is less than 1. The eigenvalue is zero for $\gamma = 1$ and $\alpha = 0$. The following lemma is useful to analyze Eq. (6.23).

LEMMA 6.1   Stability of a time-varying system

Consider the time-varying system

$$\begin{aligned} x(t+1) &= A(t)x(t) \\ y(t) &= C(t)x(t) \end{aligned} \tag{6.24}$$

Assume that there exists a symmetric matrix $P(t) > 0$ such that

$$A^T(t)P(t+1)A(t) - P(t) = -C^T(t)C(t) \tag{6.25}$$

Then Eqs. (6.24) are stable. Moreover, if the system is uniformly completely observable, that is, if there exist $\beta_1 > 0$, $\beta_2 > 0$, and $N > 1$ such that

$$0 < \beta_1 I \leq \sum_{k=t}^{t+N-1} \Phi^T(k,t)C^T(k)C(k)\Phi(k,t) \leq \beta_2 I < \infty$$

for all t and where $\Phi(k,t)$ is the fundamental matrix, then Eqs. (6.24) are also exponentially stable.

Proof: Introduce the function

$$V(t) = x^T(t)P(t)x(t)$$

Hence

$$V(t+1) - V(t) = x^T(t)A^T(t)P(t+1)A(t)x(t) - x^T(t)P(t)x(t)$$
$$= -x^T(t)C^T(t)C(t)x(t) \le 0$$

The function V can be considered a Lyapunov function for a discrete-time system. To prove stability for a discrete-time system using Lyapunov theory, we have to show that the difference

$$\Delta V(t) = V(t+1) - V(t) \le 0$$

and that the matrix $P(t)$ is positive definite. Iterating the system equations N steps gives

$$V(t+N) - V(t) = -\sum_{k=t}^{t+N-1} x^T(k)C^T(k)C(k)x(k)$$

$$= -x^T(t)\left(\sum_{k=t}^{t+N-1} \Phi^T(k,t)C^T(k)C(k)\Phi(k,t)\right)x(t)$$

$$\le -\beta_1 x^T(t)x(t) \le -\frac{\beta_1}{\lambda_{max}P(t)}V(t)$$

where $\lambda_{max}(Pt)$ is the largest eigenvalue of $P(t)$. Hence

$$V(t+N) \le \left(1 - \frac{\beta_1}{\lambda_{max}P(t)}\right)V(t) = \beta_3 V(t)$$

From Eq. (6.25) it follows that

$$P(t) = C^T(t)C(t) + A^T(t)P(t+1)A(t)$$
$$= C^T(t)C(t)$$
$$\quad + A^T(t)\left(C^T(t+1)C(t+1) + A^T(t+1)P(t+2)A(t+1)\right)A(t)$$

$$\vdots$$

$$= \sum_{k=t}^{\infty} \Phi^T(k,t)C^T(k)C(k)\Phi(k,t)$$

$$> \sum_{k=t}^{t+N-1} \Phi^T(k,t)C^T(k)C(k)\Phi(k,t) \ge \beta_1 I$$

This shows that $\lambda_{max}(P(t)) > \beta_1$ and $\beta_3 < 1$, which implies that $V(t)$ goes to zero exponentially. Furthermore,

$$P(t+N) = P(t) - \sum_{k=t}^{t+N-1} \Phi^T(k,t)C^T(k)C(k)\Phi(k,t) \le \beta_3 P(t)$$

or

$$P(t) \leq \frac{1}{1 - \beta_3} \sum_{k=t}^{t+N-1} \Phi^T(k,t) C^T(k) C(k) \Phi(k,t) \leq \frac{\beta_2}{1 - \beta_3} I$$

The matrix $P(t)$ is thus bounded from above and below. Since $V(t)$ goes to zero exponentially and $P(t)$ is bounded, it follows that the system (6.24) is exponentially stable. □

Applying this lemma to Eq. (6.23), we get the following theorem.

THEOREM 6.4 Exponential stability

The difference equation (Eq. 6.23) is globally exponentially stable if there exist positive constants β_1, β_2, and N such that for all t,

$$0 < \beta_1 I \leq \sum_{k=t}^{t+N-1} \varphi(k) \varphi^T(k) \leq \beta_2 I < \infty \tag{6.26}$$

Proof: Choose $P = I$ and

$$C(t) = \frac{\sqrt{\gamma(2\alpha + (2 - \gamma)\varphi^T \varphi)}}{\alpha + \varphi^T \varphi} \varphi^T$$

where the argument t of φ is suppressed. A straightforward calculation shows that Eq. (6.25) is satisfied, so the system is stable. To prove exponential stability, first observe that uniform observability of $(A(k), C(k))$ is equivalent to uniform observability of $((A(k) - B(k)C(k)), C(k))$. Choosing

$$B(k) = -\frac{\gamma}{\sqrt{(\gamma(2\alpha + (2 - \gamma)\varphi^T \varphi)}} \varphi$$

we find that $A(k) - B(k)C(k) = I$, and uniform asymptotic stability then corresponds to Eq. (6.26). □

Notice that Eq. (6.26) is closely related to persistent excitation. (Compare with Definition 2.1.) It is thus found that exponential convergence of the gradient algorithm is closely connected to whether the input signal to the system is persistently exciting of sufficiently high order.

It should be pointed out that condition (6.26) is a persistent excitation condition for the regressors, not the external reference signal for the system. The excitation can be provided by the command signals and by the disturbances acting on the process. Notice, however, that excitation may be lost by feedback, which can introduce relations between the variables appearing in the regression vector. We discuss this later in this section.

Recursive Least Squares

Parameter convergence for recursive least squares is first discussed for the simple model (6.21), which is linear in the parameters and for which are no disturbances. Let the parameter vector have n elements. The parameters can be calculated exactly from n data points, provided that the vectors $\varphi(1), \ldots, \varphi(n)$ are linearly independent. The least-squares estimate is given by

$$\hat{\theta}(n) = \left(\sum_{k=1}^{n} \varphi(k)\varphi^T(k) \right)^{-1} \sum_{k=1}^{n} \varphi(k)y(k)$$

$$= \left(\sum_{k=1}^{n} \varphi(k)\varphi^T(k) \right)^{-1} \sum_{k=1}^{n} \varphi(k)\varphi^T(t)\theta^0 = \theta^0 \qquad (6.27)$$

The correct state is obtained in n steps. If the estimate is instead calculated by recursive least squares, the following estimate is obtained:

$$\hat{\theta}(t) = \left(P^{-1}(0) + \sum_{k=1}^{t} \varphi(k)\varphi^T(k) \right)^{-1} \left(\sum_{k=1}^{t} \varphi(k)y(k) + P^{-1}(0)\hat{\theta}(0) \right) \qquad (6.28)$$

where $\hat{\theta}(0)$ is the initial estimate and $P(0)$ is the initial covariance of the estimator. By making $P(0)$ positive definite but arbitrarily large, the result from the recursive estimation can be made arbitrarily close to the true value.

From this analysis we obtain the following result.

THEOREM 6.5   Property of RLS

Let the recursive least squares be applied to data generated by Eq. (6.21). Let $P(0)$ be positive definite and let $\hat{\theta}(0)$ be bounded. Assume that

$$\beta(t)I \leq \sum_{k=1}^{t} \varphi(k)\varphi^T(k)$$

where $\beta(t)$ goes to infinity. Then the estimate converges to θ^0.   ☐

This discussion shows that in the deterministic case it is possible to obtain parameter estimators that converge in a finite number of steps. The key assumption is that the regressors are linearly independent, so $\sum \varphi(k)\varphi^T(k)$ is of full rank. When the parameters are changing, a least-squares estimator, in which the covariance matrix P is regularly reset to αI, is a good implementation. This procedure is called *covariance resetting*. To obtain an estimate that reacts rapidly to parameter changes, it is also possible to have several estimators in parallel, which are reset sequentially.

Results similar to Theorem 6.3 can also be established for the least-squares algorithm and several of its variants. The key is to replace function $V(t)$ in Theorem 6.3 by

$$V(t) = \tilde{\theta}^T(t)P^{-1}(t)\tilde{\theta}(t)$$

and add assumptions that guarantee that the eigenvalues of P stay bounded. One way to do this is to use the constant trace algorithm (see Section 11.5).

So far, only the general model (6.21) has been discussed. The properties of estimates of parameters of discrete-time transfer functions will now be considered. The uniqueness of the estimates is first explored. For this purpose we assume that the data is actually generated by

$$A^0(q)y(t) = B^0(q)u(t) + e(t + n) \qquad (6.29)$$

where A^0 and B^0 are relatively prime. If $e = 0$, $\deg A > \deg A^0$, and $\deg B > \deg B^0$, it follows from Theorem 2.1 that the estimate is not unique because the columns of the matrix Φ are linearly dependent. Theorem 2.10 gives conditions for uniqueness of the least-squares estimate.

The Stochastic Case

Consider the model

$$y(t) = \varphi^T(t)\theta^0 + e(t)$$

where $\{e(t)\}$ is a sequence of independent Gaussian $(0, \sigma)$ random variables. The least-squares estimator is given by Eq. (6.28). The covariance of the estimate for large t is (see Theorem 2.2)

$$P(t) = \sigma^2 \left(\sum_{k=1}^{t} \varphi(k)\varphi^T(k) \right)^{-1}$$

By taking the covariance of the estimate as a measure of the rate of convergence, it is found that under uniform persistent excitation the matrix P converges at the rate $1/t$. This implies that the estimates converge at the rate $1/\sqrt{t}$.

THEOREM 6.6   Convergence of RLS

Let the least-squares method for estimating parameters of a transfer function be applied to data generated by the model of Eq. (6.29) where $\{e(t)\}$ is a sequence of uncorrelated random variables with zero mean and variance σ^2. Assume that the estimated model has the same structure as the process generating the data, that is, the ideal case. Further assume that the input signal is persistently exciting of order $\deg A + \deg B + 1$. Then

(i) $\qquad\qquad\qquad \hat{\theta}(t) \to \theta^0 \quad$ in the mean square as $\quad t \to \infty$

(ii) $\qquad\qquad \text{var}\,(\hat{\theta} - \theta^0) \approx \dfrac{\sigma^2}{t} \left(\lim_{t\to\infty} \dfrac{1}{t}\,\Phi^T\Phi \right)^{-1}$

$\hfill\square$

Remark 1.   The estimates do not converge to the true parameters when $e(t)$ is correlated with $e(s)$ for $t \neq s$.

Remark 2.   Theorem 6.6 gives the convergence rate for the parameter error in the ideal case. More complex behavior can be obtained when the different components of the regression vector have different convergence rates (see Example 2.11).   □

Unmodeled Dynamics

So far, it has been assumed that the true process is compatible with the model used in parameter estimation. It frequently happens that the true process is more complex than the estimated model. This is often referred to as *unmodeled dynamics*. The problem is complex, and a careful analysis is lengthy; roughly speaking, the parameters will converge to a value that minimizes the least-squares criterion:

$$V(\theta) = \frac{1}{T}\sum_0^T (Ay_f(t) - Bu_f(t)) \tag{6.30}$$

where y_f and u_f are the filtered process input and output, that is,

$$y_f = H_f y$$
$$u_f = H_f u$$

and the parameter θ represents the coefficients of the polynomials A and B. The minimum exists under certain regularity conditions, and the minimizing θ is unique under the condition of persistency of excitation. The minimizing value will depend on the data filter H_f and the spectrum of the reference signal and the disturbances.

Identification in Closed Loop

When discussing parameter estimation in Chapter 2, we observed that identifiability could be lost if the input was generated by feedback from the output. The reason is that the feedback introduces dependencies in the regression vector. (Compare with Example 2.10.) Since this is very important for the behavior of direct adaptive controllers, we will investigate the problem in a little more detail. In this analysis we will consider what happens when we perform system identification to data generated by feedback. Consider a process described by

$$Ay(t) = Bu(t) + v(t) \tag{6.31}$$

with the controller

$$Ru = Tu_c - Sy$$

where polynomials R, S, and T have constant parameters. The closed-loop system is given by

$$y = \frac{BT}{AR + BS} u_c + \frac{R}{AR + BS} v$$

$$u = -\frac{AT}{AR + BS} u_c + \frac{S}{AR + BS} v$$

With a system identification experiment it is possible to determine the transfer functions

$$G_1 = \frac{BT}{AR + BS} \qquad G_2 = \frac{AT}{AR + BS}$$

$$G_3 = \frac{R}{AR + BS} \qquad G_4 = \frac{S}{AR + BS}$$

that appear in these equations. There are no problems with identifiability if the input signal u_c is persistently exciting of sufficiently high order because the polynomials A and B are then readily determined from G_1 and G_2. However, if the command signal is zero and all excitation comes from the disturbance, we can determine only the polynomial

$$A_c = AR + BS \tag{6.32}$$

To achieve identifiability, it must also be required that the signal v be persistently exciting of sufficiently high order. The question of identifiability of polynomials A and B then becomes a problem of uniquely determining A and B from Eq. (6.32) when polynomials R and S are known. If A_0 and B_0 are solutions, the general solution is

$$A = A_0 + QS \quad B = B_0 - QR$$

where Q is an arbitrary polynomial. When the model structure is specified, the highest degree of A is also given. The solution is thus unique only if polynomials R and S have sufficiently high degree. To achieve identifiability in closed loop, it is therefore important that the controller be of sufficiently high order. It is natural to assume that R and S have the same degree. Identifiability is then obtained if

$$\deg R = \deg S \geq \deg A \tag{6.33}$$

In Example 3.2, in which $\deg A = 2$, $\deg B = 1$, and $\deg R = \deg S = 1$, we do not have identifiability under closed loop with $u_c = 0$. However, if it is required that the controller has integral action as in Example 3.10, we have $\deg R = \deg S = 2$, and the condition (6.33) holds. To achieve identifiability, it must, of course, be required that the disturbance be persistently exciting.

Also observe that if a pole placement design is used, all models that are estimated will give the correct closed-loop characteristic polynomial.

Design Calculations

The design calculations are an important part of indirect adaptive systems. Theoretically, the design procedure is represented by the function χ, which maps process parameters $\hat{\theta}$ to controller parameters ϑ. The properties of χ will, of course, depend on the parameterization of the model and the design procedure chosen. The function can often be quite complicated. It is important that the map gives unique controller parameters and that there are no singularities in the map. We discuss the properties of the map in some simple cases.

Consider the process model

$$Ay = Bu \tag{6.34}$$

where it is assumed that A has degree n and B has degree $n-1$. The model thus has $2n$ parameters. If pole placement design is used, the controller parameters are given by

$$AR + BS = A_o A_m \tag{6.35}$$

where R and S have the same degree m as the observer polynomial A_o. The minimum-degree solution corresponds to $m = n - 1$, but an observer of higher order is often preferable to improve the robustness of the system. Without loss of generality, R can be monic. The controller then has $2m + 1$ parameters. The function χ is thus a map from R^{2n} to R^{2m+1}, where $m \geq n - 1$. Since Eq. (6.35) becomes singular when polynomials A and B have a common factor, it follows that the map χ has singularities. The problem with design singularities is illustrated by an example.

EXAMPLE 6.6   **Singularities for pole placement design**

Consider the model of Eq. (6.34) with

$$A(q) = q^2 + a_1 q + a_2$$
$$B(q) = b_0 q + b_1$$

In Example 3.2 a controller was designed for

$$A_m(q) = q^2 + a_{m1}q + a_{m2}$$
$$A_o(q) = q + a_o$$

In this case the controller and process parameters are

$$\vartheta = \begin{pmatrix} r_1 & s_0 & s_1 \end{pmatrix} \qquad \theta = \begin{pmatrix} a_1 & a_2 & b_0 & b_1 \end{pmatrix}$$

and the map $\chi : R^4 \to R^3$ is given by

$$
r_1 = \frac{a_o a_{m2} b_0^2 + (a_2 - a_{m2} - a_o a_{m1}) b_0 b_1 + (a_o + a_{m1} - a_1) b_1^2}{b_1^2 - a_1 b_0 b_1 + a_2 b_0^2}
$$

$$
s_0 = \frac{b_1 (a_o a_{m1} - a_2 - a_{m1} a_1 + a_1^2 + a_{m2} - a_1 a_o)}{b_1^2 - a_1 b_0 b_1 + a_2 b_0^2}
$$

$$
+ \frac{b_0 (a_{m1} a_2 - a_1 a_2 - a_o a_{m2} + a_o a_2)}{b_1^2 - a_1 b_0 b_1 + a_2 b_0^2} \tag{6.36}
$$

$$
s_1 = \frac{b_1 (a_1 a_2 - a_{m1} a_2 + a_o a_{m2} - a_o a_2)}{b_1^2 - a_1 b_0 b_1 + a_2 b_0^2}
$$

$$
+ \frac{b_0 (a_2 a_{m2} - a_2^2 - a_o a_{m2} a_1 + a_o a_2 a_{m1})}{b_1^2 - a_1 b_0 b_1 + a_2 b_0^2}
$$

The map χ is singular when the denominator in Eqs. (6.36) vanishes, that is, when

$$
b_1^2 - a_1 b_0 b_1 + a_2 b_0^2 = 0 \qquad\qquad \Box
$$

Singularities of the type in Example 6.6 will appear for practically all design methods. Since the singularities are algebraic surfaces, the parameter estimates must pass them if the algorithms are not initialized properly. There are several ways to avoid the difficulties. One possibility is to test for common factors and to cancel them if they appear, but such a procedure will require test quantities. It will also make χ discontinuous, which creates difficulties in the analysis. Another and better solution is to find design techniques such that the mapping χ is smooth. This is an open research problem, which so far has received little attention.

The following example illustrates what happens if no precautions are taken with cancellations.

EXAMPLE 6.7   **Indirect adaptive system with design singularities**

Consider the system in Example 6.6, and let the controller be an indirect adaptive system that is based on estimation of the parameters of the model. The desired dynamics A_m are chosen to correspond to a second-order system with $\omega = 1.5$ and $\zeta = 0.707$. The observer polynomial is chosen to be $A_o = z$.

Figure 6.8 shows the results obtained when the adaptive algorithm is applied to a first-order system

$$
G(s) = \frac{1}{s+1}
$$

Notice the strange behavior of the output. This would have been even worse if the control signal had not been kept bounded in the simulation. The parameter estimates converge very quickly to values such that A and B have a common factor. The Diophantine equation is then singular, as shown in Example 6.6,

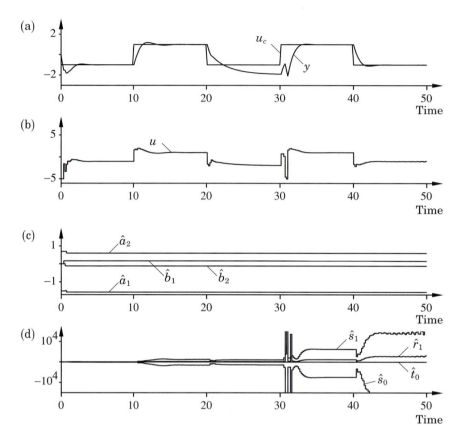

Figure 6.8 Simulation of an indirect adaptive pole placement controller based on a second-order process model of a first-order process. (a) Output and reference value. (b) Control signal. (c) Estimated process parameters. (d) Calculated controller parameters.

and the controller parameters become very large. The consequences of canceling a possible common factor and making a design for a first-order system are illustrated in Fig. 6.9. In this particular case a factor is canceled if poles and zeros are so close that

$$\left| A\left(\frac{-b_1}{b_0}\right) \right| = \left| \frac{b_1^2 - a_1 b_0 b_1 + a_2 b_0^2}{b_0^2} \right| \le 0.01 \tag{6.37}$$

The performance is now very good.   □

Summary

Parameter convergence for indirect adaptive algorithms depends critically on the assumptions of identifiability and persistency of excitation. Analysis of

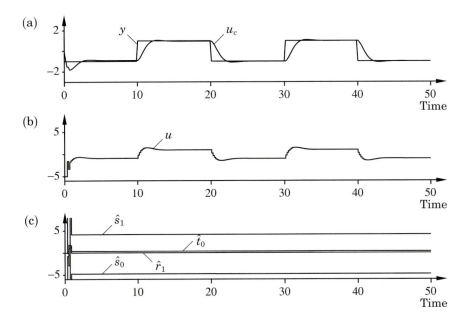

Figure 6.9 Simulation of an indirect adaptive pole placement controller based on a second-order process model. A possible common factor in the estimated process transfer function is canceled before the control law is calculated if the condition of Eq. (6.37) holds. (a) Output and reference value. (b) Control signal. (c) Calculated controller parameters.

the convergence rate of estimators shows that the convergence rate depends drastically on the underlying process being deterministic or stochastic. It also depends on the algorithm. A least-squares algorithm in the deterministic case gives convergence in a finite number of steps, provided that the input is persistently exciting. The gradient algorithms give exponential but generally much slower convergence than the least-squares algorithm. The convergence rate is much slower in the stochastic case. Analysis of the convergence rate for estimators gives only partial insight into the convergence rate of adaptive algorithms. To obtain a detailed understanding, it is necessary to consider that the input to the system is generated by feedback.

6.5   STABILITY OF DIRECT DISCRETE-TIME ALGORITHMS

Stability was discussed in connection with model-reference adaptive system in Chapter 5. It was in fact the key design issue in the MRAS. The problem was easy to resolve in the cases in which all the state variables were measured and for output feedback of systems in which the dynamics were SPR or could easily be made SPR. In these cases the MRAS has the property that arbitrarily large adaptation gains can be used.

A stability proof for a direct discrete-time adaptive control law (MRAS or STR) for a general linear system will now be given. Some simplifications will be made in the algorithm to avoid too many technicalities.

The Algorithm

Direct algorithms for adaptive control were discussed in Section 3.5. We give the proof for a simple algorithm of this type. Consider a process described by the difference equation

$$A^*(q^{-1})y(t) = B^*(q^{-1})u(t - d) \tag{6.38}$$

Let the desired response from command signal to process output be characterized by

$$A_m^*(q^{-1})y(t) = t_0 u_c(t - d)$$

This specification implies that all process zeros are canceled. Furthermore, let the observer polynomial be A_o. A direct algorithm can then be formulated as follows. Estimate parameters of the model

$$A_o^* A_m^* y(t + d) = R^* u(t) + S^* y(t) = \varphi^T(t)\theta \tag{6.39}$$

where

$$\theta = \begin{pmatrix} r_0 & r_1 & \cdots & r_k & s_0 & s_1 & \cdots & s_l \end{pmatrix}^T$$

$$\varphi(t) = \begin{pmatrix} u(t) & u(t-1) & \cdots & u(t-k) & y(t) & y(t-1) & \cdots & y(t-l) \end{pmatrix}^T \tag{6.40}$$

The parameters are estimated by using the following projection estimator:

$$\hat{\theta}(t) = \hat{\theta}(t - 1) + \frac{\gamma \varphi(t - d)}{\alpha + \varphi^T(t - d)\varphi(t - d)} e(t) \tag{6.41}$$

$$e(t) = y(t) - \varphi^T(t - d)\hat{\theta}(t - 1)$$

with $0 < \gamma < 2$ and $\alpha > 0$. This estimator is the same as Eqs. (6.22) except that φ now has index $t - d$ instead of t. The properties given in Theorem 6.3 are still valid.

The control law is

$$\hat{R}^* u(t) + \hat{S}^* y(t) = t_0 A_o^* u_c(t) \tag{6.42}$$

or, equivalently,

$$\hat{\theta}^T(t)\left(A_o^* A_m^* \varphi(t)\right) = t_0 A_o^* u_c(t) \tag{6.43}$$

where $u_c(t)$ is the desired setpoint. Notice that it must be required that $\hat{\theta}_1(t) = \hat{r}_0(t) \neq 0$; otherwise, the control law is not causal.

Preliminaries

Since the proof consists of several steps, we outline the basic idea. The properties of the estimator were given in Theorem 6.3, which proved that the estimates are bounded and that a normalized prediction error converges to zero. However, the theorem does not show that the estimates converge. By introducing the control law and the properties of the system to be controlled, it can then be established that the signals are bounded and that the controlled output converges to the command signal.

If the input and output signals of the system can be shown to be bounded, then φ given by Eqs. (6.40) is bounded. If $\varphi(t-d)$ is bounded for all t, it follows from Property (ii) of Theorem 6.3 that the prediction error $e(t)$ goes to zero. The following result is useful to establish the boundedness of φ.

L E M M A 6.2   Key technical lemma

Let $\{s_t\}$ be a sequence of real numbers and let $\{\sigma_t\}$ be a sequence of vectors such that

$$\|\sigma_t\| \leq c_1 + c_2 \max_{0 \leq k \leq t} |s_k|$$

Assume that

$$z_t = \frac{s_t^2}{\alpha_1 + \alpha_2 \sigma_t^T \sigma_t} \to 0 \tag{6.44}$$

and that

$$\lim_{t \to \infty} s(t) = 0$$

where $\alpha_1 > 0$ and $\alpha_2 > 0$. Then $\|\sigma_t\|$ is bounded.

Proof: The result is trivial if s_t is bounded. Hence assume that s_t is not bounded. Then there exists a subsequence $\{t_n\}$ such that $|s_{t_n}| \to \infty$ and $|s_t| \leq s_{t_n}$ for $t \leq t_n$. For this sequence it follows that

$$\left| \frac{s_t^2}{\alpha_1 + \alpha_2 \sigma_t^T \sigma_t} \right| \geq \frac{s_t^2}{\alpha_1 + \alpha_2 (c_1 + c_2 |s_t|)^2} \geq \frac{1}{\alpha_3 c_2^2} > 0$$

where $0 < \alpha_3 < \alpha_2$. This contradicts Eq. (6.44) and proves the statements.   ☐

Main Result

The main result can now be stated as the following theorem.

T H E O R E M 6.7   Boundedness and convergence

Consider a system described by Eq. (6.38). Let the system be controlled with the adaptive control algorithm given by Eqs. (6.40), (6.41), and (6.42) where the command signal u_c is bounded. Assume that

A1: The time delay d is known.

A2: Upper bounds on the degrees of the polynomials A^* and B^* are known.

A3: The polynomial B has all its zeros inside the unit disc.

A4: The sign of $b_0 = r_0$ is known.

Then

(i)   The sequences $\{u(t)\}$ and $\{y(t)\}$ are bounded.

(ii) $\lim\limits_{t\to\infty} \left| A_m^*(q^{-1})y(t) - t_0 u_c(t-d) \right| = 0$

Proof: Introduce the control error

$$
\begin{aligned}
\varepsilon(t) &= A_o^* \left(A_m^* y(t) - t_0 u_c(t-d) \right) = P^* y(t) - t_0 A_o^* u_c(t-d) \\
&= P^* y(t) - \theta^T(t-d)\left(P^* \varphi(t-d) \right) \\
&= P^* e(t) + P^* \left(\theta^T(t-1)\varphi(t-d) \right) - \theta^T(t-d)\left(P^* \varphi(t-d) \right) \\
&= P^* e(t) + \sum_{i=0}^{\deg P} p_i \left(\theta(t-1-i) - \theta(t-d) \right)^T \varphi(t-d-i) \qquad (6.45)
\end{aligned}
$$

where $P = A_o A_m$ has been introduced to simplify the writing. The first two equalities are trivial. The third is obtained from Eq. (6.39), the fourth from Eqs. (6.41), and the last by expanding the expression.

It now follows from properties (ii) and (iii) of Theorem 6.3 that

$$
\lim_{t\to\infty} \frac{\varepsilon(t)}{\sqrt{\alpha + \varphi^T(t-d)\varphi(t-d)}} = 0
$$

It follows from the first equality in Eq. (6.45) that

$$
A_o^* A_m^* y(t) = \varepsilon(t) + t_0 A_o^* u_c(t)
$$

Since the polynomials A_o and A_m are stable and since u_c is bounded, it follows that

$$
|y(t)| \le \alpha_1 + \beta_1 \max_{0\le k\le t} |\varepsilon(k)|
$$

Moreover, since the polynomial B is stable, it follows that

$$
|u(t-d)| \le \alpha_2 + \beta_2 \max_{0\le k\le t} |y(k)|
$$

Hence

$$
|\varphi(t-d)| \le \alpha_3 + \beta_3 \max_{0\le k\le t} |\varepsilon(k)|
$$

If we apply Lemma 6.2, it follows that $\varphi(t)$ is bounded and that $\varepsilon(t) \to 0$ as $t \to \infty$. Since the polynomial A_o^* is stable, property (ii) also follows. $\square$

Remark 1.   We used an algorithm for which the details of the proof are simple. With minor modification the results can be extended to cover many of the direct algorithms given in Section 3.5.

Remark 2. A minor modification of the algorithm is necessary to ensure that $\hat{r}_0 \neq 0$. One way to do this is as follows: If $\hat{r}_0(t) = 0$, modify γ to give $\hat{r}_0(t) \neq 0$. Theorem 6.3 will still be valid with this modification of the algorithm. Since the estimator properties enter into the proof only via Theorem 6.3, the result still holds.

Remark 3. Notice that it does not follow that the parameter estimates converge. The fact that the control error nonetheless goes to zero depends on an interplay between the estimation and the control algorithms. This property is a special feature of direct algorithms.

Remark 4. The minimum-phase property is used to conclude that u is bounded when y is bounded.

Remark 5. Notice the similarity between Eq. (6.45) and the augmented error introduced in Chapter 5. □

Discussion

It has been established that a direct adaptive controller gives a closed-loop system with bounded signals and desired asymptotic properties, provided that Assumptions A1–A4 are valid. Assumptions A1 and A2 are necessary to write down the algorithm. Knowledge of the time delay (with a resolution corresponding to the sampling period) is essential. The signals will not be bounded if d is too small. Assumption A3 implies that the sampled system is minimum-phase; it is required because all process zeros are canceled in the design procedure. The error equation will not be linear in the parameters if this is not done. Assumption A4 is essential, since b_0 is absorbed in the adaptation gain γ, to guarantee that $\hat{r}_0(t) \neq 0$ for all times. Assumption A2 implies that the adaptive control law must have a sufficient number of parameters. This means that the model used to design the adaptive controller must be at least as complex as the process to be controlled. The consequences of violating the assumptions will be discussed later.

Extensions

The results can be extended in several different directions. Similar results can also be given in the continuous-time case, in which the underlying model can be written as

$$A(p)y(t) = B(p)u(t)$$

where A and B are polynomials in the differential operator $p = d/dt$. Assumptions A1–A4 are then replaced by the following assumptions:

A1': The pole excess $\deg A - \deg B$ is known.

A2': Upper bounds on the degrees of the polynomials A and B are known.

A3': The polynomial B has all its zeros in the left half-plane.

A4': The sign of b_0 is known.

The results can also be extended to systems with disturbances generated from known dynamics.

The gradient estimation algorithm can be replaced by other, more efficient methods. Theorem 6.3 then needs to be generalized. Many types of least-squares-like algorithms can be covered by replacing the function $V = \tilde{\theta}^T \tilde{\theta}$ in Theorem 6.3 by $V = \tilde{\theta}^T P^{-1} \tilde{\theta}$ and adding assumptions that guarantee that the eigenvalues of P stay bounded. Other control laws can also be treated. One important situation that has not been treated is the case in which the control signal is kept bounded by saturation. Theorem 6.3 still holds, but Theorem 6.7 does not, since Eq. (6.42) does not hold when the control signal saturates.

Gronwall-Bellman Lemma

The essential idea in the proof of Theorem 6.7 is the separation of the adaptive controller into two parts. First, some properties of the estimator are established that are independent of how the control signal is generated. Second, properties of the controlled system are derived. Convergence and stability are derived on the basis of the key technical lemma (Lemma 6.2). This procedure can be used for many different adaptive schemes.

The key technical lemma is a simplified version of the Gronwall-Bellman lemma, which is a standard tool for proving the existence of solutions to ordinary differential equations. There are both continuous-time and discrete-time versions of this lemma.

LEMMA 6.3   Gronwall-Bellman lemma: Continuous time

If $u, v \geq 0$, if c_1 is a positive constant, and if

$$u(t) \leq c_1 \int_0^t u(s)v(s)\,ds \qquad\qquad (6.46)$$

then

$$u(t) \leq c_1 \exp\left(\int_0^t v(s)\,ds \right) \qquad\qquad \square$$

LEMMA 6.4   Gronwall-Bellman lemma: Discrete time

If $u, v \geq 0$, if c_1 is a positive constant, and if

$$u(t) \leq c_1 \sum_{k=0}^{t-1} u(k)v(k) \qquad\qquad (6.47)$$

then

$$u(t) \leq c_1 \exp\left(\sum_{k=0}^{t-1} v(k) \right) \qquad\qquad \square$$

By using the Gronwall-Bellman lemma, many direct adaptive algorithms can be analyzed in the following way:

- Show that growth conditions such as Eq. (6.46) or Eq. (6.47) hold.
- Show properties analogous to Eq. (6.44) for the signals u and v.
- Use the Gronwall-Bellman lemma to get stability.

These steps can be used as a template for proving convergence and stability for adaptive algorithms.

6.6 AVERAGING

The results in the previous sections do not permit a detailed investigation of adaptive control algorithms. For example, no information about transient behavior is available until much more detailed analysis is undertaken. The conventional methods for investigating nonlinear systems involve investigation of equilibria and analysis of the local behavior near the equilibria. Such an approach will give only local properties, although in some special cases it may be possible to proceed further and obtain global properties. The results of the analysis can then be augmented by simulations. For purposes of this discussion it is useful to write the equations of motion of the complete system in a comprehensive form such as Eqs. (6.1) and (6.2) or Eqs. (6.3). In an adaptive system it is natural to separate the states of the system and the process parameters. The process parameters are changing more slowly than the states. This separation of time scales is used in the averaging theory to gain more insight about the properties of the closed-loop system. The idea of averaging originated in the analysis of planetary motion.

The Averaged Equations

The analysis of the dynamics of adaptive systems is generally quite complicated because the complete system is often of high order. Analysis of a direct algorithm for a discrete-time second-order system with four unknown parameters using a gradient method leads to a difference equation of order 8 (two states of the system, four parameters, and two difference equations to generate the regression variables). Ten more equations are obtained if a least-squares estimation algorithm is used.

Because of the special properties of adaptive systems, however, there is an approximate method that will simplify the analysis considerably. The basic idea is that the parameters change much more slowly than the other variables of the system. This property is intrinsic to the adaptive algorithms. If this were not the case, we could hardly justify using the notion of parameters.

To describe the averaging methods, consider the adaptive system described by Eqs. (6.1) and (6.2). The rate of change of the parameter $\hat{\theta}$ can be made arbitrarily small by choosing the adaptation gain γ sufficiently small. For simplicity we use the simple gradient algorithm

$$\frac{d\hat{\theta}}{dt} = \gamma \varphi(\vartheta, \xi) e(\vartheta, \xi) \tag{6.48}$$

The product φe on the right-hand side depends on ϑ and ξ, where $\vartheta = \vartheta(\hat{\theta})$ varies slowly and ξ varies fast. The key idea in the averaging method is to approximate the product φe by

$$G(\hat{\theta}) = \text{avg} \left\{ \varphi \left(\vartheta(\hat{\theta}), \xi(\vartheta(\hat{\theta}), t) \right) e \left(\vartheta(\hat{\theta}), \xi(\vartheta(\hat{\theta}), t) \right) \right\}$$

where avg$\{\cdot\}$ denotes the average and $\xi(\vartheta(\hat{\theta}), t)$ is computed under the assumption that the parameters $\hat{\theta}$ are constant. The average can be computed in many ways. Typical examples are

$$\text{avg} \left\{ f \left(\hat{\theta}, \xi(\hat{\theta}, t), t \right) \right\} = \frac{1}{T} \int_0^T f \left(\hat{\theta}, \xi(\hat{\theta}, t), t \right) dt$$

$$\text{avg} \left\{ f \left(\hat{\theta}, \xi(\hat{\theta}, t), t \right) \right\} = \lim_{T \to \infty} \int_0^T f \left(\hat{\theta}, \xi(\hat{\theta}, t), t \right) dt$$

$$\text{avg} \left\{ f \left(\hat{\theta}, \xi(\hat{\theta}, t), t \right) \right\} = E f \left(\hat{\theta}, \xi(\hat{\theta}, t), t \right)$$

The first alternative is applicable when f is periodic with period T, and the last equation applies when ξ is a stationary stochastic process. Notice that the averaged equations can be calculated only when the signals are bounded. This implies that the closed-loop system must be stable if the parameters $\hat{\theta}$ are fixed. The calculation of $\xi(\vartheta(\hat{\theta}), t)$ is a straightforward exercise in linear system analysis. However, the expressions may be complex for high-order systems. Symbolic calculation is a useful tool for carrying out the calculations. The use of averaging thus results in the following averaged nonlinear differential equation for the parameters:

$$\frac{d\bar{\theta}}{dt} - \gamma \, \text{avg} \left\{ \varphi \left(\vartheta(\bar{\theta}), \xi(\vartheta(\bar{\theta}), t) \right) e \left(\vartheta(\bar{\theta}), \xi(\vartheta(\bar{\theta}), t) \right) \right\} = 0 \tag{6.49}$$

This equation can also be written as

$$\frac{d\bar{\theta}}{dt} - \gamma \, \text{avg} \left\{ (G_{\varphi v}(\vartheta(\bar{\theta}), p)v) \, (G_{ev}(\vartheta(\bar{\theta}), p)v) \right\} = 0 \tag{6.50}$$

Notice that the transfer functions G_{ev} and $G_{\varphi v}$ depend on the averaged parameter $\bar{\theta}$. When the averaged equations are obtained, the behavior of the state variables ξ can be obtained by linear analysis.

Several averaging theorems give conditions for $\bar{\theta}$ being close to $\hat{\theta}$. The conditions typically require smoothness of the functions involved and periodicity or near periodicity of the time functions. There are also stochastic averaging theorems. Notice that averaging analysis was used in Theorems 4.1 and 4.2.

A significant advantage of averaging theory is that it reduces the dimensions of the problem. The theorems require that the adaptation gain be small, but experience has shown that averaging often gives a good approximation, even for large adaptation gains.

When the averaging equations are obtained, analysis proceeds in the conventional manner by investigation of the equilibria of the averaged equations and linearization at the equilibria to determine the local behavior. Notice that the averaged equations may possess equilibria (i.e., solutions to $\text{avg}\{\varphi e\} = 0$) even if the exact equations do not have an equilibrium. This corresponds to the case in which the true parameters are meandering in the neighborhood of the equilibrium to the averaged equation.

Sinusoidal Driving Forces

A simple case of averaging is when the external driving force is sinusoidal, that is, $v(t) = u_0 \sin \omega t$. The signals φ and e are then given by

$$\varphi(t) = G_{\varphi v}(\vartheta, \omega)v(t)$$
$$e(t) = G_{ev}(\vartheta, \omega)v(t)$$

Notice that controller parameters ϑ depend on $\bar{\theta}$. The following result is useful for calculation of the averages.

LEMMA 6.5   Averaging for sinusoidal input

Let G_v and G_w be stable transfer functions, and let v and w denote the steady-state responses of the corresponding systems to the input $u_c = u_0 \sin \omega t$. The mean value of the product vw is then given by

$$\text{avg}(vw) = \frac{u_0^2}{2} |G_v(i\omega)|\, |G_w(i\omega)| \cos\left(\arg G_v(i\omega) - \arg G_w(i\omega)\right)$$

$$= \frac{u_0^2}{2} \operatorname{Re}\left(G_v(i\omega)G_w(-i\omega)\right)$$

Proof: The signals v and w have the amplitudes $|G_v(i\omega)|$ and $|G_w(i\omega)|$; their phase angles are $\arg G_v(i\omega)$ and $\arg G_w(i\omega)$. Integrating over one period gives the result.   □

A true parameter equilibrium exists if the equation

$$G_{ev}\left(\vartheta(\bar{\theta}), \omega\right) = 0$$

has a unique solution. To derive a necessary condition we consider the averaged equation

$$\frac{d\bar{\theta}}{dt} = \gamma \operatorname{Re}\left\{G_{\varphi v}\left(\vartheta(\bar{\theta}), \omega\right) R_v G_{ev}^T\left(\vartheta(\bar{\theta}), -\omega\right)\right\} \tag{6.51}$$

where

$$R_v = \text{avg}\left(v\,v^T\right)$$

A necessary condition for Eq. (6.51) to have a unique parameter equilibrium is that v and $\bar{\theta}$ have equal dimension and that R_v be of full rank. To have a unique parameter equilibrium for slow external driving signals, it is thus necessary that the number of estimated parameters be less than or equal to the number of external driving signals and that the external driving signals be persistently exciting. This result indicates that there may be some disadvantages to overparameterization, contrary to what is indicated in Theorem 6.7. The local stability of the equilibrium θ^0 is given by the linearized equation

$$\frac{dx}{dt} = Ax$$

where x denotes the deviation from the equilibrium $\bar{\theta} - \bar{\theta}^0$ and

$$A = G_{\varphi v}\left(\vartheta(\bar{\theta}^0), \omega\right) R_v \frac{\partial}{\partial \theta} G_{ev}^T\left(\vartheta(\bar{\theta}^0), \omega\right)$$

The preceding equations can be applied to slow or constant perturbations by setting $\omega = 0$, provided that the assumptions of averaging are fulfilled.

An Example of Averaging Analysis

Consider a process with the transfer function $kG(s)$ and an adjustable feed-forward gain. Find a feedforward gain $\hat{\theta}$ such that the input-output behavior matches the transfer function $k_0 G_m(s)$ as well as possible. It is assumed that $k > 0$ and $k_0 > 0$. The case $G_m = G$ was discussed in Chapter 5. Two different algorithms for updating the gain were proposed in Chapter 5: the MIT rule and the SPR rule. The algorithms are

$$
\begin{aligned}
\frac{d\hat{\theta}}{dt} &= -\gamma y_m e \qquad \text{(MIT)} \\[2mm]
\frac{d\hat{\theta}}{dt} &= -\gamma u_c e \qquad \text{(SPR)}
\end{aligned}
\tag{6.52}
$$

where u_c is the command signal, $y_m = k_0 G_m u_c$ is the model output, and e is the error defined by

$$e(t) = y - y_m = kG(p)\left(\hat{\theta}(t)u_c(t)\right) - k_0 G_m(p)u_c(t)$$

The analysis in Section 5.5 shows that the MIT rule gives a closed-loop system that is globally stable for any adaptation gain γ in the "ideal" case, when $G = G_m$ and G is SPR. In the presence of unmodeled dynamics it is, of course, highly unrealistic to assume that a transfer function is SPR. So far, no stability result has been given for the MIT rule. However, Example 5.5 indicates that the

MIT rule will be unstable for sufficiently high adaptation gains if the system is not SPR.

We now investigate the algorithms under nonideal conditions, using averaging. Inserting the expressions for y_m and e into the equations for the parameters, we get

$$\frac{d\hat{\theta}}{dt} + \gamma(k_0 G_m u_c) \left(kG(\hat{\theta} u_c) - k_0 G_m u_c \right) = 0$$

$$\frac{d\hat{\theta}}{dt} + \gamma u_c \left(kG(\hat{\theta} u_c) - k_0 G_m u_c \right) = 0$$

(6.53)

where the first equation holds for the MIT rule and the second holds for the SPR rule. The corresponding averaging equations are

$$\frac{d\bar{\theta}}{dt} + \gamma \left(\bar{\theta} k k_0 \operatorname{avg}\{(G_m u_c)(G u_c)\} - k_0^2 \operatorname{avg}\{(G_m u_c)^2\} \right) = 0$$

$$\frac{d\bar{\theta}}{dt} + \gamma \left(\bar{\theta} k \operatorname{avg}\{u_c(G u_c)\} - k_0 \operatorname{avg}\{u_c(G_m u_c)\} \right) = 0$$

(6.54)

The equilibrium parameters are

$$\bar{\theta}_{\text{MIT}} = \frac{k_0}{k} \frac{\operatorname{avg}\{(G_m u_c)^2\}}{\operatorname{avg}\{(G_m u_c)(G u_c)\}}$$

$$\bar{\theta}_{\text{SPR}} = \frac{k_0}{k} \frac{\operatorname{avg}\{u_c(G_m u_c)\}}{\operatorname{avg}\{u_c(G u_c)\}}$$

(6.55)

The equilibrium values correspond to the true parameters for all command signals u_c only if $G = G_m$ (i.e., there are no unmodeled dynamics). When $G \neq G_m$, the equilibrium obtained will depend on the command signal as well as on the unmodeled dynamics. Notice that the equilibrium value obtained for the MIT rule minimizes the actual mean square error.

The stability conditions for the averaged equations (Eqs. 6.54) are

$$\gamma \operatorname{avg}\{(G_m u_c)(G u_c)\} > 0 \qquad \text{(MIT)}$$
$$\gamma \operatorname{avg}\{u_c(G u_c)\} > 0 \qquad \text{(SPR)}$$

The averaged equation when the MIT rule is used will thus give a stable equilibrium for all command signals if $G_m = G$. The stability condition depends on the command signal and the process dynamics as well as on the response model.

For the SPR rule the stability condition depends only on the command signal and on the process dynamics. The equilibrium is stable for all command signals if G is SPR. For processes that are not SPR the equilibrium may well be unstable. Consider the case of a command signal composed of a constant and a sum of sinusoids:

$$u_c(t) = a_0 + 2 \sum_{k=1}^{n} a_k \sin \omega_k t$$

If Lemma 6.5 is used, the stability conditions for $\gamma > 0$ become

$$a_0^2 G_m(0) + \sum_{k=1}^{n} a_k^2 |G_m(i\omega_k)| \, |G(i\omega_k)| \cos\left\{\arg G_m(i\omega_k) - \arg G(i\omega_k)\right\} > 0$$

$$a_0^2 G(0) + \sum_{k=1}^{m} a_k^2 \operatorname{Re} G(i\omega_k) > 0$$

For a single sinusoidal command signal the MIT rule gives a stable equilibrium if the phase lags of G_m and G differ by at most 90° at the frequencies of the input signal. The SPR rule, on the other hand, gives a stable equilibrium if the phase lag of the process is at most 90°.

For command signals containing several sinusoidals the equilibrium can still be stable, provided that the command signal is dominated by components with frequencies in the range in which the phase lag of the process is less than 90°. Notice that it helps to filter the command signal so that the signals in the frequency range in which the plant has a phase shift of more than 90° are attenuated. In the MIT rule, reduction of the gain of the model can also be reduced at high frequencies. It follows from Eqs. (6.54) that the convergence rate of the parameters is strongly signal-dependent. The value of normalization as described in Section 5.3 is that the convergence rate becomes less dependent on the signal amplitudes. The preceding calculations are illustrated by an example.

EXAMPLE 6.8   **Sinusoidal command signal**

Consider a reference model with the transfer function

$$G_m(s) = \frac{a}{s + a}$$

Assume that the process has the transfer function

$$G(s) = \frac{ab}{(s + a)(s + b)}$$

Furthermore, let the command signal be a sinusoid with unit amplitude and frequency ω. Equations (6.55) give the equilibrium values

$$\bar{\theta}_{\text{MIT}} = \frac{k_0}{k} \frac{b^2 + \omega^2}{b^2}$$

$$\bar{\theta}_{\text{SPR}} = \frac{k_0}{k} \frac{a(b^2 + \omega^2)}{b(ab - \omega^2)} \qquad \omega < \sqrt{ab}$$

The stability conditions show that the MIT rule is stable for all ω, but the SPR rule is stable only if $\omega < \sqrt{ab}$. Figure 6.10 shows the estimates of the gain for the case behavior $a = 1$ and $b = 10$ when the input signals have frequencies

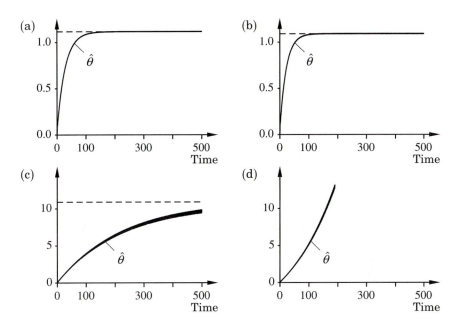

Figure 6.10 Estimated feedforward gains obtained by the MIT rule with sinusoidal input signals having frequencies (a) $\omega = 3$; (b) $\omega = 3.4$; and the SPR rule when (c) $\omega = 3$; (d) $\omega = 3.4$, for a system with $G_m = 1/(s + 1)$ and $G = 10/((s + 1)(s + 10))$. The dashed lines are the equilibrium values obtained from averaging analysis.

$\omega = 3$ and $\omega = 3.4$. The equilibrium values predicted by the averaging theory are also shown in the figure. The SPR is unstable for $\omega = 3.4 > \sqrt{10}$. Also notice the drastic difference in the equilibrium values between the different updating methods. The desired equilibrium value is $\theta_0 = k_0/k$.

The behavior is well predicted by the averaging analysis. Notice the difference in convergence rates. Initially, when $\bar{\theta} = 0$, the rates of changes are given by

$$\dot{\theta}_{\mathrm{MIT}} = \gamma k_0^2 \, \mathrm{avg}\{(G_m u_c)^2\}$$
$$\dot{\theta}_{\mathrm{SPR}} = \gamma k_0 \, \mathrm{avg}\{u_c (G_m u_c)\}$$

These expressions clearly show that the initial rates decrease with increasing frequency because $|G_m(i\omega)|$ decreases with frequency. For the SPR rule the rate decreases even more because of the phase lag between u_c and $G_m u_c$. □

In conclusion, we find that averaging analysis gives useful insights. It shows that analysis of the ideal case can be quite misleading. Even in the simple case of adjustment of a feedforward gain, unmodeled dynamics together with high-frequency excitation signals may lead to instability of the equilibrium. The equilibrium analysis also makes interesting contributions to the comparison of the MIT and SPR rules. First, the equilibrium of the MIT rule has a good physical interpretation as the parameter that minimizes the mean square error. Second, the apparent advantage of the SPR rule that very high adaptation gains can be used vanishes. In practical situations, there are always unmodeled dynamics. In the presence of unmodeled dynamics the gain must be kept small to maintain stability.

6.7   APPLICATION OF AVERAGING TECHNIQUES

In the previous sections, idealized cases were investigated. The convergence and stability analysis of self-tuning regulators were based on Assumptions A1–A4 and the premise that there are no disturbances. In Chapter 5 the stability of MRAS was proved under the SPR assumption on certain transfer functions. Assumption A2 in Theorem 6.7 implies that the model used to design the adaptive controller must be at least as complex as the process to be controlled. This is highly unrealistic because real processes are often distributed and also nonlinear.

In practice, adaptive controllers are based on simplified models. It is therefore of interest to investigate what happens when the process is more complex than assumed in the design of the controller. In this case the process is said to have *unmodeled dynamics*. If a controller is able to control processes with unmodeled dynamics and/or disturbances, we say that the controller is *robust*.

Analysis of a Simple MRAS

A simple model-reference adaptive system for a process of first order was derived in Example 5.2 by using the MIT rule. In Example 5.7 the same problem was considered, and an MRAS was obtained by using Lyapunov's stability theory. We now use averaging theory to investigate the properties of the controller. In designing the adaptive controller it is assumed that the nominal transfer function of the process is

$$G(s) = \frac{b}{s + a} \tag{6.56}$$

which is not necessarily the true transfer function of the process. The desired closed-loop system has the transfer function

$$G_m(s) = \frac{b_m}{s + a_m}$$

A model-reference adaptive control law was derived in Example 5.7 by using Lyapunov theory. A block diagram of the closed-loop system is given in

Fig. 5.11. The system is described by the equations

$$\frac{d\hat{\theta}_1}{dt} = -\gamma u_c e$$

$$\frac{d\hat{\theta}_2}{dt} = \gamma y e$$

$$e = y - y_m \tag{6.57}$$

$$y = G(p)u$$

$$y_m = G_m(p)u_c$$

$$u = \hat{\theta}_1 u_c - \hat{\theta}_2 y$$

where u_c is the reference signal, u is the process input, y is the process output, y_m is the output of the reference model, e is the error, $\hat{\theta}_1$ is the adjustable feedforward gain, and $\hat{\theta}_2$ is the adjustable feedback gain.

It is not possible to give a complete analysis of Eqs. (6.57) for general reference signals; approximations must be made even in a simple case like this. We now investigate the adaptive system when the reference signal is sinusoidal. The equilibrium points are first explored, and the behavior in their neighborhood is then investigated by averaging and linearization.

Equilibrium Values for the Parameters

It follows from Eqs. (6.57) that the parameters $\hat{\theta}_1$ and $\hat{\theta}_2$ are constant when the error e is zero. The conditions for e to be zero will now be investigated. The signal transmission from the command signal u_c to the output y is described by the transfer function

$$G_c = \frac{\hat{\theta}_1 G}{1 + \hat{\theta}_2 G}$$

and the control error becomes

$$e(t) = y(t) - y_m(t) = (G_c(p) - G_m(p))\,u_c(t)$$

Let the reference signal be $u_c = u_0 \sin \omega t$. The error e is then zero if

$$G_c(i\omega) = G_m(i\omega)$$

or

$$\hat{\theta}_1^0 G(i\omega) = \hat{\theta}_2^0 G_m(i\omega)G(i\omega) + G_m(i\omega) \tag{6.58}$$

This equation can be solved for $\hat{\theta}_1^0$ and $\hat{\theta}_2^0$ by equating the real and imaginary parts. There is a unique solution if $\text{Im}\{G(i\omega)\} \neq 0$. The solutions are easily obtained by dividing Eq. (6.58) by $G_m G$ and G, respectively, and taking imaginary parts. This gives

$$\hat{\theta}_1^0 = \frac{\text{Im}\{1/G(i\omega)\}}{\text{Im}\{1/G_m(i\omega)\}}$$

$$\hat{\theta}_2^0 = -\frac{\text{Im}\{G_m(i\omega)/G(i\omega)\}}{\text{Im}\,G_m(i\omega)}$$

(6.59)

In the nominal case we get $\hat{\theta}_1^0 = b_m/b$ and $\hat{\theta}_2^0 = (a_m - a)/b$. These equilibrium values do not depend on the frequency of the command signal. They also correspond to the desired feedback gains.

Averaging

The command signal u_c is the only external signal; hence $v = u_c$. Furthermore, $\varphi^T = \begin{pmatrix} u_c & y \end{pmatrix}$. To obtain the averaging equations, the transfer functions G_{ev} and $G_{\varphi v}$ are first calculated:

$$G_{ev} = \frac{\hat{\theta}_1 G}{1 + \hat{\theta}_2 G} - G_m$$

$$G_{\varphi v}^T = \begin{pmatrix} -1 & \dfrac{\hat{\theta}_1 G}{1 + \hat{\theta}_2 G} \end{pmatrix}$$

By using Lemma 6.5 the averaged equations can now be written as

$$\frac{d\bar{\theta}_1}{dt} = -\frac{\gamma u_0^2}{2}\,Re\left\{\frac{\bar{\theta}_1 G(i\omega)}{1 + \bar{\theta}_2 G(i\omega)} - G_m(i\omega)\right\}$$

$$\frac{d\bar{\theta}_2}{dt} = \frac{\gamma u_0^2}{2}\,Re\left\{\left(\frac{\bar{\theta}_1 G(i\omega)}{1 + \bar{\theta}_2 G(i\omega)} - G_m(i\omega)\right)\frac{\bar{\theta}_1 G(-i\omega)}{1 + \bar{\theta}_2 G(-i\omega)}\right\}$$

(6.60)

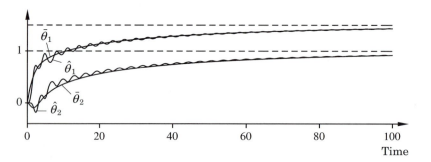

Figure 6.11   Parameter estimates and their approximation by the averaging method. The dashed lines show the equilibrium values of the gains.

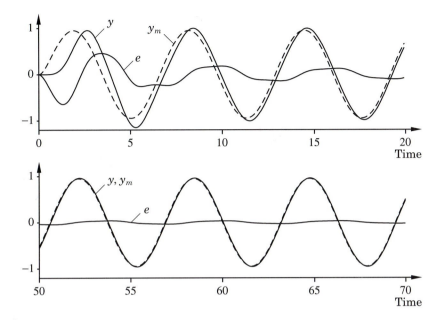

Figure 6.12   System output y (solid line) and the output of the reference model y_m (dashed line) and error e for Example 6.9 for $t=0$–20 and $t=50$–70.

Notice that these equations are valid also when G is a general transfer function, that is, G does not need to satisfy Eq. (6.56).

Accuracy of averaging

Consider the particular case of $a = 1$, $b = 2$, and $a_m = b_m = 3$. Let the adaptation gain γ be 1, and let the command signal be $u_0 \sin t$. The time histories of the parameter estimates $\hat{\theta}_1$, $\hat{\theta}_2$ and their approximations $\bar{\theta}_1$, $\bar{\theta}_2$ are shown in Fig. 6.11. The figure shows that the averaging gives a good approximation in this case. Notice that the approximation improves with time. The process output y and the output of the reference model y_m are shown in Fig. 6.12. Notice that the signals are already quite close after 10 s, although the parameters are quite far from their correct values at this time. The error $e = y - y_m$ thus appears to converge much faster than the parameters. This was seen for several different adaptive controllers in the previous chapters. Also notice that much faster convergence will be obtained with a recursive least-squares method.   □

Local Stability

The stability of the equilibrium of the averaged equations (Eqs. 6.60) will now be investigated. Straightforward but tedious calculations give the following linearized equation:

$$\frac{dx}{dt} = Ax \tag{6.61}$$

where x is a vector whose two components are the deviations of $\bar\theta_1$ and $\bar\theta_2$ from their equilibrium values and the matrix A is given by

$$A = \frac{\gamma u_0^2 |G_m|}{2\bar\theta_1^0} \begin{pmatrix} -\cos\theta_m & |G_m|\cos 2\theta_m \\ |G_m| & -|G_m|^2 \cos\theta_m \end{pmatrix} \tag{6.62}$$

where $\bar\theta_1^0$ is the equilibrium value of $\bar\theta_1$ and $\theta_m = \arctan(\omega/a_m)$. The matrix A has the characteristic equation

$$\lambda^2 + \alpha\lambda(1 + \cos^2\theta_m) + \alpha^2 \sin^2\theta_m = 0$$

where

$$\alpha = \frac{\gamma u_0^2 a_m b}{2(a_m^2 + \omega^2)}$$

The characteristic equation has its zeros in the left half-plane if $\omega \neq 0$. The equilibrium of the linearized equation (Eq. 6.61) is thus stable for all $\omega \neq 0$. The investigated MRAS has been designed by using Lyapunov theory. In the idealized case the transfer function (6.56) is SPR, and it is expected that the MRAS should have good performance.

Unmodeled Dynamics

The consequences of unmodeled dynamics are now investigated for the MRAS shown in Fig. 5.11. This system was designed on the basis of the assumption that the transfer function of the process has the form (6.56). We now investigate what happens if the process actually has a pole excess larger than 1. Before we go into details, a specific example is investigated.

EXAMPLE 6.10   **Unmodeled dynamics**

Assume that the nominal transfer function (6.56) has $a = 1$ and $b = 2$ but that the actual transfer function is

$$G(s) = \frac{458}{(s+1)(s^2 + 30s + 229)} \tag{6.63}$$

The dynamics correspond to the nominal plant $2/(s+1)$ cascaded with $229/(s^2 + 30s + 229)$. The process thus has two poles $s = -15 \pm 2i$, which were neglected in the model used to design the adaptive controller. Figure 6.13 shows the behavior of the controller parameters when the command signal is a step and

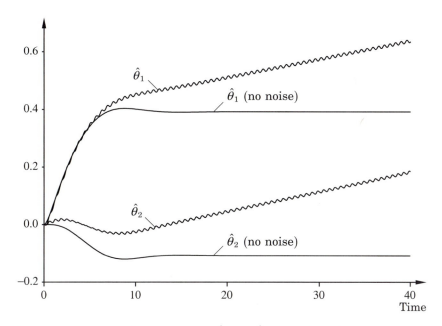

Figure 6.13 Controller parameters $\hat{\theta}_1$ and $\hat{\theta}_2$ when the adaptive control law of Eqs. (6.57) is applied to the process of Eq. (6.63). The command signal is a step, and there is sinusoidal measurement noise. The smooth curves show the behavior when there is no measurement noise.

there is a sinusoidal measurement error. Figure 6.14 shows the behavior of the parameters when the command signal is sinusoidal with different frequencies.

□

Example 6.10 shows that the presence of unmodeled dynamics will drastically change the behavior of the adaptive system. Figure 6.14 shows that the equilibrium depends on the frequency of the command signal and that it may be unstable for certain frequencies. We now attempt to understand the mechanisms that change the behavior of the system so drastically and to find suitable remedies.

Step Commands

First, the behavior illustrated in Fig. 6.13 is analyzed. The case of step commands is first investigated when there is no measurement noise. When $\omega = 0$, the equilibrium condition of Eq. (6.58) reduces to

$$\bar{\theta}_2 = \frac{1}{G_m(0)} \bar{\theta}_1 - \frac{1}{G(0)} \tag{6.64}$$

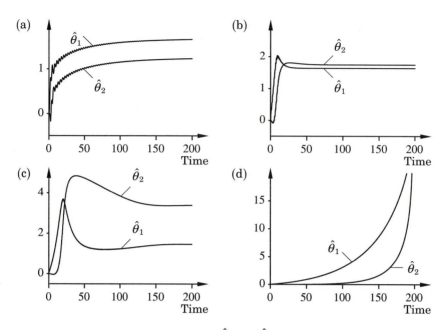

Figure 6.14   Controller parameters $\hat{\theta}_1$ and $\hat{\theta}_2$ when the adaptive control law of Eqs. (6.57) is applied to the process of Eq. (6.63) when the command signal is $u_c = \sin \omega t$ with (a) $\omega = 1$; (b) $\omega = 3$; (c) $\omega = 6$; (d) $\omega = 20$.

The equilibrium set is thus a straight line in the parameter space. The line is uniquely determined by the steady-state gains $G(0)$ and $G_m(0)$. Notice in particular that the equilibrium set is not a point. This is easily understood from the viewpoint of system identification. We wish to determine two parameters, $\hat{\theta}_1$ and $\hat{\theta}_2$. However, the excitation used is a step that is persistently exciting of first order and thus admits determination of only one parameter. (See Example 2.5.)

Averaging is now applied to obtain further insight into the behavior of the system. The averaging analysis applies to the set of parameter values such that the closed-loop system is stable for fixed parameters. To find this set, notice that the closed-loop system is a linear time-invariant system when parameters $\hat{\theta}_1$ and $\hat{\theta}_2$ are constant. The closed-loop eigenvalues are the zeros of the equation

$$1 + \hat{\theta}_2 G(s) = 0$$

A necessary condition for stability is that $1 + \hat{\theta}_2 G(s)$ has its roots in the left half-plane. This condition is also sufficient in the nominal case because the transfer function $G(s)$ is then SPR, and arbitrarily large feedback gains can be used. When there are unmodeled dynamics, the transfer function $G(s)$ is usually not SPR and the closed-loop system typically becomes unstable when $\hat{\theta}_2$ is sufficiently large.

EXAMPLE 6.11   **Step commands**

With the transfer function of Eq. (6.63) used in Example 6.10, the closed-loop characteristic equation is given by

$$(s + 1)(s^2 + 30s + 229) + 458\theta_2 = 0$$

or

$$s^3 + 31s^2 + 259s + 229 + 458\theta_2 = 0$$

This equation has all roots in the left half-plane if

$$-0.5 < \hat{\theta}_2 < 17.03 = \theta_2^{stab}$$

The averaged equations for the parameter estimates are obtained by setting $\omega = 0$ in Eqs. (6.60). If it is assumed that $G_m(0) = 1$, the equations become

$$\frac{d\bar{\theta}_1}{dt} = -\frac{\gamma u_0^2}{2}\left(\frac{\bar{\theta}_1 G(0)}{1 + \bar{\theta}_2 G(0)} - 1\right)$$

$$\frac{d\bar{\theta}_2}{dt} = \frac{\gamma u_0^2}{2}\frac{\bar{\theta}_1 G(0)}{1 + \bar{\theta}_2 G(0)}\left(\frac{\bar{\theta}_1 G(0)}{1 + \bar{\theta}_2 G(0)} - 1\right) \tag{6.65}$$

These differential equations have the equilibrium set of Eq. (6.64).

Close to the equilibrium set, the equations are described by the following linearized equation:

$$\frac{dx}{dt} = \frac{\gamma u_0^2}{2\theta_1^0}\begin{pmatrix} -1 & 1 \\ 1 & -1 \end{pmatrix}x \tag{6.66}$$

where $x_1 = \bar{\theta}_1 - \theta_1^0$ and $x_2 = \bar{\theta}_2 - \theta_2^0$. Consider a point away from the equilibrium line, that is, $x_2 = x_1 + \delta$ or $\bar{\theta}_2 = \bar{\theta}_1 - 1/G(0) + \delta$. The velocity of the state vector at that point is $\dot{x}_1 = \gamma u_0^2 \delta/\theta_1^0$, $\dot{x}_2 = -\gamma u_0^2 \delta/\theta_1^0$. The vector field of the linearized equation is thus as shown in Fig. 6.15. The vector field thus

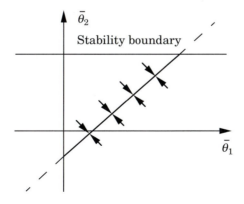

Figure 6.15   Equilibrium set and local behavior of the averaged equations.

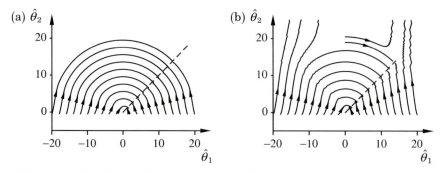

Figure 6.16 Phase plane of the controller parameters (a) in the nominal case of $G(s) = 2/(s+1)$ and (b) in the case of unmodeled dynamics Eq. (6.63). The dashed lines are the equilibrium sets of the parameters in the nominal case.

pushes the parameter toward the equilibrium for $\bar{\theta}_1 > 0$ and away from the equilibrium for $\bar{\theta}_1 < 0$. Notice that the system is not structurally stable because one eigenvalue of the linearized equation is zero. This means that we can expect drastically different properties when the system is perturbed.

It is usually difficult to go beyond the local analysis. However, in this particular case it is possible to obtain the global properties of the averaged equation. Outside the equilibrium set of Eq. (6.64), the averaged equations (Eqs. 6.65) can be divided to give

$$\frac{d\bar{\theta}_2}{d\bar{\theta}_1} = -\frac{G(0)\bar{\theta}_1}{1 + \bar{\theta}_2 G(0)}$$

This differential equation has the solution

$$\bar{\theta}_2^2 + \frac{2}{G(0)} \bar{\theta}_2 + \bar{\theta}_1^2 = \text{const}$$

The parameters of the averaged equations will thus move along circular paths with the center at $(0, -1/G(0))$. The motion is clockwise for $\bar{\theta}_2 > \bar{\theta}_1 - 1/G(0)$ and counterclockwise for $\bar{\theta}_2 < \bar{\theta}_1 - 1/G(0)$. The motion slows down and stops when the parameters reach the equilibrium set

$$\{\bar{\theta}_1, \ \bar{\theta}_2 | \bar{\theta}_1 > 0, \ \bar{\theta}_2 = \bar{\theta}_1 - 1/G(0)\}$$

The averaged equation approximates the nonlinear equations for the parameters only for parameters such that the closed-loop system is stable. In the nominal case, when the transfer function of the plant is $G(s) = 2/(s + 1)$, the stability region is $-1/G(0) < \hat{\theta}_2$. In the case of unmodeled dynamics the stability region is defined by $-1/G(0) < \hat{\theta}_2 < \hat{\theta}_2^{stab}$. This means that trajectories that start far away from the origin will escape from the stability region. Figure 6.16 shows the actual parameter paths in the nominal case and for the unmodeled dynamics given by the transfer function of Eq. (6.63) in Example 6.10. With unmodeled dynamics the trajectories will diverge if the initial values are too large. The deviation from circular arcs is due to the initial transient when $y(t)$ is different from the equilibrium value. The adaptation gain used in the example is quite large ($\gamma = 1$). The trajectories will be arbitrarily close to circles by choosing γ sufficiently small. The "jitter" in the trajectories in Fig. 6.16(b) is caused by oscillations in the parameters, not numerical errors.   □

The analysis and the simulations show that the adaptive system can be unstable if the input signal is a step and if there are unmodeled dynamics.

Measurement Noise

We now investigate the effects of measurement noise. The simulation shown in Fig. 6.13 indicates that measurement noise may cause the parameters to drift. Figure 6.17 shows parameter $\hat{\theta}_2$ as a function of parameter $\hat{\theta}_1$ with and without measurement noise. The simulation indicates that the equilibrium is lost in the presence of measurement noise. The parameters move toward a set close to the equilibrium set, oscillate rapidly in the neighborhood of this set, and drift along the set. The analysis tools developed will now be used to explain the behavior of the system. Assume that the command signal is a step with amplitude u_0 and that the measurement noise can be modeled as an additive zero mean signal n at the process output. It follows from Eqs. (6.57) that

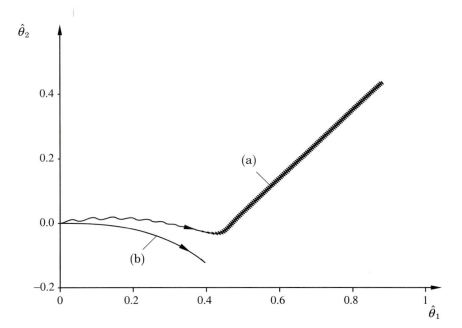

Figure 6.17 Phase plane of the controller parameters (a) with and (b) without measurement noise.

the error cannot be made identically zero by proper choice of the parameters. Hence no true equilibrium exists such that the parameters are constant. The phenomenon is a typical behavior of a system that lacks structural stability. Intuitively, the results can be explained as follows: A step input is persistently exciting of order 1 only, which means that it admits consistent estimation of one parameter only. When two parameters are adjusted, the equilibrium values of the parameters make a submanifold, not a point. Measurement errors and other disturbances may cause the parameters to drift along the equilibrium set. In the presence of unmodeled dynamics, the feedback gain may then become so large that the closed-loop system becomes unstable. By using averaging, the equilibrium set and the drift rate along the set can be determined.

The parameter values will drift also in the nominal case. However, the closed-loop system is stable for all parameter values.

Sinusoidal Command Signals

Several of the difficulties encountered with step commands are due to the fact that a step is persistently exciting of first order only. This means that the equilibrium set is a manifold and only a linear combination of the parameters can be determined. With a sinusoidal command signal that is persistently exciting of second order, two parameters can be determined consistently. It may therefore be expected that some of the difficulties will disappear. However, the simulation shown in Fig. 6.14 indicates that there are some problems with sinusoidal command signals in combination with unmodeled dynamics.

As before, it is assumed that the adaptive controller is designed as if the process were described by the transfer function

$$G(s) = \frac{b}{s + a}$$

Since the character of the unmodeled dynamics is important, it is assumed that the actual plant is described by the frequency function

$$G(i\omega) = \frac{b}{a + i\omega} r(\omega)e^{-i\phi(\omega)} \tag{6.67}$$

The functions r and ϕ represent the distortions of amplitude and phase due to unmodeled dynamics. It is assumed that the transfer function corresponding to r and ϕ has no poles in the right half-plane.

The unmodeled dynamics may change the properties of the system drastically. For example, the nominal system will be stable for all values of the feedback gain, since it is SPR. If the unmodeled dynamics are such that the additional phase lag can be large, the system with unmodeled dynamics will be unstable for sufficiently large feedback gains. The critical gain can be determined as follows. The phase lag of the plant is $\phi(\omega) + \arctan(\omega/a)$. This lag is π if

$$\frac{\omega}{a} = \tan(\pi - \phi(\omega)) = -\tan\phi(\omega)$$

or

$$\omega \cos \phi(\omega) + a \sin \phi(\omega) = 0 \tag{6.68}$$

The process gain of this frequency is

$$|G(i\omega)| = \frac{b\,r(\omega)}{\sqrt{a^2 + \omega^2}}$$

The system thus becomes unstable for the gain

$$\bar{\theta}_2 = \bar{\theta}_2^0 = \frac{\sqrt{a^2 + \omega^2}}{b\,r(\omega)} \tag{6.69}$$

where ω is the smallest value that satisfies Eq. (6.68).

Equilibrium Analysis

The possible equilibria of the parameters will first be determined. Introducing the transfer function of Eq. (6.67) into Eq. (6.59) gives (after straightforward but tedious calculations)

$$\hat{\theta}_1 = \frac{b_m}{b} \cdot \frac{(a \sin \phi(\omega) + \omega \cos \phi(\omega))}{\omega r(\omega)}$$

$$\hat{\theta}_2 = \frac{\omega(a_m - a) \cos \phi(\omega) + (\omega^2 + a a_m) \sin \phi(\omega)}{\omega b r(\omega)} \tag{6.70}$$

$$= \frac{1}{b r(\omega)} \left((\omega \sin \phi(\omega) - a \cos \phi(\omega)) + \frac{a_m}{\omega} (a \sin \phi(\omega) + \omega \cos \phi(\omega)) \right)$$

A comparison with the nominal case shows that the equilibrium will be shifted because of the unmodeled dynamics. The shift in the equilibrium depends on the frequency of the input signal as well as on the unmodeled dynamics.

It is of particular interest to determine whether there are conditions that may lead to difficulties. The feedforward gain vanishes for frequencies such that Eq. (6.68) is satisfied. This is precisely the frequency at which the process has a phase lag of 180°. The feedback gain for this frequency is

$$\hat{\theta}_2 = \frac{1}{br(\omega)} (\omega \sin \phi - a \cos \phi) = \frac{\sqrt{a^2 + \omega^2}}{br(\omega)}$$

This implies that $\hat{\theta}_2 |G(i\omega)| = 1$, that is, that the loop gain then becomes unity.

We thus find that the equilibrium values of the parameters for sinusoidal input signals will depend on the unmodeled dynamics and the frequency of the sinusoidal command signal. When the frequency is such that the plant has a phase shift of 180°, the feedforward gain is zero and the feedback gain is such that the closed-loop system is unstable. This observation is illustrated by an example.

EXAMPLE 6.12   **Sinusoidal command signal**

Consider the system in Example 6.10. The transfer function with the unmodeled dynamics is

$$G(s) = \frac{458}{(s+1)(s^2 + 30s + 229)} = \frac{458}{s^3 + 31s^2 + 259s + 229}$$

The equilibrium values of the controller gains are

$$\bar{\theta}_1 = \frac{3(259 - \omega^2)}{458}$$

$$\bar{\theta}_2 = \frac{2(137 + 7\omega^2)}{229}$$

when $a_m = b_m = 3$. The transfer function G has a phase shift of 180° at $\omega = \sqrt{259} = 16.09$. At this frequency the equilibrium values of the controller gains are $\hat{\theta}_1 = 0$ and $\hat{\theta}_2 = 3900/229 = 17.03$. The closed-loop system is unstable for this feedback gain. This explains the results shown in Fig. 6.14.

□

Summary of the MRAS Examples

The investigation of the first-order MRAS is summarized in the following table:

Inputs	Exact Model Structure	Unmodeled Dynamics
Step command	Equilibrium set is a half-line.	Equilibrium set is a line segment. Stability is lost for some initial values.
Step command + measurement noise	Solution will move toward a line and then drift along the line.	Solution will move toward a line and drift along the line until stability is lost.
Sinusoidal	Equilibrium set is a point that is independent of the frequency.	Equilibrium set is a point that depends on the frequency. The equilibrium is unstable for sufficiently high frequencies.

Several interesting conclusions can be drawn from the examples. When the input signal is not sufficiently exciting, the equilibrium is a manifold independently of the presence of unmodeled dynamics or disturbances. When there are disturbances, the estimates will drift along the manifold. In the case of unmodeled dynamics the closed-loop system may eventually become unstable. From a methodological point of view the examples give insights that can be derived from equilibrium analysis, which can be carried out with a moderate effort in many cases. We can find out if an equilibrium exists in the sense that the parameters remain constant. Notice that the averaged equations may have an equilibrium even if the exact equations do not. However, it is rarely the case that global analysis can be carried out.

6.8 AVERAGING IN STOCHASTIC SYSTEMS

The importance of averaging was illustrated in the previous sections. However, the excitation has been restricted to constant or sinusoidal inputs. In this section averaging is used on discrete-time systems with stochastic inputs. Assume that the system is described by

$$A^*(q^{-1})y(t) = B^*(q^{-1})u(t-d) + C^*(q^{-1})e(t) \tag{6.71}$$

where $e(t)$ is a zero-mean Gaussian stochastic process. Depending on the specifications, different self-tuning regulators can be used to control the system (compare Chapter 4). For simplicity it is assumed that the basic direct self-tuning algorithm (Algorithm 4.1) is used. The controller parameters are then estimated from a model of the form

$$y(t) = R^*(q^{-1})u(t-d) + S^*(q^{-1})y(t-d) \tag{6.72}$$

or

$$y(t) = \varphi(t-d)^T \theta \tag{6.73}$$

The parameters θ are estimated by using the recursive least-squares method. In applying averaging, it is appropriate to use the form

$$\hat{\theta}(t) = \hat{\theta}(t-1) + \gamma(t)R(t)^{-1}\varphi(t-d)\left(y(t) - \varphi^T(t-d)\hat{\theta}(t-1)\right)$$
$$R(t) = R(t-1) + \gamma(t)\left(\varphi(t-d)\varphi^T(t-d) - R(t-1)\right) \tag{6.74}$$

where the covariance matrix $P(t)$ is related to $R(t)$ through

$$P(t) = \gamma(t)R(t)^{-1}$$

and $\gamma(t) = 1/t$. In some cases it is convenient to replace the matrix $R(t)$ by a scalar $r(t)$. This gives shorter computation times and requires less storage, but it gives slower convergence. For stochastic approximation we obtain

$$r(t) = r(t-1) + \gamma(t)\left(\varphi(t-d)^T\varphi(t-d) - r(t-1)\right) \tag{6.75}$$

The controller is

$$u(t) = -\frac{\hat{S}^*(q^{-1})}{\hat{R}^*(q^{-1})}y(t) \tag{6.76}$$

or

$$\varphi(t)^T\hat{\theta}(t) = 0$$

The self-tuning regulator is described by Eqs. (6.73) and (6.74). The control law of Eq. (6.76) is then used on the system of Eq. (6.71). The resulting closed-loop system is a set of nonlinear, stochastic difference equations, which can be very difficult to analyze. The difficulty arises mainly from the interplay between the estimated parameters as well as the fact that these parameters are used in the controller. By using the averaging idea it is possible to derive *associated deterministic differential equations*. The convergence properties of the algorithm can then be determined by using these equations. The method was suggested by Ljung in 1977 and is sometimes called the *ODE (ordinary differential equation) approach*. Only a heuristic derivation and motivation are given here; further details can be found in the references at the end of this chapter.

A Heuristic Derivation

For sufficiently large t the step size $\gamma(t)$ in Eqs. (6.74) is small, and the correction in $\hat{\theta}(t)$ is small. As in Section 6.6, we can separate the states from the parameters and assume that the parameters are constant in evaluating the behavior of the closed-loop system. Both $R(t)$ and $\varphi(t)$ depend on the parameter estimates. Since $\hat{\theta}$ is assumed to change slowly, the behavior of the model can be approximated by

$$y(t) = \varphi^T(t - d, \bar{\theta})\bar{\theta}$$

where $\bar{\theta}$ is the averaged value of the estimates. Also, φ depends on the estimated variables through the feedback. The updating equation for R can be approximated by

$$\bar{R}(t) = \bar{R}(t - 1) + \gamma(t)\left(G(\bar{\theta}) - \bar{R}(t - 1)\right) \qquad (6.77)$$

where

$$G(\bar{\theta}) = E\left\{\varphi(t - d, \bar{\theta})\varphi^T(t - d, \bar{\theta})\right\} \qquad (6.78)$$

The expectation is taken with respect to the underlying stochastic process in Eq. (6.71) and evaluated for the fixed value of the parameters $\bar{\theta}$. In the same way the parameter update is approximated by

$$\bar{\theta}(t) = \bar{\theta}(t - 1) + \gamma(t)\bar{R}(t)^{-1}f(\bar{\theta}) \qquad (6.79)$$

where

$$f(\bar{\theta}) = E\left\{\varphi(t - d, \bar{\theta})\left(y(t) - \varphi^T(t - d, \bar{\theta})\bar{\theta}\right)\right\} \qquad (6.80)$$

Equations (6.79) and (6.77) are the averaged difference equations describing the estimator. Now let $\Delta\tau$ be a small number, and let t' be defined by

$$\Delta\tau = \sum_{k=t}^{t'} \gamma(k)$$

Then

$$\bar{\theta}(t') = \bar{\theta}(t) + \Delta\tau\bar{R}(t)^{-1}f\left(\bar{\theta}(t)\right)$$
$$\bar{R}(t') = \bar{R}(t) + \Delta\tau\left(G\left(\bar{\theta}(t)\right) - \bar{R}(t)\right)$$

With a change of time scale such that $t = \tau$ and $t' = t + \Delta\tau$, these equations can be seen as a difference approximation of the ordinary differential equations

$$\frac{d\bar{\theta}}{d\tau} = \bar{R}(\tau)^{-1}f\left(\bar{\theta}(\tau)\right) \tag{6.81}$$

$$\frac{d\bar{R}}{d\tau} = G\left(\bar{\theta}(\tau)\right) - \bar{R}(\tau) \tag{6.82}$$

If stochastic approximation is used, Eq. (6.82) is replaced by

$$\frac{d\bar{r}}{d\tau} = g\left(\bar{\theta}(\tau)\right) - \bar{r}(\tau)$$

where

$$g(\bar{\theta}) = E\left\{\varphi^T(t-d)\varphi(t-d)\right\}$$

and $\bar{R}$ is replaced by $\bar{r}$ in Eq. (6.81). These equations are called the *associated ordinary differential equations* to Eqs. (6.74) and (6.75). They are a special kind of averaged equations. First, the difference equations are replaced by differential equations; second, there is a time scaling compared with the original system. The time scaling can be interpreted as a logarithmic compression of the original time. That is, more and more steps of length $\gamma(t)$ are needed to get the step $\Delta\tau$ as the time progresses.

The arguments leading to Eqs. (6.81) and (6.82) have been heuristic. However, it can be rigorously shown that, provided that the estimates $\hat{\theta}(t)$ are "sufficiently often" in the domain of attraction of the associated differential equations, then

- Only stable stationary points of Eqs. (6.81) and (6.82) are possible convergence points for the estimates.

- The trajectories $\bar{\theta}(\tau)$ are the "asymptotic paths" of the estimates $\hat{\theta}(t)$.

The associated ODE can be used to find possible convergence points of an adaptive algorithm, $\bar{\theta}^0$ and $\bar{R}^0$. The equations can then be linearized around these stationary points. It is easily seen that the linearized equations are

$$\frac{d}{dt}\begin{pmatrix} \bar{\theta} - \bar{\theta}^0 \\ \bar{R} - \bar{R}^0 \end{pmatrix} = \begin{pmatrix} G(\bar{\theta})^{-1}\dfrac{\partial f(\bar{\theta})}{\partial\bar{\theta}} & 0 \\ X & -I \end{pmatrix}_{\bar{\theta}\,=\,\bar{\theta}^0}\begin{pmatrix} \bar{\theta} - \bar{\theta}^0 \\ \bar{R} - \bar{R}^0 \end{pmatrix}$$

where the element X is not important for the local stability. The stationary point is thus stable if the matrix

$$K = G(\bar{\theta})^{-1}\frac{\partial f(\bar{\theta})}{\partial\bar{\theta}}\bigg|_{\bar{\theta}\,=\,\bar{\theta}^0} \tag{6.83}$$

has all its eigenvalues in the left half-plane. The associated ODEs can thus be used in the following way:

1. Compute the expressions for $\varphi(t)$ and $\varepsilon(t) = y(t) - \varphi(t-d)^T \bar{\theta}$ for a fixed value of $\bar{\theta}$.

2. Compute the expected values $G(\bar{\theta})$ and $f(\bar{\theta})$.

3. Determine possible convergence points for Eqs. (6.81) and (6.82), and determine the local stability properties by using Eq. (6.83).

4. Simulate the equations.

Even if Eqs. (6.81) and (6.82) can be quite difficult to analyze in detail, it is usually easy to determine the possible stationary points. The equations can also be simulated to obtain a feel for the behavior of the convergence properties. The change in the time scale makes it more favorable to simulate the ODEs than the averaged difference equations.

Stability of Stochastic Self-Tuners

Averaging methods can be used for stability analysis of stochastic self-tuning regulators. Consider a simple self-tuner based on least-squares estimation and minimum-variance control (Algorithm 4.1 with $Q^* = P^* = 1$). Let the algorithm be applied to a system described by Eq. (6.71). The self-tuner is assumed to be compatible with the model in the sense that the time delay and the model orders are the same. The closed-loop system is globally stable if the pulse transfer function

$$G(z) = \frac{1}{C(z)} - \frac{1}{2}$$

is SPR (see Ljung (1977b)). The filter P^* that is used to filter the regressors can be interpreted as an estimate of the observer polynomial C^*. The condition for global stability is then that the transfer function

$$G(z) = \frac{P(z)}{C(z)} - \frac{1}{2}$$

is SPR. The local stability condition is that the real part of polynomial $C(z)$ is positive at all zeros of the polynomial $B(z)$ (see Holst (1979)). The method with stochastic averaging is illustrated with three examples.

EXAMPLE 6.13   Stochastic averaging

Consider the system

$$y(t) + ay(t-1) = u(t-1) + bu(t-2) + e(t) + ce(t-1)$$

with $a = -0.99$, $b = 0.5$, and $c = -0.7$. Let the estimated model be

$$y(t) = u(t-1) + r_1 u(t-2) + s_0 y(t-1)$$

and use the controller

$$u(t) = -s_0 y(t) - r_1 u(t-1)$$

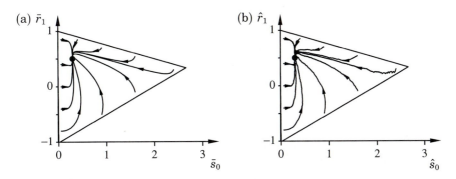

Figure 6.18 Phase plane for the controller parameters in Example 6.13 when recursive least-squares estimation is used. (a) Trajectories of the associated ODE. (b) Realizations of the difference equations. The parameter values corresponding to the minimum-variance controller are indicated by a dot.

The closed-loop system is described by

$$y(t) = \frac{(1 + cq^{-1})(1 + r_1 q^{-1})}{(1 + aq^{-1})(1 + r_1 q^{-1}) + s_0 q^{-1}(1 + bq^{-1})} e(t)$$

$$u(t) = \frac{-s_0(1 + cq^{-1})}{(1 + aq^{-1})(1 + r_1 q^{-1}) + s_0 q^{-1}(1 + bq^{-1})} e(t)$$

In this case,

$$\varphi^T(t-1) = \begin{pmatrix} u(t-2) & y(t-1) \end{pmatrix} \qquad \theta^T = \begin{pmatrix} r_1 & s_0 \end{pmatrix}$$

and

$$\varepsilon(t) = y(t)$$

Thus

$$f(\bar{\theta}) = \begin{pmatrix} r_{yu}(2) \\ r_y(1) \end{pmatrix} \qquad G(\bar{\theta}) = \begin{pmatrix} r_u(0) & r_{yu}(1) \\ r_{yu}(1) & r_y(0) \end{pmatrix}$$

where $r_y(\tau)$, $r_u(\tau)$, and $r_{yu}(\tau)$ are the covariance functions of y and u and the cross-covariance between y and u.

The stationary point is given by $f(\bar{\theta}) = 0$, which gives $r_{yu}(2) = 0$ and $r_y(1) = 0$. This is exactly the result obtained in Theorem 4.1. Figure 6.18(a) shows the phase plane of the ODE when recursive least-squares estimation is used. The stationary point corresponds to the minimum-variance controller, and the triangle indicates the stability boundary for the closed-loop system. Figure 6.18(b) shows realizations of the estimates $\hat{s}_0$ and $\hat{r}_1$ when recursive least-squares estimation has been used. The estimator is started with a very small step size. The realizations agree very well with the trajectories of the ODE. The ODEs have been simulated for $0 \leq \tau \leq 50$; 75,000 steps had to be simulated for the difference equations in Fig. 6.18(b). A forgetting factor of $\lambda = 0.99995$ was necessary to get close to the stationary point. □

EXAMPLE 6.14   **Moving-average self-tuner**

Consider an integrator with a time delay τ. (Compare Example 4.6.) For the time delay $\tau < h$ the system is described by

$$A(q) = q(q - 1)$$
$$B(q) = (h - \tau)q + \tau = (h - \tau)(q + b) = (h - \tau)B'$$
$$C(q) = q(q + c)$$

where

$$b = \frac{\tau}{h - \tau} \qquad \text{and} \qquad |c| < 1$$

The system is minimum-phase, $|b| < 1$, when $\tau < h/2$. Moving-average controllers of different orders will now be analyzed. (Compare Section 4.2.)

Case 1 $(d = 1)$

The minimum-variance strategy obtained through

$$AR + (h - \tau)B'S = B'C$$

giving

$$R(q) = q + b$$
$$S(q) = \frac{1 + c}{h - \tau} q$$

is the only possibility to get a moving average of order zero. Since the process zero is canceled, it is necessary for stability that the system be minimum-phase. The characteristic equation of K in Eq. (6.83) is in this case

$$(\lambda + 1)\left(\lambda + \frac{1}{1 - bc}\right) = 0$$

Since $|b|$ and $|c|$ are both less than 1, it follows that the eigenvalues of K are both negative.

Case 2 $(d = 2)$

Since B is of first order and C is of second order, there are several possibilities to get an output that is a moving-average process. We get the following combinations:

Case	B^+	
2(a)	$q + b$	Minimum variance
2(b)	$q + b$	Deadbeat
2(c)	1	Moving average

To investigate the equilibria, first notice that Cases 2(a) and 2(b) can give stable equilibria only if $b < 1$ (i.e., $\tau < h/2$).

Case 2(a) corresponds to the minimum-variance controller. The characteristic equation of the matrix K is

$$\lambda^2 - \lambda\left(b + c + \frac{c}{1 - bc}\right) + \frac{bc}{1 - bc} = 0$$

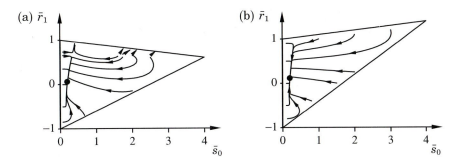

Figure 6.19 Simulation of the ODEs of the parameter estimates for the integrator when $d = 2$ and $c = -0.8$. (a) $\tau = 0.4$. (b) $\tau = 0.6$. The parameter values corresponding to the moving-average controller are indicated by dots.

Since b is nonnegative, it follows that this equation has roots in the right half-plane or at $\lambda = 0$ for all c in the interval $(-1, 1)$. The equilibrium is thus always unstable.

In Case 2(b) the characteristic equation of the matrix K is given by

$$\frac{(1 + c^2)(1 - bc)^2 + c^4(1 - b^2)}{c(c - b)(1 - bc)} \lambda^2 + \frac{1 + c^2 - bc}{c} \lambda + 1 = 0$$

This equation has all roots in the left half-plane if $b < c$.

In Case 2(c), moving-average control, the characteristic equation is

$$\lambda^2 + 2\lambda(b - c) + b(b - c) = 0$$

Since b is positive, it follows that this equation has its roots in the left half-plane if $b > c$. Notice that the moving-average controller is locally stable for $b > c$ even if $h/2 < \tau < h$, that is, when the controlled process is non-minimum-phase.

Summarizing, we find that if $d = 1$, there is only one equilibrium, which corresponds to the minimum-variance control. This equilibrium is locally stable only if $\tau < h/2$. When $d = 2$, there are three equilibria, corresponding to Cases 2(a), 2(b), and 2(c). Equilibrium 2(a) is always unstable; equilibrium 2(b) is stable if $b < c$; and equilibrium 2(c) is stable if $b > c$.

The phase portraits of the ODEs associated with the algorithm are shown in Fig. 6.19 for the case in which $d = 2$ and $c = -0.8$. When $\tau = 0.4$, there are three equilibria. They correspond to Case 2(a), which is a saddle point, Case 2(b), which is an unstable focus, and Case 2(c), which is a stable node. The stable node corresponds to the moving-average controller. The parameters are $r_1 = 0.08$ and $s_0 = 0.20$. For $\tau = 0.6$ there is only one equilibrium, which corresponds to the moving-average controller with the parameters $r_1 = 0.12$ and $s_0 = 0.20$. Figure 6.19 also shows that starting points exist for which the algorithm does not converge. The estimates are driven toward the stability boundary. □

The examples show how it is possible to use the associated ODE both to analyze the system and to get a feel for the behavior close to the stationary points as well as far away from them.

EXAMPLE 6.15   **Local instability of a minimum-variance STR**
Consider a process described by

$$y(t) - 1.6y(t-1) - 0.75y(t-2)$$
$$= u(t-1) + u(t-2) + 0.9u(t-3) + e(t) + 1.5e(t-1) + 0.75e(t-2)$$

The B polynomial has zeros at

$$z_{1,2} = -0.50 \pm 0.81i$$

Furthermore,

$$C(z_{1,2}) = -0.40 \pm 0.40i$$

The real part of C is thus negative at the zeros of B. This implies that the parameters corresponding to minimum-variance control make an unstable equilibrium for the ODEs. Furthermore, it follows from Theorem 4.2 that these parameter values are the only possible equilibrium point for the parameters. The following heuristic argument indicates that the estimates are bounded:

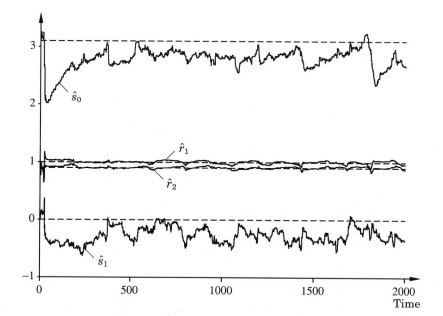

Figure 6.20   The parameter estimates when a self-tuning controller is used on the process in Example 6.15. The dashed lines correspond to the optimal minimum-variance controller.

If the parameters are such that the closed-loop system is unstable, the inputs and the outputs will be so large that they will dominate the stochastic terms in the model. The estimates will then quickly approach values that correspond to a deadbeat controller for the system, which gives a stable closed-loop system. This argument can be made rigorous (see Johansson (1988)). The estimates will thus vary in a bounded area without converging to any point. Figure 6.20 shows the parameter estimates when a direct self-tuning regulator with the controller structure

$$u(t) = -\frac{\hat{s}_0 + \hat{s}_1 q^{-1}}{1 + \hat{r}_1 q^{-1} + \hat{r}_2 q^{-2}} y(t)$$

is used. The simulation is initialized with values that correspond to the minimum-variance controller. The simulation is done by using RLS with a forgetting factor $\lambda = 0.98$. Figure 6.20 shows that the estimates try to reach the optimal values but are repelled when they get close. Notice also that the behavior is similar to that shown in Fig. 6.3. The example shows that a minimum-phase system exists in which the parameters corresponding to the minimum-variance controller are not a stable equilibrium for the self-tuning algorithm. This particular example led, in fact, to extensive research effort on the stability of stochastic self-tuners.   □

6.9 ROBUST ADAPTIVE CONTROLLERS

In the previous sections we showed that both continuous-time and discrete-time adaptive controllers perform well in idealized cases. For the discrete-time self-tuning regulator, Assumptions A1–A4 in Theorem 6.7 were necessary to prove convergence and stability. The examples indicate that the MRAS algorithm in Eqs. (6.57) is incapable of dealing with unmodeled dynamics and disturbances. The insight given by the analysis also suggests various improvements of the algorithms. In this section, different ways to improve the robustness properties are discussed.

The first and most obvious observation is that the underlying controller structure must be appropriate. A pure proportional feedback is not appropriate, since the controller gain should be reduced at high frequencies to maintain robustness. Notice that a digital control law with appropriate prefiltering gives a very effective reduction of gain at frequencies higher than the Nyquist frequency associated with the sampling. However, any use of filtering in this way requires prior information about the unmodeled dynamics.

The examples also show that Theorem 6.7, although it is of significant theoretical interest, has limited practical value. The theorem clearly will not hold if Assumption A2 is violated. This assumption will not hold in a practical case, in which there are always unmodeled dynamics. It is also not realistic to neglect disturbances. This raises the possibility that global stability can be established only under unrealistic assumptions.

Theorem 6.7 also gives poor guidelines for the choice of controller complexity. To satisfy Assumption A2, it seems logical to increase the controller complexity. However, this will impose additional requirements on the input signal to maintain persistency of excitation.

Projections, Leakage, and Dead Zones

Equilibrium analysis based on averaging shows that the equilibria depend on the unmodeled dynamics and the nature of the command signal in a complicated way. Some general conclusions can be extracted, however. If the command signal is not persistently exciting of an order that corresponds to the number of updated parameters, the equilibrium set will in general be a manifold rather than a point. For systems that are linear in the parameters, the equilibria will actually be an affine set, which means that the controller gains may be very large on some points of the set. Small amounts of measurement noise or other disturbances may then cause a loss of equilibrium and result in drift of the parameters.

Several ideas have been proposed to modify the adaptive algorithms to avoid the difficulty. One possibility is to modify the algorithm so that the parameters are *projected* into a given fixed set. However, this requires that appropriate prior knowledge be available. For example, in Example 6.11 it is sufficient to project into a set such that $0 \leq \hat{\theta}_2 \leq 17$. A convenient way to obtain a controller with a finite gain is to introduce a path parallel to the process with gain ρ. Let G_r be the transfer function of the controller. The arrangement with the parallel path is equivalent to use a controller with the transfer function

$$G_r' = \frac{G_r}{1 + \rho G_r}$$

This is clearly bounded by $1/\rho$ when G_r has high gain.

In Section 5.3 we showed that the normalization in the estimator (6.2) is important to improve the properties of the algorithms. The normalization comes automatically when least-squares methods are used. Another modification is to change the parameter updating in Eq. (6.2) to

$$\frac{d\hat{\theta}}{dt} = \gamma \frac{\varphi e}{\alpha + \varphi^T \varphi} + \alpha_1 (\theta^0 - \hat{\theta}) \tag{6.84}$$

where θ^0 is an *a priori* estimate of the parameters and $\alpha_1 > 0$ is an appropriate constant. The added term $\alpha_1(\theta^0 - \hat{\theta})$, sometimes called *leakage*, will make sure the estimates are driven toward θ^0 when they are far from θ^0. However, the modification will change the equilibrium. *A priori* knowledge is also required to choose θ^0 and α_1.

To avoid the problem of shift in equilibria, the following modification has also been suggested:

$$\frac{d\hat{\theta}}{dt} = \gamma \frac{\varphi e}{\alpha + \varphi^T \varphi} + \alpha_1 |e| (\theta^0 - \hat{\theta}) \qquad (6.85)$$

A third way to avoid the difficulty is to switch off the parameter estimation if the input signal is not appropriate. There are several ways to determine when the estimates should be switched off. A simple way is to update only when the error is large, that is, to introduce a *dead zone* in the estimator. Such an approach is discussed below. However, it is necessary to have prior knowledge to select the dead zone.

It has also been suggested that the width of the dead zone be varied adaptively. From the equilibrium analysis it appears more appropriate to use a criterion based on persistent excitation. An alternative to switching off the estimate is to introduce intentional perturbation signals so as to ensure a proper amount of excitation.

Filtering and Monitoring of Excitation

From the system identification point of view the problem of unmodeled dynamics can be interpreted as follows. In fitting a low-order model to a system with complex dynamics, the results depend critically on the frequency content of the input signal. Precautions must thus be taken to ensure that the frequency content of the input signal is concentrated to the frequency range at which the simple model is expected to fit well. This indicates that the signals should be filtered before they are entered into the parameter estimator or the parameter update law. However, filtering alone is not sufficient, since it may happen that the input signal has only frequencies outside the useful frequency range. (A typical case is the system in Example 6.12 with $u_c(t) = \sin 16.09t$.) No amount of filtering can remedy such a situation. We are then left with only two options: to switch off the estimation or to introduce intentional perturbation signals.

Effects of Disturbances

Loss of robustness due to disturbances was found for the MRAS in Section 6.6. Similar problems can be encounted for discrete-time systems. In Section 6.5 direct self-tuning regulators were discussed in the ideal case, in which there are no disturbances, but the results can be extended in different directions to cover disturbances. Consider the case in which the process is described by

$$A(q)y(t) = B(q)u(t) + v(t) \qquad (6.86)$$

where v is a bounded disturbance. To get some insight into what can happen, first consider an example. (See Egardt (1979).)

EXAMPLE 6.16   **Bounded disturbances**

Consider the system

$$y(t + 1) + ay(t) = u(t) + v(t + 1)$$

Use an adaptive control law with $A_o^* = A_m^* = 1$. (The desired response is thus $y_m(t + 1) = u_c(t)$.) The control law is

$$u(t) = -\hat{\theta}(t)y(t) + u_c(t)$$

where

$$\hat{\theta}(t + 1) = \hat{\theta}(t) + \frac{y(t)}{1 + y^2(t)} e(t + 1)$$

$$e(t + 1) = y(t + 1) - \hat{\theta}y(t) - u(t)$$

Introduce

$$\tilde{\theta} = \hat{\theta} - \theta^0$$

where $\theta^0 = -a$. The closed-loop system can be described by the equations

$$\tilde{\theta}(t + 1) = \frac{1}{1 + y^2(t)} \tilde{\theta}(t) + \frac{y(t)v(t + 1)}{1 + y^2(t)} \qquad (6.87)$$

$$y(t + 1) = -\tilde{\theta}(t)y(t) + u_c(t) + v(t + 1)$$

To show that $y(t)$ may be unbounded, we want to construct a disturbance v and a command signal u_c such that the parameter error goes to infinity. Assume that initial conditions are chosen such that $\tilde{\theta}(1) = 0$ and $y(1) = 1$. Define

$$f(t) \triangleq \left(\sqrt{t(t - 1)} - (t - 1) \right) \left(1 + \frac{1}{t - 1} \right) \qquad t = 2, 3, \ldots, T - 5$$

for some large T. Choose the following disturbance:

$$v(t) = 1 - \frac{1}{\sqrt{t - 1}} + f(t) \qquad t = 2, 3, \ldots, T - 5$$

and the following command signal:

$$u_c(t - 1) = \frac{1}{\sqrt{t}} - f(t) \qquad t = 2, 3, \ldots, T - 5$$

The signals v and u_c are bounded. A straightforward calculation gives

$$\tilde{\theta}(t) = \sqrt{t} - 1$$

$$y(t) = \frac{1}{\sqrt{t}}$$

for $t = 1, \ldots, T - 5$. Further, let

$$v(t) = 0 \qquad t = T - 4, \ldots, T$$

$$u_c(t - 1) = \begin{cases} 0 & t = T - 4 \\ 1 & t = T - 3, \ldots, T \end{cases}$$

It can then be verified that $\tilde{\theta}(t)$ and $y(t)$ for large T are approximately given by the following table.

t	$\tilde{\theta}(t)$	$y(t)$
$T - 4$	$\sqrt{T}$	-1
$T - 3$	$\dfrac{\sqrt{T}}{2}$	$\sqrt{T}$
$T - 2$	$\dfrac{1}{2\sqrt{T}}$	$-\dfrac{T}{2}$
$T - 1$	$\dfrac{1}{\sqrt{T}T^2}$	$\dfrac{\sqrt{T}}{4}$
T	$\dfrac{16}{\sqrt{T}T^3}$	1

Now choose $v(T + 1)$ and $u_c(T)$ such that $\tilde{\theta}(T + 1) = 0$ and $y(T + 1) = 1$. The state vector of Eqs. (6.87) is then equal to the initial state. By repeating the procedure for increasing values of T, a subsequence of $y(t)$ will increase as $-T/2$ and therefore is unbounded. □

Example 6.16 shows that the algorithm may behave badly even if it is assumed that the disturbances are bounded. Robustness against bounded disturbances can be obtained by using conditional updating as shown in the following theorem.

THEOREM 6.8 Conditional updating

Consider the plant (6.86) where v is a disturbance that is bounded by

$$\sup_t \left| \frac{R}{A_o A_m B} v \right| \leq C_1$$

where R is the polynomial in the feedback law and C_1 is a constant. Assume that the direct adaptive algorithm defined by Eqs. (6.41) and (6.42) is used, with the modification that the parameters are updated only when the estimation error is such that

$$|e| \geq \frac{2C_1}{2 - \max(b_0/r_0, 1)}$$

Let Assumptions A1–A3 hold, and assume in addition that $0 < b_0 < 2r_0$. Then the inputs and outputs of the closed-loop system are bounded. □

Proofs of this theorem can be found in Egardt (1979) and Goodwin and Sin (1984). The modification of the algorithm is referred to as *conditional updating* or *introduction of a dead zone in the estimator*.

Of course, the result is of limited practical value because it requires an upper bound on the disturbance, which is not known *a priori*. The bound also depends on the ratio $b_0/r_0 = b_0/\hat{b}_0$, where b_0 is the instantaneous gain. The estimate of this gain is thus essential. If $b_0/r_0 = 1$ and $A_o = A_m = 1$, it follows that $R = B$, and the condition for updating becomes

$$|e(t)| \geq 2 \sup |v(t)|$$

This means that the estimate will be updated when the estimation error is twice as large as the maximum noise amplitude.

Another modification of the algorithm also leads to bounded signals. The modification consists of using the updating law of Eqs. (6.41) if the magnitude of the estimates is less than a given bound and to project into a bounded set if Eqs. (6.41) give estimates outside the bounds. We refer to Theorem 4.4 of Egardt (1979) for details. This method will, of course, require that the bounds on the parameters be known *a priori*.

Signal Normalization

Various modifications of the adaptive algorithm are discussed in more detail in Chapter 11. Therefore only a few sketchy remarks are given here. Notice that Theorem 6.8 gives stability conditions for adaptive control applied to the model (6.86), when v is a bounded disturbance. Unmodeled dynamics can, of course, be modeled by Eq. (6.86), but v will no longer be bounded, since it depends on the inputs and outputs. By introducing the signal defined by

$$Cr(t) = \max \left(|u(t)|, |y(t)| \right)$$

where C is a stable filter, and introducing the normalized signals

$$\tilde{y} = \frac{y}{r}, \qquad \tilde{u} = \frac{u}{r}, \qquad \tilde{v} = \frac{v}{r}$$

the model of Eq. (6.86) can be replaced by

$$A\tilde{y} = B\tilde{u} + \tilde{v}$$

where $\tilde{v}$ is now bounded. By invoking Theorem 6.8, it can be established that adaptive control with a dead zone or projection gives a system with bounded signals. The detailed justification is complicated.

The Minimum-Phase Assumption

In Theorem 6.7 and for the MRAS the process is required to be minimum-phase. This assumption is used to conclude that the input signal is bounded when the output is bounded. The minimum-variance controller, which cancels the open-loop process zeros, cannot be used when the process is nonminimum-phase.

Instead, the LQG self-tuner or the moving-average controller with increased prediction horizon can be used.

It should be remarked that sampled data systems often can be non-minimum-phase because of "sampling zeros" even if the continuous-time system that is sampled is minimum-phase. These zeros are given by the following theorem.

THEOREM 6.9 Limiting sampled-data zeros

Let $G(s)$ be a rational function

$$G(s) = K\frac{(s - z_1)(s - z_2),\ldots,(s - z_m)}{(s - p_1)(s - p_2),\ldots,(s - p_n)} \tag{6.88}$$

and let $H(z)$ be the corresponding pulse transfer function. Assume that $m < n$. As the sampling period $h \to 0$, m zeros of H go to 1 as $\exp(z_ih)$, and the remaining $n - m - 1$ zeros of H go to the zeros of $B_{n-m}(z)$, where $B_k(z)$ is the polynomial

$$B_k(z) = b_1^k z^{k-1} + b_2^k z^{k-2} + \cdots + b_k^k \tag{6.89}$$

and

$$b_i^k = \sum_{l=1}^{i}(-1)^{i-l}l^k \begin{pmatrix} k+1 \\ i-l \end{pmatrix} \qquad i = 1,\ldots,k \tag{6.90}$$

The first polynomials B_k are

$$B_1(z) = 1$$
$$B_2(z) = z + 1$$
$$B_3(z) = z^2 + 4z + 1 \qquad\qquad \square$$

This theorem is proved in Åström *et al.* (1984). It implies that direct methods for adaptive control that require that the plant be minimum-phase cannot be used with too short a sampling period. When very fast sampling is required, a continuous-time representation may then be preferable. Another possibility is to describe the system in the *delta operator*, defined by

$$\delta = \frac{q - 1}{h}$$

or in *Tustin's operator*:

$$\Delta = \frac{1}{2h}\frac{q - 1}{q + 1}$$

This yields parameterizations that give a much better resolution at $q = 1$. The δ operator gives a description that is equivalent to the q operator description. The advantage of the transformation is that the δ operator description has better numerical properties when the sampling is fast. All the poles of the q

operator form are clustered around the point $q = 1$. This gives rise to numerical sensitivity. For the δ operator it can be shown that the limiting value

$$\lim_{h \to 0} \frac{B_h(\delta)}{A_h(\delta)} = \frac{B_0(\delta)}{A_0(\delta)}$$

is such that the coefficients in B_0 and A_0 are the same as the coefficients in the continuous-time transfer function. This implies that the structure of the transfer function in the δ operator is essentially the same as that of the continuous-time transfer function, provided that the sampling period is sufficiently short.

The High-Frequency Gain

For a process that has no right half-plane zeros, the standard direct discrete-time algorithm is based on the model

$$A_o^* A_m^* y(t + d) = b_0 \left(R^* u(t) + S^* y(t) \right)$$

where b_0 is the coefficient of the first nonvanishing term in the B polynomial. With some abuse of language this coefficient is called the *high-frequency gain* because it is the first nonvanishing coefficient of the impulse response. For continuous-time systems the transfer function of the process is approximately $G(s) = b_0 s^{-dh}$. In Theorem 6.7 it was required that the sign of the coefficient b_0 be known. There are several ways to deal with the parameter b_0. It may be absorbed into R and S and estimated. The polynomial R then has the form

$$R(q) = r_0 q^k + r_1 q^{k-1} + \cdots + r_k$$

The problem with this approach is that some safeguards must be taken to avoid the estimate r_0 becoming too small. Another possibility is to introduce a crude fixed estimate of b_0. The following analysis shows what happens when this is done. Let the true system be

$$y(t + 1) = b_0 \left(u(t) + \psi^T(t) \theta^0 \right)$$

and let the model be

$$y(t + 1) = r_0 \left(u(t) + \psi^T(t) \theta \right) = r_0 u(t) + \varphi^T(t) \theta$$

With zero command signal the control law becomes

$$u(t) = -\psi^T(t) \hat{\theta}(t)$$

The equation for parameter updating is

$$\hat{\theta}(t + 1) = \hat{\theta}(t) + P(t + 1)\varphi(t) e(t + 1)$$

where
$$e(t + 1) = y(t + 1) = b_0 u(t) + b_0 \psi^T(t) \theta^0$$

$$= -b_0 \psi^T(t) \left(\hat{\theta}(t) - \theta^0 \right) = -\frac{b_0}{r_0} \varphi^T(t) \left(\hat{\theta}(t) - \theta^0 \right)$$

The estimation error is thus governed by

$$\tilde{\theta}(t + 1) = \left(I - \frac{b_0}{r_0} P(t + 1) \varphi(t) \varphi^T(t) \right) \tilde{\theta}(t)$$

With a pure projection algorithm we have

$$P(t + 1) = \frac{1}{\varphi^T(t) \varphi(t)}$$

In this case the matrix in large parentheses has one eigenvalue $(1 - b_0/r_0)$ and the remaining eigenvalues 1. With least-squares updating, the averaged equation for $\bar{\theta}$ becomes

$$\tilde{\theta}(t + 1) = \left(1 - \frac{b_0}{r_0} \right) \tilde{\theta}(t)$$

Hence, to remain stable, it must be required that

$$0 < \frac{b_0}{r_0} < 2$$

If an algorithm with a fixed r_0 is used, it is convenient to absorb r_0 in the scaling of the signals. This is discussed in more detail in Chapter 11. When the parameter b_0 is estimated, it can be treated like the other parameters. However, because of the special structure of the model it is useful to use special algorithms such as the ones discussed in Section 5.8.

Universal Stabilizers

An interesting class of adaptive algorithms was discovered during attempts to investigate whether Assumption A3 is necessary. The following question was posed. Consider the scalar system

$$\frac{dy}{dt} = ay + bu \tag{6.91}$$

where a and b are constants. Does there exist a feedback law of the form

$$u = f(\hat{\theta}, y)$$
$$\frac{d\hat{\theta}}{dt} = g(\hat{\theta}, y) \tag{6.92}$$

that stabilizes the system for all values of a and b? Morse (1983) suggested that there are no rational f and g that solve the problem. Morse's conjecture was verified by Nussbaum (1983), who proved the following result.

THEOREM 6.10   Universal stabilizer

The control law of Eqs. (6.92), with

$$f(\hat{\theta}, y) = y\hat{\theta}^2 \cos \hat{\theta}$$
$$g(\hat{\theta}, y) = y^2$$

(6.93)

and $\hat{\theta}(0) = 0$, stabilizes Eq. (6.91).

Proof: The closed-loop system is described by

$$\frac{dy}{dt} = ay + by\hat{\theta}^2 \cos \hat{\theta}$$

$$\frac{d\hat{\theta}}{dt} = y^2$$

Since $\hat{\theta}(0) = 0$ and $d\hat{\theta}/dt \geq 0$, it follows that $\hat{\theta}(t)$ is nonnegative and nondecreasing. $\hat{\theta}(t)$ is also bounded, which is shown by contradiction. Hence assume that $\lim_{t \to \infty} \hat{\theta}(t) = \infty$. Multiplication of the differential equation for y by y gives

$$y \frac{dy}{dt} = ay^2 + by^2\hat{\theta}^2 \cos \hat{\theta} = a \frac{d\hat{\theta}}{dt} + b\hat{\theta}^2 \cos \hat{\theta} \frac{d\hat{\theta}}{dt}$$

Integration with respect to time gives

$$y^2(t) = y^2(0) + 2a\hat{\theta}(t) + 2b \int_0^{\hat{\theta}(t)} x^2 \cos x \, dx$$

Hence

$$\frac{y^2(t)}{\hat{\theta}(t)} = \frac{y^2(0)}{\hat{\theta}(t)} + 2a + \frac{2b}{\hat{\theta}(t)} \int_0^{\hat{\theta}(t)} x^2 \cos x \, dx$$

But

$$\frac{1}{\hat{\theta}} \int_0^{\hat{\theta}} x^2 \cos x \, dx = \hat{\theta} \sin \hat{\theta} + 2 \cos \hat{\theta} - \frac{2}{\hat{\theta}} \sin \hat{\theta}$$

Hence

$$\lim_{t \to \infty} \inf \frac{1}{\hat{\theta}} \int_0^{\hat{\theta}} x^2 \cos x \, dx = -\infty$$

This gives

$$\lim_{\hat{\theta} \to \infty} \inf \frac{y^2(t)}{\hat{\theta}(t)} = -\infty$$

which is a contradiction because $y^2/\hat{\theta}$ is nonnegative. It thus follows that

$$\lim_{t \to \infty} \hat{\theta}(t) = \theta^0 < \infty$$

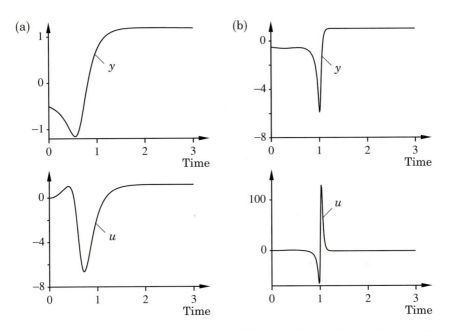

Figure 6.21   Simulation of the control law of Eqs. (6.94) applied to the plants (a) $G(s) = 1/(1 - s)$ and (b) $G(s) = 1/(s - 1)$.

Integration of the equation for $\hat{\theta}$ gives

$$\hat{\theta}(t) = \int_0^t y^2(t)\,dt$$

It then follows that

$$\lim_{t \to \infty} y(t) = 0 \qquad\qquad \Box$$

The behavior of a universal stabilizer is illustrated in Fig. 6.21. A reference value is used in the simulations, and the control law is then modified to

$$f(\hat{\theta}, y) = (u_c - y)\hat{\theta}^2 \cos\hat{\theta}$$
$$g(\hat{\theta}, y) = (u_c - y)^2 \qquad\qquad (6.94)$$

Notice that the control law of Eqs. (6.94) can be interpreted as proportional feedback with the gain $k = \hat{\theta}^2 \cos\hat{\theta}$. The behavior of the control law can be interpreted as follows. Sweep over all possible controller gains and stop when a stabilizing gain has been found. The function g can be interpreted as the rate of change of the gain sweep. The rate is large for large errors and small for small errors. The form $\cos\hat{\theta}$ makes sure that the gains can be both positive and negative. Universal stabilizers may show very violent behavior. This not surprising, since the system may be temporarily unstable during the sweep over the gains.

The control law of Eqs. (6.94) is useful because it does not contain any parameters that relate to the system that it stabilizes. It is therefore called a *universal stabilizer*. However, the control law is restricted to a first-order system. In attempting to generalize Theorem 6.10 to higher-order systems, the following question was posed. How much prior information about an unknown system is required to stabilize it? This question was answered in a general setting by Mårtensson (1985), who showed that it is sufficient to know the order of a stabilizing fixed-gain controller. If a transfer function is given, it is unfortunately a nontrivial task to find the minimal order of a stabilizing controller.

6.10   CONCLUSIONS

Analysis of adaptive systems is difficult because they are complicated. A number of different methods have been used to gain insight into the behavior of adaptive systems. The theory is useful to show fundamental limitations of the algorithms and to point out possible ways to improve them.

In this chapter a basic stability theorem has been derived on the basis of standard tools of the theory of difference equations. To show stability and convergence, it is necessary to make quite restrictive assumptions about the system to be controlled. The consequences of violating these assumptions have been analyzed.

It has been shown that analysis of equilibria and local properties around equilibria can be explored by the method of averaging. This method can also be applied to investigate global properties. Averaging can be applied in many different situations. For deterministic problems it can be used for steps and periodic signals. It can also be applied to stochastic signals. Averaging methods have also been applied to analyze what happens when adaptive systems are designed on the basis of simplified models. To apply averaging methods, it is necessary to use small adaptation gains. Unfortunately, there are no good methods to determine analytically how small the gains should be. It is also demonstrated that adaptive systems may have very complex behavior for large adaptation gains. Mechanisms that may lead to instability have been discussed. One mechanism is associated with lack of a parameter equilibrium or local instability of the equilibrium. Other mechanisms are parametric excitation and high adaptation gain. The last two mechanisms can be avoided by choosing a small adaptation gain.

PROBLEMS

6.1   Consider the indirect continuous-time self-tuning controller in Example 3.6. Collect all equations that describe the self-tuner, and show that they can be written in the form

$$\frac{d\xi}{dt} = A(\vartheta)\xi + B(\vartheta)u_c$$

$$\begin{pmatrix} e \\ \varphi \end{pmatrix} = C(\vartheta)\xi + D(\vartheta)u_c$$

$$\vartheta = \chi(\theta)$$

$$\frac{d\theta}{dt} = P\varphi e$$

$$\frac{dP}{dt} = \alpha P - P\varphi\varphi^T P$$

Give explicit expressions for all components of the vectors ξ, φ, ϑ, and θ and the matrix P.

6.2 Consider a system with unknown gain whose transfer function is SPR. Show that a closed-loop system that is insensitive to variations in the gain is easily obtained by applying proportional feedback. Carry out a detailed analysis for the case in which the transfer function is $G(s) = 1/(s+1)$.

6.3 Consider an MRAS for adjustment of a feedforward gain. Assume that the system is designed on the basis of the assumption that the process dynamics are

$$G(s) = \frac{a}{s+a}$$

(a) Investigate the behavior of the systems obtained with the SPR and MIT rules when the real system has the transfer function

$$G(s) = \frac{ab^2}{(s+a)(s+b)^2}$$

Determine in particular which frequency ranges give stable adaptation rules for sinusoidal command signals.

(b) Consider the MRAS based on the SPR rule when the reference signal is constant and when an additional constant load disturbance is acting on the input of the process. Investigate how the load disturbance influences the stationary point of the total system. Investigate the local stability properties through linearization.

6.4 Consider an MRAS for adjustment of a feedforward gain based on the MIT rule. Let the command signal be

$$u_c = a_1 \sin \omega_1 t + a_2 \sin \omega_2 t$$

and assume that the process has the transfer function

$$G(s) = \frac{1}{(s+1)^3}$$

Derive conditions for the closed-loop system to be stable.

6.5   Consider Theorem 6.7. Generalize the results to cover the case in which the polynomial B^* has isolated zeros on the unit circle.

6.6   Consider the system described by

$$y(t) = u(t - d)$$

Assume that a direct adaptive control (e.g., with $A_o^* = A_m^* = 1$) is designed according to the assumption that $d = 1$. Investigate how this controller behaves when applied to a system with $d = 2$.

6.7   Construct a proof analogous to Theorem 6.7 for continuous-time systems.

6.8   Consider the system

$$y(t) = u(t - 1) + a$$

where a is an unknown constant. Construct an adaptive control law that makes y follow a command u_c asymptotically. Prove that it converges.

6.9   Consider the continuous-time model

$$y(t) = \varphi^T(t)\theta$$

Let the parameter θ be estimated by

$$\frac{d\hat{\theta}}{dt} = \gamma \frac{\varphi(t)}{\alpha + \varphi^T(t)\varphi(t)} e(t)$$

where $\gamma > 0$ and $\alpha > 0$ are real constants and

$$e(t) = y(t) - \varphi^T(t)\hat{\theta}(t)$$

Assume that $y(t)$ is given by

$$y(t) = \varphi^T(t)\theta_0$$

Prove that

$$|\hat{\theta}(t) - \theta_0| \le |\hat{\theta}(s) - \theta_0| \le |\hat{\theta}(0) - \theta_0| \qquad t > s > 0$$

and that

$$\frac{|e(t)|}{\sqrt{\alpha + \varphi^T(t)\varphi(t)}} \to 0$$

as $t \to \infty$.

6.10 Consider the system in Example 6.12. Interpret the results as if the adaptive algorithm tried to estimate parameters a and b in the transfer function $G(s) = b/(s + a)$. Use Eqs. (6.70) to show that

$$\hat{a} = \frac{229 - 31\omega^2}{259 - \omega^2}$$

$$\hat{b} = \frac{458}{259 - \omega^2}$$

Determine the parameters for $\omega = 2.72$ and $\omega = 17.03$. Explain the results by evaluating $G(s)$ for the corresponding frequencies.

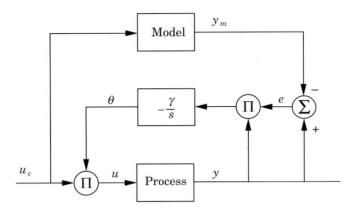

Figure 6.22 Adaptive feedforward controller in Problem 6.11.

6.11 A feedforward gain is adapted as shown in the block diagram in Fig. 6.22. The model is given by

$$\frac{dy_m}{dt} = -y_m + u_c$$

The process is not linear, however, but is given by

$$\frac{dy}{dt} = -y - ay^3 + u$$

Let $\gamma = 1$ and $u_c = 1$.

(a) What are the equilibrium points of the system?

(b) Linearize the system around the equilibrium points, and determine how the stability of the linearized system depends on the parameter a.

(c) Simulate the behavior of the nonlinear adaptive system to verify the results in part (b).

6.12 An integrator process

$$G(s) = \frac{1}{s}$$

is to be controlled by the error feedback law

$$sU(s) = (4s + \theta)(U_c(s) - Y(s))$$

where $U(s)$, $U_c(s)$, and $Y(s)$ are the Laplace transforms of the input, reference, and output signals, respectively. The desired response of the closed-loop system is given by the transfer function

$$G_m(s) = \frac{4(s + 1)}{(s + 2)^2}$$

An MRAS has been designed, giving the parameter update law

$$\frac{d\theta}{dt} = -\gamma e(t)\left(\frac{1}{(p+2)^2}u_c(t)\right)$$

where $e = y - y_m$.

(a) Find the equilibrium parameter set of the parameter update law, giving a parameter estimate that is constant for any reference input $u_c(t)$. Give an expression for the averaged nonlinear differential equation of the parameter update law.

(b) Determine the local stability of the equilibrium parameter set for a sinusoidal reference signal $u_c(t) = \sin \omega t$ by examining the characteristic polynomial of the linear differential equation obtained by linearizing the averaged differential equation around the equilibrium parameter set. Determine for what frequencies the linearized equation is stable.

6.13 Formulate the averaging equation for a discrete-time algorithm corresponding to Eq. 6.49.

6.14 Consider discrete-time adaptive control of the system

$$y(t+1) = ay(t) + but(t)$$

Derive an MRAS that gives a closed-loop system

$$y_m(t+1) = a_m y_m(t) + b_m u_c(t)$$

Use averaging methods to analyze the system when the command signal is a step and a sinusoid.

6.15 Consider Problem 6.14. Investigate the behavior of the system when the command signal is a step and when there is sinusoidal measurement noise.

6.16 Consider the MRAS given by Eqs. (6.57). Investigate the local behavior of the closed-loop system when the command signal is a sinusoid and the gradient method

$$\frac{d\hat{\theta}}{dt} = \gamma\varphi e$$

is replaced with a least-squares method of the form

$$\frac{d\hat{\theta}}{dt} = P\varphi e$$

$$\frac{dP}{dt} = -P\varphi\varphi^T P + \lambda P$$

6.17 Show that there is no constant-gain controller that can simultaneously stabilize the systems $G(s) = 1/(1+s)$ and $G(s) = 1/(1-s)$.

6.18 Show that there is a fixed-gain controller that will simultaneously stabilize the systems $G(s) = 1/(s + 1)$ and $G(s) = 1/(s - 1)$.

6.19 Consider the MRAS given by Eqs. (6.57). Make a simulation study to investigate the consequences of introducing leakage as described by Eqs. (6.84) and (6.85) in the estimation algorithm. Study sinusoidal command signals as well as step commands and measurement noise.

6.20 Consider the MRAS in Problem 6.4. Make a simulation study to investigate the consequences of using conditional updating. Study sinusoidal command signals as well as step commands and measurement noise.

6.21 Consider the system in Problem 6.4. Let the input be sinusoidal with frequency ω. Investigate the effects of sinusoidal measurement noise on the system.

6.22 Consider direct algorithms for control of the system

$$y(t + 1) = ay(t) + bu(t)$$

to give an input-output relation

$$y_m(t + 1) = a_m y(t) + b_m u_c(t)$$

Investigate by simulation the convergence rates obtained when $\hat{b}$ is fixed to different values.

6.23 Investigate the behavior of the universal stabilizer in the presence of measurement noise.

6.24 Consider a system for adjustment of a feedforward gain based on the MIT rule. Let the command signal be $u_c(t) = \sin \omega t$, and let $G(s) = 1/(s + 1)$. Simulate the parameter behavior for the MIT rule with adaptation gains $\gamma = 10$ and $\gamma = 11$. Compare the analysis in Example 6.8.

6.25 Consider the simulation shown in Fig. 6.11, which was performed with adaptation gain $\gamma = 1.0$. Repeat the simulation with different adaptation gains.

REFERENCES

A standard text on nonlinear systems is:

Guckenheimer, J., and P. Holmes, 1983. *Nonlinear Oscillations, Dynamical Systems and Bifurcations of Vector Fields*, Applied Mathematics Series. New York: Springer-Verlag.

The stability problem has been of major concern since the MRAS was proposed. Flaws in earlier stability proofs were pointed out in:

Feuer, A., and A. S. Morse, 1978. "Adaptive control of single-input single-output linear systems." *IEEE Trans. Automat. Contr.* **AC-23**: 557–569.

The proof of Theorem 6.7 follows the ideas in:

Goodwin, G. C., P. J. Ramadge, and P. E. Caines, 1980. "Discrete-time multivariable adaptive control." *IEEE Trans. Automat. Contr.* **AC-25**: 449–456.

Equivalent results for continuous-time systems are presented in:

Morse, A. S., 1980. "Global stability of parameter-adaptive control systems." *IEEE Trans. Automat. Contr.* **AC-25**: 433–439.

Narendra, K. S., Y.-H. Lin, and L. S. Valavani, 1980. "Stable adaptive controller design. Part II: Proof of stability." *IEEE Trans. Automat. Contr.* **AC-25**: 440–448.

Many variations of the stability theorem are given in:

Goodwin, G. C., and K. S. Sin, 1984. *Adaptive Filtering Prediction and Control*, Information and Systems Science Series. Englewood Cliffs, N.J.: Prentice-Hall.

Related results are presented in:

de Larminat, P., 1979. "On overall stability of certain adaptive control systems." *Preprints of the 5th IFAC Symposium on Identification and System Parameter Estimation*, pp. 1153–1159. Darmstadt, Germany.

Egardt, B., 1980a. "Stability analysis of discrete-time adaptive control schemes." *IEEE Trans. Automat. Contr.* **AC-25**: 710–716.

Egardt, B., 1980b. "Stability analysis of continuous-time adaptive control systems." *Siam J. Contr. Optimiz.* **18**: 540–558.

Gawthrop, P. J., 1980. "On the stability and convergence of a self-tuning controller." *Int. J. Contr.* **31**: 973–998.

Kumar, P. R., 1990. "Convergence of adaptive control schemes using least-squares parameter estimates." *IEEE Trans. Automat. Contr.* **AC-35**: 416–424.

A stability analysis for bounded disturbances is given in:

Egardt, B., 1979. *Stability of Adaptive Controllers.* Lecture Notes in Control and Information Sciences, vol. 20. Berlin: Springer-Verlag.

The case of mean square bounded disturbances was investigated in:

Praly, L., 1984. "Stochastic adaptive controllers with and without positivity condition." *Proceedings of the 23rd IEEE Conference on Decision and Control*, pp. 58–63. Las Vegas, Nevada.

The idea of conditional updating and projection of estimates into a bounded range is also treated in Egardt (1979). Conditional updating is also discussed in:

Peterson, B. B., and K. S. Narendra, 1982. "Bounded error adaptive control." *IEEE Trans. Automat. Contr.* **AC-27**: 1161–1168.

An elegant formalism for the growth-rate estimates in Lemma 6.2 is found in:

Narendra, K. S., A. M. Annaswamy, and R. P. Singh, 1985. "A general approach to the stability analysis of the adaptive systems." *Int. J. Contr.* **AC-41**: 193–216.

Proof of convergence for the original self-tuner based on recursive least-squares estimation and minimum-variance control is found in

Guo, L., and H.-F. Chen, 1991. "The Åström-Wittenmark self-tuning regulator revisited and ELS-based adaptive trackers." *IEEE Trans. Automat. Contr.* **AC-36**: 802–812.

Chen, H.-F., and L. Guo, 1991. *Identification and Stochastic Adaptive Control.* Boston: Birkhäuser.

The method of averaging to investigate nonlinear oscillations was developed by:

Krylov, A. N., and N. N. Bogoliubov, 1937. *Introduction to Non-linear Mechanics* (English translation 1943). Princeton, N.J.: Princeton University Press.

A simple presentation of the key ideas is given in:

Minorsky, N., 1962. *Nonlinear Oscillations.* Princeton, N.J.: Van Nostrand.

More detailed treatments are given in:

Hale, J. K., 1963. *Oscillations in Nonlinear Systems.* New York: McGraw-Hill.

Hale, J. K., 1969. *Ordinary Differential Equations.* New York: Wiley-Interscience.

Arnold, V. I., 1983. *Geometrical Methods in the Theory of Ordinary Differential Equations.* New York: Springer-Verlag.

Guckenheimer, J., and P. Holmes, 1983. *Nonlinear Oscillations, Dynamical Systems and Bifurcations of Vector Fields.* Berlin: Springer-Verlag.

Sastry, S., and M. Bodson, 1989. *Adaptive Control: Stability, Convergence, and Robustness.* Englewood Cliffs, N.J.: Prentice-Hall.

Many results on classical stability theory for ordinary differential equations are found in:

Bellman, R., 1953. *Stability Theory of Differential Equations.* New York: McGraw-Hill.

The example of nonrobustness in Example 6.8 is based on:

Rohrs, C., L. S. Valavani, M. Athans, and G. Stein, 1985. "Robustness of continuous-time adaptive control algorithms in the presence of unmodeled dynamics." *IEEE Trans. Automat. Contr.* **AC-30**: 881–889.

This initiated the discussion of the robustness problem. The analysis in Sections 6.6 and 6.7 is largely based on:

Åström, K. J., 1983. "Analysis of Rohr's counterexample to adaptive control." *Proceedings of the 22nd IEEE Conference on Decision and Control,* pp. 982–987. San Antonio, Texas.

Åström, K. J., 1984. "Interactions between excitation and unmodeled dynamics in adaptive control." *Proceedings of the 23rd IEEE Conference on Decision and Control,* pp. 1276–1281. Las Vegas, Nevada.

The idea of introducing leakage is found in:

Ioannou, P. A., and P. V. Kokotovic, 1983. *Adaptive Systems with Reduced Models.* New York: Springer-Verlag.

The idea of normalization was suggested by Praly. See, for example:

Praly, L., 1986. "Global stability of a direct adaptive control scheme with respect to a graph topology." In *Adaptive and Learning Systems: Theory and Applications,* ed. K. S. Narendra. New York: Plenum Press.

It is further explored in:

Narendra, K. S., and A. M. Annaswamy, 1987. "A new adaptive law for robust adaptation without persistent excitation." *IEEE Trans. Automat. Contr.* **AC-32**: 134–145.

Further discussion of robustness is given in:

Anderson, B. D. O., R. R. Bitmead, C. R. Johnson, Jr., P. V. Kokotovic, R. L. Kosut, I. M. Y. Mareels, L. Praly, and B. D. Riedle, 1986. *Stability of Adaptive Systems: Passivity and Averaging Analysis.* Cambridge, Mass.: MIT Press.

Goodwin, G. C., D. J. Hill, D. Q. Mayne, and R. H. Middleton, 1986. "Adaptive robust control. Convergence, stability and performance." *Proceedings of the 25th IEEE Conference on Decision and Control,* pp. 468–473. Athens, Greece.

Kreisselmeier, G., and B. D. O. Anderson, 1986. "Robust model reference adaptive control." *IEEE Trans. Automat. Contr.* **AC-31**: 127–133.

Ortega, R., and Y. Tang, 1989. "Robustness of adaptive controllers—A survey." *Automatica* **25**: 651–677.

Ydstie, B. E., 1992. "Transient performance and robustness of direct adaptive control." *IEEE Trans. Automat. Contr.* **AC-37**: 1091–1105.

Stochastic averaging was introduced in:

Ljung, L., 1977a. "Analysis of recursive stochastic algorithms." *IEEE Trans. Automat. Contr.* **AC-22**: 551–575.

The ordinary differential equations associated with a discrete time estimation problem were derived. This particular form of averaging is called the ODE method. Extensive applications of the method are given in:

Ljung, L., and T. Söderström, 1983. *Theory and Practice of Recursive Identification.* Cambridge, Mass.: MIT Press.

More recent proofs of the method are found in:

Kushner, H., 1984. *Approximation and Weak Convergence Methods of Random Processes.* Cambridge, Mass.: MIT Press.

Kushner, H., and D. Clark, 1978. *Stochastic Approximation Methods for Constrained and Unconstrained Systems,* Applied Mathematical Science Series 26. Berlin: Springer-Verlag.

Kushner, H., and A. Schwartz, 1984. "An invariant measure approach to the convergence of stochastic approximations with state dependent noise." *SIAM J. on Control and Optimization* **22**: 13–27.

Metivier, M., and P. Priouret, 1984. "Applications of a Kushner and Clark lemma to general classes of stochastic algorithms." *IEEE Trans. Information Theory* **IT-30**: 140–151.

An accessible account is also given in:

Kumar, P. R., and P. Varaiya, 1986. *Identification and Adaptive Control.* Englewood Cliffs, N.J.: Prentice-Hall.

Stochastic averaging was applied to the self-tuning regulator based on least-squares estimation and minimum-variance control in:

Ljung, L., 1977b. "On positive real transfer function and convergence of some recursive schemes." *IEEE Trans. Automat. Contr.* **AC-22**: 539–551.

More details about Example 6.14 are given in:

Åström, K. J., and B. Wittenmark, 1985. "The self-tuning regulators revisited." *Preprints of the 7th IFAC Symposium on Identification and System Parameter Estimation,* pp. xxv–xxxiii. York, U.K.

Conditions for local stability of the equilibrium are given in:

Holst, J., 1979. "Local convergence of some recursive stochastic algorithms." *Preprints of the 5th IFAC Symposium on Identification and System Parameter Estimation,* pp. 1139–1146. Darmstadt, Germany.

Analysis of stability in self-tuning regulators based on Lyapunov theory is given in:

Johansson, R., 1988. "Stochastic stability of direct adaptive control." Report TFRT-7377, Department of Automatic Control, Lund Institute of Technology, Lund, Sweden.

The fact that rapid sampling may create zeros of the pulse transfer function outside the unit disc is discussed in:

Åström, K. J., P. Hagander, and J. Sternby, 1984. "Zeros of sampled systems." *Automatica* **20**: 31–38.

Work on universal stabilizers was initiated by a discussion of whether Assumption A4 of Theorem 6.7 is necessary. See:

Morse, A. S., 1983. "Recent problems in parameter adaptive control." In *Outils et Modèles Mathematiques pour l'Automatique, l'Analyse de Systèmes et le Traitement du Signal,* ed. I. D. Landau, vol 3, pp. 733–740. Paris: Editions du CNRS.

The problem was solved for scalar systems in:

Nussbaum, R. D., 1983. "Some remarks on a conjecture in parameter adaptive control." *Syst. Contr. Lett.* **3**: 243–246.

Universal stabilizers for multivariable systems are discussed in:

Mårtensson, B., 1985. "The order of any stabilizing regulator is sufficient a priori information for adaptive stabilization." *Syst. Contr. Lett.* **6**(2): 87–91.

Mårtensson showed (to summarize roughly) that the order of a stabilizing controller is the only information required for adaptive stabilization of a multivariable system.

CHAPTER 7

STOCHASTIC
ADAPTIVE CONTROL

7.1 INTRODUCTION

In earlier chapters the adaptive control problem was approached from a heuristic point of view. The unknown parameters of the process or the regulator were estimated by using real-time estimation, and the estimated parameters were then used as if they were the true ones. The uncertainties of the parameter estimates were not taken into account in the design. This procedure gives a *certainty equivalence* controller. The model-reference adaptive controllers and the self-tuning regulators have been derived under the assumption that the parameters are constant but unknown. When the process parameters are constant, the estimation routines usually are such that the uncertainties decrease rapidly after the estimation is started. However, the uncertainties can be large at the startup or if the parameters are changing. In such cases it may be important to let the control law be a function of the parameter estimates as well as of the uncertainties of the estimates.

It would be appealing to formulate the adaptive control problem from a unified theoretical framework. This can be done by using nonlinear stochastic control theory, in which the process, its parameters, and the environment are described by using a stochastic model. The difference compared with the treatment in the previous chapters is that the parameters of the process also are described by using a stochastic model. The criterion is formulated so as to minimize the expected value of a loss function. It is difficult to find the controller that minimizes the expected loss function. Conditions for the existence of an optimal controller are not known. However, under the condition that a solution exists, it is possible to derive a functional equation by using dynamic program-

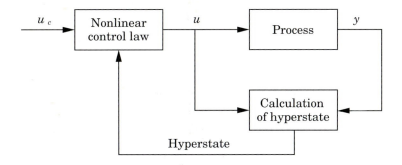

Figure 7.1   Block diagram of an adaptive regulator obtained from stochastic control theory.

ming. This equation, called the *Bellman equation*, can be solved numerically only in very simple cases. The structure of the optimal regulator is shown in Fig. 7.1. The controller is composed of two parts: an estimator and a feedback regulator. The estimator generates the conditional probability distribution of the state given the measurements. This distribution is called the *hyperstate* of the problem. The feedback regulator is a nonlinear function that maps the hyperstate into the space of control variables.

The structural simplicity of the solution is obtained at the price of introducing the hyperstate, which can be a quantity of very high dimension. Notice that the structure is similar to that of the self-tuning regulator. The self-tuning regulator can be regarded as an approximation; the conditional probability distribution is replaced by a distribution with all mass at the conditional mean value. In Fig. 7.1 there is no distinction between the parameters and the other state variables of the process. The regulator can therefore handle very rapid parameter variations. Furthermore, the averaging methods based on separation of the states of the process and the parameters (used in Chapter 6) cannot be used to analyze the system. The optimal control law has an interesting property. The control attempts to drive the output to the desired value, but it will also introduce perturbations when the estimates are uncertain. This will improve the estimates and the future control. The optimal controller achieves a correct balance between maintaining good control and small estimation errors. This is called *dual control*.

The chapter is organized in the following way. The idea with multistep decision problems is introduced in Section 7.2, where the two-armed bandit problem is introduced. A general stochastic adaptive control problem is formulated in Section 7.3, and Section 7.4 gives the derivation of the Bellman equation. The consequences of the structure of the solution are discussed, and the dual property is analyzed. Different ways to approximate the dual controller are discussed in Section 7.5. However, only very simple examples of dual controllers can be solved numerically, but the solutions give some useful indications of how suboptimal controllers can be constructed. Some examples are given in Section 7.6, and the stochastic adaptive approach is summarized in Section 7.7.

7.2   MULTISTEP DECISION PROBLEMS

The idea of decision under uncertainty is discussed in this section. There are many situations in which decisions must be taken despite uncertainties about the processes or the statistics. One example is route planning, in which the traffic will influence the time it takes to get from one point to another. Another example is testing of medical drugs. In investigating the effect of a new drug, it is necessary to plan the test, but it is also important to have the possibility to go back to a standard procedure if the patient is not responding well to the new treatment. The characteristic features of these types of problems are that there are uncertainties about the possible outcome of different control actions. Further, there is a sequence of control actions to be taken. At each time, feedback is used to update or change the procedure. One of the first stochastic adaptive problems of this kind that was solved can be represented by the classical *two-armed bandit (TAB) problem*. This is a typical problem of sequential design of statistical experiments.

The TAB problem can be described in the following way. A player is faced with two slot machines, I and II. If machine I is played the gain is one unit with probability p; machine II gives a gain of one unit with probability q. In the simplest case, p is known and q is unknown and is chosen before each game of length N according to a given probability distribution. During the game the unknown quantity q has to be estimated, and the player must decide at each step which machine to play to maximize the total gain of each game of N plays.

The two-armed bandit problem can be used to illustrate the essential ideas of multistep decision problems. One strategy that can be used is open-loop control, that is, the control sequence is chosen without any measurements being made. The decision is taken with respect to the *a priori* knowledge about p and the distribution of q. In the TAB case, machine I should be played if p is larger than the mean value of q; otherwise, machine II should be played. A second strategy is what is called *open-loop optimal feedback (OLOF) control*. This controller is derived by maximizing the multistep gain function at each step under the assumption that no further measurements will be available, that is, an open-loop control sequence is determined. The first step in the control sequence is then used, and the performance of the system is measured. On the basis of the new information (feedback) a new maximization is done (compare the receding horizon controller in Chapter 4). The first step in the OLOF control is thus the same as the first step in the open-loop control. The measurements are thus in the TAB problem used to update the estimate of the unknown probability q.

To find the optimal solution to the TAB problem, it is possible to use dynamic programming to derive the optimal strategy that maximizes the expected gain depending on the outcome of previous plays in the game. How this is done is shown for a more general case in Section 7.4. The optimal strategy for a simple TAB problem is illustrated in the following example adopted from Yakowitz (1969).

<hr>

EXAMPLE 7.1   Two-armed bandit problem

Assume that $p = 0.6$ and that q is uniformly distributed over the interval $[0, 1]$. If machine I is played all the time, the expected gain is 0.6 per play; if machine II is played all the time, the expected gain per play is 0.5. The open-loop strategy then suggests that machine I should be played all the time. However, for each game of length N there is a probability that the q has a larger value than 0.6. If infinitely many plays are available, the player can play machine II, estimate q, and then decide which machine to play for the rest of the plays.

To determine the profit of knowledge of q, assume that the player is told the value of q before each game. The player's optimal strategy is then to play the machine having the highest probability. In this case the expected gain per play is $E\{\max(p, q)\} = 0.68$. This means that the expected gain can be increased by 13% compared with the open-loop strategy if q is estimated. Table 7.1 shows the average gain per play for different values of N. As the number of plays increases, the gain will approach the maximum value 0.68. Relatively many plays are needed to get close to the optimum.

Figure 7.2 shows the state transition diagram for the optimal strategy when $N = 6$. The player is initially in the state (0, 0) and starts to play machine II to find out if machine II has better winning probability than machine I. The numbers in the circles indicate the number of times machines I and II, respectively, have given a gain of one unit. In states (0, 0) and (0, 1) there will be a switch to machine I after the player loses once; after state (0, 2) the optimal strategy allows one loss before switching to machine I.   □

Table 7.1   Average gain per play for different values of N for the two-armed bandit problem in Example 7.1.

N	Gain per play
6	0.62
10	0.64
25	0.655
100	0.6676
500	0.6755
1000	0.6773

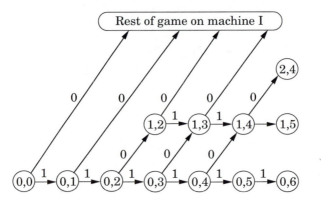

Figure 7.2   The optimal strategy for the two-armed bandit problem in Example 7.1 when $N = 6$.

7.3   THE STOCHASTIC ADAPTIVE PROBLEM

The stochastic adaptive control problem is formulated for a simple class of systems by giving the class of models, the criterion, and the admissible control strategies.

The Model

Consider the discrete-time, single-input, single-output system

$$y(t) + a_1(t)y(t-1) + \cdots + a_n(t)y(t-n) =$$
$$b_0(t)u(t-1) + \cdots + b_{n-1}(t)u(t-n) + e(t) \qquad (7.1)$$

where y, u, and e are output, input, and disturbance, respectively. The noise sequence $\{e(t)\}$ is assumed to be Gaussian with zero mean and variance R_2. Further, it is assumed that $e(t)$ is independent of $y(t-1)$, $y(t-2)$, ..., $u(t-1)$, $u(t-2)$, ..., $a_i(t)$, $a_i(t-1)$ ..., and $b_i(t)$, $b_i(t-1)$, ... It is further assumed that $b_0(t) \neq 0$ and that the system is minimum-phase for all t. The time-varying parameters

$$x(t) = \left(\begin{array}{ccccc} b_0(t) & \cdots & b_{n-1}(t) & a_1(t) & \cdots & a_n(t) \end{array} \right)^T \qquad (7.2)$$

are modeled by a Gauss-Markov process, which satisfies the stochastic difference equation

$$x(t + 1) = \Phi x(t) + v(t) \qquad (7.3)$$

where Φ is a known constant matrix and $\{v(t)\}$ is a sequence of independent, equally distributed normal vectors with zero mean value and known covariance

R_1. The initial state of the system in Eq. (7.3) is assumed to be normally distributed with mean value

$$Ex(0) = m \tag{7.4}$$

and covariance

$$\text{cov } \{x(0), x(0)\} = R_0 \tag{7.5}$$

It is assumed that $e(t)$ is independent of $v(t)$ and of $x(0)$.

The input-output relation of the system of Eq. (7.1) can be written in the compact form

$$y(t) = \varphi^T(t-1)x(t) + e(t) \tag{7.6}$$

where

$$\varphi^T(t-1) = \Big(u(t-1) \quad \ldots \quad u(t-n) \quad -y(t-1) \quad \ldots \quad -y(t-n) \Big) \tag{7.7}$$

The model is thus defined by Eqs. (7.3) and (7.6).

The Criterion

It is assumed that the purpose of the control is to keep the output of the system as close as possible to a known reference value trajectory $u_c(t)$. The deviation is measured by the criterion

$$J_N = E \left\{ \frac{1}{N} \sum_{t=1}^{N} (y(t) - u_c(t))^2 \right\} \tag{7.8}$$

where E denotes mathematical expectation. This is called an *N-stage criterion*. The loss function should be minimized with respect to $u(0), u(1), \ldots, u(N-1)$. The controller obtained for $N = 1$ is sometimes called a *myopic controller*, since it is short-sighted and looks only one step ahead. The minimizing controller will be very different if $N = 1$ or if N is large.

Admissible Control Strategies

To specify the problem completely, it is necessary to define the admissible control strategies. A control strategy is admissible if $u(t)$ is a function of all outputs observed up to and including time t, that is, $y(t), y(t-1), \ldots$ all applied control signals $u(t-1), \ldots$ and the *a priori* data. Let $\mathcal{Y}_t$ denote all values of the output up to and including $y(t)$ or, more precisely, the σ-algebra generated by $y(t), \ldots, y(0)$ and $x(0)$.

Discussion of the Problem Formulation

To get a reasonable problem, it is assumed that the noise in Eq. (7.1) is of least-squares type, that is, $C(q) = q^n$. Further, there is no extra time delay in the system. In the formulation it has been assumed that the measurements $y(t)$ are obtained at each sampling interval. It is possible to define other control problems leading to other controllers by changing the way in which the future measurements will become available. The realism of the assumption that Φ is known in Eq. (7.3) is open to question. The case $\Phi = I$ can, however, be used as a generic case to study the dual control problem.

The process of Eq. (7.1) is a nonlinear model, since the parameters as well as the old inputs and outputs are the states of the system. Notice for instance that the distributions of the parameters and the disturbances are Gaussian but $y(t)$ is not Gaussian. The problem could also be phrased in more general terms by assuming that both the model and the criterion are general nonlinear functions. In this chapter we consider the special case defined by Eqs. (7.1) and (7.8) to illustrate the ideas and the difficulties with the stochastic adaptive approach.

7.4   DUAL CONTROL

We now analyze the problem formulated in Section 7.3. The problem of estimating the parameters of Eq. (7.1) is first considered. The control problem is first solved for the case in which the parameters are known. The problem is then solved for the case in which $N = 1$ in the criterion of Eq. (7.8). The solution of the complete problem is finally discussed. The control problem is solved by using dynamic programming.

The Estimation Problem

To solve the dual control problem, it is necessary to be able to evaluate the influence of the control signal on the future outputs and to estimate and predict the behavior of the stochastic parameters. The estimation problem is defined so as to compute the conditional probability distribution of the parameters, given the measured data.

The system is written in a standard state space form, using Eqs. (7.3) and (7.6). The conditional distribution of $x(t+1)$, given $\mathcal{Y}_t$, is given by the following theorem.

THEOREM 7.1   Conditional distribution of the states

Consider the model of Eq. (7.3) with the output defined by Eq. (7.6), where $e(t)$ and $v(t)$ are independent zero mean Gaussian variables with covariances

R_2 and R_1, respectively. The initial state of the system is given by Eqs. (7.4) and (7.5).

The conditional distribution of $x(t)$, given $\mathcal{Y}_{t-1}$, is Gaussian with mean $\hat{x}(t)$ and covariance $P(t)$ satisfying the difference equations

$$\hat{x}(t+1) = \Phi\hat{x}(t) + K(t)\left(y(t) - \varphi^T(t-1)\hat{x}(t)\right)$$
$$P(t+1) = \left(\Phi - K(t)\varphi^T(t-1)\right)P(t)\Phi^T + R_1 \tag{7.9}$$
$$K(t) = \Phi P(t)\varphi(t-1)\left(R_2 + \varphi^T(t-1)P(t)\varphi(t-1)\right)^{-1}$$

with the initial conditions

$$\hat{x}(0) = m$$
$$P(0) = R_0$$

Furthermore, the conditional distribution of $y(t)$, given $\mathcal{Y}_{t-1}$, is Gaussian with mean value

$$m_y(t) = \varphi^T(t-1)\hat{x}(t)$$

and covariance

$$\sigma_y^2(t) = R_2 + \varphi^T(t-1)P(t)\varphi(t-1)$$

Proof: If $\varphi(t-1)$ is a known time-varying vector, then the theorem is identical to the Kalman filtering theorem, which can be found in standard textbooks on stochastic control. Going through the details of the proof of the Kalman filtering theorem, we find that it is still valid, since $\varphi(t-1)$ is a function of $\mathcal{Y}_{t-1}$. In other words, the vector $\varphi(t-1)$ is not known in advance, but it is known when it is needed in the computations.   □

Remark.   Notice that the conditional distribution of $y(t)$, given $\mathcal{Y}_{t-1}$, is Gaussian even if $y(t)$ is not Gaussian.   □

The estimation problem is thus easily solved for the model structure chosen. The conditional distribution of the state of the system is called the *hyperstate*. The distribution is Gaussian in the problem under consideration. It is then sufficient to consider the mean and covariance of $x(t)$. Further, some of the old inputs and outputs must be stored to compute the distribution defined in Eqs. (7.9). In the problem under consideration the hyperstate is finite-dimensional and can be characterized by the triple

$$\xi(t) = \left(\ \tilde{\varphi}(t-1)\quad \hat{x}(t)\quad P(t)\ \right) \tag{7.10}$$

where

$$\tilde{\varphi}^T(t-1) = \left(\ 0\quad u(t-2)\quad \dots\quad u(t-n)\quad -y(t-1)\quad \dots\quad -y(t-n)\ \right) \tag{7.11}$$

The vector $\tilde{\varphi}^T(t-1)$ is the same as $\varphi^T(t-1)$, except that $u(t-1)$ is replaced by a zero. The updating of the hyperstate is given by Theorem 7.1 and the definition of $\tilde{\varphi}^T(t-1)$. In the general case the conditional probability distribution is not Gaussian. This will considerably increase the computational difficulties and the storage requirements.

Systems with Known Parameters

If the parameters of the system of Eq. (7.1) are known, it is easy to determine the optimal feedback. The vector $\tilde{\varphi}^T$ defined by Eq. (7.11) is used to show the dependence of $u(t)$:

$$y(t + 1) = \varphi^T(t)x(t + 1) + e(t + 1)$$
$$= b_0(t + 1)u(t) + \tilde{\varphi}^T(t)x(t + 1) + e(t + 1)$$

The optimal feedback when $b_0(t + 1)$ and $x(t + 1)$ are known is then given by

$$u(t) = \frac{u_c(t + 1) - \tilde{\varphi}^T(t)x(t + 1)}{b_0(t + 1)} \tag{7.12}$$

Notice that $\tilde{\varphi}(t)$ is a function of the admissible data. This controller gives

$$y(t + 1) = u_c(t + 1) + e(t + 1)$$

and it minimizes Eq. (7.8), since $e(t + 1)$ is independent of $\mathcal{Y}_t$ and $u(t)$. The minimal loss is given by

$$\min J_N = R_2$$

Notice that it is necessary to assume that $b_0(t + 1) \neq 0$ and that the system is minimum-phase at every instant of time. The control signal may otherwise be unbounded.

Certainty Equivalence Control

When the parameters $x(t + 1)$ are not known, it can be tempting to replace Eq. (7.12) with

$$u(t) = \frac{u_c(t + 1) - \tilde{\varphi}^T(t)\hat{x}(t + 1)}{\hat{b}_0(t + 1)} \tag{7.13}$$

The true parameter values are replaced by the expected values, given $\mathcal{Y}_t$. The controller of Eq. (7.13) is called the *certainty equivalence controller*. Certainty equivalence control is the strategy used in the self-tuning regulators in Chapters 3 and 4 and in the model-reference adaptive systems in Chapter 5. In these controllers it was also necessary to ensure that $\hat{b}_0 \neq 0$.

Cautious Control

We now consider the special case in which $N = 1$ in Eq. (7.8). According to Theorem 7.1 the conditional distribution of $y(t + 1)$, given $\mathcal{Y}_t$, is Gaussian with mean $\varphi^T(t)\hat{x}(t+1)$ and covariance $R_2 + \varphi^T(t)P(t+1)\varphi(t)$. Then

$$E\left\{(y(t+1) - u_c(t+1))^2 \,|\, \mathcal{Y}_t\right\}$$

$$= \left(\varphi^T(t)\hat{x}(t+1) - u_c(t+1)\right)^2 + \varphi^T(t)P(t+1)\varphi(t) + R_2$$

$$= \left(\tilde{\varphi}^T(t)\hat{x}(t+1) + \hat{b}_0(t+1)u(t) - u_c(t+1)\right)^2$$
$$+ \tilde{\varphi}^T(t)P(t+1)\tilde{\varphi}(t) + u^2(t)p_{b_0}(t+1)$$
$$+ 2u(t)\tilde{\varphi}^T(t)P(t+1)\ell + R_2 \tag{7.14}$$

The first equality is obtained by using the standard formula that

$$E(\zeta^2) = m^2 + p$$

when ζ is a Gaussian variable with mean m and variance p. The column vector ℓ selects the first column of the matrix $P(t)$, that is,

$$\ell^T = \begin{pmatrix} 1 & 0 & \cdots & 0 \end{pmatrix}$$

Further, p_{b_0} is the covariance of the parameter estimate $\hat{b}_0$. Equation (7.14) is quadratic in $u(t)$. Minimization of Eq. (7.14) with respect to $u(t)$ gives the admissible one-step optimal controller

$$u(t) = \frac{\hat{b}_0(t+1)u_c(t+1) - \tilde{\varphi}^T(t)\left(\hat{b}_0(t+1)\hat{x}(t+1) + P(t+1)\ell\right)}{\hat{b}_0^2(t+1) + p_{b_0}(t+1)} \tag{7.15}$$

The minimum value of the loss function is

$$\min_{u(t)} E\left\{(y(t+1) - u_c(t+1))^2 \,|\, \mathcal{Y}_t\right\}$$

$$= \left(\tilde{\varphi}^T(t)\hat{x}(t+1) - u_c(t+1)\right)^2 + R_2 + \tilde{\varphi}^T(t)P(t+1)\tilde{\varphi}(t)$$
$$- \frac{\left(\hat{b}_0(t+1)u_c(t+1) - \tilde{\varphi}^T(t)\left(\hat{b}_0(t+1)\hat{x}(t+1) + P(t+1)\ell\right)\right)^2}{\hat{b}_0^2(t+1) + p_{b_0}(t+1)} \tag{7.16}$$

The *one-step-ahead controller*, or *cautious controller*, of Eq. (7.15) differs from Eq. (7.13) because the uncertainties of the parameter estimates are also taken into account. The controller becomes cautious when the estimates are uncertain. Notice that the cautious controller of Eq. (7.15) reduces to the certainty equivalence controller of Eq. (7.13) when $P(t+1) = 0$.

EXAMPLE 7.2   Integrator with time-varying gain

Consider an integrator in which the gain is changing. Let the process be described by

$$y(t) - y(t-1) = b(t)u(t-1) + e(t)$$

where

$$b(t + 1) = \varphi_b b(t) + R_1 v(t)$$

The errors e and v are zero-mean Gaussian white noise with the standard deviations R_2 and 1, respectively. Further, it is assumed that $u_c = 0$. The certainty equivalence controller is given by

$$u(t) = -\frac{1}{\hat{b}(t + 1)} \, y(t)$$

and the cautious controller is

$$u(t) = -\frac{\hat{b}(t + 1)}{\hat{b}^2(t + 1) + p_b(t + 1)} \, y(t)$$

The gain in the cautious controller has been reduced by a factor

$$\frac{\hat{b}^2}{\hat{b}^2 + p_b}$$

compared with the certainty equivalence controller. Notice that the gain approaches zero when the uncertainty increases.   □

Multistep Optimization

The general multistep optimization problem can be solved by using dynamic programming. The fact that the conditional distributions are Gaussian will simplify the problem.

It follows from a fundamental result of stochastic control theory (see Åström (1970), Lemma 8:3.2) that

$$\min_{u(t-1)\ldots u(N-1)} E\left\{ \sum_{k=t}^{N} (y(k) - u_c(k))^2 \right\}$$

$$= E_{\mathcal{Y}_{t-1}}\left(\min E\left\{ \sum_{k=t}^{N} (y(k) - u_c(k))^2 \,\Big|\, \mathcal{Y}_{t-1} \right\} \right)$$

and it is assumed that the minimum exists. $E(\cdot \,|\, \mathcal{Y}_{t-1})$ is a function of the hyperstate of Eq. (7.10) and t. Define

$$V\left(\xi(t), t\right) = \min_{u(t-1)\ldots u(N-1)} E\left\{ \sum_{k=t}^{N} (y(k) - u_c(k))^2 \,\Big|\, \mathcal{Y}_{t-1} \right\}$$

$V\left(\xi(t), t\right)$ can be interpreted as the minimum expected loss for the remaining part of the control horizon given the data up to $t - 1$.

Consider the situation at time $N - 1$. When $u(N - 1)$ is changed, only $y(N)$ will be influenced. This means that we have the same situation as for the one-step minimization. From Eq. (7.16) we get

$$V\left(\xi(N),\, N\right)$$
$$= \left(\tilde{\varphi}^T(N-1)\hat{x}(N) - u_c(N)\right)^2 + R_2 + \tilde{\varphi}^T(N-1)P(N)\tilde{\varphi}(N-1)$$
$$- \frac{\left(\hat{b}_0(N)u_c(N) - \tilde{\varphi}^T(N-1)\left(\hat{b}_0(N)\hat{x}(N) + P(N)\ell\right)\right)^2}{\hat{b}_0^2(N) + p_{b_0}(N)}$$

At time $N - 1$ we get

$$V\left(\xi(N-1),\, N-1\right)$$
$$= \min_{u(N-2)} E\left\{(y(N-1) - u_c(N-1))^2 + V\left(\xi(N),\, N\right)\,\middle|\,\mathcal{Y}_{N-2}\right\}$$

Notice that the minimization is done only over $u(N-2)$, since $u(N-1)$ was eliminated in the previous minimization. This recursively defines the loss at time $N - 1$, which then can be used for iteration backwards one more step of time, and so on. This dynamic programming procedure leads to a recursive equation, which defines the minimum expected loss. At time t we get

$$V\left(\xi(t),\, t\right) = \min_{u(t-1)} E\left\{(y(t) - u_c(t))^2 + V\left(\xi(t+1),\, t+1\right)\,\middle|\,\mathcal{Y}_{t-1}\right\} \qquad (7.17)$$

This functional equation is called the *Bellman equation* of the problem. The simplicity of the form of Eq. (7.17) is misleading. The equation cannot be solved analytically, but it requires extensive numerical computations to get the solution even for very simple problems.

The first term on the right-hand side of Eq. (7.17) can be evaluated in the same way as in the one-step minimization. The second term causes the difficulties in the optimization, since we have to evaluate

$$E\left\{V\left(\xi(t+1),\, t+1\right)\,\middle|\,\mathcal{Y}_{t-1}\right\}$$

The average with respect to the distribution of $y(t)$, given $\mathcal{Y}_{t-1}$, must be computed. According to Theorem 7.1 this distribution is Gaussian with mean $m_y(t)$ and variance $\sigma_y^2(t)$. This gives

$$E\left\{V\left(\xi(t+1),\, t+1\right)\,\middle|\,\mathcal{Y}_{t-1}\right\}$$
$$= \frac{1}{\sigma_y\sqrt{2\pi}}\int_{-\infty}^{\infty} V\left(\tilde{\varphi}(t),\, \hat{x}(t+1),\, P(t+1),\, t+1\right)e^{-(s-m_y)^2/(2\sigma_y^2)}\,ds \qquad (7.18)$$

where

$$\hat{x}(t+1) = \Phi\hat{x}(t) + K(t)\left(s - \varphi^T(t-1)\hat{x}(t)\right)$$

$$P(t+1) = \left(\Phi - K(t)\varphi^T(t-1)\right)P(t)\Phi^T + R_1$$

$$K(t) = \Phi P(t)\varphi(t-1)/\sigma_y^2(t)$$

$$\sigma_y^2(t) = R_2 + \varphi^T(t-1)P(t)\varphi(t-1)$$

$$\tilde{\varphi}_1(t) = u(t-1)$$

$$\tilde{\varphi}_i(t) = \tilde{\varphi}_{i-1}(t-1) \qquad i = 2,\ldots,n, n+2,\ldots,2n$$

$$\tilde{\varphi}_{n+1}(t) = s$$

These equations, together with Eq. (7.18), can be used to compute recursively the control signal and the loss as functions of the hyperstate. The control variable $u(t-1)$ influences the immediate loss (i.e., the first term on the right-hand side of Eq. (7.17)). Notice that $u(t-1)$ also influences the expected future loss, since it influences $\varphi(t-1)$, which influences $\hat{x}(t+1)$, $P(t+1)$, and $\tilde{\varphi}^T(t)$. This means that the choice of the control signal $u(t-1)$ influences the immediate loss, the future parameter estimates, their accuracy, and also the future values of the output signal. The optimal controller is a *dual controller*. It makes a compromise between the control action and the probing action.

The probing action will add an active learning feature to the controller, in contrast to the cautious and certainty equivalence controllers, in which the learning is "accidental." The optimal feedback will generate control actions that will improve the accuracy of the future estimates at the expense of the short-term loss. The cautious controller obtained when $N = 1$ will not benefit if probing is introduced; it only tries to make the loss as small as possible at the next instant of time.

Separation and Certainty Equivalence

The optimal one-step controller of Eq. (7.15) cannot be obtained by using the certainty equivalence principle, but the estimation and the control problems can be separated. As was mentioned in Section 7.1, most adaptive controllers are based on the hypothesis that the certainty equivalence principle can be used. The derivations in this section show that the separation principle also can be used in the considered problem. However, the uncertainties must also be used in the computation of the control signal. It is thus of interest to investigate whether there are classes of systems for which the certainty equivalence and separation principles hold.

One case in which the certainty equivalence principle holds is the celebrated linear quadratic Gaussian case for known systems. For adaptive controllers there are very few cases to which the certainty equivalence principle is applicable. One exception is when the unknown parameters are stochastic variables that are independent between different sampling intervals. The certainty equivalence principle also holds for the stochastic linear quadratic problem formulation, when the process noise is white but not necessarily Gaussian and when the measurement noise is additive but not necessarily white.

The separation principle is valid for much more general cases. The cautious controller and the dual controller derived in this section are obtained by using separation.

Numerical Solution

Even in the simplest cases there is no analytic solution to the Bellman equation (Eq. 7.17). It is therefore necessary to resort to numerical solution. One iteration of Eq. (7.17) involves

- Discretization of the loss V in the variables of the hyperstate,
- Evaluation of the integral in Eq. (7.18) using a quadrature formula, and
- Minimization over $u(t-1)$ for each combination of the discretized hyperstate.

Both V and u are functions of the hyperstate, so the storage requirements increase rapidly when the order of the system increases. Assume that the dimension of the hyperstate is 2 and that each variable is discretized into ten steps. Thus the loss and control tables contain 100 values each. Let the hyperstate have dimension 6, and let each variable be discretized in ten steps. The dimension of the loss and control tables is then 10^6 each. This means that only very simple problems have been solved numerically because of the "curse of dimensionality."

Discontinuity of the Control Signal

A feature of the optimal solution is that the control law can become discontinuous in situations such as that shown in Fig. 7.3. The figure shows the loss function as a function of the control signal for three different but close values of the hyperstate. If there are several local minima, the control signal can become

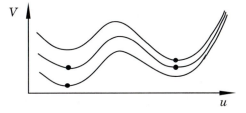

Figure 7.3 Illustration of how several local minima of the loss function can give a discontinuity in the control signal. The global minima are marked with dots. On the middle curve the local minima have the same value.

discontinuous when the global minimum changes from one local minimum to another. This can be interpreted as a change of mode for the controller. For instance, the controller may introduce probing to increase the knowledge of the unknown parameters.

7.5 SUBOPTIMAL STRATEGIES

The optimal multistep dual controller derived in Section 7.4 is of little practical use, because the numerical computations limit its applicability. The dual structure of the controller is very important, however. Many ways to make practical approximations have been suggested; this section surveys some of the possibilities. The properties of the cautious controller are first investigated, and different ways to improve this controller are then discussed.

Cautious Controllers

Minimization over only one step leads to the one-step or cautious controller of Eq. (7.15). This controller takes the parameter uncertainties into account, in contrast to the certainty equivalence controller of Eq. (7.13). However, the gain of Eq. (7.15) will decrease if the variance of $\hat{b}_0$ increases. This will give less information about b_0 in the next step, and the variance will increase further. The controller is then caught in a vicious circle, and the magnitude of the control signal becomes very small. This is called the *turn-off phenomenon*.

<hr>

EXAMPLE 7.3   Turn-off

Consider the integrator with unknown gain in Example 7.2 with $R_1 = 0.09$ and $\varphi_b = 0.95$. Figure 7.4 shows a representative simulation of the cautious controller. The control signal is small for periods of time, and the variance of the estimated gain increases during the turn-off. After some time the control activity suddenly starts again.   □

The turn-off will generally start when the control signal is small and when the parameter b_0 is small. The problem with turn-off makes the cautious controller unsuitable for control of systems with quickly varying parameters. The cautious controller can be useful if the parameters of the process are constant or almost constant, but the certainty equivalence controller with some simple safety measures can often be used in such cases also.

Classification of Suboptimal Dual Controllers

The problem of turn-off has led to many suggestions of how to derive controllers that are simple but still have some dual features. Some ways are:

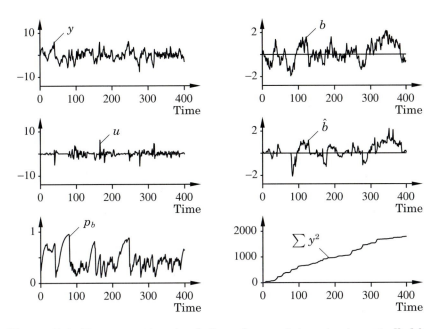

Figure 7.4   Representative simulation when an integrator is controlled by using a cautious controller. Turn-off occurs when the control signal is small.

- Adding perturbation signals to the cautious controller,
- Constraining the variance of the parameter estimates,
- Extensions of the loss function, and
- Serial expansion of the loss function.

Some of these modifications are now discussed.

Perturbation Signals

The turn-off is due to lack of excitation (compare Chapter 6). One way to increase the excitation is to add a perturbation signal. Pseudo-random binary sequences (PRBS) and white noise signals have been suggested. The perturbation can be added all the time or only when the variance is exceeding some limit. The addition of the extra signal will naturally increase the probing loss but may make it possible to improve the total performance.

Constrained One-Step Minimization

One class of suboptimal dual controllers is obtained by constrained one-step minimization. Suggested constraints are

- Limitation of the minimum value of the control signal and
- Limitation of the variance.

One method is to choose the control as

$$u(t) = \begin{cases} u_{\lim} \cdot \text{sign}(u_{\text{cautious}}) & \text{if } |u_{\text{cautious}}| < |u_{\lim}| \\ u_{\text{cautious}} & \text{if } |u_{\text{cautious}}| \geq |u_{\lim}| \end{cases}$$

This will give an extra probing signal if the cautious controller gives too small an input signal.

Different ways to constrain the minimization by using the P-matrix have been suggested. For instance, the one-step loss of Eq. (7.14) can be minimized under the constraint that

$$\text{tr} \, P^{-1}(t + 2) \geq M$$

P^{-1} is proportional to the information matrix. The constraint on the trace of P^{-1} means that the information about the parameters is always larger than some chosen value M. A similar approach is to constrain only the variance of $\hat{b}_0$ to

$$p_{b_0}(t + 2) \leq \begin{cases} \gamma \hat{b}_0^2(t + 2) & \text{if } p_{b_0}(t + 1) \leq \hat{b}_0^2(t + 1) \\ \alpha p_{b_0}(t + 1) & \text{otherwise} \end{cases}$$

These modifications of the cautious controller have the advantage that the control signal can be easily computed, but the algorithms will contain application-dependent parameters that have to be chosen by the user. Finally, the approximations will not prevent the turn-off. The extra perturbation is not activated until the turn-off occurs.

Extensions of the Loss Function

An approach that is similar to constrained minimization is to extend the loss function to prevent the shortsightedness of the cautious controller. One obvious way is to try to solve the two-step minimization problem. The derivation in Section 7.4 shows that it is not possible to get an analytic solution when $N = 2$ in Eq. (7.8).

Another approach is to extend the loss function with a function of $P(t+2)$, which will reward good parameter estimates. The following loss function can be used:

$$\min_{u(t)} E \left\{ (y(t + 1) - u_c(t + 1))^2 + \rho f(P(t + 2)) \big| \mathcal{Y}_t \right\} \tag{7.19}$$

where ρ is a fixed parameter. Since the crucial parameter is b_0, we can use

$$f(P(t + 2)) = p_{b_0}(t + 2)$$

or

$$f(P(t + 2)) = R_2 \frac{p_{b_0}(t + 2)}{p_{b_0}(t + 1)} \tag{7.20}$$

This leads to a loss function with two local minima; it is necessary to make a numerical search for the global minimum. It is possible to utilize the structure of the problem and make a serial expansion up to second order of the loss function. The expansion gives a simple noniterative suboptimal dual controller in which the increase in computations compared with a self-tuning or cautious regulator is very moderate.

Two similar approaches are to modify the loss functions to

$$\min_{u(t)} E\left\{(y(t+1) - u_c(t+1))^2 - \rho\frac{\det P(t+1)}{\det P(t+2)}\,\Big|\,\mathcal{Y}_t\right\} \tag{7.21}$$

and

$$\min_{u(t)} E\left\{(y(t+1) - u_c(t+1))^2 - \rho\varepsilon^2(t+1)\,\Big|\,\mathcal{Y}_t\right\} \tag{7.22}$$

respectively. The innovation $\varepsilon(t+1)$ is defined as

$$\varepsilon(t+1) = y(t+1) - \varphi^T(t)\hat{x}(t+1)$$

Both these loss functions lead to quadratic criteria that make it possible to derive simple analytic expressions for the control signal.

Serial Expansion of the Loss Function

The suboptimal dual controllers discussed above have been derived for the input-output model of Eq. (7.1). Suboptimal dual controllers have also been derived for state space models. One approach is to make an expansion of the loss function in the Bellman equation. Such an expansion can be done around the certainty equivalence or the cautious controllers. This approach has mainly been used when the control horizon N is rather short, usually less than 10. One reason is the quite complex computations that are involved.

Summary

There are many ways to make suboptimal dual controllers. Most of the approximations that are discussed start with the cautious controller and try to introduce some active learning. This can be done by including a term in the loss function that reflects the quality of the estimates. This term should also be a function of the control signal that is going to be determined. The suboptimal controllers should also be such that they can be used for higher-order systems without too much computation.

7.6 EXAMPLES

Some examples are used to illustrate the properties of the controllers discussed in this chapter.

EXAMPLE 7.4   Optimal dual controller

The first example is a numerically solved dual control problem from Åström and Helmersson (1982). Consider the integrator in Example 7.2. The gain is assumed to be constant but unknown, that is, $\varphi_b = 1$ and $R_1 = 0$. It is assumed that the parameter b is a random variable with a Gaussian prior distribution; the conditional distribution of b, given inputs and outputs up to time t, is Gaussian with mean $\hat{b}(t)$ and covariance $P(t)$. The hyperstate can then be characterized by the triple $(y(t), \hat{b}(t), P(t))$. The equations for updating the hyperstate are given by Eqs. (7.9).

Define the loss function

$$
V_N = \min_u E \left\{ \sum_{k=t+1}^{t+N} y^2(k) \,|\, \mathcal{Y}_t \right\}
$$

where $\mathcal{Y}_t$ denotes the data available at time t, that is, $\{y(t), y(t-1), \ldots\}$. By introducing the normalized variables

$$
\eta = y/\sqrt{R_2} \qquad \beta = \hat{b}/\sqrt{P} \qquad \mu = -u\hat{b}/y
$$

it can be shown that V_N depends on η and β only. Further introduce the normalized innovation

$$
\varepsilon(t) = \frac{y(t+1) - y(t) - \hat{b}(t)u(t)}{R_2 + u(t)^2 P(t)}
$$

For $R_2 = 1$ the Bellman equation for the problem can be written as

$$
V_N(\eta,\beta) = \min_\mu U_N(\eta,\beta,\mu)
$$

where

$$
V_0(\eta,\beta) = 0
$$

and

$$
U_N(\eta,\beta,\mu) = 1 + \eta^2(1-\mu)^2 + \frac{\mu^2\eta^2}{\beta^2} + \int_{-\infty}^{\infty} V_{N-1}(\eta_p,\beta_p)\phi(s)\,ds
$$

where ϕ is the normal probability density with zero mean and unit variance and

$$
\eta_p = \eta - \mu\eta + s\sqrt{1 + \frac{\mu^2\eta^2}{\beta^2}}
$$

$$
\beta_p = \beta\sqrt{1 + \frac{\mu^2\eta^2}{\beta^2}} - \frac{\mu\eta}{\beta}s
$$

Notice that η_p and β_p are the one-step-ahead predicted values of η and β. When the minimization is performed, the control law is obtained as

$$
\mu_N(\eta,\beta) = \arg\min_\mu U_N(\eta,\beta,\mu)
$$

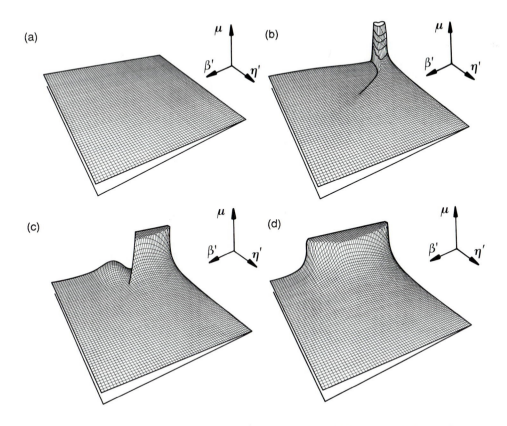

Figure 7.5   Illustration of the cautious control and dual control laws when
(a) $N = 1$; (b) $N = 3$; (c) $N = 6$; and (d) $N = 31$. The control signal is shown
as a function of $\eta' = \eta/(1 + \eta)$ and $\beta' = \beta^2/(1 + \beta^2)$. The control signal is
limited, which explains the plateaus.

The minimization can be done analytically for $N = 1$, giving

$$\mu_1(\eta, \beta) = \arg \min \left(1 + \eta^2(1 - \mu)^2 + \frac{\mu^2 \eta^2}{\beta^2} \right) = \frac{\beta^2}{1 + \beta^2}$$

The original variables give

$$u(t) = -\frac{1}{\hat{b}(t + 1)} \cdot \frac{\hat{b}^2(t + 1)}{\hat{b}^2(t + 1) + P(t + 1)} \, y(t)$$

This control law is the one-step control, or myopic control, derived in Example 7.2.

   For $N > 1$ the optimization can no longer be done analytically. Instead,
we have to resort to numerical calculations. Figure 7.5 shows the dual control

laws obtained for different time horizons N. The discontinuity of the control law corresponds to the situation in which a probing signal is introduced to improve the estimates.

The certainty equivalence controller

$$u(t) = -y(t)/\hat{b}$$

can be expressed as

$$\mu = 1$$

in normalized variables. Notice that all control laws are the same for large β, that is, if the estimate is accurate. The optimal control law is close to the cautious control for large control errors. For estimates with poor precision and moderate control errors, the dual control gives larger control actions than the other control laws. The optimal dual controller has been computed on a Vax 11/780. The normalized variables η and β are discretized into 64 values each. The control table and the loss function table are thus of dimension 64×64. One iteration of the Bellman equation takes about 6 hours of CPU time.   □

EXAMPLE 7.5   **Probing**

An interesting feature of the dual control law is that it behaves quite differently from the heuristic algorithms. The most significant feature is the probing that takes place to gain more information about the unknown parameters. The effect of probing is most significant when the output y is small. Probing can be illustrated by using the results of Example 7.4. Both the cautious and certainty equivalence control laws are continuous in y and zero for $y = 0$. However, the dual control law is very different. To show this, consider the control signal for $y = 0-$. Figure 7.6 shows the control signal for $y = 0-$ as a function of the normalized parameter precision, $\beta = \hat{b}/\sqrt{P}$, for different time horizons. All control laws give zero control signal when the parameter estimate

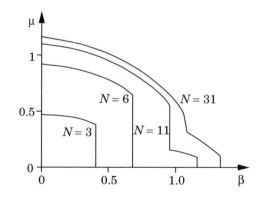

Figure 7.6 Control signal as a function of the normalized parameter precision β for optimal control laws for different time horizons.

is reasonably precise. However, for uncertain estimates, the control signal is different from zero, and the transition is discontinuous. This discontinuity can be used to define a probing zone. Notice that the probing zone increases with increasing time horizon. For $N = 31$, probing occurs when $\beta \leq 1.3$, that is, when $\hat{b} \leq 1.3\sqrt{P}$. ☐

Time-varying parameters

The system in Examples 7.2 and 7.3 will now be controlled by a suboptimal dual controller that minimizes Eq. (7.19), with $f(P(t+2))$ given by Eq. (7.20). (See Wittenmark and Elevitch (1985).) Figure 7.7 shows the same experiment using the same noise sequences as in Fig. 7.4. With the suboptimal dual controller there is no tendency toward turn-off. The simulation in Fig. 7.7 shows that the suboptimal dual controller is much better than the cautious controller. Comparisons using Monte Carlo simulations have also been done with the numerically computed optimal dual controller. The result is that the suboptimal dual controller is as good as the numerically computed optimal controller. A summary of some simulations is shown in Fig. 7.8, which shows mean values and standard deviations of the loss when the standard deviation of the parameter noise R_1 is changed. It is assumed that

$$\varphi_b = \sqrt{1 - R_1}$$

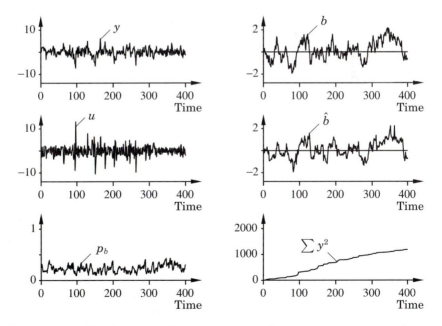

Figure 7.7 Simulation of the integrator with time-varying gain using a suboptimal dual controller. Compare Fig. 7.4.

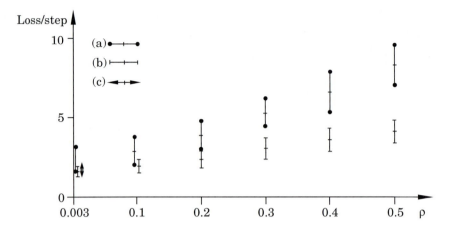

Figure 7.8 The mean values and standard deviations of the losses for Monte Carlo runs with different values of $\sqrt{R_1}$ for the system in Example 7.6. (a) Cautious controller. (b) Suboptimal dual controller from Wittenmark and Elevitch (1985). (c) Numerically computed optimal dual controller from Åström and Helmersson (1982).

For $\sqrt{R_1}$ = 0.003 there is good agreement between the suboptimal controller and the optimal dual controller that was derived under the assumption that $R_1 = 0$. The optimal dual controller from Example 7.4 corresponds to $R_1 = 0$. The controller obtained has been used also for $\sqrt{R_1}$ = 0.003.

In Eq. (7.20), $R_2/p_{b_0}(t + 1)$ is used as a normalization factor in the term added to the loss function. The reason is an attempt to preserve a property of the dual optimal controller. In Example 7.4 the loss V_N was a function of the normalized variables η and β. The loss function of Eq. (7.19) with Eq. (7.20) will also have this property for the integrator example. Simulations indicate that the normalization in Eq. (7.20) also makes the choice of ρ less crucial.   □

7.7 CONCLUSIONS

Optimal multistep controllers have been derived by using stochastic control theory. The solution is defined through the Bellman equation. This functional equation is difficult to solve even for very simple systems. The optimal solution has some interesting properties; it makes a compromise between good control and good estimation by introducing probing actions. This dual effect is of great importance, since it introduces an active learning feature into the controller. It is important to preserve this dual feature when suboptimal controllers are considered. The cautious or one-step controller does not have any active learning; the control may instead be turned off when the parameter uncertainties become too large. One important question is whether it is worth the effort to look at more elaborate control structures than the certainty equivalence controllers. The self-tuning regulators perform very well, as can be seen in Chapters 3, 4, and 6. The extra computations are not too extensive in several of the suboptimal dual controllers discussed in Section 7.5. This indicates that active learning can easily be included.

There are two situations in which dual control can pay off. One is when the time horizon is very short, in which case it is important to get good parameter estimates immediately. Areas in which this is the case are economic systems and control of missiles. The other situation in which dual features are important is when the parameters are varying rapidly and when the b_0 parameter can change sign, as in the simulations given in Section 7.6. Grinders are one type of physical process in which the gain may change sign. Grinders are common, for instance, in the mining, pulp, and paper industries.

Even if the optimal dual controller is impossible to calculate for realistic processes, it gives important hints about how to make sensible modifications of certainty equivalence and cautious controllers.

PROBLEMS

7.1 Discuss possible difficulties of extending the problem given in Section 7.3 to the case in which the system in Eq. (7.1) has an additional time delay.

7.2 Show that the cautious controller of Eq. (7.15) minimizes the loss function of Eq. (7.14) and that the minimum value of the loss function is Eq. (7.16).

7.3 Consider the process in Example 7.2, but with a constant but unknown gain b. Calculate and compare the minimum values of the loss function when

(a) the parameter b is known (i.e., the minimum-variance controller),

(b) the certainty equivalence controller is used,

(c) the cautious controller is used.

7.4 Compute the suboptimal control law that minimizes the loss function of Eq. (7.21). (*Hint*: See Goodwin and Payne (1977), p. 296.)

7.5 Compute the suboptimal control law that minimizes the loss function of Eq. (7.22). (*Hint*: See Milito *et al.* (1982).)

7.6 Assume that the process is described by one of the known models

$$y(t) = \varphi(t)\theta_i + e(t) \qquad i = 1, \ldots, m$$

but it is not known which is the correct one. Let the initial information be described by the probabilities $p_i = \mathrm{P}(\theta = \theta_i)$. Formulate the dual control problem and discuss the computational difficulties associated with the solution.

7.7   Discuss the consequences of formulating the dual control problem for the model

$$x(t + 1) = \Phi(t)x(t) + \Gamma(t)u(t)$$
$$y(t) = C(t)x(t) + e(t)$$

where Φ, Γ, and C contain some unknown parameters. For simplicity, consider the case in which the system is given in controllable canonical form, that is,

$$\Phi(t) = \begin{pmatrix} -a_1(t) & -a_2(t) & \cdots & -a_n(t) \\ 1 & 0 & \cdots & 0 \\ \vdots & & \ddots & \vdots \\ 0 & & \cdots & 1 & 0 \end{pmatrix}$$

$$\Gamma^T(t) = \begin{pmatrix} b_0(t) & \cdots & b_{n-1}(t) \end{pmatrix}$$

$$C(t) = \begin{pmatrix} 1 & 0 & \cdots & 0 \end{pmatrix}$$

REFERENCES

The basic ideas of stochastic control and dynamic programming are discussed in:

Bellman, R., 1961. *Adaptive Control Processes: A Guided Tour.* Princeton, N.J.: Princeton University Press.

Åström, K. J., 1970. *Introduction to Stochastic Control Theory.* New York: Academic Press.

Bertsekas, D., 1978. *Stochastic Optimal Control.* New York: Academic Press.

More general treatments and surveys of stochastic adaptive control are found in:

Wittenmark, B., 1975. "Stochastic adaptive control methods: A survey." *Int. J. Control* **21**: 705–730.

Bar-Shalom, Y., and E. Tse, 1976. "Concepts and methods in stochastic control." In *Control and Dynamic Systems: Advances in Theory and Applications*, ed. C. T. Leondes, Vol. 12, pp. 99–172. New York: Academic Press.

Åström, K. J., 1978. "Stochastic control problems." In *Mathematical Control Theory. Lecture Notes in Mathematics*, ed. W. A. Coppel. Berlin: Springer-Verlag.

Kumar, P. R., and P. Varaiya, 1986. *Stochastic Systems: Estimation, Identification, and Adaptive Control.* Englewood Cliffs, N.J.: Prentice-Hall.

The two-armed bandit problem is discussed, for instance, in Bellman (1961) and in:

Yakowitz, S. J., 1969. *Mathematics of Adaptive Control Processes.* New York: American Elsevier.

The dual control concept with control loss and probing loss is discussed in:

Feldbaum, A. A., 1965. *Optimal Control Theory.* New York: Academic Press.

The difference between certainty equivalence and separation is treated in:

Witsenhausen, H. S., 1971. "Separation of estimation and control for discrete time systems." *Proceedings IEEE* **59**: 1557–1566.

Because of the difficulty of solving the Bellman equation, only a few dual optimal control problems have been solved. The simplified case in which the process is described as a Markov chain is discussed in:

Åström, K. J., 1965. "Optimal control of Markov processes with incomplete state information I." *J. Math. Anal. Appl.* **10**: 174–205.

Åström, K. J., 1969. "Optimal control of Markov processes with incomplete state information II." *J. Math. Anal. Appl.* **26**: 403–406.

Sternby, J., 1976. "A simple dual control problem with an analytical solution." *IEEE Trans. Automat. Contr.* **AC-21**: 840–844.

The case in which the process is a delay and there is an unknown gain is solved numerically in:

Åström, K. J., and Wittenmark, B., 1971. "Problems of identification and control." *J. Math. Anal. Appl.* **34**: 90–113.

This reference also gives examples of the turn-off phenomenon. The integrator with unknown gain is analyzed in:

Bohlin, T., 1969. "Optimal dual control of a simple process with unknown gain." Technical Paper PT 18.196, IBM Nordic Laboratory, Lidingö, Sweden.

Åström, K. J., and Helmersson, A., 1982. "Dual control of a lower order system." *Proceedings of the National CNRS Colloque "Développement et Utilisation d'Outils et Modèles Mathématiques en Automatique, Analyse de Systèmes et Traitement du Signal."* Belle-Ile, France.

Åström, K. J., and Helmersson, A., 1986. "Dual control of an integrator with unknown gain." *Comp. & Maths. with Appls.* **12A**(6): 653–662.

The computational problems of the optimal solution have led to many different suggestions for suboptimal dual controllers. Extra perturbation to avoid turn-off is discussed in:

Wieslander, J., and B. Wittenmark, 1971. "An approach to adaptive control using real-time identification." *Automatica* **7**: 211–217.

Jacobs, O. L. R., and J. W. Patchell, 1972. "Caution and probing in stochastic control." *Int. J. Control* **16**: 189–199.

Constrained minimization of the one-step loss function is treated in:

Alster, J., and P. R. Bélanger, 1974. "A technique for dual adaptive control." *Automatica* **10**: 627–634.

Hughes, D. J., and O. L. R. Jacobs, 1974. "Turn-off, escape and probing in non-linear stochastic control." *Preprint IFAC Symposium on Stochastic Control.* Budapest.

Mosca, E., S. Rocchi, and G. Zappa, 1978. "A new dual active control algorithm." *Preprints 17th IEEE Conference on Decision and Control*, pp. 509–512. San Diego, Calif.

Different extensions of the one-step loss function are discussed in:

Wittenmark, B., 1975. "An active suboptimal dual controller for systems with stochastic parameters." *Automatic Control Theory and Application* **3**: 13–19.

Goodwin, G. C., and R. L. Payne, 1977. *Dynamic System Identification: Experiment Design and Data Analysis*. New York: Academic Press.

Sternby, J., 1977. "Topics in dual control." Ph.D. thesis TFRT-1012, Department of Automatic Control, Lund Institute of Technology, Lund, Sweden.

Milito, R., C. S. Padilla, R. A. Padilla, and D. Cadorin, 1982. "An innovations approach to dual control." *IEEE Trans. Automat. Contr.* **AC-27**: 132–137.

Wittenmark, B., and C. Elevitch, 1985. "An adaptive control algorithm with dual features." *Preprints 7th IFAC Symposium on Identification and System Parameter Estimation*, pp. 587–592. York, U.K.

Linearization of the loss function is found in Bar-Shalom and Tse (1976) and in:

Wenk, C. J., and Bar-Shalom, Y., 1980. "A multiple model adaptive control algorithm for stochastic systems with unknown parameters." *IEEE Trans. Automat. Contr.* **AC-25**: 703–710.

Bar-Shalom, Y., P. Mookerjee, and J. A. Molusis, 1982. "A linear feedback dual controller for a class of stochastic systems." *Proceedings of the National CNRS Colloque "Développement et Utilisation d'Outils et Modèles Mathématiques en Automatique, Analyse de Systèmes et Traitement du Signal."* Belle-Ile, France.

A discussion of an industrial example in which dual control can be useful is found in:

Dumont, G., and K. J. Åström, 1988. "Wood chip refiner control." *IEEE Control Systems Magazine* **8**(2): 38–43.

CHAPTER 8

AUTO-TUNING
———

8.1 INTRODUCTION

Adaptive schemes like MRAS and STR require *a priori* information about the
process dynamics. It is particularly important to know time scales, which are
critical for determining suitable sampling intervals and filtering. The importance of *a priori* information was overlooked for a long time but became apparent in connection with the development of general-purpose adaptive controllers.
Several manufacturers were forced to introduce a *pre-tune mode* to help in ob-
taining the required prior information. The importance of prior information
also appeared in connection with attempts to develop techniques for automatic
tuning of simple PID regulators. Such regulators, which are standard building
blocks for industrial automation, are used to control systems with a wide range
of time constants.

From the user's point of view it would be ideal to have an auto-tuning
function in which the regulator can be tuned simply by pushing a button. Although conventional adaptive schemes seemed to be ideal tools to provide automatic tuning, they were found to be inadequate because they required prior
knowledge of time scales. Special techniques for automatic tuning of simple
regulators were therefore developed. These techniques are also useful for providing pre-tuning of more complicated adaptive systems. In this chapter we
describe some of these techniques. They can be characterized as crude robust
methods that provide ballpark information. They are thus ideal complements
to the more sophisticated adaptive methods. An overview of industrial PID
controllers with auto-tuning is given in Section 12.3.

The chapter is organized as follows: The standard PID controller is discussed in Section 8.2. Different auto-tuning techniques are given in Section
8.3. Transient and frequency response methods for tuning are developed in

375

Sections 8.4 and 8.5, respectively, and Section 8.6 is devoted to analysis of relay oscillations. Conclusions are presented in Section 8.7.

8.2   PID CONTROL

The PID controllers are the standard tool for industrial automation. The flexibility of the controller makes it possible to use PID control in many situations. The controllers can also be used in cascade control and other controller configurations. Many simple control problems can be handled very well by PID control, provided that the performance requirements are not too high. The PID algorithm is packaged in the form of standard regulators for process control and is also the basis of many tailor-made control systems. The textbook version of the algorithm is

$$u(t) = K_c \left(e(t) + \frac{1}{T_i} \int_0^t e(s)\, ds + T_d \frac{de}{dt} \right) \tag{8.1}$$

where u is the control variable, e is the error defined as $e = u_c - y$ where u_c is the reference value, and y is the process output. The algorithm that is actually used contains several modifications. It is standard practice to let the derivative action operate only on the process output. It may be advantageous to let the proportional part act only on a fraction of the reference value. The derivative action is replaced by an approximation that reduces the gain at high frequencies. The integral action is modified so that it does not keep integrating when the control variable saturates (*anti-windup*). Precautions are also taken so that there will not be transients when the regulator is switched from manual to automatic control or when parameters are changed.

If the nonlinearity of the actuator can be described by the function f, a reasonably realistic PID regulator can be described by

$$\begin{aligned} u(t) &= f(v(t)) \\ v(t) &= P(t) + I(t) + D(t) \end{aligned} \tag{8.2}$$

where

$$\begin{aligned} P(t) &= K_c \left(\beta u_c(t) - y(t) \right) \\ \frac{dI}{dt} &= \frac{K_c}{T_i} \left(u_c(t) - y(t) \right) + \frac{1}{T_t} \left(v(t) - u(t) \right) \\ \frac{T_d}{N} \frac{dD}{dt} &= -D - K_c T_d \frac{dy}{dt} \end{aligned} \tag{8.3}$$

The last term in the expression for dI/dt is introduced to get anti-windup when the output saturates. This guarantees that the integral part I is bounded. The parameter T_t is a time constant for resetting the integral action when the actuator saturates. The essential parameters to be adjusted are K_c, T_i, and T_d.

The parameter N can be fixed; a typical value is $N = 10$. The tracking time constant is typically a fraction of the integration time T_i.

8.3 AUTO-TUNING TECHNIQUES

Several ways to do auto-tuning have been proposed. The most common method is to make a simple experiment on the process. The experiment can be done in open loop or closed loop. In the open-loop experiments the input of the process is excited by a pulse or a couple of steps. A simple process model, for instance of second order, is then estimated by using recursive least squares or some other recursive estimation method. If a second-order process model is estimated, then the PID controller can be used to make pole placement. The speed and the damping of the system are then the design parameters. A popular design method is to choose the controller zeros such that they cancel the two process poles. This gives good responses to setpoint changes, while the response to load disturbances is determined by the open-loop dynamics. The transient response method for automatic tuning of PID regulators is used in products from Yokogawa, Eurotherm, and Honeywell. It is used for pre-tuning in adaptive controllers from Leeds and Northrup and Turnbull Control.

The tuning experiments can also be done in closed loop. A typical example of this is the self-oscillating method of Ziegler and Nichols or its variants. The relay auto-tuner based on self-oscillation is used in products from SattControl and Fisher-Rousemount. In these regulators the tuning is initiated simply by pushing the tuning button. One advantage of making experiments in closed loop is that the output of the process may be kept within reasonable bounds, which can be difficult for processes with integrators if the experiment is done in open loop.

The auto-tuning function is often a built-in feature in standard stand-alone PID controllers. Automatic tuning can also be done by using external equipment. The tuner is then connected to the process and performs an experiment, usually in open loop. The tuner then suggests parameter settings, which are transferred to the PID controller either manually or automatically. Since the external tuner must be able to work with PID controllers from different manufacturers, it is important that the tuner have detailed information about the implementation of the PID algorithm in specific cases.

Another method for auto-tuning is to use an expert system to tune the controller. This is done during normal operation of the process. The expert system waits for setpoint changes or major load disturbances and then evaluates the performance of the closed-loop system. Properties such as damping, period of oscillation, and static gain are estimated. The controller parameters are then changed according to the built-in rules, which mimic the behavior of an experienced control engineer. Pattern recognition or expert system is used in controllers from Foxboro and Fenwal.

8.4   TRANSIENT RESPONSE METHODS

Several simple tuning methods for PID controllers are based on transient response experiments. Many industrial processes have step responses of the type shown in Fig. 8.1, in which the step response is monotonous after an initial time. A system with a step response of the type shown in Fig. 8.1 can be approximated by the transfer function

$$G(s) = \frac{k}{1 + sT} e^{-sL} \tag{8.4}$$

where k is the static gain, L is the apparent time delay, and T is the apparent time constant. The parameter a is given by

$$a = k \frac{L}{T} \tag{8.5}$$

The Ziegler-Nichols Step Response Method

A simple way to determine the parameters of a PID regulator based on step response data was developed by Ziegler and Nichols and published in 1942. The method uses only two of the parameters shown in Fig. 8.1, namely, a and L. The regulator parameters are given in Table 8.1. The Ziegler-Nichols tuning rule was developed by empirical simulations of many different systems. The rule has the drawback that it gives closed-loop systems that are often too poorly damped. Systems with better damping can be obtained by modifying the numerical values in Table 8.1. By using additional parameters it is also possible to determine whether the Ziegler-Nichols rule is applicable. If the time constant T is also determined, an empirical rule is established that the

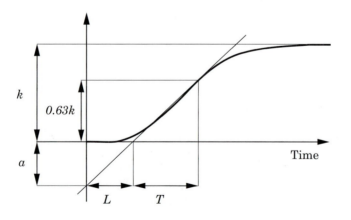

Figure 8.1   Unit step response of a typical industrial process.

Table 8.1 Regulator parameters obtained by the Ziegler-Nichols step response method.

Controller	K_c	T_i	T_d
P	$1/a$		
PI	$0.9/a$	$3L$	
PID	$1.2/a$	$2L$	$L/2$

Ziegler-Nichols rule is applicable if $0.1 < L/T < 0.6$. For large values of L/T it is advantageous to use other tuning rules or control laws that compensate for dead time. For small values of L/T, improved performance may be obtained with higher-order compensators. It is also possible to use more sophisticated tuning rules based on three parameters.

Characterization of a Step Response

The parameters k, L, and T can be determined from a graphical construction such as the one indicated in Fig. 8.1. It may be useful to take averages of several steps if the signals are noisy. There are also methods based on area measurements that can be used. One method of this type is illustrated in Fig. 8.2. The area A_0 is first determined. Then

$$T + L = \frac{A_0}{k} \tag{8.6}$$

The area A_1 under the step response up to time $T + L$ is then determined, and T is then given by

$$T = \frac{eA_1}{k} \tag{8.7}$$

where e is the base of the natural logarithm. The essential drawbacks of the method are that it may be difficult to know the size of the step in the control

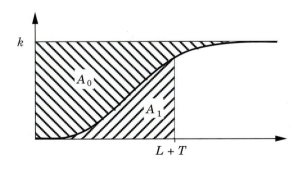

Figure 8.2 Area method for determining L and T.

signal and to determine whether a steady state has been reached. The step should be so large that the response is clearly noticeable above the noise but not so large that production is disturbed. Disturbances will also influence the result significantly.

On-line Refinement

If a reasonable regulator tuning is obtained, the damping and natural frequency of the closed-loop system can also be determined from a closed-loop transient response. The regulator tuning can then be improved.

8.5 METHODS BASED ON RELAY FEEDBACK

The main drawback of the transient response method is that it is sensitive to disturbances because it relies on open-loop experiments. The relay-based methods avoid this difficulty because the required experiments are performed in closed loop.

The Key Idea

The basic idea is the observation that many processes have limit cycle oscillations under relay feedback. A block diagram of such a system is shown in Fig. 8.3. The input and output signals obtained when the command signal u_c is zero are shown in Fig. 8.4. The figure shows that a limit cycle oscillation is established quite rapidly. We can intuitively understand what happens in the following way: The input to the process is a square wave with frequency ω_u. By a Fourier series expansion we can represent the input by a sum of sinusoids with frequencies ω_u, $3\omega_u$, and so on. The output is approximately sinusoidal, which means that the process attenuates the higher harmonics effectively. Let the amplitude of the square wave be d; then the fundamental component has the amplitude $4d/\pi$. Making the approximation that all higher harmonics can

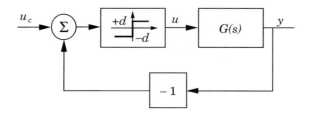

Figure 8.3   Linear system with relay control.

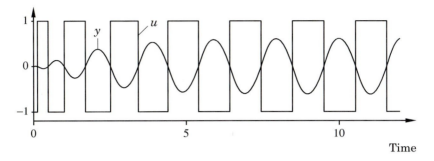

Figure 8.4   Input and output of a system with relay feedback.

be neglected, we find that the process output is a sinusoid with frequency ω_u and amplitude

$$a = \frac{4d}{\pi} \, |G(i\omega_u)|$$

To have an oscillation, the output must also go through zero when the relay switches. Moreover, the fundamental component of the input and the output must have opposite phase. We can thus conclude that the frequency ω_u must be such that the process has a phase lag of 180°. The conditions for oscillation are thus

$$\arg G(i\omega_u) = -\pi \qquad \text{and} \qquad |G(i\omega_u)| = \frac{a\pi}{4d} = \frac{1}{K_u} \qquad (8.8)$$

where K_u can be regarded as the equivalent gain of the relay for transmission of sinusoidal signals with amplitude a. For historical reasons this parameter is called the *ultimate gain*. It is the gain that brings a system with transfer function $G(s)$ to the stability boundary under pure proportional control. The period $T_u = 2\pi/\omega_u$ is similarly called the *ultimate period*. An experiment with relay feedback is thus a convenient way to determine the ultimate period and the ultimate gain. Notice also that an input signal whose energy content is concentrated at ω_u is generated automatically in the experiment.

The Ziegler-Nichols Closed-Loop Method

Ziegler and Nichols have devised a very simple heuristical method for determining the parameters of a PID controller based on the critical gain and the critical period. The controller settings are given in Table 8.2. These parameters give a closed-loop system with quite low damping. Systems with better damping can be obtained by slight modifications of the numbers in the table. A modified method of this type is ideally matched to the determination of K_u and T_u by the relay method. This gives the relay auto-tuner shown in Fig. 8.5. When tuning is demanded, the switch is set to T, which means that relay feedback is activated and the PID regulator is disconnected. When a stable limit

Table 8.2   Regulator parameters obtained by the Ziegler-Nichols closed-loop method.

Controller	K_c	T_i	T_d
P	$0.5K_u$		
PI	$0.4K_u$	$0.8T_u$	
PID	$0.6K_u$	$0.5T_u$	$0.12T_u$

cycle is established, the PID parameters are computed, and the PID controller is then connected to the process. Naturally, the method will not work for all systems. First, there will not be unique limit cycle oscillations for an arbitrary transfer function. Second, PID control is not appropriate for all processes. Relay auto-tuning has empirically been found to work well for a large class of systems encountered in process control.

The Method of Describing Function

The approximative method used to derive the conditions for relay oscillations given by Eqs. (8.8) is called the *method of harmonic balance*. We will now describe a slight variation of the method that can be used to obtain additional insight. This is called the *describing function method*. It can be described as follows: Consider a simple feedback system composed of a linear part with the transfer function $G(s)$ and feedback with an ideal relay as shown in Fig. 8.3. The conditions for limit cycle oscillations can be determined approximately by investigating the propagation of sinusoidal signals around the loop. There will be higher harmonics because of the relay, but they will be neglected. The propagation of a sine wave through the linear system is described by the complex number $G(i\omega)$. Similarly, the propagation of a sine wave through

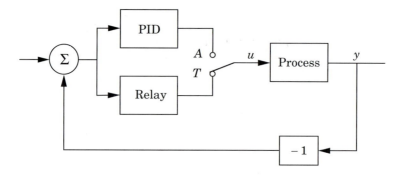

Figure 8.5   Block diagram of a relay auto-tuner.

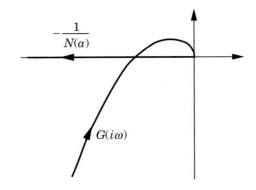

Figure 8.6 Nyquist curve $G(i\omega)$ and the describing function $N(a)$ for a relay.

the nonlinearity can also be characterized by a complex number $N(a)$, which depends on the amplitude of the signal at the input of the nonlinearity. $N(a)$ is called the describing function of the nonlinearity. The condition for oscillation is then that the signal comes back with the same amplitude and phase as it passes the closed loop. This gives the condition

$$G(i\omega)N(a) = -1$$

This condition can be represented graphically by also plotting the curve $N(a)$ in the Nyquist diagram. (See Fig. 8.6.) For the relay the nonlinearity is

$$N(a) = \frac{4d}{a\pi}$$

because a is the input signal amplitude and the fundamental component of the output has amplitude $4d/\pi$. A possible oscillation is at the intersection of the curves. The frequency is read from the Nyquist curve and the amplitude from the describing function.

EXAMPLE 8.1   **Relay oscillation**

Consider a system with relay feedback as in Fig. 8.3 with

$$G(s) = \frac{K\alpha}{s(s+1)(s+\alpha)}$$

$K = 5$, $\alpha = 10$, $d = 1$, and $u_c = 0$. This was the system used to generate Fig. 8.4. Simple calculations show that

$$\arg G(i\omega_u) = -\frac{\pi}{2} - \tan^{-1}\omega_u - \tan^{-1}\frac{\omega_u}{\alpha}$$

$$= -\frac{\pi}{2} - \tan^{-1}\frac{\omega_u(\alpha+1)}{\alpha - \omega_u^2} = -\pi$$

This implies that the Nyquist curve intersects the negative real axis for $\omega_u = \sqrt{\alpha}$. The approximative analysis thus gives the following estimate of the period:

$$T_u = \frac{2\pi}{\sqrt{\alpha}} = \frac{6.28}{\sqrt{\alpha}} = 1.99$$

Using Eqs. (8.8) gives $a = 4d|G(i\omega_u)|/\pi = 0.58$. From the simulations it can be determined that the true values are $T_u = 2.07$ and $a = 0.62$, which show that the describing function method gives fair but not very accurate estimates in this example.   □

Several refinements of the method are useful. The amplitude of the limit cycle oscillation can be specified by introducing a feedback that adjusts the relay amplitude. A hysteresis in the relay is useful to make the system less sensitive to noise. The parameters T_u and K_u can be used to determine the parameters of a PID regulator. The method can be made insensitive to disturbances by comparing and averaging over several periods of the oscillation.

<hr>

EXAMPLE 8.2   Auto-tuning of cascaded tanks

The properties of a relay auto-tuner are illustrated by an example. The process to be controlled consists of three cascaded tanks. The level of the lower tank is measured, and the control variable is the voltage to the amplifier driving the pump for the inlet. The signals are noisy. The relay in the auto-tuner has a hysteresis, which is determined automatically on the basis of measurements of the process disturbances. The relay amplitude is also adjusted automatically to keep a specified amplitude of the limit cycle. The limit cycle is judged to be stationary by measuring the periods and amplitudes of two positive half-periods. Figure 8.7 shows the process inputs and outputs in one experiment, illustrating the effect of amplitude adjustment. When the tuning is finished,

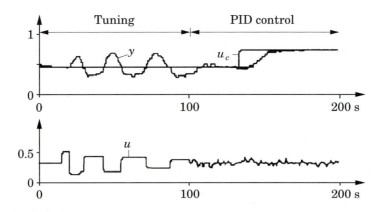

Figure 8.7 Results obtained by applying an auto-tuner to level control of three cascaded tanks.

the regulator is switched to PID control automatically. A change of the setpoint shows that the tuning has been successful.   □

Improved Estimates and Pre-tuning

So far, only two parameters, K_u and T_u, have been extracted from the relay experiment. Much more information can be obtained. By changing the setpoint during the relay experiment it is possible to determine the static process gain k. The product kK_u can then be used to assess the appropriateness of PID control with Ziegler-Nichols tuning. A common rule is that the Ziegler-Nichols method can be used if $2 < kK_u < 20$. Values that are lower than 2 indicate that a control law that admits dead-time compensation should be used. Large values of kK_u indicate that improved performance can be obtained with a more complex control algorithm. The relay experiment can also be used to estimate a discrete-time transfer function by using standard system identification methods.

The relay method is ideally suited as a pre-tuner for a more sophisticated adaptive controller. A model such as Eq. (8.4) is very useful to select the sampling period and achievable closed-loop response for an MRAS or an STR. It provides a PID controller that can serve as a backup controller. If the static gain is also determined, the quantity kK_u can be used to assess the process dynamics. The ultimate period can be used to obtain an estimate of an appropriate sampling period. Parameter estimates that can serve as initial values in the recursive parameter estimator can be obtained by applying a parameter estimation method to the data from the relay experiments. If an adaptive controller based on a pole placement design is used, the ultimate period can also be used to find appropriate values of desired closed-loop bandwidths.

8.6   RELAY OSCILLATIONS

Since limit cycling under relay feedback is a key idea of relay auto-tuning, it is important to understand why a linear system oscillates under relay feedback and when the oscillation is stable. It is also important to have methods for determining the period and the amplitude of the oscillations. Consider the system shown in Fig. 8.3. Introduce the following state space realization of the transfer function $G(s)$:

$$\frac{dx}{dt} = Ax + Bu$$
$$y = Cx$$

(8.9)

The relay can be described by

$$u = \begin{cases} d & \text{if } e > 0 \\ -d & \text{if } e < 0 \end{cases}$$

(8.10)

where $e = u_c - y$. We have the following result.

THEOREM 8.1   Limit cycle period

Assume that the system defined in Fig. 8.3 and by Eqs. (8.9) and (8.10) has a symmetric limit cycle with period T. The period T is then the smallest value of $T > 0$ that satisfies the equation

$$C(I + \Phi)^{-1}\Gamma = 0 \qquad (8.11)$$

where

$$\Phi = e^{AT/2}$$

and

$$\Gamma = \int_0^{T/2} e^{As}\, ds\, B$$

Proof: Let t_k denote the times when the relay switches. Since the limit cycle is symmetric, it follows that

$$t_{k+1} - t_k = T/2$$

Assume that the control signal u is d over the interval (t_k, t_{k+1}). Integration of Eqs. (8.9) over the interval gives

$$x(t_{k+1}) = \Phi x(t_k) + \Gamma d$$

Since the limit cycle is symmetric, it also follows that

$$x(t_{k+1}) = -x(t_k)$$

Hence

$$x(t_k) = -(I + \Phi)^{-1}\Gamma d$$

Since the output $y(t)$ must be zero at t_k, it follows that

$$y(t_k) = Cx(t_k) = -C(I + \Phi)^{-1}\Gamma d = 0$$

which gives Eq. (8.11).   □

Remark 1.   The condition of Eq. (8.11) can also be written as

$$H_{T/2}(-1) = 0 \qquad (8.12)$$

where $H_{T/2}(z)$ is the pulse transfer function obtained when sampling the system of Eqs. (8.9) with period $T/2$.

Remark 2.   The result that the period is given by Eq. (8.12) also holds for linear systems with a time delay, provided that $T/2$ is larger than or equal to the delay.

Remark 3.   Similar conditions can also be derived for relays with hysteresis.
                                                                                  □

Comparison with the Describing Function

Having obtained the exact formula of Eq. (8.11) for T, it is possible to investigate the precision of the describing function approximation. Consider the symmetric case and introduce $h = T/2$. The pulse transfer function obtained in sampling the system of Eqs. (8.9) with period h is given by

$$H_h(e^{sh}) = \frac{1}{h} \sum_{n=-\infty}^{\infty} \frac{1}{s + in\omega_s} \left(1 - e^{-h(s+in\omega_s)}\right) G(s + in\omega_s)$$

where $\omega_s = 2\pi/h$. Put $sh = i\pi$:

$$H_h(-1) = \sum_{-\infty}^{\infty} \frac{2}{i(\pi + 2n\pi)} G\left(i\frac{\pi + 2n\pi}{h}\right)$$

$$= \sum_{0}^{\infty} \frac{4}{\pi(1 + 2n)} \text{Im}\left(G\left(i\frac{\pi + 2n\pi}{h}\right)\right) = 0$$

The first term of the series gives

$$H_h(-1) \approx \frac{4}{\pi} \text{Im}\left(G\left(i\frac{\pi}{h}\right)\right) = \frac{4}{\pi} \text{Im}\left(G\left(i\frac{2\pi}{T}\right)\right) = 0$$

which is the same result for calculation of T obtained from the describing function analysis. This implies that the describing function approximation is accurate only if $G(s)$ has low-pass character. An example illustrates determination of the period of oscillation.

EXAMPLE 8.3   Limit cycle period

Consider the same process as in Example 8.1. To apply Theorem 8.1, the system $G(s)$ is sampled with period h. The pulse transfer function is

$$H_h(z) = \frac{Kh}{(z - 1)} - \frac{K\alpha(1 - e^{-h})}{(\alpha - 1)(z - e^{-h})} + \frac{K\left(1 - e^{-\alpha h}\right)}{\alpha(\alpha - 1)(z - e^{-\alpha h})}$$

Hence

$$H_h(-1) = -\frac{Kh}{2} + \frac{K\alpha(1 - e^{-h})}{(\alpha - 1)(1 + e^{-h})} - \frac{K(1 - e^{-\alpha h})}{\alpha(\alpha - 1)(1 + e^{-\alpha h})}$$

$$= -\frac{Kh}{2} + \frac{K\alpha}{\alpha - 1}\left(\frac{1 - e^{-h}}{1 + e^{-h}} - \frac{1}{\alpha^2}\frac{1 - e^{-\alpha h}}{1 + e^{-\alpha h}}\right) = 0$$

Numerical search for the value of h that satisfies this equation gives $h = 1.035$. This gives $T_u = 2.07$, which agrees with the simulation in Fig. 8.4.   □

Stable periodic solutions will not be obtained for all systems. A double integrator under pure relay control, for example, will give periodic solutions with an arbitrary period.

8.7   CONCLUSIONS

In this chapter we have described simple robust methods that can be used to get crude estimates of process dynamics. The methods can be used for automatic tuning of simple regulators of the PID type or as pre-tuners for more sophisticated adaptive control algorithms. Two types of methods have been discussed: a transient method based on open-loop step tests and a closed-loop method based on relay feedback.

PROBLEMS

8.1   Consider a process characterized by the transfer function

$$G(s) = \frac{k}{1 + sT} e^{-sL}$$

Show that parameters T and L are exactly given by Eqs. (8.6) and (8.7).

8.2   Consider a process with the transfer function

$$G(s) = \prod_{k=1}^{n} \frac{1}{(1 + sT_k)} e^{-sL}$$

Show that Eq. (8.6) gives

$$T + L = \sum_{k=1}^{n} T_k + L$$

8.3   Consider a process described by the transfer function

$$G(s) = \frac{k}{s} e^{-sL}$$

Determine a proportional regulator that gives an amplitude margin $A_m = 2$. Show that it is identical to the setting obtained by applying the Ziegler-Nichols rule in Table 8.2.

8.4   Determine the period of the limit cycle obtained when processes with transfer functions

(a) $G(s) = \dfrac{k}{s} e^{-sL}$ (b) $G(s) = \dfrac{1}{(s + 1)^3}$ (c) $G(s) = \dfrac{1}{s^2}$

are provided with relay feedback. Use both the approximate and exact methods.

8.5   Consider a process with the transfer function given in Problem 8.3. Determine a proportional regulator obtained with the Ziegler-Nichols method given in Table 8.2.

REFERENCES

The PID regulator is very common. It is the standard tool for solving most process control problems. Various aspects of PID control are discussed in:

Smith, C. L., 1972. *Digital Computer Process Control.* Scranton, Pa.: Intext Educational Publishers.

Shinskey, F. G. , 1979. *Process-Control Systems Application Design Adjustment.* New York: McGraw-Hill.

Desphande, P. B., and R. H. Ash, 1981. *Elements of Computer Process Control with Advanced Control Applications.* Research Triangle Park, N.C.: Instrument Society of America.

The Ziegler-Nichols tuning rules were presented in:

Ziegler, J. G., and N. B. Nichols, 1942. "Optimum settings for automatic controllers." *Trans. ASME* **64**: 759–768.

Tuning rules based on three parameters k, T, and L are presented in:

Cohen, G. H., and G. A. Coon, 1953. "Theoretical consideration of retarded control." *Trans. ASME* **15**: 827–834.

A discussion of many different tuning rules for PID controllers is found in:

McMillan, G. K., 1983. *Tuning and Control Loop Performance.* Research Triangle Park, N.C.: Instrument Society of America.

Interesting views on PID control versus more advanced controls for process control applications are found in:

McMillan, G. K., 1986. "Advanced control algorithms: Beware of false prophecies." *Intech* January: 55–57.

The relay auto-tuner was presented in:

Åström, K. J., and T. Hägglund, 1984. "Automatic tuning of simple regulators with specifications on phase and amplitude margins." *Automatica* **20**: 645–651.

It is also patented:

Hägglund, T., and K. J. Åström, 1985. "Method and an apparatus in tuning a PID regulator." U.S. Patent Number 4549123.

A detailed treatment of PID control is given in:

Åström, K. J., and T. Hägglund, 1995. *PID Control.* Research Triangle Park, N.C.: Instrument Society of America.

Discussions of automatic tuning of simple regulators are found in:

Åström, K. J., and T. Hägglund, 1988. *Automatic Tuning of PID Regulators.* Research Triangle Park, N.C.: Instrument Society of America.

Åström, K. J., T. Hägglund, C. C. Hang, and W. K. Ho, 1993. "Automatic tuning and adaptation for PID controllers: A survey." *Control Eng. Practice* **1**: 699–714.

Tsypkin pioneered the research in relay feedback. His results on relay oscillations are treated in detail in:

Tsypkin, Y. Z., 1984. *Relay Control Systems.* Cambridge, U.K.: Cambridge University Press.

CHAPTER 9

GAIN SCHEDULING

———

9.1 INTRODUCTION

In many situations it is known how the dynamics of a process change with the operating conditions of the process. One source for the change in dynamics may be nonlinearities that are known. It is then possible to change the parameters of the controller by monitoring the operating conditions of the process. This idea is called *gain scheduling*, since the scheme was originally used to accommodate changes in process gain only. Gain scheduling is a nonlinear feedback of special type; it has a linear controller whose parameters are changed as a function of operating conditions in a preprogrammed way. The idea of relating the controller parameters to auxiliary variables is old, but the hardware needed to implement it easily was not available until recently. To implement gain scheduling with analog techniques, it is necessary to have function generators and multipliers. Such components have been quite expensive to design and operate. Gain scheduling has thus been used only in special cases, such as in autopilots for high-performance aircraft. Gain scheduling is easy to implement in computer-controlled systems, provided that there is support in the available software.

Gain scheduling based on measurements of operating conditions of the process is often a good way to compensate for variations in process parameters or known nonlinearities of the process. It is controversial whether a system with gain scheduling should be considered an adaptive system or not, because the parameters are changed in an open-loop or preprogrammed fashion. If we use the informal definition of adaptive controllers given in Section 1.1, gain scheduling can be regarded as an adaptive controller. Gain scheduling is a very useful technique for reducing the effects of parameter variations. In fact it is the foremost method for handling parameter variations in flight control systems. There are also many commercial process control systems in which gain scheduling can be used to compensate for static and dynamic nonlinearities. Split-range controllers that use different sets of parameters for different ranges of the process output can be regarded as a special type of gain-scheduling controllers.

Section 9.2 gives the principle of gain scheduling. Different ways to design systems with gain scheduling are treated in Section 9.3, and Section 9.4 gives a method based on nonlinear transformations. Section 9.5 describes some applications of gain scheduling. Conclusions are given in Section 9.6.

9.2   THE PRINCIPLE

It is sometimes possible to find auxiliary variables that correlate well with the changes in process dynamics. It is then possible to reduce the effects of parameter variations simply by changing the parameters of the controller as functions of the auxiliary variables (see Fig. 9.1). Gain scheduling can thus be viewed as a feedback control system in which the feedback gains are adjusted by using feedforward compensation. The concept of gain scheduling originated in connection with the development of flight control systems. In this application the Mach number and the dynamic pressure are measured by air data sensors and used as scheduling variables.

A main problem in the design of systems with gain scheduling is to find suitable scheduling variables. This is normally done on the basis of knowledge of the physics of a system. In process control the production rate can often be chosen as a scheduling variable, since time constants and time delays are often inversely proportional to production rate. (Compare Example 1.5.)

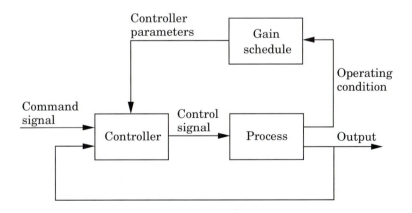

Figure 9.1   Block diagram of a system in which influences of parameter variations are reduced by gain scheduling.

When scheduling variables have been determined, the controller parameters are calculated at a number of operating conditions by using some suitable design method. The controller is thus tuned or calibrated for each operating condition. The stability and performance of the system are typically evaluated by simulation; particular attention is given to the transition between different operating conditions. The number of entries in the scheduling tables is increased if necessary. Notice, however, that there is no feedback from the performance of the closed-loop system to the controller parameters.

It is sometimes possible to obtain gain schedules by introducing nonlinear transformations in such a way that the transformed system does not depend on the operating conditions. The auxiliary measurements are used together with the process measurements to calculate the transformed variables. The transformed control variable is then calculated and retransformed before it is applied to the process. The controller thus obtained can be regarded as being composed of two nonlinear transformations with a linear controller in between. Sometimes the transformation is based on variables that are obtained indirectly through state estimation. Examples are given in Sections 9.4 and 9.5.

One drawback of gain scheduling is that it is an open-loop compensation. There is no feedback to compensate for an incorrect schedule. Another drawback of gain scheduling is that the design may be time-consuming. The controller parameters must be determined for many operating conditions, and the performance must be checked by extensive simulations. This difficulty is partly avoided if scheduling is based on nonlinear transformations.

Gain scheduling has the advantage that the controller parameters can be changed very quickly in response to process changes. Since no estimation of parameters occurs, the limiting factors depend on how quickly the auxiliary measurements respond to process changes.

9.3   DESIGN OF GAIN-SCHEDULING CONTROLLERS

It is difficult to give general rules for designing gain-scheduling controllers. The key question is to determine the variables that can be used as scheduling variables. It is clear that these auxiliary signals must reflect the operating conditions of the plant. Ideally, there should be simple expressions for how the controller parameters relate to the scheduling variables. It is thus necessary to have good insight into the dynamics of the process if gain scheduling is to be used. The following general ideas can be useful:

- Linearization of nonlinear actuators,
- Gain scheduling based on measurements of auxiliary variables,
- Time scaling based on production rate, and
- Nonlinear transformations.

The ideas are illustrated by some examples.

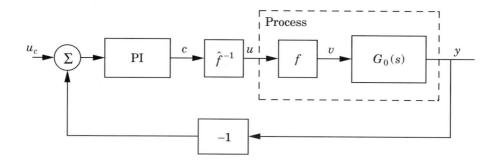

Figure 9.2   Compensation of a nonlinear actuator using an approximate inverse.

EXAMPLE 9.1   **Nonlinear actuator**

Consider the system with a nonlinear valve in Example 1.4. The nonlinearity is assumed to be

$$v = f(u) = u^4 \qquad u \geq 0$$

Let $\hat{f}^{-1}$ be an approximation of the inverse of the valve characteristic. To compensate for the nonlinearity, the output of the controller is fed through this function before it is applied to the valve (see Fig. 9.2). This gives the relation

$$v = f(u) = f\left(\hat{f}^{-1}(c)\right)$$

where c is the output of the PI controller. The function $f(\hat{f}^{-1}(c))$ should have less variation in gain than f. If $\hat{f}^{-1}$ is the exact inverse, then $v = c$.

Assume that $f(u) = u^4$ is approximated by two lines (see Fig. 9.3): one connecting the points (0, 0) and (1.3, 3) and the other connecting (1.3, 3) and

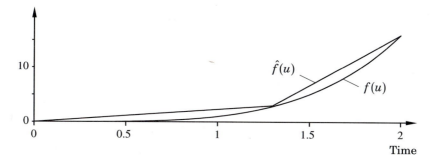

Figure 9.3   The nonlinear valve characteristic $v = f(u) = u^4$ and a two-line approximation $\hat{f}(u)$.

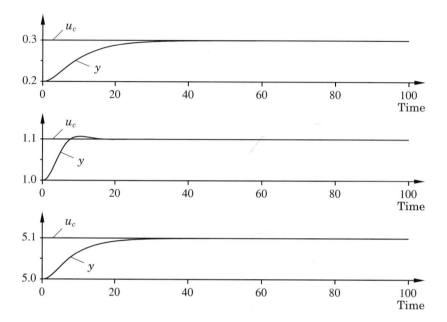

Figure 9.4 Simulation of the system in Example 9.1 with nonlinear valve and compensation using an approximation of the valve characteristic. Compare Fig. 1.9.

(2, 16). Then

$$\hat{f}^{-1}(c) = \begin{cases} 0.433c & 0 \le c \le 3 \\ 0.0538c + 1.139 & 3 \le c \le 16 \end{cases}$$

Figure 9.4 shows step changes in the reference signal at three different operating conditions when the approximation of the inverse of the valve characteristic is used between the controller and the valve. (Compare with the uncompensated system in Fig. 1.9.) There is considerable improvement in the performance of the closed-loop system. By improving the inverse it is possible to make the process even more insensitive to the nonlinearity of the valve. □

Example 9.1 shows a simple and very useful idea to compensate for known static nonlinearities. In practice it is often sufficient to approximate the nonlinearity by a few line segments. There are several commercial single-loop controllers that can make this kind of compensation. DDC packages usually include functions that can be used to implement nonlinearities.

The resulting controller in Example 9.1 is nonlinear and should (in its basic form) not be regarded as gain scheduling. In Example 9.1 there is no measurement of any operating condition apart from the controller output. In other situations the nonlinearity is determined from measurement of several variables. However, a gain-scheduling controller should contain measurement of a variable that is related to the operating point of the process. Gain scheduling based on an auxiliary signal is illustrated in the following example.

EXAMPLE 9.2   **Tank system**

Consider a tank in which the cross section A varies with height h. The model is

$$V = \int_0^h A(\tau)\,d\tau$$

$$\frac{dV}{dt} = A(h)\frac{dh}{dt} = q_i - a\sqrt{2gh}$$

where V is the volume, q_i is the input flow, and a is the cross section of the outlet pipe. Let q_i be the input, and let h be the output of the system. The linearized model at an operating point, q_{in}^0 and h^0, is given by the transfer function

$$G(s) = \frac{\beta}{s + \alpha}$$

where

$$\beta = \frac{1}{A(h^0)} \qquad \alpha = \frac{q_{in}^0}{2A(h^0)h^0} = \frac{a\sqrt{2gh^0}}{2A(h^0)h^0}$$

A good PI control of the tank is given by

$$u(t) = K\left(e(t) + \frac{1}{T_i}\int e(\tau)\,d\tau\right)$$

where

$$K = \frac{2\zeta\omega - \alpha}{\beta}$$

and

$$T_i = \frac{2\zeta\omega - \alpha}{\omega^2}$$

This gives a closed-loop system with natural frequency ω and relative damping ζ. Introducing the expressions for α and β gives the following gain schedule:

$$K = 2\zeta\omega A(h^0) - \frac{q_{in}^0}{2h^0}$$

$$T_i = \frac{2\zeta}{\omega} - \frac{q_{in}^0}{2A(h^0)h^0\omega^2}$$

The numerical values are often such that $\alpha \ll 2\zeta\omega$. The schedule can then be simplified to

$$K = 2\zeta\omega A(h^0)$$

$$T_i = \frac{2\zeta}{\omega}$$

In this case it is thus sufficient to make the gain proportional to the cross section of the tank.   □

Example 9.2 illustrates that it can sometimes be sufficient to measure one or two variables in the process and use them as inputs to the gain schedule. Often, it is not as easy as in Example 9.2 to determine the controller parameters as a function of the measured variables. The design of the controller must then be redone for different working points of the process. Some care must also be exercised if the measured signals are noisy. They may have to be filtered properly before they are used as scheduling variables.

The next example illustrates that gains, delays, and time constants are often inversely proportional to the production rate of the process. This fact can be used to make time scaling.

EXAMPLE 9.3   **Concentration control**

Consider the concentration control problem in Example 1.5. The process is described by Eq. (1.3). Assume that we are interested in manipulating the concentration in the tank, c, by changing the inlet concentration, c_{in}. For a fixed flow the dynamics can be described by the transfer function

$$G(s) = \frac{1}{1 + sT} e^{-s\tau}$$

where

$$T = V_m/q \qquad \tau = V_d/q$$

If $\tau < T$, then it is straightforward to determine a PI controller that performs well when q is constant. However, it is difficult to find universal values of the controller parameters that will work well for wide ranges of q. This is illustrated in Fig. 1.11, which shows the step responses of a fixed-gain controller for varying flows. Since the process has a time delay, it is natural to look for sampled data controllers. Sampling of the model with sampling period $h = V_d/(dq)$, where d is an integer, gives

$$c(kh + h) = ac(kh) + (1 - a)u(kh - dh)$$

where

$$a = e^{-qh/V_m} = e^{-V_d/(V_m d)}$$

Notice that the sampled data model has only one parameter, a, that does not depend on q. A constant-gain controller can easily be designed for the sampled data system.

The gain scheduling is realized simply by having a controller with constant parameters, in which the sampling rate is inversely proportional to the flow rate. This will give the same response, independent of the flow, in looking at the sampling instants, but the transients will be scaled in time. Figure 9.5 shows the output concentration and the control signals for three different flows. To

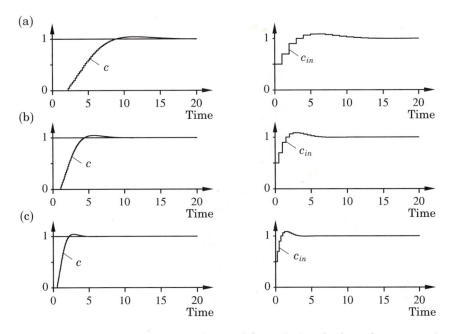

Figure 9.5 Output concentration and control signal when the process in Example 9.3 is controlled by a fixed digital controller but the sampling interval is $h = 1/(2q)$. (a) $q = 0.5$; (b) $q = 1$; (c) $q = 2$.

implement this gain-scheduling controller, it is necessary to measure not only the concentration but also the flow. Errors in the flow measurement will result in jitter in the sampling period. To avoid this, it is necessary to filter the flow measurement.

The Ziegler-Nichols transient response method discussed in Section 8.4 is based on a model with a time delay and a first-order system. Table 8.1 gives

$$K_c = \frac{0.9\tau}{T} = \frac{0.9V_d}{V_m}$$

$$T_i = 3\tau = \frac{3V_d}{q}$$

That is, the integration time is inversely proportional to the flow q. This is the same effect as is obtained with the discrete-time controller when the sampling period is inversely proportional to q. □

In Examples 9.1 and 9.2 it was possible to determine the schedules exactly. The behavior of the closed-loop system does not depend on the operating conditions. In other cases it is possible to obtain only approximate relations for different operating conditions. The design then has to be repeated for several operating conditions to create a table. It is also necessary to interpolate between the values of the table to obtain a smooth behavior of the closed-loop system. This can lead to extensive calculations and simulations before the full gain schedule is obtained.

The gain schedule is usually obtained through simulations of a process model, but it is also possible to build up the gain table on-line. This might be done by using an auto-tuner or an adaptive controller. The adaptive system is used to get the controller parameters for different operating points. The parameters are then stored for later use when the system returns to the same or a neighboring operating point.

9.4   NONLINEAR TRANSFORMATIONS

It is of great interest to find transformations such that the transformed system is linear and independent of the operating conditions. The process in Example 9.3 is one example in which this can be done by time scaling. The obtained sampled model is independent of the flow because the time is scaled as

$$t_s = \frac{V_d}{q} t$$

This means that the key variable is distance traveled by a particle instead of time. All processes associated with material flows—rolling mills, band transporters, flows in pipes, and so on—have this property.

A system of the form

$$\frac{dx(t)}{dt} = f\left(x(t)\right) + g\left(x(t)\right)u(t)$$

can also be transformed into a linear system, provided that all states of the system can be measured and a generalized observability condition holds. (Compare Section 5.10.) The system is first transformed into a fixed linear system. The transformation is usually nonlinear and depends on the states of the process. A controller is then designed for the transformed model, and the control signals of the model are retransformed into the original control signals. The result is a special type of nonlinear controller, which can be interpreted as a gain-scheduling controller. Knowledge about the nonlinearities in the model is built into the controller. The method with nonlinear transformations is illustrated by an example.

EXAMPLE 9.4   **Nonlinear transformation of a pendulum**

Consider the system

$$\frac{dx_1}{dt} = x_2$$

$$\frac{dx_2}{dt} = -\sin x_1 + u \cos x_1 \tag{9.1}$$

$$y = x_1$$

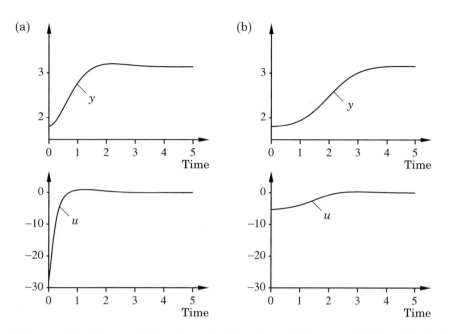

Figure 9.6 The pendulum described by Eqs. (9.1), controlled by (a) the nonlinear controller of Eq. (9.3) and (b) the fixed-gain controller of Eq. (9.4). The desired characteristic equation (Eq. 9.2) is defined by $p_1 = 2.8$ and $p_2 = 4$.

which describes a pendulum, where the acceleration of the pivot point is the input and the output y is the angle from a downward position. Introduce the transformed control signal

$$v(t) = -\sin x_1(t) + u(t) \cos x_1(t)$$

This gives the linear equations

$$\frac{dx}{dt} = \begin{pmatrix} 0 & 1 \\ 0 & 0 \end{pmatrix} x + \begin{pmatrix} 0 \\ 1 \end{pmatrix} v$$

Assume that x_1 and x_2 are measured, and introduce the control law

$$v(t) = -l_1' x_1(t) - l_2' x_2(t) + m' u_c(t)$$

The transfer function from u_c to y is

$$\frac{m'}{s^2 + l_2' s + l_1'}$$

Let the desired characteristic equation be

$$s^2 + p_1 s + p_2 \qquad\qquad (9.2)$$

which can be obtained with

$$l'_1 = p_2 \qquad l'_2 = p_1 \qquad m' = p_2$$

Transformation back to the original control signal gives

$$u(t) = \frac{v(t) + \sin x_1(t)}{\cos x_1(t)} = \frac{1}{\cos x_1(t)}\left(-p_2 x_1(t) - p_1 x_2(t) + p_2 u_c(t) + \sin x_1(t)\right)$$

$$(9.3)$$

The controller is thus highly nonlinear. Figure 9.6 shows the output and the control signal when the controller of Eq. (9.3) is used and when a fixed-gain controller

$$u(t) = -l_1 x_1(t) - l_2 x_2(t) + m u_c(t) \tag{9.4}$$

is used. The parameters l_1, l_2, and m are chosen to give the characteristic equation (Eqs. 9.2) when the system is linearized around $x_1 = \pi$, that is, the upright position.

Notice that Eq. (9.3) can be used for all angles except for $x_1 = \pm\pi/2$, that is, when the pendulum is horizontal. The magnitude of the control signal increases without bounds when x_1 approaches $\pm\pi/2$. The linearized model is not controllable at this operating point. □

The following example illustrates how to use the method of nonlinear transformations for a second-order system.

EXAMPLE 9.5   **Nonlinear transformation of a second-order system**

Consider the system

$$\frac{dx_1}{dt} = f_1(x_1, x_2)$$

$$\frac{dx_2}{dt} = f_2(x_1, x_2, u)$$

$$y = x_1$$

Assume that the state variables can be measured and that we want to find a feedback such that the response of the variable x_1 to the command signal is given by the transfer function

$$G(s) = \frac{\omega^2}{s^2 + 2\zeta\omega s + \omega^2} \tag{9.5}$$

Introduce new coordinates z_1 and z_2, defined by

$$z_1 = x_1$$

$$z_2 = \frac{dx_1}{dt} = f_1(x_1, x_2)$$

and the new control signal v, defined by

$$v = F(x_1, x_2, u) = \frac{\partial f_1}{\partial x_1} f_1 + \frac{\partial f_1}{\partial x_2} f_2 \tag{9.6}$$

These transformations result in the linear system

$$\frac{dz_1}{dt} = z_2$$

$$\frac{dz_2}{dt} = v$$

(9.7)

It is easily seen that the linear feedback

$$v = \omega^2(u_c - z_1) - 2\zeta\omega z_2$$

(9.8)

gives the desired closed-loop transfer function of Eq. (9.5) from u_c to $z_1 = x_1$ for the linear system of Eqs. (9.7). It remains to transform back to the original variables. It follows from Eqs. (9.6) and (9.8) that

$$F(x_1, x_2, u) = \frac{\partial f_1}{\partial x_1} f_1 + \frac{\partial f_1}{\partial x_2} f_2 = \omega^2(u_c - x_1) - 2\zeta\omega f_1(x_1, x_2)$$

Solving this equation for u gives the desired feedback. It follows from the implicit function theorem that a condition for local solvability is that the partial derivative $\partial F/\partial u$ is different from zero. □

The generalization of Example 9.5 requires a solution to the general problem of transforming a nonlinear system into a linear system by nonlinear feedback. Conditions and examples are given in the references at the end of this chapter. Figure 9.7 shows the general case when the full state is measured. There is a nonlinear transformation

$$u = g_1(x, v)$$
$$z = g_2(x)$$

that makes the relation between v and z linear. A state feedback controller from z is then computed that gives v. The control signal v is then transformed into

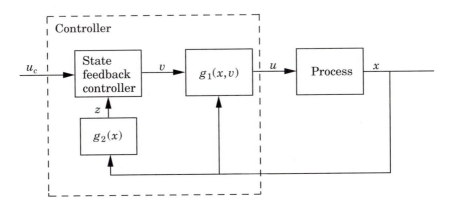

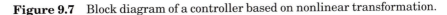

Figure 9.7 Block diagram of a controller based on nonlinear transformation.

the original control signal u. Feedback linearization requires good knowledge about the nonlinearities of the process. Uncertainties will give a transformed system that is not linear, although it may be easier to control than the original system.

A simple version of the problem also occurs in control of industrial robots. In this case the basic equation can be written as

$$J\frac{d^2\varphi}{dt^2} = T_e$$

where J is the moment of inertia, φ is an angle at a joint, and T_e is a torque, which depends on the motor current, the torque angles, and their first two derivatives. The equations are thus in the desired form, and the nonlinear feedback is obtained by determining the currents that give the desired torque. The problem is therefore called the *torque transformation*.

9.5   APPLICATIONS OF GAIN SCHEDULING

Gain scheduling is a very useful method. It requires good knowledge about the process and that some auxiliary variables can be measured. A great advantage with the method is that the controller adapts quickly to changing conditions.

This section contains examples of some cases in which it is advantageous to use gain scheduling in some of the forms that have been presented above. The examples include ship steering, pH control, combustion control, engine control, and flight control.

Ship Steering

Autopilots for ships are normally based on feedback from a heading measurement, using a gyrocompass, to a steering engine, which drives the rudder. It is common practice to use a control law of the PID type with fixed parameters. Although such a controller can be made to work reasonably well, its performance is poor in heavy weather and when the speed of the ship is changed. The reason is that the ship dynamics change with the speed of the ship and that the disturbances change with the weather. There is a growing awareness that autopilots can be improved considerably by taking these changes into account. This is illustrated by analysis of some simple models.

The ship dynamics are obtained by applying Newton's equations to the motion of the ship. For large ships the motion in the vertical plane can be separated from the other motions. It is customary to describe the horizontal motion by using a coordinate system fixed to the ship (see Fig. 9.8). Let V be the total velocity, let u and v be the x and y components of the velocity, and let r be the angular velocity of the ship.

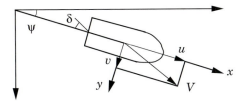

Figure 9.8   Coordinates and notations used to describe the equations of motion of ships.

In normal steering, the ship makes small deviations from a straight-line course. It is thus natural to linearize the equations of motion around the solution $u = u_0$, $v = 0$, $r = 0$, and $\delta = 0$. The natural state variables are the sway velocity v, the turning rate r, and the heading ψ. The following equations are obtained:

$$
\begin{aligned}
\frac{dv}{dt} &= (u/l)a_{11}v + ua_{12}r + (u^2/l)b_1\delta \\
\frac{dr}{dt} &= (u/l^2)a_{21}v + (u/l)a_{22}r + (u^2/l^2)b_2\delta \\
\frac{d\psi}{dt} &= r
\end{aligned}
\tag{9.9}
$$

where u is the constant forward velocity and l is the length of the ship.

The parameters in the state equation (Eqs. 9.9) are surprisingly constant for different ships and different operating conditions (see Table 9.1). The transfer function from rudder angle δ to heading ψ is easily determined from Eqs. (9.9). The following result is obtained:

$$
G(s) = \frac{K(1 + sT_3)}{s(1 + sT_1)(1 + sT_2)}
\tag{9.10}
$$

where

$$
\begin{aligned}
K &= K_0 u/l \\
T_i &= T_{i0} l/u \qquad i = 1, 2, 3
\end{aligned}
\tag{9.11}
$$

The parameters K_0 and T_{i0} are also given in Table 9.1. Notice that they may change considerably even if the parameters of the state model do not change much. In many cases the model can be simplified to

$$
G(s) = \frac{b}{s(s + a)}
\tag{9.12}
$$

where

$$
\begin{aligned}
b &= b_0 \left(\frac{u}{l}\right)^2 = b_2 \left(\frac{u}{l}\right)^2 \\
a &= a_0 \left(\frac{u}{l}\right)
\end{aligned}
\tag{9.13}
$$

Table 9.1   Parameters of models for different ships.

Ship	Mine-sweeper	Cargo	Tanker Full	Ballast
Length (m)	55	161	350	
a_{11}	-0.86	-0.77	-0.45	-0.43
a_{12}	-0.48	-0.34	-0.43	-0.45
a_{21}	-5.2	-3.39	-4.1	-1.98
a_{22}	-2.4	-1.63	-0.81	-1.15
b_1	0.18	0.17	0.10	0.14
b_2	-1.4	-1.63	-0.81	-1.15
K_0	2.11	-3.86	0.83	5.88
T_{10}	-8.25	5.66	-2.88	-16.91
T_{20}	0.29	0.38	0.38	0.45
T_{30}	0.65	0.89	1.07	1.43
a_0	-0.14	0.19	-0.28	-0.06
b_0	-1.4	-1.63	-0.81	-1.15

This model is called the *Nomoto model*. Its gain b can be expressed approximately as follows:

$$b = c \left(\frac{u}{l}\right)^2 \frac{Al}{D} \qquad (9.14)$$

where D (in cubic meters) is the displacement, A (in square meters) is the rudder area, and c is a parameter whose value is approximately 0.5. The parameter a will depend on trim, speed, and loading. Its sign may change with the operating conditions.

A ship is influenced by disturbances due to wind, waves, and currents. The effects of these can be described as additional forces. Reasonable models have constant, periodic, and random components. The disturbances due to waves are typically periodic. The period may vary with the speed of the ship and its orientation relative to the waves.

The effects of parameter variations can be seen from the linearized models in Eqs. (9.9), (9.10), and (9.12). First, consider variations in the speed of the ship. It follows from Eqs. (9.11) and (9.13) that the gain is proportional to the square of the velocity and that the time constants are inversely proportional to the velocity. A reduction to half-speed thus reduces the gain to a quarter of its value and doubles the time constants.

The gain is essentially determined by the ratio of the rudder forces to the moment of inertia. Thus the relative water velocity at the rudder is what determines the gain. This velocity is influenced by waves and currents. The relative velocity may decrease drastically when there are large waves coming from behind and the ship is riding on the waves. The relative velocity may be very small or even zero. Controllability is then lost because there is no rudder force. The situation is even worse if the waves are not hitting the ship straight from behind, because the waves will then generate torques that tend to turn the ship.

The ship dynamics are also influenced by other factors. The hydrodynamic forces, and consequently also the parameters a_{ij} and b_j in the linearized model of Eqs. (9.9), depend on trim loading and water depth. This may be seen from Table 9.1, which gives parameters for a tanker under different loading conditions. Some consequences of the parameter variations are illustrated by an example.

EXAMPLE 9.6   **Ship steering**

Assume that the ship steering dynamics can be approximated by the Nomoto model of Eq. (9.12) and that a controller of PD type with the transfer function

$$G_r(s) = K(1 + sT_d)$$

is used. The loop transfer function is

$$G(s)G_r(s) = \frac{Kb(1 + sT_d)}{s(s + a)}$$

The characteristic equation of the closed-loop system is

$$s^2 + s(a + bKT_d) + bK = 0$$

The relative damping is

$$\zeta = \frac{1}{2}\left(\frac{a}{\sqrt{bK}} + T_d\sqrt{bK}\right)$$

The damping will depend on the speed of the ship. Assume that the model of Eq. (9.12) has the values a_{nom} and b_{nom} at the nominal speed u_{nom}. The variable u_{nom} is the nominal velocity used to design the feedback. Assume that u is the actual constant velocity. Using the speed dependence of a and b given by Eqs. (9.13) gives

$$a = a_{\text{nom}}\frac{u}{u_{\text{nom}}}$$

$$b = b_{\text{nom}}\left(\frac{u}{u_{\text{nom}}}\right)^2$$

This gives the damping

$$\zeta = \frac{1}{2}\left(\frac{a_{\text{nom}}}{\sqrt{Kb_{\text{nom}}}} + \frac{u}{u_{\text{nom}}}T_d\sqrt{Kb_{\text{nom}}}\right)$$

Consider an unstable tanker with

$$a_{\text{nom}} = -0.3$$
$$b_{\text{nom}} = 0.8$$
$$K = 2.5$$
$$T_d = 0.86$$

This gives $\zeta = 0.5$ and $\omega = 1.4$ at the nominal velocity. Furthermore,

$$\omega = 1.4u/u_{\text{nom}}$$
$$\zeta = -0.11 + 0.61u/u_{\text{nom}}$$

The closed-loop characteristic frequency and damping will thus decrease with decreasing velocity. The closed-loop system becomes unstable when the speed of the ship decreases to $u = 0.17u_{\text{nom}}$.

By scaling the parameters of the autopilot according to speed, it is possible to obtain closed-loop performance that is less sensitive to speed variations. The scaling of the parameters of the controller depends on the control goal. One design criterion is time invariance; that is, the time response of the ship should always be the same. If true time invariance is desired, the controller gains should be inversely proportional to the square of the speed. Path invariance is another criterion. In this case the path on the map is always the same. The gains should then be inversely proportional to the velocity of the ship. The gains are limited at low speed to avoid large rudder motions. □

pH Control

Control of pH (the concentration of hydrogen ions) is a well-known control problem that presents difficulties due to large variations in process dynamics. The problem is similar to the simple concentration control problem in Example 9.3. The main difficulty arises from a static nonlinearity between pH and concentration. This nonlinearity depends on the substances in the solution and on their concentrations.

The pH number is a measure of the concentration or, more precisely, the activity of hydrogen ions in a solution. It is defined by

$$\text{pH} = -\log[\text{H}^+] \tag{9.15}$$

where $[\text{H}^+]$ denotes the concentration of hydrogen ions. The formula (9.15) is, strictly speaking, not correct, since $[\text{H}^+]$ has the dimension of concentration, which is measured in the unit M = mol/l. The correct version of Eq. (9.15) is thus $\text{pH} = -\log([\text{H}^+]/f_{\text{H}})$, where f_{H} is a constant with the dimension liters per mole. The formula of Eq. (9.15) will be used here, however, because it is universally accepted in textbooks of chemistry.

Water molecules are dissociated (split into hydrogen and hydroxyl ions) according to the formula

$$\text{H}_2\text{O} \rightleftharpoons \text{H}^+ + \text{OH}^-$$

In chemical equilibrium the concentration of hydrogen H^+ (or rather H_3O^+) and hydroxyl OH^- ions are given by the formula

$$\frac{[H^+][OH^-]}{[H_2O]} = \text{constant} \tag{9.16}$$

Only a small fraction of the water molecules are split into ions. The water activity is practically unity, and we get

$$[H^+][OH^-] = K_w \tag{9.17}$$

where the equilibrium constant K_w has the value $10^{-14}\,[(\text{mol/l})^2]$ at $25°C$. The main nonlinearity of the pH control problem will now be discussed.

EXAMPLE 9.7   **Titration curve for a strong acid-base pair**

Consider neutralization of m_A mol of hydrochloric acid HCl by m_B mol of sodium hydroxide NaOH in a water solution. The following reaction takes place:

$$HCl + NaOH \rightarrow H^+ + OH^- + Na^+ + Cl^-$$

Let the total volume be V. The concentration of chloride ions is then

$$[Cl^-] = x_A = m_A/V$$

and the concentration of sodium ions is given by

$$[Na^+] = x_B = m_B/V$$

because the acid and the base are completely ionized. Since the number of positive ions equals the number of negative ions, it follows that

$$x_A + [OH^-] = x_B + [H^+]$$

The concentration of hydroxyl ions can be related to the hydrogen ion concentration by Eq. (9.17). Hence

$$x = x_B - x_A = [OH^-] - [H^+] = \frac{K_w}{[H^+]} - [H^+] = 10^{pH-14} - 10^{-pH} \tag{9.18}$$

Solving for $[H^+]$ gives

$$[H^+] = \sqrt{x^2/4 + K_w} - x/2$$
$$[OH^-] = \sqrt{x^2/4 + K_w} + x/2$$

This gives

$$pH = f(x) = -\log\left(\sqrt{x^2/4 + K_w} - x/2\right) \tag{9.19}$$

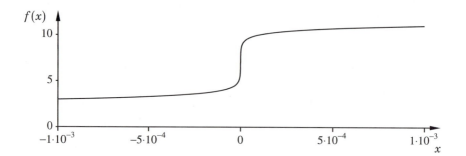

Figure 9.9   Titration curve of Eq. (9.19) for neutralization of a 0.001 M solution of HCl with a 0.001 M solution of NaOH.

The graph of the function f is called the *titration curve*. It is the fundamental nonlinearity for the neutralization problem. An example of the titration curve is shown in Fig. 9.9, which shows that there is considerable variation in the slope of the titration curve. The abscissa of the titration curve in Fig. 9.9 is given in terms of the concentration difference $x_B - x_A$. The x-axis can also be recalibrated into the amount of the reagent.

The derivative of the function f is given by

$$f'(x) = \frac{10 \log e}{2\sqrt{x^2/4 + K_w}} = \frac{10 \log e}{10^{pH-14} + 10^{-pH}} \tag{9.20}$$

The derivative has its largest value $f' = 2.2 \cdot 10^6$ for pH $= 7$. It decreases rapidly for larger and smaller values of pH. For pH $= 4$ and 10 we have $f' = 4.3 \cdot 10^3$. The gain can thus vary by several orders of magnitude. □

Figure 9.9 shows that the pH of a strong acid that is almost neutralized may change very rapidly if only a small amount of base is added. The reason for this is that strong acids and bases are completely dissociated. A weak acid is not completely dissociated, so it can absorb hydrogen ions by converting them to undissociated acid. It can also create hydrogen ions by dissociating acid molecules. This means that a weak acid or a weak base has an ability to resist changes in pH. This property is called *buffering*. The titration curve of a solution that contains weak acids or bases will therefore be less steep than the titration curves of strong acids or bases.

Example 9.7 shows that there will be a severe nonlinearity in the system due to the titration curve. An additional example illustrates the difficulties in controlling such a system.

EXAMPLE 9.8   pH control

Consider the problem of controlling the pH of an acid effluent that is fed to a stirred tank with volume V (in liters) and neutralized with NaOH. Let c_A (in moles per liter) be the concentration of acid in the influent stream, and let q

(in liters per second) be the flow of the effluent. Let c_B (in moles per liter) be the concentration of the reagent. Assume that the reagent concentration is so high that the reagent flow u (in liters per second) is negligible in comparison with q. The system is modeled by a linear dynamic model, which describes the mixing dynamics as if there were no reactions, and a static nonlinear titration curve, which gives pH as a function of the concentrations. Let x_A and x_B be the concentrations of acid and base in the tank if there were no chemical reactions. Mass balances then give

$$\frac{dx_A}{dt} = \frac{q}{V}(c_A - x_A)$$
$$\frac{dx_B}{dt} = \frac{u}{V}c_B - \frac{q}{V}x_B$$

(9.21)

The pH is given by Eq. (9.19). It is further assumed that the dynamics of the pH sensor and the pump together can be described by the transfer function

$$G(s) = \frac{1}{(1 + sT)^2}$$

A simple calculation indicates the difficulties in the control problem. Assuming proportional control with gain k, the linearized loop transfer function from the error in pH to pH becomes

$$G_0(s) = \frac{c_B k f'}{q(1 + sT_m)(1 + sT)^2}$$

where T_m is the mixing time constant

$$T_m = V/q$$

and f' is the slope of the titration curve given by Eq. (9.20). The critical gain for stability is

$$k_c = \frac{q}{f'c_B T}(2 + T/T_m)(1 + T/T_m) \approx \frac{2q}{f'c_B T}$$

where the approximation holds for $T \ll T_m$. Since the slope of the titration curve varies drastically with pH, the critical gain will vary accordingly. Some values for different values of the pH of the mixture are:

pH	Critical gain
7	0.009
8	0.046
9	0.46
10	4.6

To make sure that the closed-loop system is stable for small perturbations around an equilibrium of pH = 7, the gain should thus be less than 0.009. A reasonable value of the gain for operation at pH = 8 is $k = 0.01$, but this

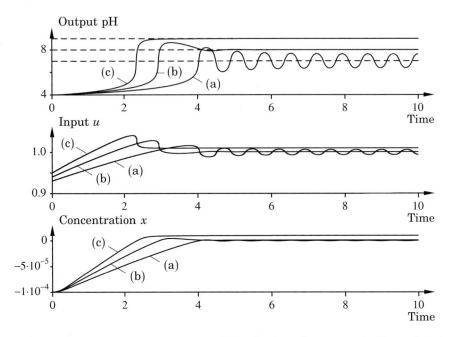

Figure 9.10   Output pH and control signal when the process in Example 9.8 is controlled by using a PI controller when pH_{ref} is (a) 7; (b) 8; (c) 9.

gain will give an unstable system at pH = 7 and is too low for a reasonable response at pH = 9. Figure 9.10 shows PI control with gain 0.01 and reset time 1. The process is started at equilibrium pH = 4. The reference value is then changed to 7, 8, and 9.

The calculations and the simulation illustrate the key problems with pH control. The difficulties are compounded by the presence of time delays and flow variations. One way to get around the problem is to use the concentration x as the output rather than pH. Figure 9.11 shows a possible control scheme in which the measured pH and the reference value of pH are transformed into equivalent concentrations. This means that the variable x is computed for the

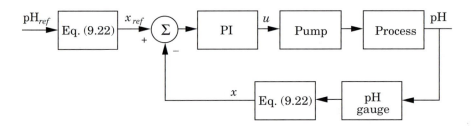

Figure 9.11   Control configuration for the pH control problem in Example 9.8.

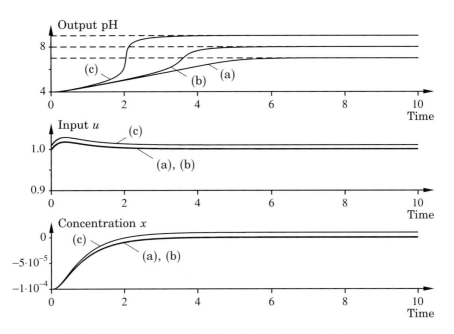

Figure 9.12   The same experiment as in Fig. 9.10, but with the controller structure in Fig. 9.11. The gain of the controller is 1000, and the reset time is 1. (a) $pH_{ref} = 7$; (b) $pH_{ref} = 8$; (c) $pH_{ref} = 9$.

measured pH by the formula

$$x = f^{-1}(\text{pH}) = 10^{\text{pH}-14} - 10^{-\text{pH}} \tag{9.22}$$

The transfer function from u to x is

$$\frac{c_B}{q(1 + sT_m)(1 + sT)^2}$$

which is independent of the operating point. Figure 9.12 shows the same experiments as in Fig. 9.10, but with the control modification shown in Fig. 9.11. It should be noted that the nonlinear compensation with Eq. (9.22) can be used, since a strong acid-base pair is controlled. The more general problem of mixtures of many weak acids and bases does not have an easy linearizing transformation. It is then necessary to measure the concentrations of the components or to make an on-line measurement of the titration curve. Some other form of adaptation can then be reasonable.   □

Combustion Control

In combustion control of a boiler it is important to adjust the oxygen content of the flue gases. The flow of combustion air depends on the burn rate in the

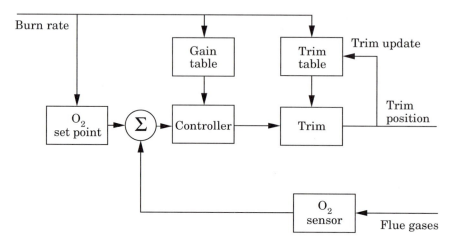

Figure 9.13   Adaptive feedforward and gain scheduling in an oxygen trim controller.

boiler. The measurement signal is the oxygen content in the exhaust stack, and the control signal is the trim position, which controls the flow of combustion air. There is a significant time delay between the input to the burner and the oxygen sensor in the exhaust stack. With a conventional controller there is then a loss of efficiency before the correct trim position is reached after a change in the burn rate. One configuration based on adaptive feedforward and gain scheduling is shown in Fig. 9.13. The working range of the boiler is divided into regions. For each region there is a memory (digital integrator). All integrators are zero initially. When the boiler starts to operate, the trim control will adjust the oxygen setpoint. When the setpoint level is achieved, the appropriate integrator is set to the correct trim position. A trim profile will be built up as the boiler works over its range. When the boiler returns to a position at which the integrator is set, the stored trim value is instantly fed to the trim drive actuator, thus eliminating the lag from the control loop. If the fuel changes, the trim profile is updated automatically. The controller thus works with an adaptive feedforward compensation from the burn rate. There is also a gain scheduling of the loop gain of the controller to get tight control under all firing conditions. This gain schedule is built up in commissioning the controller.

Fuel-Air Control in a Car Engine

A schematic drawing of a microcomputer control system for a car engine is shown in Fig. 9.14. The accelerator is connected to the throttle valve. The fuel injection is governed by a table lookup controller. The control variable, which

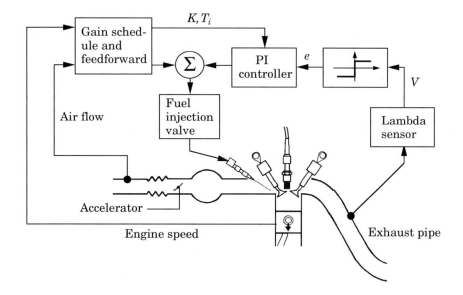

Figure 9.14 Schematic diagram of a microcomputer engine control system.

is the opening time for the fuel injection valve, is controlled by a combination of feedforward and feedback. The feedforward signal is a nonlinear function of engine speed and load. The load is represented by the air flow, which can be measured by using a hot wire anemometer. In one common system the table has 16×16 entries with linear interpolation. There is also feedback in the system from an exhaust oxygen sensor. The fuel-air ratio is measured by using a zirconium oxide catalytic sensor called the *lambda sond*. This sensor gives

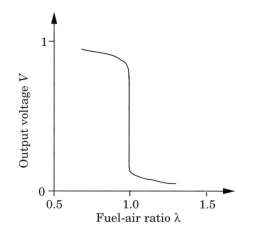

Figure 9.15 The characteristic of a lambda sond.

an output that changes drastically when the fuel-air ratio is 1. A typical sensor characteristic is shown in Fig. 9.15. The lambda sond is positioned after the exhaust manifold in an excess oxygen environment, where the exhaust gas from all the cylinders is mixed. This creates a delay in the feedback loop. Notice the feedforward path via the table discussed earlier in the paragraph. The feedback has a special form; continuous control cannot be used because of the strongly nonlinear characteristics of the lambda sond. The error signal is formed by normalizing the output of the lambda sensor as follows:

$$e = \begin{cases} 1 & \text{if } V > 0.5 \\ -1 & \text{if } V \le 0.5 \end{cases}$$

The error signal is thus positive if the fuel-air ratio is low (lean mixture) and negative when the ratio is high (rich mixture). The error signal is sent to a PI controller whose gain and integration time are set from the scheduling table. The values are set on the basis of load (air flow) and engine speed. The gain

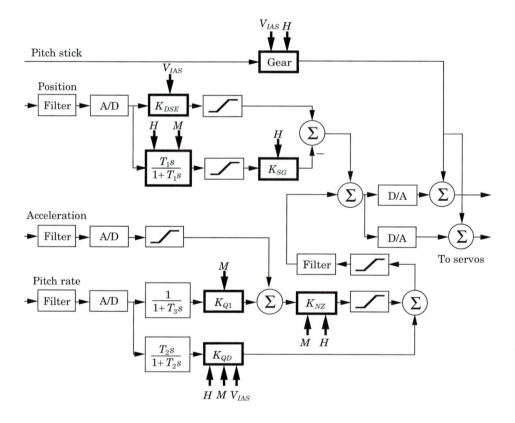

Figure 9.16 Simplified block diagram of the pitch control of the autopilot for a supersonic aircraft. The highlighted blocks show the parts of the autopilot where gain scheduling is used.

schedule is implemented simply by adding entries for the gain and integration time to the table used for feedforward of the nominal control variable. Because of the relay characteristic, there will be an oscillation in the fuel-air ratio. This is beneficial, because the catalytic sensor needs a variation to operate properly. The amplitude and the frequency of the oscillation are determined by the parameters of the controller.

Flight Control Systems

Figure 9.16 shows a block diagram of the pitch channel of a flight control system for a supersonic aircraft. The pich stick signal is the command signal from the pilot. Position, acceleration, and pitch rate are feedback signals. There are three scheduling variables: height H, indicated airspeed V_{IAS}, and Mach number M. The parameters of the controller that are scheduled are drawn as boxes; the arrows indicate the scheduling variables. The schedule for the gain K_{QD} is given by

$$K_{QD} = K_{QD_{IAS}} + (K_{QD_H} - K_{QD_{IAS}})MF$$

where $K_{QD_{IAS}}$ is a function of indicated airspeed V_{IAS} (shown in Fig. 9.17) and K_{QD_H} is a function of height (also shown in Fig. 9.17). The variable MF is given by

$$MF = \frac{1}{s+1}K_{MF}$$

where K_{MF} is a function of the Mach number and s is the Laplace transform variable.

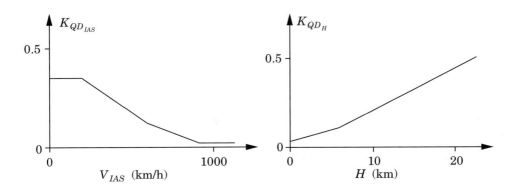

Figure 9.17   Scheduling functions. The function $K_{QD_{IAS}}$ is also different for different flight modes.

9.6   CONCLUSIONS

Gain scheduling is a good way to compensate for known nonlinearities. With such a scheme the controller reacts quickly to changing conditions. One drawback of the method is that the design may be time-consuming if it is not possible to use nonlinear transformations or auto-tuning. Another drawback is that the controller parameters are changed in open loop, without feedback from the performance of the closed-loop system. This makes the method impossible to use if the dynamics of the process or the disturbances are not known accurately enough.

Example 9.3 and the ship steering example in Section 9.5 show that it is often useful to introduce normalized variables. The processes then become constant in the new variables, and the gain scheduling of the controllers is easily derived.

PROBLEMS

9.1   Simulate the tank system in Example 9.2. Let the tank area vary as

$$A(h) = A_0 + h^2$$

Further assume that $a = 0.1A_0$.

(a) Study the behavior of the closed-loop system when the full gain schedule is used and when the modified gain schedule is used.

(b) Study the sensitivity of the system to changes in the parameters of the process.

(c) Study the sensitivity of the closed-loop system to noise in the measurement of the level.

9.2   Consider the concentration control problem in Example 9.3. Design a fixed sampled-data controller with a fixed sampling period for the system. Compare it with a controller based on the time-scaled model.

9.3   A model of a ship is given in Section 9.5. Show that the two scalings suggested in Example 9.6 correspond to the time invariance and the path invariance behavior of the ship.

9.4   The simulations in Example 9.8 are done by using the model of Eqs. (9.21) and (9.19) with $q = 1000$, $V = 1000$, $T = 0.1$, and $K_w = 10^{-14}$. The controller is a PI controller with gain 0.01 and reset time 1. Verify the simulations in Fig. 9.10 and Fig. 9.12.

9.5   Consider the ship steering problem in Example 9.6. Simulate the closed-loop system, and determine the sensitivity with respect to the speed of the ship.

9.6   The controller in Example 9.3 gives a control that is equal when measured in terms of the number of sampling intervals but not when measured in terms of time. Suggest and test possibilities to get the same time responses independent of the flow through the tank.

REFERENCES

The use of gain scheduling in aircraft control is discussed in:

Stein, G., 1980. "Adaptive flight control: A pragmatic view." In *Applications of Adaptive Control*, eds. K. S. Narendra and R. V. Monopoli. New York: Academic Press.

A typical application of gain scheduling and compensation of nonlinearities in the process industry is given in:

Whatley, M. J., and D. C. Pott, 1984. "Adaptive gain improves reactor control." *Hydrocarbon Processing* May: 75–78.

Nonlinear transformations in a general context were originally discussed by using geometric control theory in:

Krener, A. J., 1973. "On the equivalence of control systems and the linearization of nonlinear systems." *SIAM J. Control* 11: 670.

Brockett, R. W., 1978. "Feedback invariants for nonlinear systems." *Preprint 7th IFAC World Congress*, pp. 1115–1120. Helsinki, Finland.

Necessary and sufficient conditions under which transformations from nonlinear to linear systems exist are given in:

Su, R., 1982. "On the linear equivalents of nonlinear systems." *Systems & Control Letters* 2: 48.

Hunt, L. R., R. Su, and G. Meyer, 1983. "Design for multiinput systems." In *Differential Geometric Control Theory Conference*, eds. R. W. Brockett, R. S. Millman, and H. J. Sussman, pp. 268–298. Boston: Birkhauser.

Linearizing control is discussed, for instance, in:

Isidori, A., 1989. *Nonlinear Control Systems: An Introduction* (2nd ed.). Berlin: Springer-Verlag.

Nijmeijer, H., and A. J. van der Schaft, 1991. *Nonlinear Dynamical Control Systems*. New York: Springer-Verlag.

A neat application for design of a flight control system for a helicopter was made by:

Meyer, G., R. Su, and L. R. Hunt, 1984. "Application of nonlinear transformations to automatic flight control." *Automatica* 20: 103–107.

Applications of the same idea in simpler setting are given in:

Orava, P. J., and A. J. Niemi, 1974. "State model and stability analysis of a pH control process." *Int. J. Control* 20: 557–567.

Källström, C. G., K. J. Åström, N. E. Thorell, J. Eriksson, and L. Sten, 1979. "Adaptive autopilots for tankers." *Automatica* **20**: 241–254.

Niemi, A. J., 1981. "Invariant control of variable flow processes." *Proceedings of the 8th IFAC World Congress*, pp. 2687–2692. Kyoto, Japan.

ROBUST AND SELF-OSCILLATING SYSTEMS

10.1   WHY NOT ADAPTIVE CONTROL?

In previous chapters we showed that adaptive control can be very useful and can give good closed-loop performance. However, that does not mean that adaptive control is the universal tool that should always be used. A control engineer should be equipped with a variety of tools and the knowledge of how to use them. A good guideline is to use the simplest control algorithm that satisfies the specifications. Robust high-gain control should definitely be considered as alternatives to adaptive control algorithms. Section 10.2 treats robust high-gain control. The self-oscillating adaptive system (SOAS) is presented in Section 10.3. This is a special class of adaptive systems with strong ties to high-gain control and auto-tuning. Relay feedback is a key ingredient of the SOAS. Another class of switching systems, variable-structure systems, is discussed in Section 10.4. Variable-structure systems have been developed mainly in the Soviet Union and can be regarded as a generalization of the SOAS. Conclusions are given in Section 10.5.

10.2   ROBUST HIGH-GAIN FEEDBACK CONTROL

Some design methods deal explicitly with process uncertainties. One powerful method has been developed by Horowitz. This procedure, which has its origin in Bode's classical work on feedback amplifiers, is based on several ideas. The specifications are expressed in terms of the transfer function from command

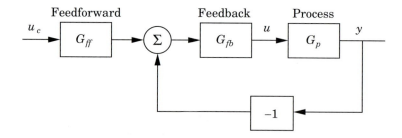

Figure 10.1   A two-degree-of-freedom system.

signal to process output. The plant is characterized by its nominal transfer function. For each frequency it is also assumed that the process uncertainty is known in terms of variations in amplitude and phase. A solution is determined in terms of a controller with a feedback G_{fb} and a feedforward G_{ff}, as shown in Fig. 10.1. Such a configuration is called a *two-degree-of-freedom* system because there are two transfer functions to be determined.

Several other design methods can be used to design robust controllers. One technique is based on LQG design. By adjusting the weighting matrices in the LQG problem, a *loop transfer recovery (LTR)* is achieved. This design procedure can cope with phase uncertainty at high frequencies. The key idea is to keep the loop gain less than 1 at high frequencies, where the phase error is large.

In Horowitz's procedure the feedback transfer function G_{fb} is first determined such that the closed-loop uncertainty is within the specified limits. The nominal value of the transfer function is then modified by the feedforward compensation G_{ff}. The method is based on graphical constructions using the Nichols chart. It gives a high-order linear compensator that can cope with the specified plant uncertainty. The procedure attempts to keep the loop gain as low as possible. A key idea in the Horowitz design method is the observation that a system in which the Nyquist curve is close to a straight line through the origin can tolerate a significant change of gain. The response time will change with the gain, but the shape of the response will remain invariant. For minimum-phase systems with a known pole excess it is always possible to find a frequency range in which the phase is constant. By proper compensation it is then possible to obtain a loop gain at which the Nyquist curve is close to a straight-line segment. The assumption that the pole excess is known implies that the phase of the system is known for high frequencies. This is not always a realistic assumption. The Horowitz design method was originally developed for structured perturbations but has also been extended to unstructured uncertainties.

The main step in the procedure is to determine the tolerances for the gain in the closed-loop transfer function. The plant uncertainties are specified as gain and phase variations of the plant transfer function at different frequencies. The given tolerances and uncertainties are used to calculate constraints for the open-loop transfer function. The feedback compensator G_{fb} is then designed such that the compensated open-loop system satisfies the tolerances. This is usually an iterative procedure, which can conveniently be done graphically by using a Nichols chart. Finally, the prefilter G_{ff} is designed such that the closed-loop specifications are fulfilled. This may be done by using the Bode diagram.

The major drawback of the method is that it is impossible to know *a priori* whether the desired closed-loop specifications are attainable. It is thus a trial-and-error method, but the iterations give the designer insight into the tradeoffs between different specifications, such as closed-loop sensitivity, complexity of the controller, and measurement noise amplification.

EXAMPLE 10.1   **An industrial robot arm**

A simple model of a robot arm is used in this example. The transfer function from the control input (motor current I) to measurement output (motor angular velocity ω) is

$$G_p(s) = \frac{k_m(J_a s^2 + ds + k)}{J_a J_m s^3 + d(J_a + J_m)s^2 + k(J_a + J_m)s}$$

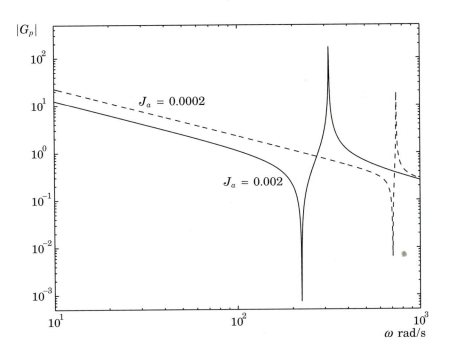

Figure 10.2   Bode plots for the robot arm in Example 10.1 for $J_a = 0.0002$ and $J_a = 0.002$.

with $J_a \in [0.0002, 0.002]$, $J_m = 0.002$, $d = 0.0001$, $k = 100$, and $k_m = 0.5$. The moment of inertia J_a of the robot arm varies with the arm angle. Bode plots of the plant gain for the extreme values of the arm inertia J_a are given in Fig. 10.2. The purpose of the control system is to control the angular velocity step responses at various arm angles. The aim is to get a closed-loop system with a bandwidth between 15 and 40 Hz. The disturbance rejection specification has been set to 6 dB. A feedback compensator that satisfies the specifications is

$$G_{fb}(s) = \frac{125(1 + s/50)(1 + s/300)}{s(1 + s/800)(1 + s/5000)}$$

This compensator is essentially a PI controller with a lead filter. The final prefilter has the transfer function

$$G_{ff}(s) = \frac{1 + s/1000}{(1 + s/26)(1 + s/200)(1 + s/200)}$$

Simulated responses are shown in Figs. 10.3 and 10.4.

To make a comparison, an adaptive controller is also designed for the process. In this particular problem the essential uncertainty is in one parameter only, the moment of inertia. It is then natural to try to make a special adaptive design in which only this parameter is estimated.

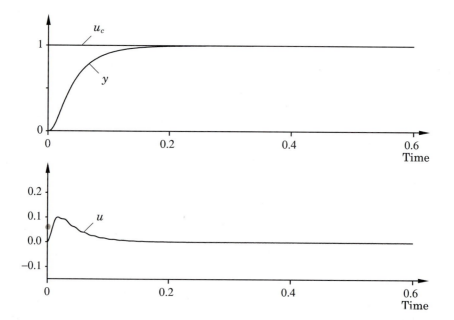

Figure 10.3 Simulation of the step response with the arm inertia $J_a = 0.0002$ for the robust system.

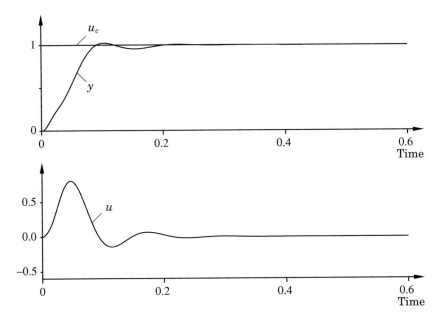

Figure 10.4 Simulation of the step response with the arm inertia J_a = 0.002 for the robust system.

The adaptive controller is designed on the basis of a simplified model. If we neglect the elasticity in the robot arm, the system can be described by

$$J \frac{d\omega}{dt} = k_m I \tag{10.1}$$

where $J = J_a + J_m$ is the total moment of inertia and k_m the current gain of the motor. The plant of Eq. (10.1) can be controlled adequately with a PI controller. The controller parameters can be chosen to be

$$K = \frac{2\zeta_0 \omega_0 J}{k_m}$$

$$T_i = \frac{2\zeta_0}{\omega_0}$$

This gives the following characteristic equation for the closed-loop system:

$$s^2 + 2\zeta_0 \omega_0 s + \omega_0^2 = 0$$

The controller parameters are thus related to the model by simple equations. Notice that the integration time T_i does not depend on the moment of inertia of the robot arm and that the controller gain K should be proportional to the moment of inertia.

A root-locus calculation indicates that the design based on the simplified model will work well if

$$\omega_0 < \omega_{crit} = \zeta_0 \left(\frac{kJ_m}{J_a^2} \right)^{1/2}$$

The most critical case occurs for $J_a = 0.002$. It implies that ω_0 must be less than 200 rad/s.

The fact that the design is based on a simplified model limits the closed-loop bandwidth. A fast response to command signals can still be obtained by use of feedforward compensation. For this purpose, let the desired response to angular velocity commands be given by

$$G_m(s) = \frac{\omega_m^2}{s^2 + 2\zeta_0 \omega_m s + \omega_m^2}$$

The feedforward controller can now be designed such that the closed-loop system gets the desired response.

An adaptive system can be obtained simply by estimating the total moment of inertia by applying recursive least squares to the model of Eq. (10.1) and feeding the estimate into the above design equation. To estimate the parameters of the continuous-time model of Eq. (10.1), it is necessary to introduce

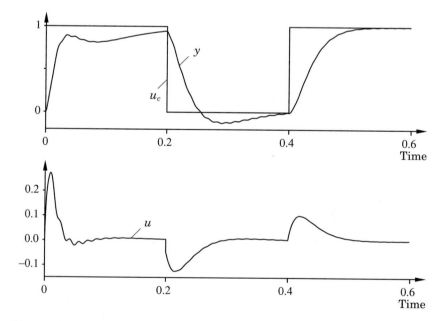

Figure 10.5   Simulation of the tailored adaptive systems response with the arm inertia $J_a = 0.0002$. The controller is initially tuned for $J_a = 0.002$.

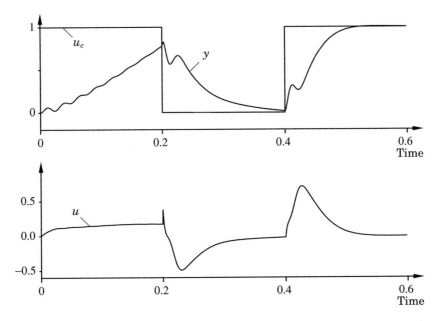

Figure 10.6 Simulation of the tailored adaptive systems response with the arm inertia $J_a = 0.002$. The controller is initially tuned for $J_a = 0.0002$.

filtering. This is done by integrating Eq. (10.1) over the time interval $(t, t+h)$:

$$\omega(t+h) - \omega(t) = \frac{k_m}{J} \int_t^{t+h} I(s)\,ds$$

A least-squares estimator of J is easily constructed from this equation. This estimate is then used in the PI control law. Simulations of the system are shown in Figs. 10.5 and 10.6. The parameter h was chosen to be 0.1 s. The figures show that the system adapts to a good response after two transients. Notice the magnitudes of the control signal for the cases of low and high inertia.

The controller structures for the robust and adaptive cases are quite similar by design. The feedback part of the robust controller is essentially a PI controller with a lead-lag filter. The parameters are $K = 2.5$ and $T_i = 0.02$. The lead-lag filter increases the controller gain to 6.7 at frequencies around 500 rad/s. The feedback part of the adaptive controller is also a PI controller, but the parameters are adjustable. They range from $K = 0.15$ and $T_i = 0.07$ for $J_a = 0.0002$ to $K = 1.05$ and $T_i = 0.07$ for $J_a = 0.002$. The feedback gain in the adaptive controller is thus 40 times smaller than the gain of the robust controller. This means that the effects of measurement noise are also much smaller for the adaptive controller. Both systems are designed to give the same response time to command signals. Notice, however, that feedforward is used in very different ways in the two systems. In the robust design, it is used to decrease the response time to command signals; in the adaptive design, it is used to increase the response time. The reason is that the bandwidth of the closed inner loop is large in the robust design, to take care of the plant variations, whereas the adaptive design allows a low closed-loop bandwidth, since the uncertainty is eliminated. The responses of the adaptive system are better over the full parameter range when the parameters are adapted, but it will take some time for the parameters to adapt. The robust controller will have a better response when the parameters of the process are changing rapidly from one constant value to another.   □

Comparison between Robust and Adaptive Control

The robust design method will generally give systems that respond more quickly when the parameters change, but it is important that the range of parameter variation be known. The adaptive controller responds more slowly but can generally handle larger parameter variations. The adaptive controller will give better responses to command signals and load variations when controller parameters have converged, provided that the model structure is sufficiently correct. The controllers designed by Horowitz's method will generally have high-loop gains, which make them more sensitive to noise.

10.3   SELF-OSCILLATING ADAPTIVE SYSTEMS

A system that is insensitive to parameter variations can be obtained by using a two-degree-of-freedom configuration with a high-gain feedback and a feedforward compensator (compare Section 10.2). This section introduces an adaptive technique to keep the gain in the feedback loop high by using a relay feedback. Relays combine the properties of high gain and inexpensive implementations. However, relays often introduce oscillations into the system.

The idea of the self-oscillating adaptive system (SOAS) originated in work at Honeywell on adaptive flight control in the late 1950s. The inspiration came from work on nonlinear systems by Flügge-Lotz at Stanford. Systems based on the idea were flight-tested in the F-94C, the F-101, and the X-15 aircraft. (See Fig. 1.2.) The idea has also been applied in process control, but the SOAS has not found widespread use. One reason is that substantial modifications of the basic scheme are necessary to make the systems work well. A characteristic feature of the SOAS is that there is a limit cycle oscillation. The system thus represents a type of adaptive control in which there are intentional perturbations, which excite the system all the time. The SOAS is one of the simplest systems with this property. The SOAS is based on three useful ideas: model-following, automatic generation of test signals, and use of a relay with a dither signal as a variable gain. The key result is that the loop gain is automatically adjusted to give an amplitude margin $A_m = 2$.

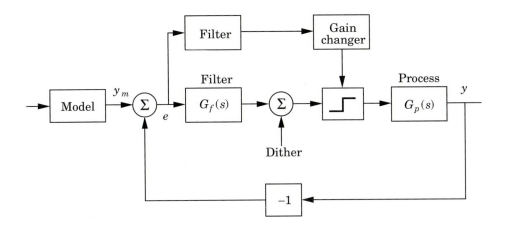

Figure 10.7   Block diagram of a self-oscillating adaptive system (SOAS).

Principles of the SOAS

Since we want to emphasize the ideas, we will limit the discussion to the basic version of the system. A block diagram of an SOAS is shown in Fig. 10.7. This is a two-degree-of-freedom system. There is a high-gain feedback loop around the process. The desired response to command signals is obtained by the reference model. Ideally, the high-gain loop will make the process output y follow the model output y_m. The response of the closed-loop system will be relatively insensitive to the variations in process dynamics because of the high loop gain. The system is thus a typical model-following design. The special feature is that the high-gain loop is nonlinear. The high-gain loop is supplemented with a feedforward model and a device to change the gain of the relay. The model determines the closed-loop response and the gain changer limits the amplitude of the limit cycle oscillation.

The High-Gain Loop

The feedback compensator contains a lead filter $G_f(s)$ and a relay. The relay is motivated by the desire to have as high a gain as possible. Because of the relay, there will be a limit cycle, whose amplitude is kept at a tolerable limit by adjusting the relay amplitude by a separate feedback loop. The relay gives a high gain for small inputs, and the gain decreases with increasing input amplitude. The key difficulty in the design of an SOAS is to find a suitable compromise between the limit cycle amplitude and the response speed. A low relay amplitude gives a limit cycle with a low amplitude but also a slow response speed. A large relay amplitude gives a rapid response but also a large amplitude of the limit cycle oscillation. The relations can to some extent be influenced by the lead filter.

Properties of the Basic SOAS

The limit cycle in a system with relay feedback was discussed in Section 8.6. This will now be used to analyze the self-oscillating adaptive system. Consider the system shown in Fig. 10.7 without the gain changer.

The relay is used to introduce a limit cycle oscillation in the system. The period and the amplitude of the oscillation can be determined by the methods discussed in Section 8.6. When the reference signal is changed, or when there are disturbances, there will also be other signals in the system, which will be superimposed on the limit cycle oscillations. The signals that appear in the system will thus be of the form

$$s(t) = a \sin \omega t + b(t)$$

where $a \sin \omega t$ denotes the limit cycle oscillation. The key to understanding the SOAS is to find out how signals of this type propagate in the system. It is straightforward to determine the transmission of the signal through the linear subsystems; the signal propagation through the relay is the main difficulty. This analysis will be simplified considerably if it is assumed that $b(t)$ varies much more slowly than $\sin \omega t$. Furthermore, assume that $b(t)$ is smaller than a. This should be true at least in steady state, since $b(t)$ is the difference between the model output and the process output.

The Dual-Input Describing Function

It is assumed that $b(t)$ varies so slowly that it can be approximated by a constant. The input signal to the relay is thus of the form

$$u(t) = a \sin \omega t + b$$

The relay input and output are shown in Fig. 10.8. The relay output can be expanded in a Fourier series

$$y(t) = bN_B + aN_A \sin \omega t + aN_{A_2} \sin 2\omega t + \cdots \tag{10.2}$$

where the numbers N_A and N_B are given by

$$N_B = \frac{1}{2\pi b} \int_0^{2\pi} y(t)\, dt = \frac{d(\pi + \alpha) - d(\pi - 2\alpha) + \alpha d}{2\pi b}$$

$$= \frac{4\alpha d}{2\pi b} = \frac{2\alpha d}{\pi b} = \frac{2d}{\pi b} \sin^{-1}\left(\frac{b}{a}\right)$$

$$N_A = \frac{1}{\pi a} \int_0^{2\pi} y(t) \sin \omega t\, dt = \frac{2d}{\pi a} \int_\alpha^{\pi - \alpha} \sin \omega t\, dt$$

$$= \frac{4d}{\pi a} \cos \alpha = \frac{4d}{\pi a} \sqrt{1 - (b/a)^2}$$

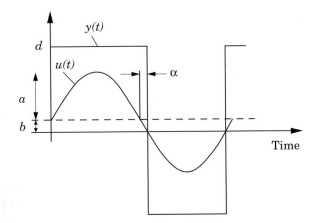

Figure 10.8 Relay inputs and outputs.

Small values of b/a give the approximations

$$N_A \approx \frac{4d}{\pi a} \qquad N_B \approx \frac{2d}{\pi a}$$

Notice that

$$N_A \approx 2N_B \qquad (10.3)$$

The transmission of the constant level b and of the first harmonic $\sin \omega t$ are thus characterized by the equivalent gains N_B and N_A. Since the linear parts will normally attenuate high frequencies more than low frequencies, a reasonable approximation is often obtained by considering only the constant part and the first harmonic. The number N_B, which describes the propagation of a constant signal, is called the *dual-input describing function*, by analogy with the ordinary describing function that describes the propagation of sinusoids through static nonlinearities. Notice that the describing function N_B depends on a (the amplitude of the sinusoidal oscillation). This dependence is the key to understanding how the SOAS works. The dual-input describing function can be used to characterize the transmission of slowly varying signals. A detailed analysis of the accuracy of the approximation is fairly complicated. Let it therefore suffice to mention some rules of thumb for using the approximation. The ratio a/b should be greater than 3, and the ratio of the limit cycle frequency to the signal frequency should also be greater than 3. It is strongly recommended that the analysis be supplemented by simulation.

Main Result

The tools for explaining how the SOAS works are now available. Consider the system in Fig. 10.7. From Section 8.5 the period of the limit cycle is given by

Eqs. (8.8) when the describing function method is used. The amplitude of the limit cycle at the relay input is also given by Eqs. (8.8):

$$N_A|G(i\omega_u)| = 1 \tag{10.4}$$

The transmission of a sinusoidal signal through a relay can thus be approximately described by an equivalent gain, which is inversely proportional to the signal amplitude at the relay input. The amplitude thus automatically adjusts so that the loop gain is unity at the frequency ω_u.

Now consider the propagation of slowly varying signals superimposed on the limit cycle oscillations. The propagation of the signals through the linear parts of the system can be described by the transfer function $G(s)$. If the signals vary slowly in comparison with the limit cycle oscillations, the propagation through the relay is approximately described by the dual-input describing function N_B. The propagation of slowly varying signals is thus approximately described by the loop transfer function

$$G_0(s) = N_B(a)G(s)$$

It follows from Eqs. (10.3) and (10.4) that

$$|G_0(i\omega_u)| = N_B(a)|G(i\omega_u)| = \frac{1}{2} N_A|G(i\omega_u)| = 0.5$$

We thus obtain the following important result, which describes the operation of the SOAS.

RESULT 10.1   Amplitude margin of the SOAS

The SOAS automatically adjusts itself so that the response to reference signals is approximately described by the closed-loop transfer function

$$G_c(s) = \frac{kG(s)}{1 + kG(s)}$$

where the gain k is such that the amplitude margin is 2.   □

This result explains the adaptive properties of the SOAS. The result can also be stated in the following way: The relay acts as a variable gain. The magnitude of the gain depends on the amplitude of the sinusoidal signal at the relay input. This gain is automatically set by the limit cycle oscillation to such a value that the loop gain becomes 0.5 at the frequency of the limit cycle.

The result is illustrated by an example.

EXAMPLE 10.2   A basic SOAS

Assume that the linear parts are characterized by the transfer function

$$G(s) = \frac{K\,\alpha}{s(s+1)(s+\alpha)}$$

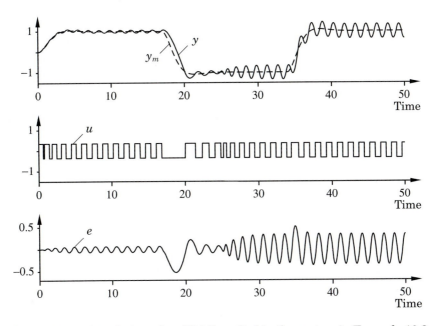

Figure 10.9   Simulation of an SOAS applied to the system in Example 10.2. The dashed line shows the desired response y_m.

From Example 8.1 the period of the limit cycle is approximately given by

$$\omega_u = \sqrt{\alpha}$$

The magnitude of the transfer function at this frequency is

$$|G(i\omega_u)| = \frac{K}{\alpha + 1}$$

If the relay amplitude is d, it follows that the amplitude of the limit cycle oscillation at the relay input is approximately given by

$$e_0 = \frac{Kd}{1 + \alpha}$$

The limit cycle amplitude is thus inversely proportional to α. A simulation of the system is shown in Fig. 10.9. The feedforward transfer function is a second-order system with the damping 0.7 and the natural frequency 1 rad/s. The nominal values of the parameters are $K = 3$, $d = 0.35$, and $\alpha = 20$. The approximate analysis gives a limit cycle with period $T = 1.4$ and amplitude 0.05. The process gain is suddenly increased by a factor of 5 at $t = 25$. Notice the rapid adaptation. However, the amplitude of the oscillation will also increase by a factor of 5. If the value of d is chosen such that the error would be 0.05 for the higher value of K, then the system becomes too slow for small K. $\square$

Design of an SOAS

The self-oscillating adaptive system is a simple nonlinear feedback system that is capable of adapting rapidly to gain variations. The system has a continuous limit cycle oscillation. This is not suitable when valves or other mechanical parts are used as actuators. However, an SOAS may conveniently be used with thyristors as actuators. The presence of the limit cycle oscillation may also cause other inconveniences. Since the system will automatically adjust to an amplitude margin $A_m = 2$, it is also necessary that the characteristics of the process be such that this design principle gives suitable closed-loop properties. The key problem in the design of the SOAS is the compromise between the limit cycle amplitude and the response speed. This compromise is influenced by the selection of the linear compensator, $G_f(s)$, and of the relay amplitude. (Compare Fig. 10.7.) The design for an SOAS can be described by the following procedure.

Step 1:   The relay amplitude is first determined such that the desired control authority (tracking rate, force, speed, etc.) is obtained. This can be estimated by analyzing the response of the process to constant control signals.

Step 2:   When the relay amplitude is specified, the desired limit cycle frequency can be determined from the condition

$$d\,|G_p(i\omega_u)| = e_0$$

where e_0 is the tolerable limit cycle amplitude in the error signal and $G_p(s)$ is the transfer function of the process. It is necessary to check that the frequency obtained is reasonable. For example, the frequency ω_u may become so high that the process dynamics become uncertain.

Step 3:   The final step is to determine the transfer function G_f of the linear compensator such that

$$\arg G_f(i\omega_u) + \arg G_p(i\omega_u) = -\pi$$

A large phase lead may be necessary, but this may not be realizable because of noise sensitivity.

Step 4:   Check that the linear closed-loop system with the loop gain $G_0 = KG_fG_p$ will work well when the gain K is adjusted so that the amplitude margin is 2. If this is not the case, the compensator G_f must be modified.   □

Notice that it is necessary to have an estimate of the magnitude of the process transfer function in Steps 1 and 2. Knowledge of the phase curve of the process transfer function is necessary in the third step. Also notice that it may not be possible to resolve the compromises in all steps. It is then necessary to add additional loops for changing the gain.

Gain Changers

External feedback loops, which adjust the relay amplitude, may be used to resolve the compromise between a high tracking rate and a small limit cycle amplitude. The so-called up-logic used in the first SOAS can be described as follows:

$$d = \begin{cases} d_1 & \text{if } |e| > e_l \\ d_2 + (d_1 - d_2)e^{-(t-t_0)/T} & \text{if } |e| < e_l \end{cases}$$

The time t_0 is the last time that $|e| < e_l$. The relay amplitude is thus increased to d_1 when the error exceeds a limit e_l. The relay amplitude then decreases to a lower level d_2 when the error is less than e_l. This gain changer increases the relay amplitude and the response rate when large reference signals are applied.

Another type of gain changer has been used to control the amplitude of the limit cycle. The limit cycle amplitude at the process output is measured by a band-pass filter and a rectifier. The relay amplitude is then adjusted to keep the limit cycle amplitude constant at the process output.

Dither Signals

In some applications it is desirable to avoid the limit cycle. One idea that has been used successfully is to introduce a variable gain after the relay. The gain is adjusted so that the limit cycle vanishes. In the early applications it was difficult to implement multiplications. A trick that was used to implement the multiplication is illustrated in Fig. 10.10. A high-frequency triangular wave is added to the signal before the relay. With low-pass filtering, the average effect

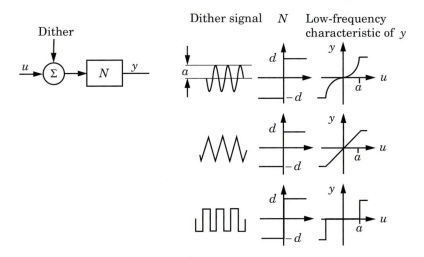

Figure 10.10 The principle of using a dither signal.

of the additive triangular signal is the same as multiplication by a constant. The constant is inversely proportional to the amplitude of the triangular wave. The triangular wave is called a *dither signal*. Use of a dither signal is an illustration of the idea that an oscillation may be quenched by another high-frequency oscillation.

EXAMPLE 10.3   SOAS with lead network and gain changer

The relay control in Example 10.2 gave an error amplitude of about $e_0 = 0.03$. Assume that we want to decrease the amplitude by a factor of 3 while maintaining $d = 0.35$. This gives a new oscillation frequency ω'_u such that

$$d\,|G(i\omega'_u)| = 0.01$$

or $\omega'_u = 10$ rad/s. To get this oscillating frequency, a lead network G_f is added such that

$$\arg G_f(i\omega'_u) + \arg G_p(i\omega'_u) = -\pi$$

Figure 10.11 shows a simulation of the system in Example 10.2 with the compensation network

$$G_f(s) = 1.2\,\frac{s+5}{s+15}$$

As in Fig. 10.9, the gain is increased by a factor of 5 at $t = 25$. It is seen that the lead network decreases the amplitude of the oscillation while maintaining

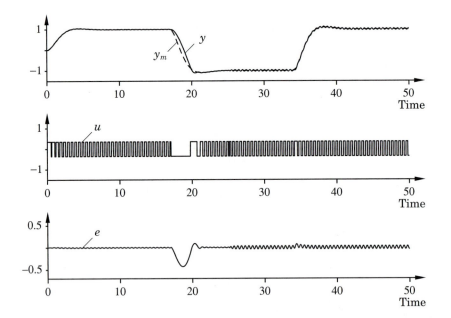

Figure 10.11   Simulation of the system in Example 10.3 using an SOAS with a lead network. The dashed line shows the desired response y_m.

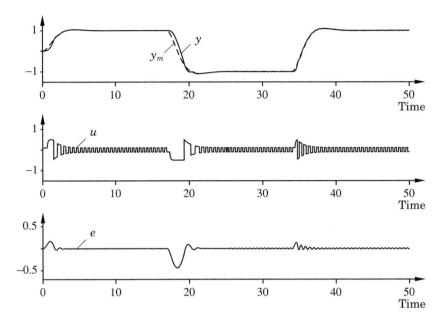

Figure 10.12   Simulation of the system in Example 10.3 using an SOAS with a lead network and a gain changer. The dashed line shows the desired response y_m.

the response speed. To speed up the response, we can introduce the up-logic for the gain. Figure 10.12 shows a simulation in which $d_1 = 0.5$, $d_2 = 0.1$, and $e_l = 0.1$. The error signal is decreased, but there is still an oscillation. The behavior of the closed-loop system can be sensitive to the choice of the parameters in the gain changer. Too large a value of d_1 will cause the error to be larger than e_0, and there will be no decrease in d nor in the amplitude of the error. The oscillation can be quenched by adding a dither signal at the input of the process.   □

The examples show how the properties of the SOAS can be changed by using lead filters, gain changers, and dither signals.

Externally Excited Adaptive Systems

A system that is closely related to the SOAS is obtained by injecting a high-frequency sinusoid to measure the gain of the process and to set the controller gain. Such a system is called an *externally excited adaptive system* (EEAS) and gives the designer more freedom than the SOAS because the frequency of the excitation can be chosen more easily. This system is used for track-keeping in compact disc players. The main source for the parameter variation is a gain variation in the laser diode system.

Summary

The basic SOAS is simple to implement and can cope with large gain changes in the process. Result 10.1 shows that the SOAS will automatically adjust itself so that the amplitude margin is 2. However, the limit cycle in the SOAS is noticeable and can be disturbing. The introduction of lead network, gain changer, and dither can decrease the amplitude of the oscillation. The EEAS is a similar system in which a high-frequency signal is introduced externally.

10.4   VARIABLE-STRUCTURE SYSTEMS

In Section 10.2 we showed how fixed robust controllers can be obtained by increasing the complexity of the controller. Another way to obtain a robust controller is to use a special version of on-off control called a *variable-structure system (VSS)*. The key idea is to apply strong control action when the system deviates from the desired behavior. The name "variable-structure" alludes to the fact that the controller structure may be changed.

Sliding Modes

One way to change the structure of the system is to use different controllers in different parts of the state space of the system. Consider the case in which the control law switches on the surface

$$\sigma(x) = 0$$

Assume that the closed-loop system is described by

$$\frac{dx}{dt} = \begin{cases} f^+(x) & \sigma(x) > 0 \\ f^-(x) & \sigma(x) < 0 \end{cases} \tag{10.5}$$

Two situations may occur, which for the two-dimensional case are shown in Fig. 10.13. In Case (a) the trajectories will pass the switching curve and continue into the other region. However, the dynamics are different in the two regions. In Case (b) the vector fields will drive the state toward the surface $\sigma(x) = 0$. The control will change rapidly from one value to another on the switching surface. This is called *chattering*. The net effect is that the state will move toward the surface $\sigma(x) = 0$ and then slide along the surface. This is called *sliding mode*. This sliding motion can be described as follows: Let f_n denote the projection of f on the normal of the surface $\sigma(x) = 0$. Introduce a number α such that

$$\alpha f_n^+ + (1 - \alpha)f_n^- = 0$$

The sliding motion is then given by

$$\frac{dx}{dt} = \alpha f^+ + (1 - \alpha)f^-$$

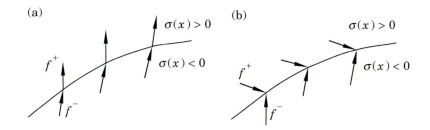

Figure 10.13   The trajectories of Eqs. (10.5) at the switching surface. (a) No sliding mode; (b) sliding mode.

This was formally shown by A. F. Filippov in the early 1960s. The control law giving the sliding motion is sometimes called the equivalent control law. If the switching is not ideal, then the trajectory will move back and forth over the switching surface. This is the case, for instance, when there is hysteresis in the switching. However, on average the motion will be along the switching surface.

Stability and Robustness

In a variable-structure system we attempt to find a switching surface such that the closed-loop system behaves as desired. We now construct a variable-structure controller. For this purpose we assume that the system that we want to control is described by the nonlinear equation

$$\frac{d^n y}{dt^n} = f_1\left(y, \frac{dy}{dt}, \ldots, \frac{d^{n-1}y}{dt^{n-1}}\right) + g_1\left(y, \frac{dy}{dt}, \ldots, \frac{d^{n-1}y}{dt^{n-1}}\right)u$$

If we introduce the state

$$x = \left(\begin{array}{ccccc} \dfrac{d^{n-1}y}{dt^{n-1}} & \dfrac{d^{n-2}y}{dt^{n-2}} & \cdots & \dfrac{dy}{dt} & y \end{array}\right)^T \qquad (10.6)$$

the system can be written as

$$\frac{dx}{dt} = \begin{pmatrix} 0 & 0 & \cdots & 0 & 0 \\ 1 & 0 & \cdots & 0 & 0 \\ \vdots & & & & \vdots \\ 0 & 0 & \cdots & 1 & 0 \end{pmatrix} x + \begin{pmatrix} f_1(x) + g_1(x)u \\ 0 \\ \vdots \\ 0 \end{pmatrix}$$

$$= f(x) + g(x)u$$

$$y = \begin{pmatrix} 0 & 0 & \cdots & 0 & 1 \end{pmatrix} x \qquad (10.7)$$

where $f(x)$ and $g(x)$ are vectors. The system is nonlinear but is affine in the control signal. Further, it is assumed that all the states can be measured and that the state vector has the special form given in Eq. (10.6). For simplicity it is assumed that the purpose of the control is to find a controller such that $x = 0$ is an asymptotically stable solution. The problem with constant reference signals is considered in Problem 10.14 at the end of the chapter.

There are three important questions that must be answered for VSS:

- Will the trajectories starting at any point hit the switching line?
- Is there a sliding mode?
- Is the sliding mode stable?

There are partial answers to these questions in the literature on VSS. For the special type of system defined by Eqs. (10.7) it is easy to derive a controller that makes the sliding mode stable.

Let the switching surface be

$$\sigma(x) = p_1 x_1 + p_2 x_2 + \cdots + p_n x_n = p^T x = 0 \tag{10.8}$$

Using the definition of the state vector, we find that

$$\sigma(x) = p_1 y^{(n-1)} + p_2 y^{(n-2)} + \cdots + p_n y = 0$$

The dynamic behavior on the sliding surface can be specified by a proper choice of the numbers p_i. The motion is determined by a differential equation of order $n - 1$. It will be stable if the polynomial

$$P(s) = p_1 s^{n-1} + p_2 s^{n-2} + \cdots + p_n \tag{10.9}$$

has all its roots in the left-half plane.

To determine a control law that keeps the system on $\sigma(x) = 0$, we introduce the Lyapunov function $V(x) = \sigma^2(x)/2$. The time derivative of V is given by

$$\frac{dV}{dt} = \sigma(x)\dot{\sigma}(x) = x^T pp^T \dot{x}$$
$$= x^T p \left(p^T f(x) + p^T g(x) u(t) \right)$$

Choose the control law

$$u(t) = -\frac{p^T f}{p^T g} - \frac{\mu}{p^T g} \operatorname{sign}(\sigma(x)) \tag{10.10}$$

Then

$$\frac{dV}{dt} = -\mu\sigma(x)\operatorname{sign}(\sigma(x)) \tag{10.11}$$

which is negative definite. This implies that $\sigma(x) = 0$ is asymptotically stable. Notice that there is a discontinuity in the control signal when the switching surface is passed.

Assume that the system has initial values such that $\sigma(x) = \sigma_0 > 0$, and let t_σ be the time when the switching surface is reached the first time. From Eq. (10.11) we find that

$$\dot{\sigma}(x) = -\mu$$

Integrating this equation from 0 to t_σ gives

$$0 - \sigma_0 = -\mu(t_\sigma - 0)$$

which gives $t_\sigma = \sigma_0/\mu$. Using the same arguments for $\sigma_0 < 0$ shows that $t_\sigma = |\sigma_0|/\mu$. With the control law given by Eq. (10.10) the state will thus reach the switching surface in finite time. The subspace $\sigma(x) = 0$ is asymptotically stable, and the state will stay on the switching surface once it is reached. The motion along the surface is determined by Eq. (10.9).

Uncertainties in f and g can be handled if μ is sufficiently large. Assume that the design of the control law is based on the approximate values $\hat{f}$ and $\hat{g}$ instead of the true ones. Then

$$\frac{dV}{dt} = \sigma \left(\frac{p^T \left(f\hat{g}^T - \hat{f}g^T \right) p}{p^T \hat{g}} - \mu \frac{p^T g}{p^T \hat{g}} \, \text{sign}(\sigma) \right)$$

The right-hand side is negative if μ is sufficiently large, provided that $p^T \hat{g}$ and $p^T g$ have the same sign. The system will thus be insensitive to uncertainties in the process model.

One way to design a variable-structure system is to first transform the system to the form given by Eqs. (10.7). This is possible for controllable linear systems without zeros and for some classes of nonlinear systems. A stable switching surface in the transformed variables is then determined. The system and the switching criteria can then be transformed back to the original state variables. Notice, however, that all states must be measured.

Smooth Control Laws

The control law (10.10) has the drawback that the relay chatters. One way to avoid this is to make the relay characteristics smoother. To do this, introduce a boundary layer around the switching surface

$$B(t) = \left\{ x(t) \,\middle|\, |\sigma(x(t))| \leq \varepsilon \right\} \quad \varepsilon \geq 0$$

The parameter ε can be interpreted as a measure of the thickness of the boundary layer. The sign function in Eq. (10.10) is now replaced by the saturation function

$$\text{sat}(\sigma, \varepsilon) = \begin{cases} 1 & \sigma > \varepsilon \\ \sigma/\varepsilon & -\varepsilon \leq \sigma \leq \varepsilon \\ -1 & \sigma < -\varepsilon \end{cases}$$

The control law is then

$$u(t) = \frac{p^T f}{p^T g} - \frac{\mu}{p^T g} \, \text{sat}(\sigma(x), \varepsilon) \tag{10.12}$$

The width of the boundary layer will influence the tracking performance and the bandwidth of the closed-loop system.

EXAMPLE 10.4   **Second-order VSS**

Consider the unstable system

$$\frac{dx}{dt} = \begin{pmatrix} 1 & 0 \\ 1 & 0 \end{pmatrix} x + \begin{pmatrix} 1 \\ 0 \end{pmatrix} u = Ax + Bu$$

$$y = \begin{pmatrix} 0 & 1 \end{pmatrix} x$$

which has the transfer function

$$G(s) = \frac{1}{s(s-1)}$$

To design a variable-structure controller we determine the closed-loop dynamics by choosing the switching line

$$\sigma(x) = p_1 x_1 + p_2 x_2 = x_1 + x_2$$

Along the sliding line $\sigma = 0$ we have

$$\sigma(x) = x_1 + x_2 = \frac{dy}{dt} + y = 0$$

Since the system is in controllable form, the closed-loop behavior is independent of the system parameters at the sliding mode. The sliding mode controller

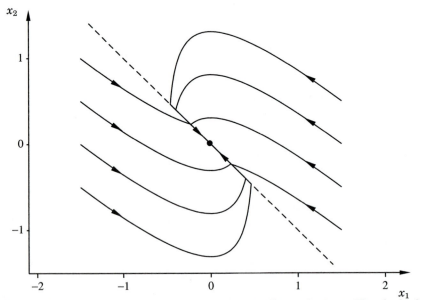

Figure 10.14 Phase portrait of the system in Example 10.4. The dashed line shows $\sigma(x) = 0$.

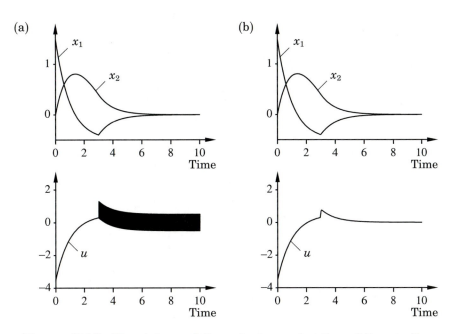

Figure 10.15 The states and the output as a function of time in Example 10.4. The initial conditions are $x_1(0) = 1.5$ and $x_2(0) = 0$. The controllers are (a) Eq. (10.10) with $\mu = 0.5$; (b) Eq. (10.12) with $\mu = 0.5$ and $\varepsilon = 0.01$.

(10.10) is now

$$u(t) = -\frac{p^T A x}{p^T B} - \mu \operatorname{sign}(\sigma(x))$$

$$= - \begin{pmatrix} 2 & 0 \end{pmatrix} x(t) - \mu \operatorname{sign}(\sigma(x))$$

The phase plane of the system is shown in Fig. 10.14 when $\mu = 0.5$. The input and the output for one initial value are shown in Fig. 10.15(a). The trajectories hit the switching line $\sigma = 0$ and stay on it. This implies that the control signal will chatter. Using the control law (10.12),

$$u(t) = - \begin{pmatrix} 2 & 0 \end{pmatrix} x(t) - \mu \operatorname{sat}(\sigma(x), \varepsilon)$$

with $\varepsilon = 0.01$, gives the behavior shown in Fig. 10.15(b). The control signal is now smooth, but the differences in the state trajectories are negligible. □

Summary

Variable-structure systems are related to the self-oscillating adaptive systems (SOAS). In variable-structure systems we want the system to get into a sliding mode to obtain insensitivity to parameter variations. The control signal of variable-structure systems will chatter in the sliding mode. The chatter can be avoided by smoothing the relay characteristics. The amplitude of the control signal is determined by the magnitude of the state variables or the error. With this modification the variable-structure system can be regarded as an SOAS in which the relay amplitude depends on the states. The switching condition is a linear function of the error in the SOAS, while in variable-structure systems it is a nonlinear function of the states.

The theory on VSS can be extended to controllers, in which the feedback is done from a reduced number of state variables. However, the conditions will become more complex than those discussed in this section. There will be more constraining conditions on the choice of the switching plane. Since the conditions for the existence of a sliding mode depend on the process and the switching plane, there have been attempts to make adaptive VSS by adaptation on $\sigma(x)$.

One main drawback of variable-structure systems is the problem of choosing the switching plane. It also requires measurement of all state variables. Another drawback is the chatter in the control signal in the sliding mode.

10.5   CONCLUSIONS

Robust high-gain control can be very effective for systems with structured parameter variations, where the range of the variations is known. If the parameter bounds are uncertain, high-gain design methods will lead to a complex and conservative design. Relay feedback is an extreme form of high-gain systems. In this chapter we have described different ways to use relay feedback to obtain systems that are insensitive to parameter variations. Self-oscillating adaptive systems and variable-structure systems are two applications of this idea. The SOAS can be designed to work quite well, but it requires engineering effort and some knowledge of the process to get a satisfactory performance of the closed-loop system. These drawbacks have resulted in lack of interest in the SOAS. However, the ideas behind SOAS have become useful in connection with the auto-tuning of simple controllers, as discussed in Chapter 8.

PROBLEMS

10.1   Determine whether each of the following plants can be stabilized by a linear fixed-parameter compensator when $a \in [-1, 1]$:
(a) a/s;   (b) $1/(s + a)$;   (c) $1/(1 + as)$;   (d) $a/(1 + s)$;
(e) $a/(1 - s)$.

10.2   Consider the process

$$G_p(s) = e^{-sT} \qquad T \in [0, 1]$$

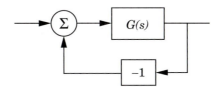

Figure 10.16 The system in Problem 10.3.

(a) Show that the process can be controlled by a controller of the structure in Fig. 10.1 with

$$G_{fb}(s) = \frac{0.6(1 + s/1.3)}{s(1 + s/2)}$$

$$G_{ff}(s) = \frac{1 + s}{1 + s/3}$$

(b) Simulate the behavior for changes in the command signal and step disturbances at the output.

(c) Discuss how to make a self-tuning regulator based on pole placement for the process.

10.3 Consider the linear closed-loop system shown in Fig. 10.16 with the same $G(s)$ as in Example 10.2 and with $\alpha = 20$. Determine the gain K so that the amplitude margin is $A_m = 2$. Simulate the system and determine its step response. Compare this with the step response of the corresponding SOAS in Example 10.2.

10.4 Consider a linear plant with the transfer function

$$G(s) = \frac{k}{s(s + 1)^2}$$

where the gain k may vary in the range $0.1 \le k \le 10$. Determine the relay amplitude d and a suitable lead network so that the limit cycle amplitude at the process output is less than 0.05 and the rise time to a step of unit amplitude is never less than 0.5. Simulate the resulting design and verify the results.

10.5 Consider the system in Example 10.2. Experiment with a gain changer of the up-logic type. Investigate how a dither signal will influence the performance of the closed-loop system.

10.6 Consider the system in Problem 10.4. Design a gain changer that keeps the limit cycle amplitude at 0.01 for the whole operating range.

10.7 Consider a system with the transfer function

$$G(s) = \frac{k}{s + 1}$$

where the gain k may change in the range 0.1 to 10. Design a servo using the SOAS principle so that the closed-loop transfer function is

$$G(s) = \frac{1}{s^2 + s + 1}$$

independent of the process gain.

10.8 Consider the system in Example 10.4 and assume that the controller (10.10) is used with $\mu = 0.5$. Assume that the process is changed such that

$$\frac{dx_1}{dt} = ax_1 + bu$$

Determine values of a and b such that the closed-loop system is still asymptotically stable.

10.9 Assume that the nth-order single-input, single-output system

$$\frac{dx}{dt} = Ax + Bu$$

is in companion form and that the control law is of the form

$$u = -\sum_{i=1}^{n} l_i x_i - \mu \operatorname{sign}(\sigma(x))$$

Derive the necessary and sufficient conditions for the existence of a sliding mode. When will the sliding mode be stable?

10.10 Consider the process in Problem 1.9. Design a robust controller for the system. Investigate the disturbance rejection of the closed-loop system.

10.11 Consider the process in Problem 1.10. Design a robust controller for the system. Investigate the disturbance rejection of the closed-loop system.

10.12 Design an SOAS for the system in Problem 1.9, and investigate its properties.

10.13 Design an SOAS for the system in Problem 1.10, and investigate its properties.

10.14 Consider the process in Eqs. (10.7). Assume that the reference value is constant y_r. The desired state is then

$$x_d = \begin{pmatrix} 0 & \cdots & 0 & y_r \end{pmatrix}$$

Determine a controller such that $x = x_d$ is an asymptotically stable solution. (*Hint:* Introduce the state error $\tilde{x} = x - x_d$, and consider the Lyapunov function $V(\tilde{x}) = \sigma^2(\tilde{x})/2$.)

REFERENCES

The robust high-gain design is closely related to early ideas on feedback amplifiers. See:

Bode, H. W., 1945. *Network Analysis and Feedback Amplifier Design.* New York: Van Nostrand.

The basic robust design method for SISO systems for specifications in the frequency domain is discussed in:

Horowitz, I. M., and M. Sidi, 1972. "Synthesis of feedback systems with large plant ignorance for prescribed time-domain tolerances." *Int. J. Control* **16**: 287–309.

Horowitz, I. M., 1973. "Optimum loop transfer function in single-loop minimum-phase feedback systems." *Int. J. Control* **18**: 97–113.

Horowitz, I. M., and U. Shaked, 1975. "Superiority of transfer function over state-variable methods in linear time-invariant feedback system design." *IEEE Trans. Automat. Contr.* **AC-20**: 84–97.

Horowitz, I. M., and M. Sidi, 1978. "Optimum synthesis for non-minimum phase feedback systems with plant uncertainty." *Int. J. Control* **27**: 361–386.

This last paper presents criteria for determining whether a given set of performance specifications are achievable, and, if so, a synthesis procedure is included for deriving the optimum design, which is defined as that with an effectively minimum-loop transmission bandwidth. The theory on QFT is summarized in:

Horowitz, I. M., 1993. *Quantitative Feedback Design Theory (QFT)*, Vol 1. Boulder, Colo.: QFT Publications.

Robust design of processes with unstructured uncertainties is treated in:

Doyle, J. C., and G. Stein, 1981. "Multivariable feedback design: Concepts for a classical/modern synthesis." *IEEE Trans. Automat. Contr.* **AC-26**: 4–16.

Morari, M., and J. C. Doyle, 1986. "A unifying framework for control system design under uncertainty and its implications for chemical process control." In *Chemical Process Control: CPCIII*, eds. M. Morari and T. McAvoy, Proceedings of the 3rd International Conference on Chemical Process Control. New York: Elsevier.

Research on relay systems was very active in the 1950s and 1960s. An authoritative treatment by one of the key contributors is:

Tsypkin, Y. Z., 1984. *Relay Control Systems.* Cambridge, U.K.: Cambridge University Press.

The method of harmonic balance and describing function is extensively treated in:

Gelb, A., and W. E. Vander Velde, 1968. *Multiple-Input Describing Functions and Nonlinear System Design*, pp. 18, 273, 308, 317, and 588. New York: McGraw-Hill.

This book also contains applications of SOAS. The background of SOAS and more about the design rules can be found in:

Lozier, J. C., 1950. "Carrier-controlled relay servos." *Elec. Eng.* **69**: 1052–1056.

Schuck, O. H., 1959. "Honeywell's history and philosophy in the adaptive control field." In *Proceedings of the Self Adaptive Flight Control Symposium,* ed. P. C. Gregory. Wright Patterson AFB, Ohio: Wright Air Development Center.

Horowitz, I. M., 1964. "Comparison of linear feedback systems with self-oscillating adaptive systems." *IEEE Trans. Automat. Contr.* **AC-9**: 386–392.

Horowitz, I. M., J. W. Smay, and A. Shapiro, 1974. "A Synthesis Theory for Self-Oscillating Adaptive Systems (SOAS)." *Automatica* **10**: 381–392.

A modified version of the SOAS was flight-tested extensively in the experimental rocket plane X-15. Experiences from that are summarized in:

Thompson, M. O., and J. R. Welsh, 1970. "Flight test experience with adaptive control systems." *Proceedings of the Agard Conference on Advanced Control Systems Concepts*, vol. 58, pp. 141–147. Agard, Neuilly-sur-Seine, France.

The general conditions for stability of the limit cycle in relay systems are still unknown. Some guidance is given by the stability conditions in:

Åström, K. J., and T. Hägglund, 1984. "Automatic tuning of simple regulators." *Proceedings of the IFAC 9th World Congress.* Vol. 3, pp. 267–272. Budapest.

Additional results on relay oscillations are found in:

Amsle, B. E., and R. E. Gorozdos, 1959. "On the analysis of bi-stable control systems." *IEEE Trans. Automat. Contr.* **AC-4**: 46–58.

Gille, J. C., M. J. Pelegrin, and P. Decaulne, 1959. *Feedback Control Systems.* New York: McGraw-Hill.

Atherton, D. P., 1975. *Nonlinear Control Engineering: Describing Function Analysis and Design.* London: Van Nostrand Reinhold.

Atherton, D. P., 1982. "Limit cycles in relay systems." *Electronics Letters* **1**(21).

A procedure for designing externally excited adaptive systems (EEAS) is given in:

Horowitz, I. M., J. W. Smay, and A. Shapiro, 1957. "A synthesis theory for the externally excited adaptive system (EEAS)." *IEEE Trans. Automat. Contr.* **AC-2**: 101–107.

Variable-structure systems are treated in:

Emelyanov, S. V., 1967. *Variable Structure Control Systems.* Munich: Oldenburger Verlag.

Itkis, U., 1976. *Control Systems of Variable Structure.* New York: Halsted Press, Wiley.

Utkin, V. I., 1977. "Variable structure systems with sliding modes." *IEEE Trans. Automat. Contr.* **AC-22**: 212–222.

Slotine, J.-J. E., and W. Li, 1991. *Applied Nonlinear Control.* Englewood Cliffs, N.J.: Prentice-Hall.

The sufficient conditions of the existence of a sliding mode is found in:

Gough, N. E., Z. M. Ismail, and R. E. King, 1984. "Analysis of variable structure systems with sliding modes." *Int. J. Systems Sci.* **15**(4): 401–409.

Adaptive variable-structure systems are discussed in:

Young, K.-K. D., 1978. "Design of variable structure model-following control systems." *IEEE Trans. Automat. Contr.* **AC-23**: 1079–1085.

Zinober, A. S. I., 1981. "Adaptive variable structure systems." In *Proceedings of the Third IMA Conference on Control Theory*, eds. J. E. Marshall, W. D. Collins, C. J. Harris, and D. H. Owens. London: Academic Press.

An overview and a presentation of several industrial applications of variable-structure systems are given in the survey paper:

Utkin, V. I., 1987. "Discontinuous control systems: State of the art in theory and applications." *Preprints of the IFAC 10th World Congress.* Vol. 1, pp. 75–94. Munich.

CHAPTER 11

PRACTICAL ISSUES AND IMPLEMENTATION

11.1 INTRODUCTION

The previous chapters were devoted mainly to development of algorithms and analysis of adaptive systems. In this chapter we discuss practical implementation of adaptive controllers. The presentation will be guided by theoretical considerations, but since the issues are quite complicated, theory can cover only part of the problems. Several issues will therefore be solved in an ad hoc manner and verified by extensive experimentation and simulation.

Since an ordinary digital controller is an integral part of an adaptive controller, it is essential to master implementation of conventional controllers. Some aspects of this are covered in Section 11.2. The discussion includes computational delay, sampling, prefiltering and postfiltering, and integrator windup. Automatic design of a controller is another important part of an adaptive controller. In this book we have mostly used simple design methods based on pole placement. In Section 11.3 we discuss how this design method can be modified in several ways to accommodate more complex specifications and to make it more robust. The design technique requires the solution of a Diophantine equation. Efficient numerical methods for doing this are discussed in Section 11.4.

Parameter estimation is another important part of an adaptive controller. Implementation of estimators is discussed in Section 11.5. This includes choice of model structure, data filters, and excitation. The ability of an adaptive controller to track time-varying parameters is an important issue. There are several ways to do this. Two techniques, exponential forgetting and covariance resetting, are discussed in detail. It turns out that exponential forgetting in combination with poor excitation can give rise to an undesirable effect called covariance windup. This phenomenon is discussed in detail together with several ways of avoiding it. The techniques discussed include constant trace algorithms, directional forgetting, and leakage. A technique to make the estimator less sensitive to outliers is also discussed in Section 11.5. It is important to have numerically efficient methods for recursive estimation. Square root algorithms, which are numerically superior to the conventional algorithms, are discussed in Section 11.6.

In Section 11.7 we discuss the interaction between estimation and control. We show that difficulties can arise if integral action is implemented inappropriately. We also show how the criteria for control and estimation can be made compatible by appropriate choices of the data filter and the experimental conditions.

Some prototype algorithms are given in Section 11.8. To implement a control system successfully, it is necessary to consider all situations that may occur in practice. Conventional controllers typically have two operating modes: manual and automatic. Adaptive controllers have many more modes. This is discussed briefly in Section 11.9, which covers startup, shut-down, and switching between different modes. Supervision of adaptive algorithms is also discussed in that section.

11.2   CONTROLLER IMPLEMENTATION

An ordinary controller is an important part of an adaptive controller. Compare with the block diagram in Fig. 1.1. It is, of course, important that the controller be implemented in a good way. Implementation of digital controllers is treated in books on digital control. Some aspects are summarized in this section.

Computational Delay

Because the analog-to-digital (A-D) and digital-to-analog (D-A) conversions and the computations take time, there will always be a delay between the measurement and the time the control signal is applied to the process. This delay, which is called the *computational delay*, depends on how the control law is implemented in the computer. Two ways are illustrated in Fig. 11.1. In Case (a) the measured variable at time t_k is used to compute the control signal applied at time t_{k+1}. In Case (b) the control signal is applied as soon as the computations are finished. The disadvantage of Case (a) is that the control action is delayed unnecessarily. In Case (b) the disadvantage is that the time delay may change, depending on the load on the computer or changes in the program. In both cases it can be necessary to include the computational delay in the design of the controller.

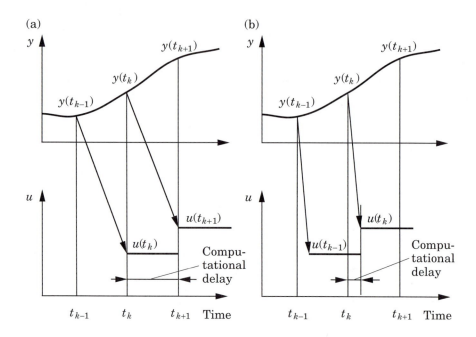

Figure 11.1 Two ways to synchronize inputs and outputs. (a) The signals measured at time t_k are used to compute the control signal applied at t_{k+1}. (b) The control signal is applied as soon as it is computed.

In Case (b) it is desirable to make the delay as small as possible. This can be done by performing as few operations as possible between the A-D and D-A conversions. Assume that the regulator has the form

$$u(t) + r_1 u(t - 1) + \ldots + r_k u(t - k)$$
$$= t_0 u_c(t) + \ldots + t_m u_c(t - m) - s_0 y(t) - \ldots - s_l y(t - l)$$

This equation can be written as

$$u(t) = t_0 u_c(t) - s_0 y(t) + u'(t - 1)$$

where

$$u'(t - 1) = t_1 u_c(t - 1) + \ldots + t_m u_c(t - m) - s_1 y(t - 1) - \ldots - s_l y(t - l)$$
$$- r_1 u(t - 1) - \ldots - r_k u(t - k)$$

Notice that the signal $u'(t - 1)$ contains information that is available at time $t - 1$. An implementation of the control algorithm that exploits this to make the computational delay as small as possible is the following:

1. Perform A-D conversion of $y(t)$ and $u_c(t)$.
2. Compute $u(t) = t_0 u_c(t) - s_0 y(t) + u'(t - 1)$.

3. Perform D-A conversion of $u(t)$.

4. Compute

$$u'(t) = t_1 u_c(t) + \ldots + t_m u_c(t - m + 1) - s_1 y(t) - \ldots - s_l y(t - l + 1)$$
$$- r_1 u(t) - \ldots - r_k u(t - k + 1)$$

The computational delay can thus be significantly reduced by a proper implementation of the controller. In a proper implementation the delay can be reduced. Apart from the two multiplications and the two additions, $u(t)$ must be tested for limitations and anti-reset windup must be done in a proper way. Since the computational delay appears in the same way as a time delay in the process dynamics, it is important to take it into account in designing a control system. A common rule of thumb is that the time delay can be neglected if it is less than 10% of the sampling period. For high-performance systems it should always be taken into account. Since the time delay is not known until the algorithm has been coded, the control design may have to be repeated. For an adaptive system it is important that the model structure is chosen so that the computational delay can be accommodated. In multitasking systems it may also happen that the computational delay varies with time.

Sampling and Pre- and Postfiltering

The choice of sampling rate is an important issue in digital control. The sampling rate influences many properties of a system such as following of command signals, rejection of load disturbances and measurement noise, and sensitivity to unmodeled dynamics. Selection of sampling rates is thus an essential design issue.

One rule of thumb that is useful for deterministic design methods is to let the sampling interval h be chosen such that

$$\omega_o h \approx 0.2 - 0.6$$

where ω_o is the natural frequency of the dominating poles of the closed-loop system. This corresponds to 12–60 samples per undamped natural period. The sampling frequency is $\omega_s = 2\pi/h$.

In all digital systems it is important that signals are filtered before they are sampled. All components of the signal with frequencies above the Nyquist frequency, $\omega_N = \omega_s/2 = \pi/h$ should be eliminated. If this is not done, a signal component with frequencies $\omega > \omega_N$ will appear as low-frequency components with the frequency

$$\omega_a = |((\omega + \omega_N) \bmod \omega_s) - \omega_N|$$

This phenomenon is called *aliasing*, and the prefilters introduced before a sampler are called anti-aliasing filters. Suitable choices of anti-aliasing filters

Table 11.1 Damping and natural frequency of second-, fourth-, and sixth-order Butterworth, ITAE, and Bessel filters. The filters have the bandwidth ω_B.

Order	Butterworth ω/ω_B	ζ	ITAE ω/ω_B	ζ	Bessel ω/ω_B	ζ
2	1	0.71	1	0.71	1.27	0.87
4	1	0.38	1.48	0.32	1.59	0.62
	1	0.92	0.83	0.83	1.42	0.96
6	1	0.26	1.30	0.32	5.14	0.49
	1	0.71	0.98	0.60	4.57	0.82
	1	0.97	0.79	0.93	4.34	0.98

are second- or fourth-order Butterworth, ITAE (integral time absolute error), or Bessel filters. They consist of one or several cascaded filters of the form

$$G_f(s) = \frac{\omega^2}{s^2 + 2\zeta\omega s + \omega^2}$$

Let ω_B be the desired bandwidth of the filter. The damping ζ and the frequency ω for filters of different orders are given in Table 11.1. The Bessel filter has the interesting property that its phase curve is approximately linear, which implies that the waveform is also approximately invariant.

The prefilter introduces additional dynamics into the system that have to be taken into account in the control design. The Bessel filter can be approximated with a time delay. Assume that the bandwidth of the filter is chosen to be

$$|G_{aa}(i\omega_N)| = \beta$$

where $G_{aa}(s)$ is the transfer function of the filter and $\omega_N = \pi/h$ is the Nyquist frequency. Parameter β is the attenuation of the filter at the Nyquist frequency. Table 11.2 gives the approximate time delay T_d as a function of β. The table

Table 11.2 The approximate time delay T_d due to the anti-aliasing filter as a function of the desired attenuation β at the Nyquist frequency for a fourth-order Bessel filter. h is the sampling period.

β	ω_N/ω_B	T_d/h
0.05	3.1	2.1
0.1	2.5	1.7
0.2	2.0	1.3
0.5	1.4	0.9
0.7	1.0	0.7

also gives ω_N as a function of filter bandwidth ω_B. The relative delay increases with attenuation. For reasonable values of the attenuation the delay is more than one sampling period. This means that the dynamics of the filter must be taken into account in the control design. We illustrate this by an example.

EXAMPLE 11.1   The effect of the anti-aliasing filter

Consider a process described by

$$G(s) = \frac{1}{s(s+1)}$$

A pole placement controller is designed to give a closed-loop system whose dominant poles are given by $\omega_m = 1$ rad/s and $\zeta_m = 0.7$. The digital controller has a sampling period of $h = 0.5$. To illustrate the effect of aliasing, we assume that the output of the system is disturbed by a sinusoidal signal; that is, the measured signal is

$$y_m(t) = y(t) + a_d \sin(\omega_d t)$$

with $a_d = 0.1$. This signal is filtered through a fourth-order Bessel filter with bandwidth ω_B. Figure 11.2 shows a number of simulations of the system that

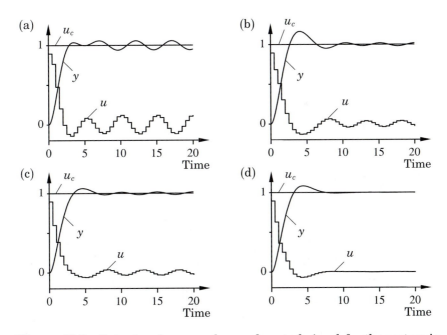

Figure 11.2   Output, reference value, and control signal for the system in Example 11.1. The measurement disturbance has the frequency $\omega_d = 11.3$ rad/s. (a) $\omega_B = 25$ rad/s; (b) $\omega_B = 6.28$ rad/s; (c) $\omega_B = 6.28$ rad/s and the regulator compensated for a delay of $0.7h$; (d) $\omega_B = 2.51$ rad/s and the regulator compensated for a delay of $1.7h$.

illustrate the effects of aliasing and filtering. Figure 11.2(a) shows setpoint, process output, and control signal. Since the bandwidth of the filter is $\omega_B = 25$, the measured signal is not attenuated much by the filter. The disturbance with frequency $\omega_d = 11.3$ is aliased to 1.3 rad/s because of the sampling with Nyquist frequency $\omega_N = 2\pi$. The aliased disturbance is clearly visible in the process input and output. In Fig. 11.2(b) the bandwidth of the prefilter is reduced to $\omega_B = 6.28$ rad/s. This bandwidth is not sufficiently small to give a substantial reduction of the disturbance. Notice also that the overshoot has increased a little because of the delay in the prefilter. In Fig. 11.2(c) the dynamics of the prefilter have been taken into account by adding a time delay of $0.7h$ in the model. The overshoot is then reduced, but the effect of the disturbance is similar to Fig. 11.2(b). In Fig. 11.2(d) the filter bandwidth has been reduced to $\omega_B = 2.51$. The disturbance is now reduced significantly. We have also taken the dynamics of the filter into account as a time delay of $1.7h$ in designing the controller. The aliased disturbance is now barely noticeable in the figure.   □

Example 11.1 shows that it is important to use an anti-aliasing filter and that the filter has to be considered in the design. For a Bessel filter, however, it is sufficient to approximate the filter by a time delay. The additional dynamics cause no principle problems for an adaptive controller because all parameters are estimated. However, inclusion of the prefilter dynamics will increase the model order significantly. In the particular case of Example 11.1 the model will increase from second order to sixth order. This means that the number of parameters that we have to estimate increases from 4 to 12. A simple way to reduce the number of parameters is to approximate the prefilter by a delay. It is then sufficient to estimate five parameters of the model

$$y(t) + a_1 y(t - h) + a_2 y(t - 2h) =$$
$$b_0 u(t - dh) + b_1 u(t - dh - h) + b_2 u(t - dh - 2h)$$

where the value of d depends on the bandwidth of the filter (see Table 11.2).

It is cumbersome and costly to change the bandwidth of an analog prefilter. This poses problems for systems in which the sampling rate has to be changed. A nice implementation in such a case is to use dual rate sampling. A high fixed sampling rate is used together with a fixed analog prefilter. A digital filter is then used to filter the signal at a slower rate when that is needed. This implies that fewer parameters have to be estimated.

The output of a D-A converter is a piecewise constant signal. This means that the control signal fed to the actuator is a piecewise constant signal that changes stepwise at the sampling instants. This is adequate for many processes. However, for some systems, such as hydraulic servos for flight control and other systems with poorly damped oscillatory modes, the steps may excite these modes. In such a case it is advantageous to use a filter that smooths the signal from the D-A converter. Such a filter is called a *postsampling filter*. The postsampling filter may be a simple continuous-time filter with a response time that is short in comparison with the sampling time. Special D-A converters that give a smooth signal have also been constructed. Another solution to the problem is to use a system with dual rate sampling. The primary control system should then be designed so that the output is piecewise linear between the sampling instants. A fast sampling can then be used to generate an approximation to this signal, possibly followed by an analog postsampling filter.

Controller Windup

Linear theory is adequate to deal with many problems in control system design. There is, however, one nonlinear problem that we must deal with in almost all practical control systems, and that is actuator saturation. The feedback will be broken when the actuator saturates. Large deviation may then occur if the process or the controller is unstable. A simple case in which this occurs is when the controller has integral action. The phenomenon was first observed in connection with PID control. It is therefore often called *integrator windup* because the integral term "winds up" when the actuator is saturated. Since integral action was also called reset, the phenomenon is also called *reset windup*. It is necessary to include a scheme for avoiding windup in systems in which the process and/or the controller is unstable.

There are many different ways to introduce anti-reset windup. One simple way is based on the interpretation of a controller as a combination of a state estimator and state feedback. Such a system is shown in Fig. 11.3. The controller is composed of two components, a state estimator and a state feedback. The state estimator determines an estimate of the state based on the process input and output. The state feedback generates the control signal based on the estimated state. It is intuitively clear from the figure that the state estimator will perform poorly when the actuator saturates because it is uses a wrong

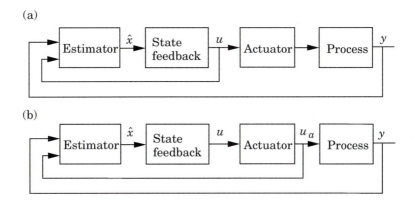

Figure 11.3 Block diagrams of controllers based on state feedback and state estimation with a process having a nonlinear actuator.

value of the control signal applied to the process. This interpretation also suggest that the problem can be avoided by feeding back the actual process input, u_a, or an estimate of it as in Fig. 11.3(b).

Since we are using polynomial representations of the controller in this book, we also give a polynomial interpretation of the scheme. Consider the controller

$$R(q)u(t) = T(q)u_c(t) - S(q)y(t)$$

where the polynomial $R(q)$ is assumed to be monic. The controller can be written in observer form as

$$A_o(q)u(t) = T(q)u_c(t) - S(q)y(t) + (A_o(q) - R(q))u(t)$$

where $A_o(q)$ is the observer polynomial. Let the saturating actuator be described by the nonlinear function $f(u)$. A controller that avoids windup is then given by

$$
\begin{aligned}
A_o(q)v(t) &= T(q)u_c(t) - S(q)y(t) + (A_o(q) - R(q))u(t) \\
u(t) &= f(v(t))
\end{aligned}
\tag{11.1}
$$

A similar scheme can be used when the saturation is dynamic. Notice that the controller responds with the observer dynamics when the feedback is broken. A particularly simple case is when $A_o^* = 1$, which corresponds to a deadbeat

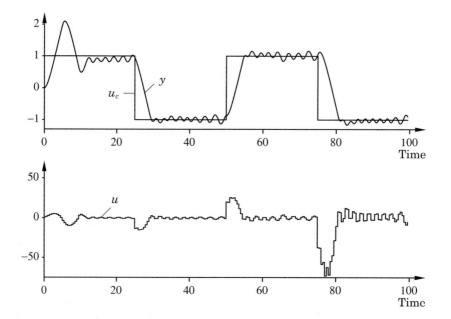

Figure 11.4   Simulation of adaptive control of the unstable process with a saturating actuator in Example 11.2.

observer. The controller is then

$$u(t) = f\Big(T^*(q^{-1})u_c(t) - S^*(q^{-1})y(t) + (1 - R^*(q^{-1}))u(t)\Big)$$

We illustrate windup by an example.

EXAMPLE 11.2   Windup and how to avoid it

Consider the simple example of adaptive control in Example 3.5. The process has the transfer function

$$G(s) = \frac{1}{s(s+1)}$$

Assume that there is an actuator that saturates when the magnitude of the control signal is 0.5. Figure 11.4 shows the behavior of the system if no precautions are taken in the controller. The figure clearly shows the detrimental effects of actuator saturation. The process runs open loop when the actuator saturates, and the output is drifting because the process has an integrator. This will also happen with a controller with fixed parameters. With an adaptive controller the saturation also causes the gain parameters (b_0 and b_1) to be underestimated. The controller gain is then too high, and the system be-

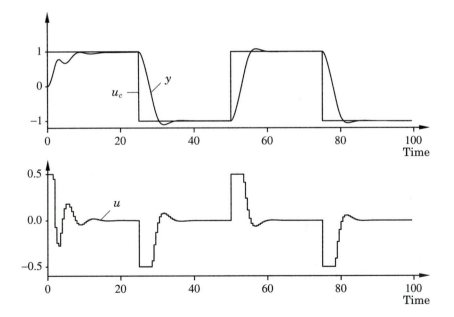

Figure 11.5 Simulation of an adaptive controller for a process with a saturating actuator with a controller having windup protection. Compare with Fig. 11.4, which shows the same simulation for a system without windup protection.

comes unstable. Windup is thus much more serious in adaptive control than in a controller with constant gain.

Figure 11.5 shows a simulation corresponding to Fig. 11.4 when the modification (11.1) is introduced to avoid windup. In this case there are clearly no difficulties. The control signal remains within the bounds $-0.5 < u < 0.5$ all the time. If the control signal saturates over a longer period of time, the adaptation should be switched off.   □

Windup will always cause difficulties. In cases like Example 11.2 the phenomenon is serious because the process is unstable. If the controller is unstable, the control law given by Eqs. (11.1) will automatically reset the controller state with a speed corresponding to the observer dynamics.

11.3   CONTROLLER DESIGN

Automatic control design is another important part of an adaptive controller. Fairly simplistic design methods were used in developing the adaptive controllers in Chapters 3, 4, and 5. Simple methods were used to keep the overall complexity of the system reasonable. To achieve this, it was sometimes necessary to assume that the processes were minimum-phase systems. This was the case, for example, for model-reference adaptive systems. Since control design is done automatically in closed loop, it is also necessary to introduce safeguards to make sure that all conditions required for the design method are fulfilled. For instance, it may be necessary to test whether the estimated process model is minimum-phase or whether there are common factors in the estimated polynomials. Direct adaptive controllers have the advantage that the design step is eliminated, since the parameters of the regulator are estimated directly. Notice, however, that several assumptions are made implicitly in using the direct algorithms. Direct methods are also restricted to special classes of systems.

Design Procedures

Many different design procedures can be used for adaptive control. Feedback by itself can make a closed-loop system insensitive to variations in process dynamics. There are also special so-called robust design methods that take process uncertainty into account explicitly. In deriving an adaptive controller it seems appealing to base it on a robust design method. It is also of interest to try to combine robust and adaptive control. The estimator should then provide estimates of the model and its uncertainty. The design method should take the uncertainty into account. Unfortunately, control and estimation theory has not yet progressed to the state in which such estimation and control procedures are available. Many of the robust design methods do also require manual interaction. Such procedures cannot be used in an adaptive controller. The pole placement design procedure is quite useful in practice, in spite of its simplicity. However, it can be improved significantly by some simple modifications that give more robust closed-loop systems. Some ways to do this are discussed in this section.

Specifications

To obtain a robust controller, it is very important that specifications be chosen in a sensible way. With a pole placement design, this means that the desired closed-loop poles have to be chosen with care. Poles that are too fast will give controllers that are very sensitive. This can be understood from the following expression, which gives a sufficient condition for stability of a pole placement design:

$$|H(z) - H_0(z)| < \left| \frac{H(z)T(z)}{H_m(z)S(z)} \right|$$

In this expression, H is the pulse transfer function of model used to design the controller, H_0 is the pulse transfer function of the true plant, $H_m = B_m/A_m$ is the desired response, and S and T are the controller polynomials. The inequality should hold on the unit circle. The condition implies that high model precision is required for those frequencies at which the desired closed-loop system has significantly higher gain than the model.

A reasonable way to determine the closed-loop poles is to make the observation that it is difficult to obtain a crossover frequency that is significantly higher than the frequency at which the plant has a phase lag of $180° - 270°$. Notice that these frequencies can conveniently be determined by a relay feedback experiment.

Youla Parameterization

The Diophantine equation is a key element of pole placement design. This equation has many solutions. If the polynomials R^0 and S^0 are solutions of the equation

$$AR^0 + BS^0 = A_c^0$$

it follows that the polynomials R and S given by

$$\begin{aligned} R &= XR^0 + YB \\ S &= XS^0 - YA \end{aligned} \tag{11.2}$$

satisfy the equation

$$AR + BS = XA_c^0$$

If a controller characterized by the polynomials R^0 and S^0 gives a closed-loop system with the characteristic polynomial A_c^0, then the controller

$$(XR^0 + YB)u = -(XS^0 - YA)y \tag{11.3}$$

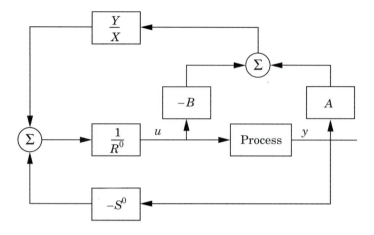

Figure 11.6   Block diagram of the closed-loop system with the controller (11.4).

gives a closed-loop system with the characteristic polynomial XA_c^0. The system is stable if the polynomial X is a stable polynomial but the polynomial Y can be chosen arbitrarily. It thus follows that if the controller $R^0 u = -S^0 y$ stabilizes the system $Ay = Bu$, then all controllers that stabilize the system are given by Eq. (11.3). The equation is called the *Youla parameterization* of all controllers that stabilize the system. Equation (11.3) can also be written as

$$ u = -\frac{S^0}{R^0} y + \frac{Y}{XR^0}(Ay - Bu) \tag{11.4} $$

This control law is illustrated by the block diagram in Fig. 11.6.

Robust Pole Placement

The Youla parameterization can be used to impose extra conditions on the controller. This idea was used in Section 3.6 to obtain controllers that have integral action. We now use it to improve the robustness of the controller. One way to do this is to require that the controller have small gain for those frequencies at which the process is very uncertain. For example, it is useful to require that the controller should have zero gain at the Nyquist frequency. This is accomplished by the condition $S(-1) = 0$. We can also require that the controller gain be zero at frequency ω_0. This is equivalent to requiring that the polynomial

$$ q^2 - 2\cos(\omega_0 h)q + 1 $$

be a factor of $S(q)$. To satisfy such requirements, the order of the closed-loop system must thus be increased. The additional poles introduced are specified by the polynomial X. We illustrate this by an example.

EXAMPLE 11.3   **Robust pole placement**

Assume that we have obtained a controller R^0, S^0 that gives a closed-loop system with the characteristic polynomial A_c^0 and that we want to improve its robustness by requiring that $S(-1) = 0$. To do this, we introduce one more closed-loop pole. We choose

$$X(q) = q - x_0$$

with $|x_0| < 1$. The polynomial Y can also be of first order. Equation (11.2) gives

$$S(q) = (q - x_0)S^0(q) - (q - y_0)A(q)$$

Requiring that $S(-1) = 0$ gives

$$y_0 = -1 + \frac{(1 + x_0)S^0(-1)}{A(-1)}$$

The robust controller is then characterized by

$$R(q) = (q - x_0)R^0(q) + (q - y_0)B(q)$$
$$S(q) = (q - x_0)S^0(q) - (q - y_0)A(q)$$

Notice that it is possible to proceed recursively to make the controller more and more complex.   □

Decoupled Command Signal and Disturbance Responses

Consider the process

$$Ay = Bu + v$$

and the controller

$$Ru = Tu_c - Sy$$

The closed-loop system is characterized by

$$y = \frac{BT}{A_c}u_c + \frac{R}{A_c}v$$

$$u = \frac{AT}{A_c}u_c - \frac{S}{A_c}v \qquad (11.5)$$

where $A_c = AR + BS$ is the closed-loop characteristic polynomial. Assume that no process zeros are canceled, and factor the characteristic polynomial as $A_c = A_oA_m$. If we choose $T = T'A_o$, Eqs. (11.5) become

$$y = \frac{BT'}{A_m}u_c + \frac{R}{A_oA_m}v$$

$$u = \frac{AT'}{A_m}u_c - \frac{S}{A_oA_m}v$$

The command signal response is governed by the dynamics of A_m, but the disturbance response is governed by the dynamics of $A_o A_m$. In this sense it is thus coupling between the command signal response and the disturbance response. In some cases it is desirable that the dynamics of the command signal responses and the disturbance responses are completely decoupled. This can be achieved by requiring that $T = T' A_o$ and $R = R' A_m$. The closed-loop system is then characterized by

$$y = \frac{BT'}{A_m} u_c + \frac{R'}{A_o} v$$

$$u = \frac{AT'}{A_m} u_c - \frac{S}{A_o A_m} v$$

The Diophantine equation then becomes

$$AA_m R' + BS = A_o A_m$$

To have a causal controller, we must require that $\deg A_m \geq \deg A$. The minimum-degree causal solution to this equation is such that $\deg S = \deg A + \deg A_m - 1$. Furthermore, $\deg R = \deg A_m + \deg A_o - \deg A$. To have a causal controller, we must thus require that $\deg S \leq \deg R$. This implies that

$$\deg A + \deg A_m - 1 \leq \deg A_m + \deg A_o - \deg A$$

Hence $\deg A_o \geq 2 \deg A - 1$. We thus find that the minimum-degree solution that decouples the response to command signals and disturbances is such that

$$\deg A_m = n \qquad \deg A_o = 2n - 1 \qquad \deg R = \deg S = 2n - 1$$

where $n = \deg A$.

A design of this type can be very useful when there are very noisy measurements and a fast setpoint response is desired.

11.4   SOLVING THE DIOPHANTINE EQUATION

Several of the design methods discussed earlier involve the solution of a Diophantine equation

$$AR + BS = A_c \tag{11.6}$$

Efficient methods for solving this equation are needed. The equation is linear in the polynomials R and S. A solution always exists if A and B are relatively prime. However, the equation has many solutions. This is easily seen: If R^0 and S^0 are solutions, then

$$R = R^0 + BQ$$

$$S = S^0 - AQ$$

are also solutions, where Q is an arbitrary polynomial. A particular solution can be specified in several different ways. Since a controller must be causal, the condition $\deg S \leq \deg R$ must hold. This condition will restrict the number of solutions significantly. An efficient way to solve the equation is to use a classical algorithm of Euclid.

Euclid's Algorithm

This algorithm finds the greatest common divisor G of two polynomials A and B. If one of the polynomials, say B, is zero, then G is equal to A. If this is not the case, the algorithm is as follows. Put $A_0 = A$ and $B_0 = B$ and iterate the equations

$$A_{n+1} = B_n$$
$$B_{n+1} = A_n \bmod B_n \tag{11.7}$$

until $B_{n+1} = 0$. The greatest common divisor is then $G = B_n$. When A and B are polynomials, $A \bmod B$ means the remainder when A is divided by B. This is in full agreement with the case when A and B are numbers. Backtracking, we find that G can be expressed as

$$AX + BY = G \tag{11.8}$$

where the polynomials X and Y can be found by keeping track of A_n div B_n in Euclid's algorithm. This establishes the link between Euclid's algorithm and the Diophantine equation. The extended Euclidean algorithm gives a convenient way to determine X and Y as well as the minimum-degree solutions U and V to

$$AU + BV = 0 \tag{11.9}$$

Equations (11.8) and (11.9) can be written as

$$F \begin{pmatrix} A \\ B \end{pmatrix} = \begin{pmatrix} X & Y \\ U & V \end{pmatrix} \begin{pmatrix} A \\ B \end{pmatrix} = \begin{pmatrix} G \\ 0 \end{pmatrix} \tag{11.10}$$

The matrix F can thus be viewed as the matrix, which performs row operations on $\begin{pmatrix} A & B \end{pmatrix}^T$ to give $\begin{pmatrix} G & 0 \end{pmatrix}^T$. A convenient way to find F is to observe that

$$\begin{pmatrix} X & Y \\ U & V \end{pmatrix} \begin{pmatrix} A & 1 & 0 \\ B & 0 & 1 \end{pmatrix} = \begin{pmatrix} G & X & Y \\ 0 & U & V \end{pmatrix}$$

The extended Euclidean algorithm can be expressed as follows: Start with the matrix

$$M = \begin{pmatrix} A & 1 & 0 \\ B & 0 & 1 \end{pmatrix}$$

If we assume that $\deg A \geq \deg B$, then calculate $Q = A$ div B, multiply the second row of M by Q, and subtract from the first row. Then apply the same procedure to the second row and repeat until the following matrix is obtained:

$$\begin{pmatrix} G & X & Y \\ 0 & U & V \end{pmatrix}$$

A nice feature of this algorithm is that possible common factors in A and B are determined automatically. The essential difficulty in implementing the algorithm is to find a good way to test for a polynomial being zero.

Solving the Diophantine Equation

By using the extended Euclidean algorithm it is now straightforward to solve the Diophantine equation

$$AR + BS = A_c \tag{11.11}$$

This is done as follows: Determine the greatest common divisor G and the associated polynomials X, Y, U, and V using the extended Euclidean algorithm. To have a solution to Eq. (11.11), G must divide A_c. A particular solution is given by

$$\begin{aligned} R^0 &= X A_c \operatorname{div} G \\ S^0 &= Y A_c \operatorname{div} G \end{aligned} \tag{11.12}$$

and the general solution is

$$\begin{aligned} R &= R^0 + QU \\ S &= S^0 + QV \end{aligned} \tag{11.13}$$

where Q is an arbitrary polynomial. The minimum-degree solution is obtained by choosing $Q = -S^0 \operatorname{div} V$. This implies that $S = S^0 \operatorname{mod} V$.

Relations to Ordinary Linear Equations

By equating coefficients of equal order, the Diophantine equation given by Eq. (11.11) can be written as a set of linear equations:

$$
\begin{pmatrix}
1 & 0 & \cdots & 0 & b_0 & 0 & \cdots & 0 \\
a_1 & 1 & \ddots & \vdots & b_1 & b_0 & \ddots & \vdots \\
a_2 & a_1 & \ddots & 0 & b_2 & b_1 & \ddots & 0 \\
\vdots & \vdots & \ddots & 1 & \vdots & \vdots & \ddots & b_0 \\
a_n & \vdots & & a_1 & b_n & \vdots & & b_1 \\
0 & a_n & & \vdots & 0 & b_n & & \vdots \\
\vdots & \ddots & \ddots & \vdots & \vdots & & \ddots & \ddots \\
0 & \cdots & 0 & a_n & 0 & \cdots & 0 & b_n
\end{pmatrix}
\begin{pmatrix}
r_1 \\ \vdots \\ r_k \\ s_0 \\ \vdots \\ s_l
\end{pmatrix}
=
\begin{pmatrix}
a_{c\,1} - a_1 \\ \vdots \\ a_{c\,n} - a_n \\ a_{c\,n+1} \\ \vdots \\ a_{c\,k+l+1}
\end{pmatrix}
\tag{11.14}
$$

$$\underbrace{\hphantom{aaaaaaaa}}_{k \text{ columns}} \quad \underbrace{\hphantom{aaaaaaaa}}_{l + 1 \text{ columns}}$$

The matrix on the left-hand side is called the *Sylvester matrix*; it occurs frequently in applied mathematics. It has the property that it is nonsingular if and only if the polynomials A and B do not have any common factors. If there are no common factors, a unique solution to Eq. (11.14) exists. Notice, however, the nonuniqueness with respect to the orders of R and S. Different choices of k and l will give different R and S, as discussed above. The solution to Eq. (11.14) can be obtained by Gaussian elimination. This method does not use the special structure of the Sylvester matrix.

11.5 ESTIMATOR IMPLEMENTATION

There are many issues that have to be considered in the implementation of an estimator. This section can be summarized by the following motto:

"Use only good relevant data, treat it carefully, and don't throw away useful information."

The key issue is that we want to obtain a model that is relevant for the task of control system design and that we want to track changes in the model. The tasks are influenced by many factors. In this section we discuss selection of model structure, filtering and excitation, parameter tracking, estimator windup, and robustness modifications.

Model Structure

The real physical processes that we try to control may have complicated dynamics. They may be nonlinear or infinite dimensional. One reason for the success of automatic control is that good control can often be based on relatively simple dynamical models. Such models can work very well under specific operating conditions, but the parameters of the model will depend on the operating conditions. In adaptive control it is attempted to fit a simple linear model on line and to adjust the parameters. For this purpose it is of paramount importance to understand what happens in fitting complicated dynamics with simple models. One fundamental fact is that the result obtained is crucially dependent on the nature of the input signal. This is illustrated by the following example.

EXAMPLE 11.4   **Fitting low-order models to high-order systems**

Consider a process with transfer function $G(s)$. Assume that one attempts to model the system by a first-order system with transfer function

$$\hat{G}(s) = \frac{b}{s + a}$$

If the input signal is sinusoidal with frequency ω_o, it is possible to get a perfect fit with finite values of the parameters if $\text{Im}\{G(i\omega_o)\} \neq 0$. Straightforward calculations show that $\hat{G}(i\omega_o) = G(i\omega_o)$ if the parameters are chosen to be

$$a = -\frac{\omega_o \, \text{Re}\{G(i\omega_o)\}}{\text{Im}\{G(i\omega_o)\}}$$

$$b = -\frac{\omega_o |G(i\omega_o)|^2}{\text{Im}\{G(i\omega_o)\}}$$

The transfer function of the model will then fit the data perfectly, but the parameters obtained depend on ω_o. The parameter values may change significantly with the frequency of the input signal. □

An interesting property of adaptive systems is that the parameters are estimated in closed loop. This implies that the simple model in the adaptive controller is fitted with the actual signals generated by the feedback. It explains intuitively the self-tuning property.

Another important observation is that the difficulty in on-line parameter estimation increases significantly with the number of parameters in the model. With many parameters the requirements on excitation also increase. For this purpose it is useful to try to reduce the number of unknown parameters as much as possible. This can be done by using *a priori* knowledge. This is often expressed in continuous-time models. The results are also strongly application dependent. Examples of this are given among the problems in the end of this chapter. Models consisting of low-order dynamics and time delays have proved very useful in process control. Such models can be represented by pulse transfer functions of the form

$$H_1(z) = \frac{b_0 z + b_1}{z^d(z + a_1)} \tag{11.15}$$

or

$$H_2(z) = \frac{b_0 z^2 + b_1 z + b_2}{z^d(z^2 + a_1 z + a_2)} \tag{11.16}$$

where the time delay τ is between dh and $dh + h$. Equation (11.16) can also represent second-order oscillatory systems. More b parameters can be included if the time delay is uncertain. In many cases it is known *a priori* that there are integrators in the model. This leads to transfer functions that contain the factor $z - 1$ in the denominator.

Data Filters and Excitation

Assume that the process is described by the discrete-time model

$$y(t) = G_0(q)u(t) + v(t) \tag{11.17}$$

Notice that possible anti-aliasing filters appear as part of the process $G_0(q)$. The disturbance $v(t)$ can be the sum of deterministic, piecewise deterministic,

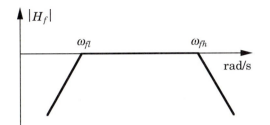

Figure 11.7   Amplitude curve for the data filter $H_f(q)$.

and stochastic disturbances. The signal has low-frequency and high-frequency components. In stochastic control problems it is important to design a controller that is tuned to a particular disturbance spectrum. In that case it is, of course, important to estimate the disturbance characteristics. In a deterministic problem we are concerned primarily with the term $G_0(q)u(t)$ in the above equation, and we are not particularly interested in the detailed character of the disturbance $v(t)$. In the following discussion we consider this case.

The presence of the disturbance $v(t)$ will, of course, create difficulties in the parameter estimation. However, the effect of $v(t)$ can be reduced by filtering. Assume that we introduce a data filter with the transfer function H_f and that we apply this filter to Eq. (11.17). Then

$$y_f(t) = G_0(q)u_f(t) + v_f(t) \tag{11.18}$$

where

$$y_f(t) = H_f(q)y(t) \qquad u_f(t) = H_f(q)u(t) \qquad \text{and} \qquad v_f(t) = H_f(q)v(t)$$

By a proper choice of the data filter we may now make the relative influence of the disturbance term smaller in Eq. (11.18) than in Eq. (11.17). The filtering should also emphasize the frequency ranges that are of primary importance for control design. The disturbance $v(t)$ typically has significant components with low frequencies. Low-frequency components should thus be reduced. Very high frequencies should similarly be attenuated. One reason for this is that if the model

$$A(q)y_f(t) = B(q)u_f(t)$$

is fitted by least squares, it is desirable that $A(q)v_f(t)$ be white noise. Since filtering with A implies that high frequencies are amplified, it means that $v_f(t)$ should not contain high frequencies. The data filter will therefore typically have band-pass character, as shown in Fig. 11.7. The center frequency is typically around the crossover frequency of the system.

In Section 3.5 we suggested using a filter with the transfer function

$$H_f(z) = \frac{1}{A_o(z^{-1})A_m(z^{-1})}$$

This filter is a typical low-pass filter that does not attenuate low frequencies. In Section 11.7 we will present other ways to choose the data filter. A typical data filter is given by

$$H_f(q) = \frac{(1-\alpha)(q-1)}{q-\alpha}$$

Some ways to choose the data filter will be discussed later.

It has been emphasized many times that it is necessary for the input signal to be persistently exciting of sufficiently high order to estimate parameters reliably. Taking into account that we are fitting low-order models to high-order systems, it is also necessary that persistency of excitation be achieved with signals in a frequency band where model accuracy is required.

Parameter Tracking

The key property of an adaptive controller is its ability to track variations in process dynamics. To do so, it is necessary to discount old data, a process that involves compromises. If parameters are constant, it is desirable to base the estimation on many measurements to reduce the effects of disturbances. If parameters are changing, however, it can be very misleading to use a long data record, since the parameters may not be the same. There are many ways to accommodate this problem. The best solutions are obtained if the nature of parameter variations is known. There are two prototype situations. One case is when parameters are slowly drifting; the other is when parameters are constant for long periods and jump from one value to another. Many attempts have been made to deal with the problem of parameter tracking, and there is a substantial literature. Most work is based on the assumption of detailed descriptions of the nature of parameter variations. A typical example is that the parameters are Markov processes with known transition probabilities. Such detailed information about the parameter variations is rarely available, and we therefore give some heuristic ways to deal with parameter tracking.

Exponential Forgetting

Exponential forgetting is a way to discard old data. It is based on the assumption that the least-squares loss function is replaced by a loss function in which old data is discounted exponentially. It follows from Theorem 2.4 that the recursive least-squares estimate with exponential forgetting is given by

$$\hat{\theta}(t) = \hat{\theta}(t-1) + K(t)\left(y(t) - \varphi^T(t)\hat{\theta}(t-1)\right)$$

$$K(t) = P(t-1)\varphi(t)\left(\lambda + \varphi^T(t)P(t-1)\varphi(t)\right)^{-1} \qquad (11.19)$$

$$P(t) = \frac{1}{\lambda}\left(I - K(t)\varphi^T(t)\right)P(t-1)$$

Table 11.3 Relations between the ratio T_f/h and the coefficient λ.

T_f/h	λ
1	0.37
2	0.61
5	0.82
10	0.90
20	0.95
50	0.98
100	0.99

where the sampling period h was chosen as the time unit. The forgetting factor is given by

$$\lambda = e^{-h/T_f}$$

where T_f is the time constant for the exponential forgetting. To make an assessment of reasonable values of the forgetting factor, we give the values of the forgetting factor for different ratios T_f/h in Table 11.3.

It is possible to generalize the method with exponential forgetting and have different forgetting factors for different parameters. However, this requires information about the nature of the changes in different parameters. Another modification is to modify Eqs. (11.19) so that only the diagonal elements are divided by λ.

Tracking of a time-varying parameter is illustrated by an example.

EXAMPLE 11.5 **Tracking parameters of a time-varying system**

Consider a process described by the differential equation

$$\frac{dy}{dt} = -y(t) + K_p(t)u$$

where the process gain is time varying. The process is controlled by an indirect adaptive controller that estimates parameters of the discrete-time model

$$y(kh + h) + ay(kh) = bu(kh)$$

and designs a controller with integral action using robust pole placement with

$$A_m(q) = q + a_m = q - e^{-h/T_m}$$

and

$$A_o(q) = q + a_o = q - e^{-h/T_o}$$

Straightforward computations give a controller of the form

$$u(kh) = t_0 u_c(kh) + t_1 u_c(kh - h) - s_0 y(kh) - s_1 y(kh - h) + u(kh - h)$$

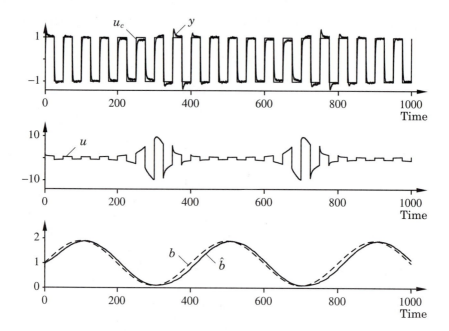

Figure 11.8   Tracking time-varying parameters.

where

$$t_0 = \frac{1 + a_m}{b}$$

$$t_1 = a_o t_0$$

$$s_0 = \frac{1 + a_o + a_m - a}{b}$$

$$s_1 = \frac{a_o a_m + a}{b}$$

Since only the process gain is unknown, we will estimate only the parameter b. In the simulation we will also assume that the gain varies sinusoidally between the values 0.1 and 1.9 with the period 400. In Fig. 11.8 we show a simulation of the system when the command signal is a square wave with period 50, there is measurement noise with the standard deviation 0.02, and the forgetting factor is $\lambda = 0.95$. Notice that the gain variation is clearly noticeable in the shape of the control signal, which changes significantly over one step. Figure 11.8 shows that the estimated gain lags the true gain. The forgetting factor is $\lambda = 0.95$, and the sampling period is $h = 0.5$. The time constant associated with the exponential forgetting is then $T_f = 10$ s, which is a crude estimate of the time lag in the estimator. Notice also that the lag is different for increasing and decreasing gains, a feature that indicates the nonlinear nature of the problem. The forgetting factor can be decreased to reduce the tracking lag. The estimates will then have more variation. To illustrate this, we simulate

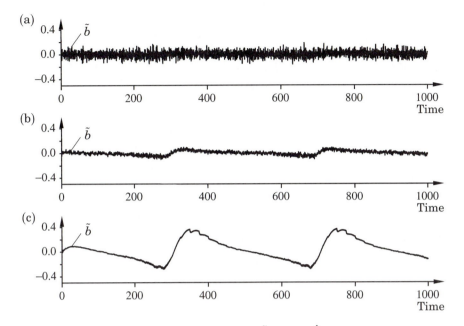

Figure 11.9 Parameter tracking error $\tilde{b} = b - \hat{b}$ for different forgetting factors: (a) $\lambda = 0.1$; (b) $\lambda = 0.7$; (c) $\lambda = 0.95$.

the same system as in Fig. 11.8 with different forgetting factors. The results are shown in Fig. 11.9. The figure shows that the forgetting factor $\lambda = 0.95$ is too large because the systematic tracking error is too large. The forgetting factor $\lambda = 0.1$, on the other hand, is too small, and the systematic error is small, but the random component is large. In this particular case the value $\lambda = 0.7$ is a reasonable compromise. The reason for the low value of λ is that the parameter variations are quite rapid. □

Covariance Resetting

In some situations the parameters are constant over long periods of time and change abruptly occasionally. Exponential forgetting, which is based on the assumption of a behavior that is homogeneous in time, is less suitable in this case. In such a situation it is more appropriate to reset the covariance matrix of the estimator to a large matrix when the changes occur. This is called *covariance resetting*. We illustrate this method by an example.

EXAMPLE 11.6   **Covariance resetting**

Consider the same system as in Example 11.5, but assume now that the parameter is piecewise constant. In Fig. 11.10 we show the results obtained

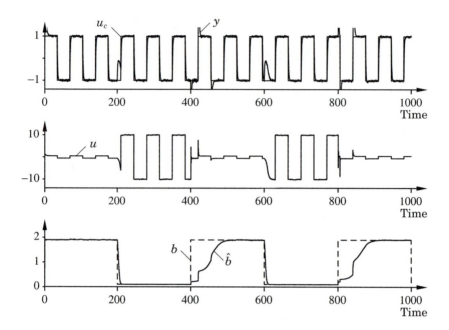

Figure 11.10 Tracking piecewise constant parameters using exponential forgetting when $\lambda = 0.95$.

with exponential forgetting with $\lambda = 0.95$. The figure shows clearly that the estimate of the process gain responds quite slowly when the gain changes. Notice also the strong asymmetry in the response of the estimate when the gain changes. It takes much longer for the estimate of the gain to increase than to decrease. The reason for this is the large difference in excitation. Also notice the stepwise nature of the estimates. Good excitation is obtained only when the command signal changes. In Fig. 11.11 we show the same system as in Fig. 11.10 with $\lambda = 1$ and covariance resetting. The covariance matrix is reset by reducing λ to 0.0001 when the parameter changes. Notice the drastic difference in the tracking rate.   □

The example clearly illustrates the advantage of using covariance resetting when the parameters change abruptly. To use this effectively, it is necessary to detect the changes in the parameters. There are many ways to do this by analyzing residuals or parameter changes. It is also possible to reset the covariance periodically.

Parallel Estimators and Other Schemes

There are many other ways to deal with parameter tracking. One possibility is to have several parallel estimators with different forgetting factors and to choose the one in which the estimates have the smallest residuals. It is also

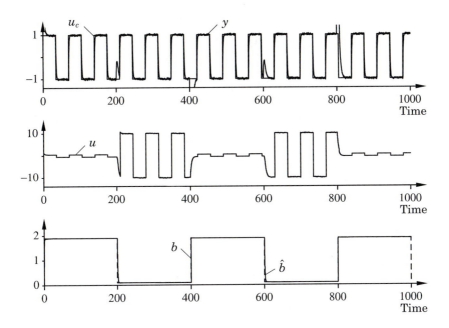

Figure 11.11   Tracking piecewise constant parameters using covariance resetting.

possible to have several parallel estimators that are reset periodically in a staggered way. There are also other schemes in which the forgetting factor is made signal dependent.

Estimator Windup

Exponential forgetting works well only if the process is properly excited all the time. There are problems with exponential forgetting when the excitation is poor. To understand this, we first consider the extreme case in which there is no excitation at all, that is, $\varphi = 0$. The equations for the estimate then become

$$\theta(t+1) = \theta(t)$$

$$P(t+1) = \frac{1}{\lambda} P(t)$$

The equation for the estimate θ is thus unstable with all eigenvalues equal to 1, and the equation for the P-matrix is unstable with all eigenvalues equal to $1/\lambda$. In this case the estimate will thus remain constant, and the P-matrix will grow exponentially if $\lambda < 1$. Since the estimator gain is $P\varphi$, the gain of the estimator will also grow exponentially. This means that the estimates may change very drastically whenever φ becomes different from zero. The phenomenon is called *estimator windup* in analogy with integrator windup.

A similar situation occurs if the regression vector is different from zero but restricted to a subspace. We illustrate this by an example.

EXAMPLE 11.7 **A constant regression vector causes windup**

Consider a process with the transfer function

$$G(s) = \frac{\beta}{s + \alpha}$$

with an indirect adaptive controller based on estimation of parameters a and b in the discrete-time model

$$y(kh + h) + ay(kh) = bu(kh)$$

The control design is the same as in Example 11.5. The controller has integral action. The parameters have the values $\alpha = 1$ and $\beta = 1$, the sampling period is $h = 0.5$ s, there is measurement noise with a standard deviation of 0.05, the setpoint is piecewise constant, and the forgetting factor is $\lambda = 0.95$. To illustrate the effect of poor excitation, the setpoint will be kept constant for long periods of time.

The parameters are in R^2. To have excitation, the regression vectors should also span R^2 persistently. When the setpoint is constant, the input and the

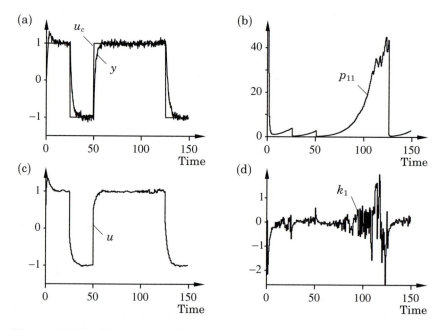

Figure 11.12 Illustration of estimator windup due to poor excitation. (a) Output y and setpoint u_c; (b) covariance p_{11}; (c) control signal u; and (d) estimator gain k_1.

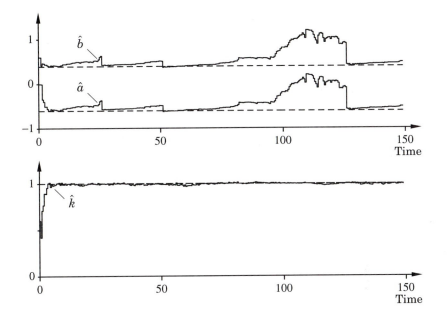

Figure 11.13   Parameter estimates in the case of estimator windup due to poor excitation. The dashed lines show the correct values of the parameters and of the gain of the process.

output settle to constant values after a transient. The regression vector then becomes

$$\varphi(t) = \begin{pmatrix} -u_c & au_c/b \end{pmatrix}$$

This vector lies in a one-dimensional subspace of R^2 and is thus not persistently exciting. The simulation shown in Fig. 11.12 illustrates the behavior of the system. The process output tracks the command signal quite well, and the control signal is also quite reasonable. Figure 11.12 shows that the element p_{11} of the P-matrix grows approximately exponentially during the periods when the command signal is constant. The deviations are due to the measurement noise that gives some excitation. The other elements of the P-matrix behave similarly. The estimator gains also grow significantly. Exponential growth of the P-matrix and the associated increase in the estimator gains are clearly noticeable in Fig. 11.12. The parameter estimates will change significantly, as shown in Fig. 11.13. The estimates are very inaccurate at the end of the periods when the command signal has been constant. In Fig. 11.13 we also show the estimate of process gain calculated from

$$\hat{k} = \frac{\hat{b}}{1 + \hat{a}}$$

This estimate is very good for the whole period because this variable is well excited. This is why the controller behaves reasonably well in spite of the poor estimates of a and b.   □

To get further insight into the windup phenomenon, we make a simplified analysis of the behavior shown in Example 11.7. For this purpose we assume that the regression vector is constant, that is, $\varphi(t) = \varphi_0$. The inverse of the P-matrix is given by

$$P^{-1}(t+1) = \lambda^t P_0^{-1} + \sum_{k=1}^{t} \lambda^{t-k} \varphi_0 \varphi_0^T = \lambda^t P_0^{-1} + \frac{1-\lambda^t}{1-\lambda} \varphi_0 \varphi_0^T$$

Using the matrix inversion lemma (Lemma 2.1), we find after some calculations that the covariance matrix can be written as

$$P(t+1) = \frac{1}{\lambda^t} \left(P_0 - \frac{P_0 \varphi_0 \varphi_0^T P_0}{\lambda^t \alpha(t) + \varphi_0^T P_0 \varphi_0} \right)$$

where

$$\alpha(t) = \frac{1-\lambda}{1-\lambda^t}$$

Furthermore, we find that

$$P(t+1)\varphi_0 = a(t) \frac{P_0 \varphi_0}{\lambda^t a(t) + \varphi_0^T P_0 \varphi_0} \tag{11.20}$$

The P-matrix can be decomposed as

$$P(t+1) = \tilde{P}(t) + \beta(t)\varphi_0 \varphi_0^T$$

where

$$\tilde{P}(t) = \frac{1}{\lambda^t} \left(P_0 - \frac{P_0 \varphi_0 \varphi_0^T P_0}{\lambda^t \alpha(t) + \varphi_0^T P_0 \varphi_0} \right) - \beta(t)\varphi_0 \varphi_0^T$$

and

$$\beta(t) = \alpha(t) \frac{\varphi_0^T P_0 \varphi_0}{(\varphi_0^T \varphi_0)^2 (\lambda^t \alpha(t) + \varphi_0^T P_0 \varphi_0)}$$

The matrix $\tilde{P}(t)$ is of rank $n-1$ with $\tilde{P}(t)\varphi_0 = 0$. Since $|\lambda| < 1$ we have as $t \to \infty$

$$a(t) \to 1 - \lambda$$

$$b(t) \to \frac{1-\lambda}{(\varphi_0^T \varphi_0)^2}$$

In the decomposition of $P(t+1)$ we thus find that the matrix $\tilde{P}$ goes to infinity as λ^{-t} and that $\beta(t)\varphi_0 \varphi_0^T$ goes to a constant $(1-\lambda)\varphi_0 \varphi_0^T / (\varphi_0^T \varphi)^2$.

Intuitively, the result of the calculation can be interpreted as follows: When the regression vector is constant, we obtain information only about the component of the parameter that is parallel to the regression vector. This component can be estimated reliably with exponential forgetting. The "projection" of the

P-matrix in this direction converges to $1 - \lambda$, and the "orthogonal" part of the *P*-matrix goes to infinity as λ^{-t}. Estimator windup is thus obtained by exponential forgetting combined with poor excitation. There are several ways to avoid estimator windup. We now discuss some of these techniques.

Conditional Updating

One possibility to avoid windup in the estimator is to update the estimate and the covariance only when there is excitation. The algorithms obtained are called algorithms with conditional updating or dead zones. A correct detection of excitation should be based on calculation of covariances or spectra as discussed in Section 2.4. Simpler conditions are often used in practice. Common tests are based on the magnitudes of the variations in process inputs and outputs or other signals such as ε and $\varphi^T P \varphi$. Notice that the quantity $\varphi^T P \varphi$ is dimension free.

If the regression vector is constant, it follows from Eq. (11.20) that

$$\varphi_0^T P(t) \varphi_0 = a(t) \frac{\varphi_0^T P_0 \varphi_0}{\lambda^t a(t) + \varphi_0^T P_0 \varphi_0}$$

As $t \to \infty$, it follows that $a(t) \to 1 - \lambda$. If $\varphi^T P \varphi$ is used as a test quantity, it is thus natural to normalize it by $1 - \lambda$. The effect of conditional updating is illustrated by an example.

EXAMPLE 11.8   **Conditional updating**

Consider the system in Example 11.7, but modify the estimator to provide conditional updating. In this particular case the estimate is updated if the test quantity

$$\varphi(t)^T P(t) \varphi(t) > 2(1 - \lambda)$$

Figure 11.14 shows a simulation that is comparable to Fig. 11.12. Notice that the exponential growth is now avoided. The elements of the *P*-matrix remain bounded, and the estimator gains are well behaved.   □

The selection of the condition for updating is critical. If the criterion is too stringent, the estimates will be poor because updating is done too infrequently. If the criterion is too liberal, we get covariance windup.

Constant-Trace Algorithms

Another way to keep the *P*-matrix bounded is to scale the matrix at each iteration. A popular scheme is to scale it in such a way that the trace of the matrix is constant. An additional refinement is to also add a small unit matrix.

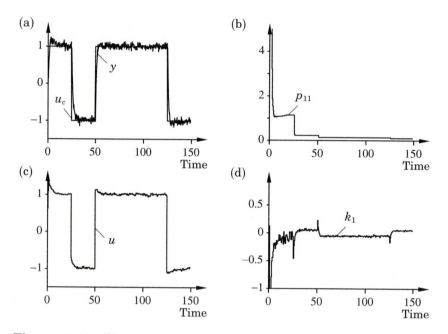

Figure 11.14   Illustration of how estimator windup can be avoided with conditional updating. Compare with Fig. 11.12.

This gives the so-called *regularized constant-trace algorithm*:

$$\hat{\theta}(t) = \hat{\theta}(t-1) + K(t) \left(y(t) - \varphi^T(t)\hat{\theta}(t-1) \right)$$

$$K(t) = P(t-1)\varphi(t) \left(\lambda + \varphi(t)^T P(t-1)\varphi(t) \right)^{-1}$$

$$\bar{P}(t) = \frac{1}{\lambda} \left(\bar{P}(t-1) - \frac{P(t-1)\varphi(t)\varphi^T(t)P(t-1)}{1 + \varphi(t)^T P(t-1)\varphi(t)} \right) \qquad (11.21)$$

$$P(t) = c_1 \frac{\bar{P}(t)}{tr\left(\bar{P}(t)\right)} + c_2 I$$

where $c_1 > 0$ and $c_2 \geq 0$. Typical values for the parameters can be

$$c_1/c_2 \approx 10^4$$

$$\varphi^T \varphi \cdot c_1 \gg 1$$

The constant-trace algorithm may also be combined with conditional updating.

Directional Forgetting

Another way to forget old data is based on the fact that one observation gives a projection of the parameter on the regression vector. Exponential forgetting can then be done only in the "direction" of the regression vector. This approach is called *directional forgetting*. To derive the equations, we observe that the inverse of the P-matrix with exponential forgetting is given by

$$P^{-1}(t+1) = \lambda P^{-1}(t) + \varphi(t)\varphi^T(t)$$

In directional forgetting we start with the formula

$$P^{-1}(t+1) = P^{-1}(t) + \varphi(t)\varphi^T(t)$$

The matrix $P^{-1}(t)$ is decomposed as

$$P^{-1}(t) = \tilde{P}^{-1}(t) + \gamma(t)\varphi(t)\varphi^T(t) \tag{11.22}$$

where $\tilde{P}^{-1}(t)\varphi(t) = 0$. This gives

$$\gamma(t) = \frac{\varphi^T(t)P^{-1}(t)\varphi(t)}{(\varphi^T(t)\varphi(t))^2}$$

Exponential forgetting is then applied only to the second term of Eq. (11.22), which corresponds to the direction where new information is obtained. This gives

$$P^{-1}(t+1) = \tilde{P}^{-1}(t) + \lambda\gamma(t)\varphi(t)\varphi^T(t) + \varphi(t)\varphi^T(t)$$

which can be written as

$$P^{-1}(t+1) = P^{-1}(t) + \left(1 + (\lambda - 1)\frac{\varphi^T(t)P^{-1}(t)\varphi(t)}{(\varphi^T(t)\varphi(t))^2}\right)\varphi(t)\varphi^T(t)$$

There are several variations of the algorithms. The forgetting factor is sometimes made a function of the data. One method has the property that the P-matrix is driven toward a matrix proportional to the identity matrix when there is poor excitation.

Leakage

Another way to avoid estimator windup, called *leakage*, was discussed in Section 6.9. In continuous time the estimator was modified as shown in Eq. (6.84) by adding the term $\alpha(\theta^0 - \theta)$. This means that the parameters will converge to θ^0 when no useful information is obtained, that is, when $e = 0$. A similar modification can also be made in discrete-time estimators. When a least-squares type of algorithm is used, it is also common to add a similar term to the P equation to drive it toward a specified matrix.

Robust Estimation

The least-squares estimate is optimal if the disturbances are Gaussian and such that the equation error is white noise. In practice the least-squares estimate has some drawbacks because the assumptions are violated. It is a direct consequence of the least-squares formulation that a single large error will have a drastic influence on the result because the errors are squared in the criterion. This is a consequence of the Gaussian assumption that implies that the probability of large errors is very small. Estimators with very different properties are obtained if it is assumed that the probability for large errors is not negligible. Without going into technicalities, we remark that the estimators will be replaced by equations such as

$$\hat{\theta}(t) = \hat{\theta}(t-1) + P(t)\varphi(t-1)f(\varepsilon(t))$$

$$\frac{d\hat{\theta}}{dt} = P\varphi f(\varepsilon)$$

where the function $f(\varepsilon)$ is linear for small ε but increases more slowly than linear for large ε. A typical example is

$$f(\varepsilon) = \frac{\varepsilon}{1 + a|\varepsilon|}$$

The net effect is to decrease the consequences of large errors. The estimators are then called *robust*.

11.6   SQUARE ROOT ALGORITHMS

It is well known in numerical analysis that considerable accuracy may be lost when a least-squares problem is solved by forming and solving the normal equations. The reason is that the measured values are squared unnecessarily. The following procedure for solving the least-squares problem is much better conditioned numerically. Start with Eq. (2.4):

$$\mathcal{E} = Y - \Phi\theta$$

An orthogonal transformation Q, that is, $Q^T Q = QQ^T = I$, does not change the Euclidean norm of the error

$$\tilde{\mathcal{E}} = Q\mathcal{E} = QY - Q\Phi\theta$$

Choose the transformation Q so that $Q\Phi$ is upper triangular. The above equation then becomes

$$\begin{pmatrix} \tilde{e}^1 \\ \tilde{e}^2 \end{pmatrix} = \begin{pmatrix} \tilde{y}^1 \\ \tilde{y}^2 \end{pmatrix} - \begin{pmatrix} \tilde{\Phi}_1 \\ 0 \end{pmatrix}\theta$$

where $\tilde{\Phi}_1$ is upper triangular. It then follows that the least-squares estimate is given by

$$\tilde{\Phi}_1\theta = \tilde{y}^1$$

and the error is $(\tilde{e}^2)^T\tilde{e}^2$. This way of computing the estimate is much more accurate than solving the normal equation, particularly if $\|\mathcal{E}\| \ll \|Y\|$. The method based on orthogonal transformation is called a *square root method* because it works with Φ or the square root of $\Phi^T\Phi$. There are several numerical methods that can be used to find an orthogonal transformation Q, for example, Householder transformations or the QR method. We will not discuss these methods further, because we are primarily interested in recursive methods.

Representation of Conditional Mean Values

Recursive square root methods can naturally be explained by using probabilistic arguments. Some preliminary results on conditional mean values for Gaussian random variables will first be developed. We can now show the following result.

THEOREM 11.1   Conditional mean values and covariances

Let the vectors x and y be jointly Gaussian random variables with mean values

$$E \begin{pmatrix} y \\ x \end{pmatrix} = \begin{pmatrix} m_y \\ m_x \end{pmatrix} \tag{11.23}$$

and covariance

$$\text{cov} \begin{pmatrix} y \\ x \end{pmatrix} = \begin{pmatrix} R_y & R_{yx} \\ R_{xy} & R_x \end{pmatrix} = R \tag{11.24}$$

where $R_{xy} = R_{yx}^T$. Further assume that $\dim x = n$ and $\dim y = p$. The conditional mean value of x, given y, is Gaussian with mean

$$E(x|y) = m_x + R_{xy}R_y^{-1}(y - m_y) \tag{11.25}$$

and covariance

$$\text{cov}(x|y) = R_{x|y} = R_x - R_{xy}R_y^{-1}R_{yx} \tag{11.26}$$

A nonnegative matrix R can be decomposed as

$$R = \rho \begin{pmatrix} I & 0 \\ K & L_x \end{pmatrix} \begin{pmatrix} D_y & 0 \\ 0 & D_x \end{pmatrix} \begin{pmatrix} I & 0 \\ K & L_x \end{pmatrix}^T \tag{11.27}$$

where D_x and D_y are diagonal matrices and L_x is lower triangular. Then

$$R_{xy}R_y^{-1} = K \tag{11.28}$$

and

$$R_{x|y} = \rho L_x D_x L_x^T \tag{11.29}$$

Proof: We first show that the vector z defined by

$$z = x - m_x - R_{xy}R_y^{-1}(y - m_y) \tag{11.30}$$

has zero mean, is independent of y, and has the covariance

$$R_z = R_x - R_{xy}R_y^{-1}R_{yx} \tag{11.31}$$

The mean value is zero. Furthermore,

$$Ez(y - m_y)^T = E\left\{(x - m_x)(y - m_y)^T - R_{xy}R_y^{-1}(y - m_y)(y - m_y)^T\right\}$$
$$= R_{xy} - R_{xy}R_y^{-1}R_y = 0$$

The variables z and y are thus uncorrelated. Since they are Gaussian, they are also independent. It now follows that

$$\begin{pmatrix} y - m_y \\ x - m_x \end{pmatrix} = \begin{pmatrix} I & 0 \\ R_{xy}R_y^{-1} & I \end{pmatrix} \begin{pmatrix} y - m_y \\ z \end{pmatrix}$$

The joint density function of x and y is

$$f(x, y) = (2\pi)^{-(n+p)/2}(\det R)^{-1/2}$$
$$\exp\left\{-\frac{1}{2}\left(z^T R_z^{-1}z + (y - m_y)^T R_y^{-1}(y - m_y)\right)\right\}$$

The density function of y is

$$f(y) = (2\pi)^{-p/2}(\det R_y)^{-1/2}\exp\left\{-\frac{1}{2}(y - m_y)^T R_y^{-1}(y - m_y)\right\}$$

where p is the dimension of y. The conditional density is then

$$f(x|y) = \frac{f(x, y)}{f(y)} = (2\pi)^{-n/2}(\det R_y)^{1/2}(\det R)^{-1/2}\exp\left\{-\frac{1}{2}z^T R_z^{-1}z\right\}$$

where n is the dimension of x. But

$$\det R = \det\begin{pmatrix} R_y & R_{yx} \\ R_{xy} & R_x \end{pmatrix} = \det\begin{pmatrix} R_y & R_{yx} \\ 0 & R_x - R_{xy}R_y^{-1}R_{yx} \end{pmatrix}$$
$$= \det R_y \cdot \det\left(R_x - R_{xy}R_y^{-1}R_{yx}\right) = \det R_y \cdot \det R_z$$

Hence

$$f(x|y) = (2\pi)^{-n/2}(\det R_z)^{-1/2}e^{-(1/2)z^T R_z^{-1}z}$$

where z is given by Eq. (11.30) and R_z by Eq. (11.31).

The first part of the theorem is thus proved. To show the second part, notice that Eq. (11.27) is

$$R = \rho\begin{pmatrix} D_y & D_y K^T \\ K D_y & L_x D_x L_x^T + K D_y K^T \end{pmatrix}$$

Identification of the different terms gives

$$R_y = \rho D_y$$
$$R_{xy} = \rho K D_y$$
$$R_x = \rho\left(L_x D_x L_x^T + K D_y K^T\right)$$

Hence

$$R_{xy}R_y^{-1} = K$$

and

$$R_{x|y} = R_x - R_{xy}R_y^{-1}R_{yx} = \rho L_x D_x L_x^T$$

Remark. It follows from the theorem that the calculation of the conditional mean of a Gaussian random variable is equivalent to transforming the joint covariance matrix of the variables to the form of Eq. (11.27). Notice that this form may be viewed as a square root representation of R. □

Application to Recursive Estimation

The basic step in recursive estimation can be described as follows: Let θ be Gaussian $N(\theta^0, P)$. Assume that a linear observation

$$y = \varphi^T \theta + e$$

is made, where e is normal $N(0, \sigma^2)$. The new estimate is then given as the conditional mean $E(\theta|y)$. The joint covariance matrix of y and θ is

$$R = \begin{pmatrix} \varphi^T P \varphi & \varphi^T P \\ P\varphi & P \end{pmatrix} + \begin{pmatrix} \sigma^2 & 0 \\ 0 & 0 \end{pmatrix}$$

The symmetric nonnegative matrix P has a decomposition $P = LDL^T$, where L is a lower triangular matrix with unit diagonal and D is a nonnegative diagonal matrix. The matrix R can then be written as

$$
\begin{aligned}
R &= \begin{pmatrix} \varphi^T LDL^T \varphi + \sigma^2 & \varphi^T LDL^T \\ LDL^T \varphi & LDL^T \end{pmatrix} \\
&= \begin{pmatrix} 1 & \varphi^T L \\ 0 & L \end{pmatrix} \begin{pmatrix} \sigma^2 & 0 \\ 0 & D \end{pmatrix} \begin{pmatrix} 1 & 0 \\ L^T \varphi & L^T \end{pmatrix}
\end{aligned}
\tag{11.32}
$$

If this matrix can be transformed to

$$R = \begin{pmatrix} 1 & 0 \\ K & \tilde{L} \end{pmatrix} \begin{pmatrix} \tilde{\sigma}^2 & 0 \\ 0 & \tilde{D} \end{pmatrix} \begin{pmatrix} 1 & K^T \\ 0 & \tilde{L}^T \end{pmatrix} \tag{11.33}$$

Theorem 11.1 can be used to obtain the recursive estimate as

$$\hat{\theta} = \theta^0 + K \left(y - \varphi^T \theta \right)$$

with covariance

$$P = \tilde{L}\tilde{D}\tilde{L}^T$$

The algorithm can thus be described as follows.

ALGORITHM 11.1   Square root RLS

Step 1:   Start with L and D as a representation of P.

Step 2:   Form the matrix of Eq. (11.32), where φ is the regression vector.

Step 3:   Reduce this to the lower triangular form of Eq. (11.33).

Step 4:   The updating gain is K, and the new P is represented by $\tilde{L}$ and $\tilde{D}$.   □

It now remains to find the appropriate transformation matrices. A convenient method is dyadic decomposition.

Dyadic Decomposition

Given vectors

$$a = \begin{pmatrix} 1 & a_2 & \ldots & a_n \end{pmatrix}^T$$

$$b = \begin{pmatrix} b_1 & b_2 & \ldots & b_n \end{pmatrix}^T$$

and scalars α and β, find new vectors

$$\tilde{a} = \begin{pmatrix} 1 & \tilde{a}_2 & \ldots & \tilde{a}_n \end{pmatrix}^T$$

$$\tilde{b} = \begin{pmatrix} 0 & \tilde{b}_2 & \ldots & \tilde{b}_n \end{pmatrix}^T$$

such that

$$\alpha a a^T + \beta b b^T = \tilde{\alpha}\tilde{a}\tilde{a}^T + \tilde{\beta}\tilde{b}\tilde{b}^T \tag{11.34}$$

If this problem can be solved, we can perform the composition of Eq. (11.33) by repeated application of the method.

Equation (11.34) can be written as

$$\alpha \begin{pmatrix} 1 \\ a_2 \\ \vdots \\ a_n \end{pmatrix} \begin{pmatrix} 1 & a_2 & \ldots & a_n \end{pmatrix} + \beta \begin{pmatrix} b_1 \\ b_2 \\ \vdots \\ b_n \end{pmatrix} \begin{pmatrix} b_1 & b_2 & \ldots & b_n \end{pmatrix}$$

$$= \tilde{\alpha} \begin{pmatrix} 1 \\ \tilde{a}_2 \\ \vdots \\ \tilde{a}_n \end{pmatrix} \begin{pmatrix} 1 & \tilde{a}_2 & \ldots & \tilde{a}_n \end{pmatrix} + \tilde{\beta} \begin{pmatrix} 0 \\ \tilde{b}_2 \\ \vdots \\ \tilde{b}_n \end{pmatrix} \begin{pmatrix} 0 & \tilde{b}_2 & \ldots & \tilde{b}_n \end{pmatrix} \tag{11.35}$$

Equating the $(1,1)$ elements gives

$$\alpha + \beta b_1^2 = \tilde{\alpha} \tag{11.36}$$

Equating the $(1, k)$ elements for $k > 1$ gives

$$\alpha a_k + \beta b_1 b_k = \tilde{\alpha}\tilde{a}_k \qquad (11.37)$$

Adding and subtracting $\beta b_1^2 a_k$ give

$$(\alpha + \beta b_1^2)a_k + \beta b_1 b_k - \beta b_1^2 a_k = \tilde{\alpha}\tilde{a}_k$$

Hence

$$\tilde{a}_k = a_k + \frac{\beta b_1}{\tilde{\alpha}}\left(b_k - b_1 a_k\right) \qquad (11.38)$$

The numbers $\tilde{\alpha}$ and $\tilde{a}_k$ can thus be determined. It now remains to compute $\tilde{\beta}$ and $\tilde{b}_k$. Equating the (k, l) elements of Eq. (11.35) for $k, l > 1$ gives

$$\alpha a_k a_l + \beta b_k b_l = \tilde{\alpha}\tilde{a}_k\tilde{a}_l + \tilde{\beta}\tilde{b}_k\tilde{b}_l$$

$$= \frac{(\alpha a_k + \beta b_1 b_k)(\alpha a_l + \beta b_1 b_l)}{\tilde{\alpha}} + \tilde{\beta}\tilde{b}_k\tilde{b}_l$$

where Eq. (11.37) has been used to eliminate $\tilde{a}_k\tilde{a}_l$. Inserting the expression in Eq. (11.36) for $\tilde{\alpha}$ gives, after some calculations,

$$(b_k - b_1 a_k)\,(b_l - b_1 a_l) = \frac{\tilde{\alpha}\tilde{\beta}}{\alpha\beta}\,\tilde{b}_k\tilde{b}_l$$

```
PROCEDURE
DyadicReduction(VAR a,b:col; VAR alpha,beta:REAL;
   i0,i1,i2 :CARDINAL);
CONST
   mzero = 1.0E-10;
VAR
   i              : CARDINAL;
   w1,w2,b1,gam : REAL;
BEGIN
   IF beta<mzero THEN beta:=0.0; END;
   b1 := b[i0];
   w1 := alpha;
   w2 := beta*b1;
   alpha := alpha + w2*b1;
   IF alpha > mzero THEN
     beta := w1*beta/alpha;
     gam := w2/alpha;
     FOR i:=i1 TO i2 DO
       b[i] := b[i] - b1*a[i];
       a[i] := a[i] + gam*b[i];
     END;
   END;
END DyadicReduction;
```

Figure 11.15   Dyadic decomposition.

These equations have several solutions. A simple one is

$$\tilde{b}_k = b_k - b_1 a_k$$

$$\tilde{\beta} = \frac{\alpha\beta}{\tilde{\alpha}}$$

A solution to the dyadic decomposition problem of Eq. (11.34) is given by the equations

$$\tilde{\alpha} = \alpha + \beta b_1^2$$

$$\tilde{\beta} = \frac{\alpha\beta}{\tilde{\alpha}}$$

$$\gamma = \frac{\beta b_1}{\tilde{\alpha}}$$

$$\tilde{b}_k = b_k - b_1 a_k \qquad k = 2, \ldots, n$$

$$\tilde{a}_k = a_k + \gamma \tilde{b}_k \qquad k = 2, \ldots, n$$

The algorithm in Fig. 11.15 is an implementation of the dyadic decomposition.

```
PROCEDURE
LDFilter(VAR theta,d:col; VAR l:matr; phi:col;
   lambda:REAL; n:CARDINAL);
VAR
   i,j  : CARDINAL;
   e,w  : REAL;
BEGIN
   d[0]  := lambda;
   e     := phi[0];
   FOR i:=1 TO n DO
      e:=e-theta[i]*phi[i];
      w:=phi[i];
      FOR j:=i+1 TO n DO w:=w+phi[j]*l[i,j]; END;
      l[0,i]:=0.0;
      l[i,0]:=w;
   END;
   FOR i:=n TO 1 BY -1 DO (* Notice backward loop *)
      DyadicReduction(l[0],l[i],d[0],d[i],0,i,n);
   END;

   FOR i:=1 TO n DO
      theta[i]:=theta[i]+l[0,i]*e;
      d[i]:=d[i]/lambda;
   END;
END LDFilter;
```

Figure 11.16   LD decomposition.

In this code, the type

```
col = ARRAY[0..maxindex] OF REAL;
```

has been introduced. By using the procedure `DyadicReduction` it is now straightforward to write a procedure that implements Algorithm 11.1. Such a procedure is given in Fig. 11.16. The algorithm performs one step of a recursive least-squares estimation. Starting from the current estimate θ, the covariance represented by its LD decomposition, and the regression vector, the procedure generates updated values of the estimate and its covariance. The data type

```
matr = ARRAY[0..maxindex] OF col;
```

is used in the program. The starting values can be chosen to be $L = I$ and $d = [\beta_0, \beta_0, \ldots, \beta_0]$. This gives $LD = \beta_0 I$.

11.7   INTERACTION OF ESTIMATION AND CONTROL

Parameter estimation and control design were treated as two separate subjects in the previous sections of this chapter. In an adaptive controller there are, of course, strong interactions between estimation and control. Some consequences of this interaction are discussed in this section.

Computational Delay

The updating of the estimated parameters and the design are done at each sampling instant. The timing of computations of the controller was discussed in Section 11.2. We pointed out that it is important to have as short a computational delay as possible. The dual time scale of the adaptive control problem implies that the process parameters are assumed to vary slowly. This means that the parameter estimates from the previous sampling instant can be used for calculating the control signal. There will thus be no extra time delay due to the adaptation, provided that the parameter update and the controller design are done after the control signal is sent out to the process.

Integral Action

Practically all controllers need integral action to ensure that calibration errors and load disturbances do not give steady-state errors. In Section 3.6 we showed how the design procedure could easily be modified to give controllers with integral action. In that section it was also shown that a particular adaptive controllers automatically gave zero steady-state error. This situation occurs quite frequently. It is also easy to check whether a particular self-tuner has this ability by investigating possible stationary solutions. A typical example is the following.

| **EXAMPLE 11.9** | **Obtaining integral action automatically** |

Consider the simple direct moving-average self-tuning controller described in Chapter 4, which is based on least-squares estimation and minimum-variance control. The estimation is based on the model

$$y(t + d) = R^*(q^{-1})u(t) + S^*(q^{-1})y(t)$$

and the regulator is

$$u(t) = -\frac{S^*}{R^*}\,y(t)$$

The conditions for a stationary solution are that

$$\hat{r}_y(\tau) = \lim_{N\to\infty} \frac{1}{N} \sum_{k=1}^{N} y(k + \tau)y(k) = 0 \qquad \tau = d,\ldots,d+l$$

$$\hat{r}_{yu}(\tau) = \lim_{N\to\infty} \frac{1}{N} \sum_{k=1}^{N} y(k + \tau)u(k) = 0 \qquad \tau = d,\ldots,d+k$$

where k and l are the degrees of the R^* and S^* polynomials, respectively. These conditions are not satisfied unless the mean value of y is zero. When there is an offset, the parameter estimates will get values such that $R^*(1) = 0$, that is, there is an integrator in the controller. However, the convergence to the integrator may be slow. □

A second way to explicitly eliminate steady-state errors is to base an adaptive controller on estimation of parameters in the model

$$A(q)y(t) = B(q)u(t) + v$$

where v is a constant that is estimated. The control design should also be modified by introducing a feedforward from the estimated disturbance. This approach has the drawback that an extra parameter has to be estimated. Furthermore, it is necessary to have different forgetting factors on the bias estimate and the other estimates; otherwise, the convergence to a new level will be very slow. Finally, if the bias is estimated in this way, it is not possible to use the self-tuner as a tuner, since there will be no reset when the estimation is switched off. This is a simple example that shows the drawbacks of mixing the functions of the feedback loop and the adaptation loop. A much better way is to design a controller with integral action, for example, by using the methods discussed in Section 3.6. A data filter of band-pass character should also be used so that the disturbance v does not influence estimation too much. We will also show how a similar approach can be used for a direct self-tuner.

Compatible Criteria for Identification and Control

So far, we have treated identification and control as two different tasks. The criterion for the identification (least squares) was chosen largely on an ad hoc basis. It is clearly desirable to try to find a criterion for identification that matches the final use of the model. This is in general a very complicated problem. We therefore discuss a simplified case. Consider a process described by the model

$$A(q)y(t) = B(q)u(t) \tag{11.39}$$

where u is the control signal and y is the measured variable. Let the controller be

$$R(q)u(t) = T(q)u_c(t) - S(q)y(t) \tag{11.40}$$

where u_c is the setpoint and $R(q)$, $S(q)$, and $T(q)$ are polynomials. The polynomials $R(q)$ and $S(q)$ satisfy the Diophantine equation

$$A(q)R(q) + B(q)S(q) = A_m(q)A_o(q) \tag{11.41}$$

where the desired closed-loop polynomial is $A_o(q)A_m(q)$. This equation has many solutions. It is customary to choose the simplest one that gives a causal controller, but it is also possible to introduce an auxiliary condition. Integral action is obtained by finding a solution such that $R(1) = 0$. High-frequency roll-off is obtained by requiring that $S(-1) = 0$.

The polynomial $T(q)$ is given by

$$T(q) = t_0 A_o(q) \tag{11.42}$$

where $t_0 = A_m(1)/B(1)$. If $R(1) = 0$, it also follows that $T(1) = S(1)$. Combining Eqs. (11.39) and (11.40), we get

$$\begin{aligned} y(t) &= t_0 \frac{B(q)}{A_m(q)} u_c(t) \\ u(t) &= t_0 \frac{A(q)}{A_m(q)} u_c(t) \end{aligned} \tag{11.43}$$

Polynomials $A_o(q)$ and $A_m(q)$ are typically chosen to give good rejection of disturbances and insensitivity to modeling errors and measurement noise.

It is desirable to formulate the adaptive control problem in such a way that the goals for control and identification are compatible. If this is done, it means that a model is fitted in such a way that it matches the ultimate use of of the model.

Consider the situation in which the goal is to control a plant with transfer function P_0. A controller is designed by using pole placement based on the approximate model whose transfer function is $P = B/A$. To compute the control law, the parameters of the polynomials A and B are estimated by using least squares, and the controller is then determined by the pole placement method. Let u_0 and y_0 denote the inputs and outputs that are obtained in controlling the actual plant, and let u and y denote the corresponding signals when the controller controls the design model. The control performance error can then be defined as

$$e_{cp} = y_0 - y$$

We have the following result.

T H E O R E M 11.2   Compatibility of identification and control

The control performance error e_{cp} is identical to the least-squares estimation error if identification is performed in closed loop and if the transfer function of the data filter is chosen to be

$$H_f = \frac{R}{A_o A_m} \tag{11.44}$$

Proof: The proof is a straightforward calculation. The output of the true system is given by

$$y_0 = \frac{P_0 T}{R + P_0 S} u_c \tag{11.45}$$

and the control signal is

$$u_0 = \frac{T}{R + P_0 S} u_c \tag{11.46}$$

The corresponding signals for the nominal plant are obtained simply by omitting the index 0 on y_0, u_0, and P_0. The control performance error then becomes

$$e_{cp} = \left(\frac{P_0 T}{R + P_0 S} - \frac{PT}{R + PS} \right) u_c = \frac{RT(P_0 - P)}{(R + P_0 S)(R + PS)} u_c$$

$$= \frac{R(P_0 - P)}{R + PS} u_0 = \frac{AR(P_0 - P)}{A_o A_m} u_0 \tag{11.47}$$

where the first equality follows from Eqs. (11.43), (11.45), and (11.46). The second equality is obtained by a simple algebraic manipulation. The third follows from Eq. (11.46), and the last equality follows from Eq. (11.41). The least-squares estimation error is given by

$$e = H_f(Ay_0 - Bu_0)$$

It follows from Eq. (11.47) that e and e_{cp} are identical if estimation is based on closed-loop data and if the data filter is chosen to be Eq. (11.44).   ☐

Remark 1.   Notice that the denominator of the filter (11.44) is given by $A_o A_m$, which are given by the specifications.

Remark 2.   Notice that for a controller with integral action the filter (11.44) is a bandpass filter.

Remark 3.   Notice that only the numerator of the filter has to be adapted.   ☐

This result gives a rational way of choosing the data filter for a servo problem.

11.8   PROTOTYPE ALGORITHMS

In this section we present some prototype algorithms for adaptive control. Guidelines for the coding of the algorithms are given. The algorithms can easily be expanded to a variety of controllers.

Algorithm Skeleton

All adaptive algorithms discussed in this chapter have the following form:

```
1     Analog_Digital_conversion
2     Compute_control_signal
3     Digital_Analog_conversion
4     If estimate then
5        begin{estimate}
6           Covariance_update
7           Parameter_update
8           If tune then
9              begin{tune}
                  th_design:=th_estimated
10                Design_calculations
11             end{tune}
12          end{estimate}
13    Organize_data
14    Compute_as_much_as_possible_of_control_signal
```

Row 1 implements the conversion of the measured output signal, the reference signal, and possible feedforward signal. All the converted signals are supposed to be filtered through appropriate anti-aliasing filters, as discussed in Section 11.2. Row 3 sets the control signal to the process. Rows 14 and 2 contain the calculations of the control signal, which are independent of whether the parameters are estimated or not. Notice the division of the calculations of the control signal to avoid overly long computation times. All calculations that are possible to do in advance are done in Row 14. Only calculations that contain the last measurements are done in Row 2.

Rows 4–13 contain calculations that are specific for an adaptive algorithm. There are two logical variables, estimate and tune, which control whether the parameters are going to be estimated and whether the controller is going to be redesigned, respectively. The estimation is done in Rows 5–7, and the design calculations are done in Row 10. Row 13 organizes the data such that the algorithm is always ready to start estimation when the operator wishes.

The various adaptive algorithms discussed in this section differ only in the design calculations. The estimator part can be the same for all algorithms. One important part of the algorithms that will not be discussed here is the operator interface. This is usually a significant part of an adaptive control system, but it is very hardware-dependent, so it is difficult to discuss in general terms. We now discuss the calculations in Rows 4–13 in more detail.

Parameter_update

We assume that the estimated model has the form

$$y(t) = \varphi^T(t)\theta$$

where the components in the regression vector φ are lagged and filtered inputs and outputs. The ordering, the number of lags, and so on depend on the specific model; these details are easily sorted out for the chosen model structure. Rows 6 and 13 of the algorithm contain the bookkeeping of the φ vector (i.e., the usual shift of some parts of the vector and supplement of the latest measurements and outputs). This part of the algorithm should also include the data filtering discussed in Sections 11.2, 11.3, and 11.7. For simplicity it is assumed that the estimation and the covariance update are done by using ordinary recursive least squares (Eqs. 11.19). The calculations can be organized as in the listing below, where eps is the residual, th_estimated is the parameter vector, P is the covariance matrix, phi is the data vector, and lambda is the forgetting factor.

```
"Compute residual
eps = y - phi'*th_estimated
"Update estimate
w = P*phi
den = lambda + phi'*w
gain = w/den
th_estimated = th_estimated + gain*eps
"Update covariance
P = (P - w*w'/den)/lambda
```

The prime is the transpose, and $*$ is matrix multiplication. This skeleton can easily be transferred to any preferred programming language.

Organize_data

This part of the code filters the process input and output by H_f, and it updates the regression vector $\varphi(t)$ and the other states of the system. If $\varphi(t)$ is updated at each sampling period, it is possible to update the estimates irregularly.

Design_calculations

When a direct algorithm such as Algorithm 3.3 is used, the controller parameters are the same as the estimated parameters, and there are no calculations that have to be done in the design block. In the indirect methods a polynomial equation has to be solved. The solution of the Diophantine equation is discussed in Section 11.4. Some care must be taken because of difficulties with possible common factors in the estimated model polynomials.

Compute_control_signal

The computation of the control signal to minimize the computational delay was discussed in Section 11.2, along with the anti-reset windup.

Summary

The program skeleton in this section can now be supplemented with details to become a complete adaptive control algorithm. These details will depend on what algorithm is chosen to be implemented and on which programming language is chosen.

11.9   OPERATIONAL ISSUES

Simple controllers typically have two operating modes, manual and automatic. It is also possible to change the parameters during operation. It is a nontrivial task to deal with the operation of a conventional controller. The current practice has developed over a long period of time. Adaptive controllers can operate in many more ways. It is a difficult problem to find a good solution to the operational problems. The problem will also vary widely with the application area. In this section we discuss a controller for industrial process control, which is designed to operate in many widely different environments.

Operating Modes

An adaptive controller has at least three operating modes: manual, constant-parameter control, and adaptation. The controllers that are used in constant-parameter mode may be of several types: PID, relay, or a general linear controller. In this mode the controller parameters must also be loaded and stored. Parameter estimation may also be initiated in the constant-parameter mode. Estimation can also be enhanced by introducing extra perturbations. The nature of these perturbations must also be specified.

Initialization

There are several ways to initialize a self-tuning algorithm, depending on the available *a priori* information about the process. In one case, nothing is known about the process. The initial values of the parameters in the estimator can then be chosen to be zero or such that the initial controller is a proportional or integral controller with low gain. Auto-tuning, discussed in Chapter 8, is a convenient way to initialize the algorithm, because it generates a suitable input signal and safe initial values of the parameters. This also gives a rational way of choosing the sampling interval.

The inputs and outputs of the process should be scaled so that they are of the same magnitude. This will improve the numerical conditions in the estimation and the control parts of the algorithm. The initial value of the covariance matrix can be 1–100 times a unit matrix if the elements in the φ vector are scaled to approximately unity. These values are usually not crucial, since the estimator will get reasonable values in a very short period of time. Our experience is that 10–50 samples are sufficient to get a very good controller when the system is excited. During the initial phase it can be advantageous to add a perturbation signal to speed up the convergence of the estimator.

The situation is different if the process has been controlled before with a conventional or an adaptive controller. The initial values should then be such that they correspond to the controller used before. Furthermore, the P-matrix should be sufficiently small.

Sometimes it is important to have disturbances be as small as possible, owing to the startup of the self-tuning algorithm. There are then two precautions that can be taken. First, the estimator can be used for some sampling periods before the self-tuning algorithm is allowed to put out any control actions. During that time a safe, simple controller should be used. It is also possible and desirable to limit the control signal. The allowable magnitude can be very small during the first period of time and can then be increased when better parameter estimates are obtained. The drawback of having small input signals is that the excitation of the process will be poor, and it will take longer to get good parameter estimates.

Supervision

An adaptive controller should also contain facilities for supervision. Basic statistics such as the mean, standard deviation, and maximum and minimum values should be computed for the process input and output. These values should be averaged over the basic period of the loop. Since excitation is so important, it should be monitored. The estimation error also gives useful information about the behavior of the loop. Common factors in the process model should be detected. This will indicate that the model structure should be changed. For special algorithms it is also possible to determine whether the controller behaves as expected. For example, for minimum variance or moving average controllers this can be determined by monitoring the covariance of the process output.

11.10   CONCLUSIONS

Practical aspects on implementation of adaptive controllers have been discussed in this chapter. There are many things to consider, since adaptive controllers are quite complicated devices. The following are some of the important issues:

- Analog anti-aliasing filters must be used. They are typically second- or fourth-order filters that effectively eliminate signal components with frequencies above the Nyquist frequency π/h, where h is the sampling period.

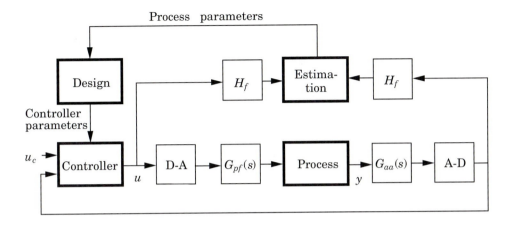

Figure 11.17   Block diagram of an adaptive control system with added filters. G_{aa} is the anti-aliasing filter, G_{pf} is the postsampling filter, and H_f is the data filter for the estimation.

The dynamics of the filters should be taken into account in the control design.

- Inputs and outputs should be filtered by a bandpass filter with H_f before these signals are sent to the parameter estimator. These filters will remove low-frequency disturbances such as levels and ramps. High frequency disturbances are also removed by H_f. The lower limit of the passband should be at least one decade below the desired crossover frequency. Known sinusoidals can also be removed by using notch filters.

- The postsampling filter G_{pf} is used to avoid excitation of high-frequency resonance modes in the process.

- Low-order models are typically used. They are estimated with algorithms having time variable exponential forgetting, regularized constant trace, or directional forgetting. The estimator should also contain a dead zone. Finally, the estimator may contain a "switch," which detects whether the system is sufficiently excited. The "switch" can measure the power in different frequency bands and thus control whether the estimator should be active or not. Square root algorithms are preferable, particularly if there is a high signal to noise ratio.

- The design method for the controller should be robust against unmodeled dynamics. Level and ramp disturbances are eliminated by introducing integrators in the controller. The control signal should be limited, and the controller should include anti-reset windup.

A block diagram of a reasonably realistic adaptive controller is given in Fig. 11.17. Adaptive controllers also contain parameters. Guidelines for choosing these have been given in this chapter.

PROBLEMS

11.1   How should the disturbance annihilation filter $H_f(q)$ be chosen if $v(t)$ in Eq. (11.17) is a sinusoidal?

11.2   Consider the data filter

$$H_f(q) = \frac{(1 - \alpha)(1 - q)}{q - \alpha}$$

Discuss how the choice of the parameter α influences elimination of constant disturbances.

11.3   Plot the Bode diagram for a fourth-order Bessel filter, and compare it with a pure time delay. Consider the cases in Table 11.2.

11.4   Determine how the behavior of the anti-reset windup controller of Eqs. (11.1) is influenced by the filter A_o.

11.5   Complete the algorithm skeleton for the cases of

(a) a direct self-tuner (Algorithm 3.3).

(b) an indirect self-tuner without zero cancellation (Algorithm 3.2).

11.6   Perform a transformation of a second- and fourth-order Bessel filter, $\omega_B = 1$, into band-pass filters, using the transformation

$$s \to \frac{s^2 + \omega_l \omega_h}{s(\omega_h - \omega_l)}$$

where ω_l and ω_h are the lower and upper cutoff frequencies, respectively. Use $\omega_l = 100$ and $\omega_h = 1000$ rad/s. Compare the band-pass characteristics by using Bode diagrams.

11.7   Use Euclid's algorithm to compute the greatest common divisor of

$$A = q^3 - 2q^2 + 1.45q - 0.35$$
$$B = q^2 - 1.1q + 0.3$$

Also determine the polynomials X and Y in Eq. (11.8).

11.8   Consider the Diophantine equation

$$AR + BS = A_c$$

and let

$$A(q) = (q - 1)(q - 0.9)$$

Use the method in Section 11.4, and compute the resulting controller when (a) $B(q) = q - 0.6$; (b) $B(q) = q - 0.9$. Assume that the desired closed characteristic polynomial is

$$A_m(q) = q^2 - q + 0.7$$

11.9 One way to solve the Diophantine equation is to multiply it by a persistently exciting signal, such as white noise. Introduce the filtered signals

$$v_a(t) = \frac{\hat{A}}{A_m A_o} \, v(t) \qquad v_b(t) = \frac{\hat{B}}{A_m A_o} \, v(t)$$

The Diophantine equation then becomes

$$Rv_a + Sv_b = v$$

The coefficients of the R and S polynomials can now be determined by using the method of least squares, and one iteration can be done at each sampling instance. Discuss the merits and drawbacks of this approach. (*Hint*: What is the convergence rate?)

REFERENCES

Implementation issues for adaptive controllers are discussed in:

Wittenmark, B., and K. J. Åström, 1980. "Simple self-tuning controllers." In *Methods and Applications in Adaptive Control*, ed. H. Unbehauen, pp. 21–30. Berlin: Springer-Verlag.

Åström, K. J., 1983. "Analysis of Rohrs' counter example to adaptive control." *Preprints of the 22nd IEEE Conference on Decision and Control*, pp. 982–987. San Antonio, Tex.

Wittenmark, B., and K. J. Åström, 1984. "Practical issues in the implementation of self-tuning control." *Automatica* **20**: 595–605.

Clarke, D. W., 1985. "Implementation of self-tuning controllers." In *Self-tuning and Adaptive Control: Theory and Applications*, eds. C. J. Harris and S. A. Billings. London: Peter Peregrinus.

Isermann, R., and K.-H. Lachmann, 1985. "Parameter adaptive control with configuration aids and supervision functions." *Automatica* **21**: 623–638.

Middleton, R. H., G. C. Goodwin, D. J. Hill, and D. Q. Mayne, 1988. "Design issues in adaptive control." *IEEE Trans. Automat. Contr.* **AC-33**: 50–58.

Wittenmark, B., 1988. "Adaptive control: Implementation and application issues." In *Adaptive Control Strategies for Industrial Use*, eds. S. L. Shah and G. Dumont, pp. 103–120, Proceedings of a Workshop, Kananaskis, Canada. New York: Springer-Verlag.

Different design methods and their properties are treated in:

Lennartson, B., and T. Söderström, 1986. "An investigation of the intersample variance for linear stochastic control." *Preprints of the 25th IEEE Conference on Decision and Control*, pp. 1770–1775. Athens.

Åström, K. J., and B. Wittenmark, 1990. *Computer Controlled Systems*, 2nd ed. Englewood Cliffs, N.J.: Prentice-Hall.

Aspects on implementation of estimation routines are found in:

Bierman, G. J., 1977. *Factorization Methods for Discrete Sequential Estimation.* New York: Academic Press.

Ljung, L., and T. Söderström, 1983. *Theory and Practice of Recursive Identification.* Cambridge, Mass.: MIT Press.

Goodwin, G. C., and K. S. Sin, 1984. *Adaptive Filtering, Prediction and Control.* Englewood Cliffs, N.J.: Prentice-Hall.

Hägglund, T., 1985. "Recursive estimation of slowly time varying parameters." *Preprints of the 7th IFAC Symposium on Identification and System Parameter Estimation,* pp. 1137–1142. York, U.K.

Kulhavý, R., 1987. "Restricted exponential forgetting in real-time identification." *Automatica* **23**: 589–600.

Ljung, L., and S. Gunnarsson, 1990. "Adaptation and tracking in system identification: A survey." *Automatica* **26**: 7–21.

Different ways to introduce integrators in adaptive controllers are treated in:

Wellstead, P. E., and P. Zanker, 1982. "Techniques for self-tuning." *Optimal Control Applications & Methods* **3**: 305–322.

Design of low-pass and band-pass filters can be studied in:

Rabiner, L. R., and B. Gold, 1975. *Theory and Application of Digital Signal Processing.* Englewood Cliffs, N.J.: Prentice-Hall.

The dyadic decomposition method described in Section 11.6 is based on:

Gentleman, W. M., 1973. "Least squares computations by Givens transformations without square roots." *J. Inst. Math. Appl.* **12**: 329–336.

Peterka, V., 1987. "Algorithms for LQG self-tuning control based on input-output delta models." In *Adaptive Systems in Control and Signal Processing 1986,* eds. K. J. Åström and B. Wittenmark, IFAC Proceedings. Oxford, U.K.: Pergamon Press.

The use of the Diophantine equation in control is surveyed in:

Kučera, V., 1993. "Diophantine equations in control: A survey." *Automatica* **29**: 1361–1375.

CHAPTER 12

COMMERCIAL PRODUCTS
AND APPLICATIONS

12.1 INTRODUCTION

There have been a large number of applications of adaptive feedback control over the past 30 years. Experiments with adaptive flight-control systems were done in 1960. Industrial experiments with self-tuning regulators were performed in 1972. Full-scale experiments with adaptive autopilots for ship steering were done in 1973. Special adaptive systems have been in continuous use for a long time. Some process control loops have been running continuously since 1974. There are also a number of special products that have been operating for a long time. Commercial systems for ship steering have been in continuous operation since 1980.

Systems implemented by using minicomputers appeared in the early 1970s. However, not until the 1980s did adaptive techniques start to have real impact on industry. The number of applications increased drastically with the advent of the microprocessor, which made the technology cost-effective. Because of this, adaptive controllers are also entering the marketplace even in single-loop controllers. Several commercial products based on adaptive techniques were introduced in the early 1980s, and second- and third-generation versions have been introduced in some cases.

Adaptive techniques are used in a number of products. Gain scheduling is the standard method for design of flight control systems for high-performance aircraft, and it is also used in robotics and process control. The self-oscillating adaptive system is used in several missiles. There are several commercial adaptive systems for ship steering, motor drives, and industrial robots. Adaptive techniques are used both in single-loop controllers and in general-purpose process control systems in the process industry. Most industrial processes are controlled by PID controllers, and a large industrial plant may have thousands of them. Many instrument engineers and plant personnel are used to select, install, and operate such controllers. In spite of this, many controllers are poorly tuned. One reason is that simple, robust methods for automatic tuning have not been available. Adaptive methods are now available for automatic tuning of PID controllers. This is in fact one of the fastest-growing areas of application for adaptive control.

However, adaptive techniques are still not widely used; the technology is not mature. Because of involvement of commercial enterprises in adaptive control, it is not always possible to find out precisely what is being done. Various ideas are hidden in proprietary information that is carefully guarded.

This chapter is organized as follows. An overview of some applications is given in Section 12.2. A number of commercial products that use adaptation are presented in Sections 12.3 and 12.4. Some specific applications are presented in more detail in the sections that follow. Ship steering, automobiles, and ultrafiltration are areas given special attention.

12.2 STATUS OF APPLICATIONS

A large number of experiments with adaptive control have been performed since the mid-1950s. The experiments have had different purposes: to verify ideas, to find out how adaptive systems perform, to compare different approaches, and to find out when they are suitable. The early experiments, which used analog implementations, were plagued by hardware problems. When digital process computers became available, they were natural tools for experimentation. Experiments with adaptive control required substantial programming, since adaptation was not part of the standard software. Applications proliferated with the advent of the microprocessor, which is a convenient tool for implementing adaptive systems. Adaptive techniques now appear both in single-loop controllers and as standard elements of large process control systems. There are tailor-made controllers for special purposes that use adaptive techniques.

Feasibility Studies

A number of feasibility studies have been performed to evaluate the usefulness of adaptive control. They cover a wide range of control problems, such as autopilots for missiles, ships, and aircraft; engine control; motion control; machine tools; industrial robots; power systems; distillation columns; chemical reactors; pH control; furnaces; heating; and ventilation. There are also applications in the biomedical area. The feasibility studies have shown that there are cases in which adaptive control is very useful and others in which the benefits are marginal. Some industrial products also use adaptive techniques. There are both general-purpose controllers and controllers for special applications.

Auto-tuning

Simple controllers with two or three parameters can be tuned manually if there is not too much interaction between the adjustments of different parameters, but manual tuning is not possible for more complex controllers. Traditionally, tuning of complex controllers has taken the route of modeling or identification and controller design. This is often a time-consuming and costly procedure, which can be applied only to important loops or to systems that are to be manufactured in large quantities.

All adaptive techniques can be used to provide automatic tuning. In such applications the adaptation loop is simply switched on. Perturbation signals may be added to improve the parameter estimation. The adaptive controller is run until the performance is satisfactory; then the adaptation loop is disconnected, and the system is left running with fixed controller parameters. The particular methods for automatic tuning of PID controllers that were discussed in Chapter 8 have been found to be particularly attractive because they require little prior information and are closely related to standard industrial practice.

Auto-tuning can be considered a convenient way to incorporate automatic modeling and design in a controller. It simplifies the use of the controller, and it widens the class of problems in which systematic design methods can be used cost-effectively. This is particularly useful for design methods such as feedforward that depend critically on good models.

Automatic tuning can be applied to simple PID controllers as well as to more complicated systems. It is very convenient to introduce tuning into a DDC package because the tuning algorithm can serve many loops. Auto-tuning can also be included in single-loop controllers. For example, it is possible to obtain standard controllers in which the mode switch has three positions: manual, automatic, and tuning. A well-designed auto-tuner is very easy to use, even for unskilled personnel. Experience has shown it to be useful both for commissioning of new systems and for routine maintenance. Auto-tuners can also be used to enhance the skill of the instrument engineers. Automatic tuning will probably also be a useful feature of more complicated controllers.

Automatic Construction of Gain Schedules

Gain scheduling is a very useful technique, but it has the drawback that it may be quite time- and cost-consuming to build a schedule. Auto-tuning can conveniently be used to build gain schedules. A scheduling variable is first determined. The parameters that are obtained when the system is running in one operating condition are then stored in a table together with the scheduling variable. The gain schedule is obtained when the process has operated at a variety of operating conditions that covers the operating range.

True Adaptive Control

The adaptive techniques may, of course, also be used for genuine adaptive control of systems with time-varying parameters. There are many ways to do this. The operator interface is important, since adaptive controllers also have parameters that must be chosen. Controllers without any externally adjusted parameters can be designed for specific applications, in which the purpose of control can be stated *a priori*. The ship steering autopilot discussed in Section 12.6 is a typical example. In many cases, however, it is not possible to specify the purpose of control *a priori*. It is at least necessary to tell the controller what it is expected to do. This can be done by introducing dials that give the desired properties of the closed-loop system. Such dials are characterized as *performance-related*. New types of controllers can be designed by using this concept. For example, it is possible to have a controller with one dial, labeled with the desired closed-loop bandwidth. Another possibility would be to have a controller with a dial that is labeled with the weighting between state deviation and control action in an LQG problem. Adaptation can also be combined with gain scheduling. A gain schedule can be used to get the parameters quickly into the correct region, and adaptation can then be used for fine-tuning.

Adaptive Feedforward

In many applications it is possible to measure some of the disturbances acting on the process. Feedforward control is very useful when there are measurable disturbances. With feedforward it is possible to decrease the influence of disturbances substantially. However, feedforward control, being an open-loop compensation, requires good models of process dynamics. Identification and adaptation therefore appear to be prerequisites for effective use of feedforward compensation. Until now, very little research and development have been done on adaptive feedforward, even if it was used in the early applications of self-tuning regulators.

Abuses of Adaptive Control

An adaptive controller is more complex than a fixed-gain controller, since it is nonlinear. Before we attempt to use an adaptive controller, it may therefore be useful to investigate whether the problem can be solved with a robust constant-gain controller, as discussed in Chapter 10. As was pointed out in Chapter 1, it is not possible to judge the need for adaptation from the variations in the open-loop dynamics. The open-loop responses may vary much while the closed-loop responses are close and vice versa.

The complexity of the controller has to be balanced against the engineering effort required to make the system operational. Experience has shown that only a modest effort is required to make a standard adaptive system work well.

12.3   INDUSTRIAL ADAPTIVE CONTROLLERS

A number of industrial products incorporate adaptive control techniques. The products can be divided into

- Tuning tools for standard controllers,
- Adaptive standard process controllers,
- General-purpose toolboxes for adaptive control, and
- Special-purpose adaptive controllers.

Because of the large number of different products, it is possible to give only some examples from the different categories.

Tuners for Standard Process Controllers

There are many products for tuning of standard controllers of PID type. Leeds and Northrup announced a PID controller with a self-tuning option in 1981. SattControl in Sweden announced auto-tuning for PID controllers in a small DDC system in 1984 and a single-loop controller with auto-tuning in 1986. Practically all PID controllers that come on the market today have some kind of built-in automatic tuning or adaptation. There are four main solutions for the tuners for standard controllers:

- A parametric model approach,
- A nonparametric model approach,
- External tuning devices, and
- Tuning tools in distributed control systems.

The main idea in the parametric model controllers is to make an experiment, usually in open loop, and estimate a first- or second-order model with time delay. The input signals are usually steps, but pulses or pseudo-random binary sequence (PRBS) signals are also used. The parameters of a PI or PID controller are then determined by using empirical tuning rules or a pole placement technique. Typical products in this category are Protonic from Hartman & Braun and UDC 6000 from Honeywell.

In the nonparametric model approach, a point on the Nyquist curve is generally estimated by using relay feedback. Compare the auto-tuning discussed in Chapter 8. On the basis of this information a modified set of Ziegler-Nichols tuning rules are used to determine the parameters of the controller. SattControl ECA40 and Fisher-Rosemount DPR900 are typical of this category.

The tuning aids discussed above are built-in features in the standard controllers. The operator initiates tuning by pushing a button or giving a command. The external tuning tools are special types of equipment that are connected to the process for the tuning or commissioning and then removed. The experiments are usually done with the process in open loop. The external tuner then determines suitable controller parameters. The new parameters are often entered manually by the operator. Since the external tuner can be used for different types of standard controllers, it must have detailed knowledge about the parameterization and implementation of algorithms from different manufacturers. Examples of external tuning tools are Supertuner from Toyo Systems in Japan, Protuner from Techmation in Arizona, PIDWIZ from BST Control in Illinois, and SIEPID from Siemens in Germany.

Tuning tools have also been introduced in distributed control systems. Because of the available computing power, it is possible to have very good human-machine interfaces and several options for tuning. Honeywell has a system called Looptune; Fisher-Rosemount Systems has a product called Intelligent Tuner.

Adaptive Standard Process Controllers

The tuners discussed above do not tune the controllers continuously but only on demand from the operator. However, there are also standard controllers with adaptation, which can follow changes in the parameters of the process. The adaptive standard controllers can be divided into

- A parametric model approach,
- A nonparametric model approach, and
- A pattern recognition approach.

The model-based adaptive controller usually estimates a first- or second-order model with time delay using a recursive least-squares algorithm. A pole placement controller with PID structure can then be determined. Examples are the Bailey Controls CLC04 and Yokogawa SLPC-181, -281.

One example of a nonparametric adaptive controller is SattControl ECA 400. (See Fig. 1.23.) It is a development of the relay-based auto-tuner. One point of the Nyquist curve is estimated continuously by using band-pass filtering. The parameters of the controller are then determined by using a modified version of the Ziegler-Nichols tuning rules.

Expert systems or pattern recognition have also been used for adaptive tuning of standard controllers. The first was the Foxboro EXACT, which was announced in October 1984. This controller is described in more detail in the text that follows. In 1987, Yokogawa announced adaptive PID controllers, SLPC-171 and SLPC-271, which have features similar to those of Foxboro's

EXACT. Another controller in this category is Fenwal 570. The Honeywell UDC 6000 controller uses step response analysis for automatic tuning and a rule base for adaptation. These controllers are designed to capture the skill of an experienced control engineer in rules. About 100–200 rules are typically implemented. The controllers are waiting for changes in the reference value or large upsets of the process. On the basis of the response and the tuning rules, the parameters of the controller are modified to increase the performance of the closed-loop system.

Several of the adaptive standard controllers, for example, Fisher DPR 910 and SattControl ECA400, have adaptive feedforward and the possibility to build up gain scheduling tables automatically. These features are very useful and can improve the performance considerably.

Standard controllers with more sophisticated control algorithms are now appearing on the market. One example is U.A.C. (Universal Adaptive Controller) from Process Automation Systems in British Columbia, which is based on predictive control. The controller can also handle multivariable systems.

General-Purpose Toolboxes for Adaptive Control

There is often a need to use more elaborate control algorithms than the standard PID controllers. It is then necessary to estimate higher-order models and to have the possibility to use different design algorithms. To cover these situations, general toolboxes for adaptive control have been developed. The adaptive algorithms are usually modules or blocks in more general packages for direct digital control (DDC). Asea Brown Boveri presented a general-purpose adaptive controller in 1982. First Control Systems in Sweden introduced an adaptive controller in 1986. It is also possible to implement adaptive control in modern distributed control systems.

PLC Implementations

Adaptive controllers can also be implemented in ordinary programmable logic controller (PLC) systems. Such solutions are used by manufacturing companies with competent in-house expertise. For example, 3M has implemented adaptive controllers in this way. The first installation was made in 1987. Currently, there are about 200 adaptive loops in operation. A wide range of processes are controlled. The systems are implemented on a variety of platforms such as General Electric, Modicon, Measurex, Square-D, and Reliance. Programming is done in Basic or C. The applications include standard loops for temperature, pressure, position, and humidity and more specialized loops associated with 3M proprietary processes. The adaptive algorithms that are used are based on estimation of parameters in models having the structure

$$A^*(q^{-1})y(t) = B_1^*(q^{-1})u(t-d) + B_2^*(q^{-1})v(t-d)$$

where v is a measurable disturbance. Polynomial A^* has degree one or two, but polynomials B_1^* and B_2^* may have higher degree to cope with variable time delay. The parameters are estimated by a special gradient technique. The control design is a modified minimum-variance strategy.

Special-Purpose Adaptive Controllers

For many processes, extensive process knowledge is available. To make good control, it is advantageous to use as much *a priori* knowledge as possible. Structures of the model and knowledge of integrators or time constants can be used to design the controller and to facilitate the tuning. For instance, special-purpose adaptive controllers have been developed for ships, pulp digesters, motor drives, ultrafiltration, and cement raw material mixing.

12.4   SOME INDUSTRIAL ADAPTIVE CONTROLLERS

Some representative commercial products and their features are described in this section. Special emphasis is put on properties such as estimation, prior information, and industrial experiences. The section ends with a discussion of some general aspects of industrial use of adaptive controllers.

SattControl ECA40 and Fisher Control DPR 900

This is the original auto-tuner based on relay oscillations, as described in Chapter 8. It was first introduced in a small (about 45 loops) DDC system for industrial process control SDM20. In this application the tuner can be connected to tune any loop in the system. Relay auto-tuning is also available in single-loop PID controllers (SattControl ECA40 and Fisher Control DPR900). In these controllers, tuning is done on demand by pushing a button on the front panel, so-called *one-button tuning*. The controllers are also provided with facilities for gain scheduling. There is a table with three controller settings.

Parameter Estimation.   The ultimate period and the ultimate gain are determined by an experiment with relay feedback. The fluctuations in the output signal are measured, and the hysteresis of the relay is set slightly wider than the noise band. The initial relay amplitude is fixed. The amplitude and period are measured for each half-period. A feedback adjusts the relay amplitude so that the limit cycle oscillation has a given amplitude. When two successive half-periods are sufficiently close, PID parameters are computed, and PID control is initiated automatically.

Control Design.   When the ultimate gain and the ultimate period are known, the parameters of a PID controller can be determined by a modified Ziegler-

Nichols rule. There is also a limited amount of logic to determine whether derivative action is needed.

Prior Information.   A major advantage of the auto-tuner is that no parameters have to be set *a priori*. To use the tuner, the process is simply brought to an equilibrium by setting a constant control signal in manual mode. The tuning is then activated by pushing the tuning button. The controller is automatically switched to automatic mode when the tuning is complete. Different control objectives may be obtained by modifying the parameters in the Ziegler-Nichols rule. One mode is chosen by default, but the user can request a slower or an extra-fast response.

Industrial Experiences.   The system has been considered very easy to use, even by inexperienced personnel. Both the auto-tuning and gain-scheduling features have been found to be very useful. In many applications the auto-tuner has contributed significantly to improved tuning. It has also been demonstrated that commissioning time can be shortened significantly by using automatic tuning and that the standard controller can be applied to processes having a wide range of time scales. Simplicity is the major advantage of the auto-tuner. This has proved particularly useful for plants that do not have qualified instrument engineers and for operation during the night shift, when instrument engineers are not available. It is also easy to explain the auto-tuner to the instrument engineers. The properties of the auto-tuner are illustrated by an example.

EXAMPLE 12.1   **Level control**

Figure 12.1 shows the behavior of the controller when it is used to control the level of a vessel in a pulp mill. A controller with pure proportional action was

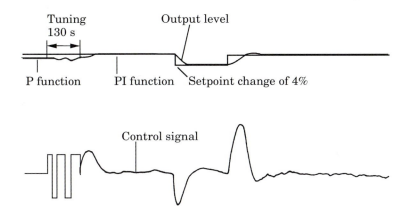

Figure 12.1   Results obtained when using the SattControl ECA40 for level control in a pulp mill.

used originally, resulting in the steady-state error shown in the figure. The tuning took about two minutes and resulted in a PI controller. This example illustrates the usefulness of the logic for selecting control action. Figure 12.1 also shows the control signal.   □

EXACT: The Foxboro Adaptive Controller

This controller is based on analysis of the transient response of the closed-loop system to setpoint changes or load disturbances and traditional tuning methods of the Ziegler-Nichols type.

Parameter Estimation. Assuming controller parameters such that the closed-loop system is stable, a typical response of the control error to a step or impulse disturbance is shown in Fig. 12.2. Heuristic logic is used to detect that a proper disturbance has occurred and to detect the peaks e_1, e_2, and e_3 and period T_p. The estimation process is simple, but it is based on the assumption that the disturbances are steps or short pulses. The algorithm can give wrong estimates if the disturbances are two short pulses because T_p will then be estimated to be the distance between the pulses.

Control Design. The control design is based on specifications on damping, overshoot, and the ratios T_i/T_p and T_d/T_p, where T_i is the integration time, T_d is the derivative time, and T_p is the period of oscillation. The damping is defined as

$$d = \frac{e_3 - e_2}{e_1 - e_2}$$

and the overshoot as

$$o = -\frac{e_2}{e_1}$$

In typical cases, both d and o must be less than 0.3. Empirical rules are used to calculate the controller parameters from T_p, d, and o. These rules are based on

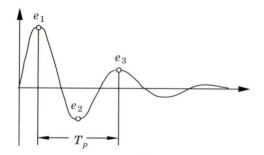

Figure 12.2   Typical response of control error to step or impulse disturbances.

traditional tuning rules of the Ziegler-Nichols type, augmented by experiences from controller tuning.

Prior Information.   The tuning procedure requires prior information about the controller parameters K_c, T_i, and T_d. It also requires information on the time scale of the process. This is used to determine the maximum time the heuristic logic waits for the second peak. Some measure of the process noise is also needed to set the tolerances in the heuristic logic. Some parameters may also be set optionally: damping d, overshoot o, maximum derivative gain, and bounds on the controller parameters.

Pre-tuning.   The tuning procedure requires reasonable controller parameters to be known so that a closed-loop system with a well-damped response is obtained. There is a pre-tune mode that can be used if the prior information that is needed is not available. A step test is done in which the user specifies the step size. Initial estimates of the controller parameters are determined from the step, and the time scale and the noise level are also determined. The pre-tune mode can be invoked only when the process is in steady state.

Industrial Experiences.   Thousands of units of EXACT controllers are in use today. The system is also available in Foxboro's system for distributed process control. Users from a large number of installations have reported favorably, citing the ease with which controllers can be well tuned and the ability to shorten commissioning time. It is also mentioned that derivative action can often yield significant benefits.

Eurotherm Temperature Controller

Temperature control is traditionally done with simple PID controllers, which are cheaper than conventional industrial controllers. Auto-tuning is now also used in such simple systems. One example is controllers produced by Eurotherm in the United Kingdom. A modified relay tuning is used in those controllers. Full control power is used until an artificial setpoint is reached. Two half-periods of a relay tuning are then used, and the controller parameters are calculated from the transient. The controller also has facilities for automatic on-line tuning based on transient response analysis.

In temperature control loops, there are usually different dynamics depending on whether the temperature is increasing or decreasing. This nonlinearity can be handled by using gain scheduling.

Asea Brown Boveri (ABB) Adaptive Controller

The Asea Brown Boveri (ABB) adaptive controller was first marketed under the name Novatune. It is an adaptive controller that is incorporated as a part of ABB Master, a distributed system for process control. The system is

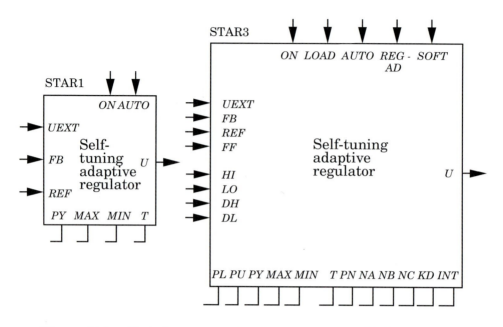

Figure 12.3   Block diagrams of the adaptive modules STAR1 and STAR3, available in the ABB adaptive controller.

block-oriented, which means that the process engineer creates a system by selecting and interconnecting blocks of different types. The system has blocks for conventional PID control, logic, and computation. Three different blocks, called STAR1, STAR2, and STAR3, are adaptive controllers. The adaptive controllers are self-tuning regulators based on least-squares estimation and minimum-variance control. The controllers all use the same algorithm; they differ in the controller complexity and the prior information that must be supplied in using them.

The ABB adaptive controller differs from the controllers that were discussed previously in that it is not based on the PID structure. Instead, its algorithm is based on a general pulse transfer function. It also admits dead-time compensation and feedforward control. The ABB adaptive controller system may be viewed as a toolbox for solving control problems.

Principle. The ABB adaptive controller is a direct self-tuning regulator similar to Algorithm 4.1 in Section 4.3. The parameters of a discrete-time model are estimated by using recursive least squares. The control design is a minimum-variance controller, which is extended to admit positioning of one pole and a penalty on the control signal. The block diagrams in Fig. 12.3 show two of the adaptive modules. The ABB adaptive controller system has three adaptive modules: STAR1, STAR2, and STAR3. STAR3 is the most complicated. The simpler ones have fewer inputs and have default values on some of the parameters in STAR3. In the block diagram the input signals are shown on the left and top sides of the box, the output signals on the right, and the parameters on the bottom. The parameters can be changed at configuration time. The parameters *PL*, *T*, and *PN* can also be changed on-line.

The simplest module, STAR1, has three input signals: the manual input *UEXT*, the measured value *FB*, and the setpoint *REF*. It has three parameters. The variable *PY* is the smallest relevant change in the feedback signal; the adaptation is inhibited for changes less than *PY*. The parameters *MAX* and *MIN* denote the bounds on the control variable, and *T* is the sampling period.

The module STAR2 has more input signals. It admits a feedforward signal *FF*. There are also four signals, *HI*, *LO*, *DH*, and *DL*, that admit dynamic changes on the bounds of the control variable and its rate of change. There are also additional parameters: *PN*, for a penalty on the control variable, and *KD*, which specifies the prediction horizon. The module also has two additional mode switches: *REGAD*, which turns off adaptation when false, and *SOFT*, which allows a soft start.

The module STAR3 has an additional function *LOAD*, which admits parameters stored in an EEPROM to be loaded. It also has several additional parameters, which admit positioning of one pole *PL* and specification of controller structure *NA*, *NB*, *NC*, and *INT*.

Parameter Estimation.   The parameter estimation is based on the model

$$(1 - PLq^{-1})y(t + KD) - (1 - PL)y(t)$$
$$= A^*(q^{-1})\Delta y(t) + B^*(q^{-1})\Delta u(t) + C^*(q^{-1})\Delta v(t)$$

where A^*, B^*, and C^* are polynomials in the delay operator q^{-1}, y is the measured variable, u is the control signal, v is a feedforward signal, and Δ is the difference operator $1 - q^{-1}$. (Compare with Algorithm 3.6.) The integers *NA*, *NB*, and *NC* give the number of coefficients in the polynomials A^*, B^*, and C^*, respectively. The number *PL* is the desired pole location for the optional pole. When parameter *INT* is zero, a similar model without differences is used. The parameters are estimated by using recursive least squares with a forgetting factor $\lambda = 0.98$. Parameter estimation is suspended automatically when the changes in the control signal and the process output are less than *PU* and *PY*. The parameter updating may also be suspended on demand through the switch *REGAD*. In combination with other modules in the ABB adaptive controller system, this constitutes a convenient way to obtain robust estimation.

Control Design.   The control law is given by

$$\left(\rho + B(q^{-1})\right)\Delta u(t) = (1 - PL)(u_c(t) - y(t)) - A^*(q^{-1})\Delta y(t) - C^*(q^{-1})\Delta v(t)$$

where ρ is a penalty factor related to *PN*. Since the algorithm is a direct self-tuner, the controller parameters are obtained directly from the estimated parameters.

Industrial Experiences.   The ABB adaptive controller has been applied to a wide range of process control problems in the steel, pulp, paper, and petro-chemical industries, wastewater treatment, and climate control. Some applications have given spectacular improvement of performance compared to PID control. This is particularly the case for processes with time delay, and in applications in which adaptive feedforward can be used. It has also been used to make special-purpose systems for special application areas such as paper winding and climate control. Some ABB adaptive controller applications are described in more detail in Section 12.5. The essential drawback of the ABB adaptive controller is that it is based on a direct self-tuner. This means that the sampling period and the parameter KD have to be chosen with care. It may, for example, be difficult to use very short sampling periods.

Firstloop: The First Control Adaptive Controller

The adaptive system Firstloop was developed by First Control Systems, a small company founded by members of the Novatune team. Firstloop is a small controller module with up to eight self-tuning regulators. The system is a toolbox with modules for adaptive control, logic, filtering square root functions, and operator communication. An interesting feature is that the adaptive controller is the *only* controller available in the system. However, by choosing the number of parameters of the estimated model, it is possible to get different controller structures—for instance, a PID controller. The adaptive controller can tune ten parameters with a sampling period of 20–50 ms. The software admits easy configuration of a control system. The First Control

Figure 12.4   The MicroController from First Control. (With courtesy of First Control Systems AB.)

controller is shown in Fig. 12.4. Firstline is a distributed process control system with a block-oriented language for control design. The adaptive controller is incorporated as a standard function module. We will describe the adaptive control module in detail.

Principle.   The adaptive control unit used in Firstloop and Firstline is based on recursive estimation of a transfer function model and a control law based on indirect pole placement. The controller also admits feedforward. The main advantage of using an indirect pole placement algorithm is that the system can be applied to nonminimum-phase systems and systems with time-varying time delays. This also implies that short sampling periods can be used. (Compare the discussion in Section 6.9.) The adaptive module comes in two versions, a standard module and an expert module. The standard module is intended for use by ordinary instrument engineers who are not specialists in adaptive control. The expert module shown in Fig. 12.5 is intended for specialists in adaptive control. Many parameters are given default values in the standard module. The variables that must be specified are shown in Fig. 12.5. The signal connections are measured value *MV*, setpoint *SP*, external control signal *UE*, feedforward *FF1*, *FF2*, and controller output *U*. The mode switches *ON*, *AUTO*, and *ADAPT* are for on/off, auto/normal, and adaptation on/off, respectively.

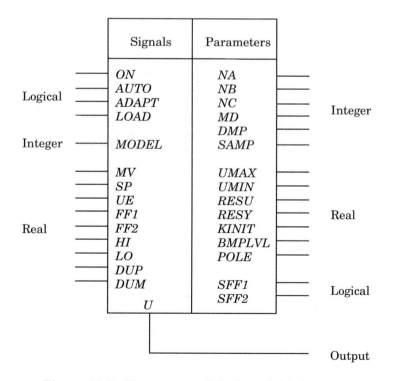

Figure 12.5   The expert module STREGX in Firstloop.

Parameters *UMAX* and *UMIN* define the actuator range. Variables *HI*, *LO*, *DUP*, and *DUM* specify the limits on the control signal that are used internally in the controller. The performance-related parameters are *POLE*, which gives the desired closed-loop pole, and *BMPLVL*, which gives the admissible initial change of the control variable at mode switches.

The desired closed-loop pole is the major variable to be selected. The choice of this variable clearly requires knowledge of the time scales of the process. The recommended rule of thumb is to start with a large value and gradually decrease it.

Parameter Estimation.   The parameters of a transfer function model are estimated. Systems with variable time delay can be captured, provided that a large number of *b* parameters are used. Up to 15 parameters can be estimated in the model. The number of parameters in the model is specified by *NA*, *NB*, and *NC*. Common factors in the pulse transfer function are canceled automatically.

Control Design.   The control design is based on pole placement. The desired response is characterized as a first-order system with delay. The remaining poles are positioned at the origin. The design of the algorithm is based on solving the Diophantine equation by a method that cancels common factors in the estimated polynomials. An LQG-based algorithm is also available. The details of the control design are proprietary.

Safety Network.   The algorithm is provided with extensive safety logic. Adaptation is interrupted when variations in measured signals and control signals are too small. The limits are given with the parameters *RESU* and *RESY*. Adaptation is also interrupted when the control error is below a certain limit, and there are safeguards to ensure that the influence of a single measurement error or sudden large disturbance is limited. (Compare Section 6.9.) Measured values that result in large model errors are also given a low weight automatically. The details of the safety logic are not available. Different models can be stored for use in different situations. The controller is initialized by a model number equal to *MODEL* when *LOAD* changes from false to true.

Industrial Experiences.   Firstloop and Firstline are used in a number of high-performance process control systems. They include control of pulp mills, paper machines, rolling mills, and pilot plants for chemical process control.

Discussion

The products described give an idea of how adaptive techniques are used in commercial products. Additional insight can be derived by analyzing the existing products and trends. Experience from the applications clearly indicates the need for tuning and adaptation; there are undoubtedly many control loops that are poorly tuned. This results in loss of energy, quality, and effective production time. It is also of interest that many different techniques are used, and there are also promising adaptive algorithms that have not yet reached the marketplace. A few specific issues will be discussed in more detail.

Computing Power. Industrial use of adaptive methods has been possible because of the availability of microprocessors. Most of the commercial systems are based on 8-bit processors, with their inherent limitations in addressing capability. This applies to all the PID auto-tuners and the first version of the ABB adaptive controller that used less than 64 kbyte of memory. With 16-bit processors and larger address spaces it is possible to use more sophisticated algorithms and better human-machine communication. The PID auto-tuners typically run with sampling rates of 10–50 Hz.

Intentional Perturbation Signals. To estimate parameters, it is necessary to have data with variations in the control signal. Such variations can be generated naturally or introduced intentionally. Natural perturbations can occur because of disturbances or poorly tuned controllers. Intentional perturbations can be introduced when natural perturbations are not present, as suggested by dual control theory. This method is used in several of the auto-tuning schemes. If prior information about the system dynamics is available, it is possible to find signals that are optimal for the purpose of estimating parameters. Relay feedback automatically generates an input signal having a lot of energy at the frequency at which the process has a phase lag of 180°. Although intentional perturbation signals are both useful and justified by theory, they are often controversial. It should be remembered, however, that poorly tuned controllers may also be considered perturbations.

Controller Structures. Different controller structures are used in the commercial systems. There are both PID controllers and general transfer function systems that admit feedforward and compensation for dead time. The main advantage of the PID structure is that it is close to current industrial practice. Within the PID family there are cases in which derivative action is of little benefit. Systems like the SattControl ECA40 can determine this and choose PI action automatically. However, there is no system that can choose the controller structure generally, although it seems possible to design such systems.

The benefits of feedforward control from measurable disturbances have been known for a long time. Experience with the ABB adaptive controller and Firstloop clearly shows the benefit of adaptive feedforward control. Since feedforward control critically depends on a good model, adaptation is almost a prerequisite for feedforward control. Adaptive controllers like the ABB adaptive controller and Firstloop use a controller structure that is a general transfer function model like

$$R(q)u(t) = T_1(q)u_c(t) + T_2(q)v(t) - S(q)y(t) \qquad (12.1)$$

where u is the control variable, u_c is the command signal, v is a measured disturbance, and y is the controlled output. The polynomials R, S, T_1, and T_2 can be chosen so that the controller corresponds to a PID controller. However, the controller modeled by Eq. (12.1) can also be much more general than a PID controller. It can incorporate many classical features such as filtering, disturbance models, Smith predictors, and notch filters. For more demanding control problems the general transfer function controller thus has significant advantages over the PID controller. However, more expertise in control engineering is needed to understand and interpret the parameters of a controller like Eq. (12.1). Since the PID controller is so common, we can expect it to coexist with more general controllers for a long time.

Multivariable Control.   Multivariable control problems can be handled to a limited extent by using the feedforward feature in the ABB adaptive controller and Firstloop. None of the commercial systems admit truly multivariable adaptive control. Up to now there have not been many applications of adaptive control to true multivariable systems. This situation can be expected to change significantly because of the substantial interest in model predictive control.

Pre-tuning.   It is interesting to note that many schemes have been provided with a pre-tuning feature. In some cases it appears that this was added afterwards. The reason is undoubtedly that too much user expertise is required for the standard algorithms. The selection of sampling periods or the equivalent time scales is a typical example. It appears that the relay method for automatic tuning would be an ideal method for pre-tuning.

Tuning Automatically or on Demand.   The existing products include systems in which tuning is initialized on demand from the operator or automatically. Users of both schemes have documented their experiences. It appears that there are a number of processes for which controllers should be retuned for different operating conditions. In many cases there are measurable signals that correlate well with the operating conditions. In these cases it seems that the combination of on-demand automatic tuning with gain scheduling is a good solution. This will give systems that change parameters faster than systems with adaptation. Of course, it is convenient to have tuning initiated automatically, but it is difficult to give general guidelines for when tuning should be initiated. The simple schemes that are currently in use are often based on simple level detection. Further research is required to find conditions for retuning; this is discussed further in Section 13.4.

An analysis of the division of labor between human and machine gives another viewpoint on the question of on-demand or automatic tuning. When tuning is done on demand of the operator, the ultimate responsibility for tuning clearly remains with the operator or the instrument engineer. This responsibility is carried even further in some systems, in which the instrument engineer has to acknowledge the tuned values before they are used. A good solution would be a system in which the responsibility and the tuning techniques could be moved from the operator to the computer system. Ideally, the system should also allow the operator to learn more about control in general and the particular process in question. Experimental architectures that allow this are available, but not in commercial systems.

Requirements for the User

The requirements for the user are very different for the various commercial systems. The PID controllers in which tuning is initiated automatically require very little. Controllers with on-demand tuning require somewhat more knowledge on the part of the user. Systems such as the ABB adaptive controller and Firstloop can be regarded as toolboxes for solving control problems that are more demanding. They also allow complex control systems to be configured. This is clearly illustrated by the experiences from ABB adaptive controller installations. The system was designed by a very qualified team that included several first-rate Ph.D.s. The design team was also responsible for many of the initial installations, which were extremely successful.

More recent versions of the toolbox systems are much easier to use. Moderate-sized systems have been successfully implemented by instrument engineers with little knowledge of advanced control. There are several reasons for the increased user-friendliness of the systems. The safety logic has been improved significantly; modules in which many parameters are given default values have been designed; and computer-based configuration tools, with a lot of knowledge built in, have been developed. The toolboxes thus allow a user to get started quickly with a modest knowledge of adaptive control, and they also make it possible for a user to construct more advanced systems when more knowledge is acquired.

12.5   PROCESS CONTROL

There are many applications of adaptive control in the field of industrial process control. Some typical examples are discussed in this section. The applications give insight into how adaptive control can be used in practice.

Temperature Control in a Distillation Column

Although the SattControl auto-tuner has been used mostly for conventional loops for control of flow, pressure, and level, it has also been applied to more difficult problems. One example is temperature control in a distillation column. This is a conventional control loop in which the temperature in a tray of a distillation column is measured and the boil-up is manipulated. This control loop was part of a process system with many loops. There had been severe problems with the temperature control for a long time, and several attempts had been made to tune the loop. Figure 12.6 shows a recording of the temperature. The figure shows that the loop is oscillatory with the controller tuning that was used ($K_c = 8$, $T_i = 2000$, and $T_d = 0$). Also notice the long period of the oscillation. The controller was switched to manual at time 11:30, and the temperature then started to drift. Auto-tuning was initiated at time 14:00.

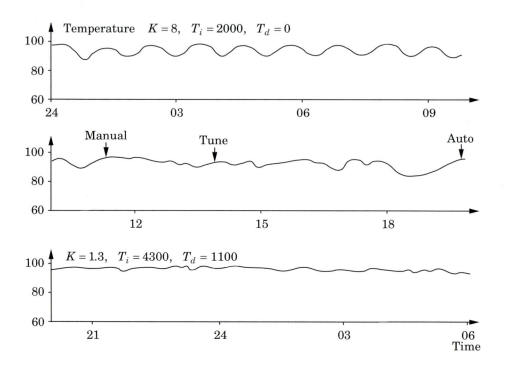

Figure 12.6 Application of the SattControl ECA40 to temperature control in a distillation column.

The tuning phase was completed after six hours at time 20:00, when the controller was automatically set to automatic control. The controller parameters obtained were $K_c = 1.3$, $T_i = 4300$, and $T_d = 1100$. Notice that the whole tuning procedure is fully automatic. The only action taken by the operator was to initiate tuning at time 14:00. The temperature variations during tuning are not larger than those obtained with the conventional controller settings. The example shows that the auto-tuner can cope with a process having drastically different time scales than those normally used.

Chemical Reactor Control

Chemical reactors are typically nonlinear. Characteristics such as catalyst activity change with time, as does the raw material. There are often inherent time delays, which may vary with production level. Poor control can result in lower product quality, damage to the catalyst, or even explosions in exothermic reactors. Chemical reactors are therefore potential candidates for adaptive control. The process in this application consists of two parallel chemical reactors in which ethylene oxide is produced by catalytic oxidation of ethylene. The process is exothermic and time-variable because of changes in catalyst activ-

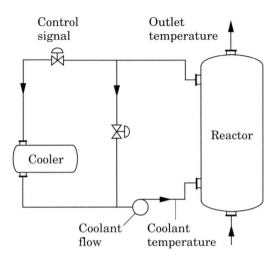

Figure 12.7 Schematic diagram of the reactor.

ity. It is essential to keep the temperature accurately controlled; a reduction of temperature variations improves the yield and prolongs the life of the catalyst. Stable steady-state operation is also a first step toward plant optimization.

The plant was equipped with a conventional control system that used PID controllers to control flow and temperature. The plant personnel were dissatisfied with the system because it was necessary to switch the controllers to manual control in case of many major disturbances, which could happen several times per day.

A schematic diagram of the process is shown in Fig. 12.7. The reactor is cooled by circulating oil to a cooler. The temperature of the coolant at the inlet to the reactor is the primary controlled variable, and the reactor outlet temperature and the coolant flow are also measured. The control signal is the flow to the cooler. The dynamics relating temperatures and flow to valve openings have variable delays and gains.

Disturbances in the process are caused by variations in the incoming gas and load changes. Large disturbances occur with changes in production level or with "shutdowns" caused by failure in surrounding process equipment. During shutdowns it is most important to maintain the process temperature as long as possible so that the production can be restarted easily. With the conventional control system, temperature fluctuations were around ±0.5°C during normal operation and up to ±2°C during larger disturbances. With adaptive control the variations were reduced to ±0.1°C during normal operation and ±0.5°C during large upsets.

The adaptive control system was implemented by using the ABB adaptive controller and the STAR3 module with feedforward from the reactor outlet temperature. By using the other modules in the system, it was also straight-forward to handle the dual valves and to reset to manual mode for startup and shutdown. The system has been in continuous operation since 1982 on a reactor at Berol Kemi AB, which produces 30,000 tons per year. The operational experiences with the system have been very good. With adaptive control, it was possible to reduce the temperature fluctuations significantly. The controllers are now kept in automatic mode most of the time, even during production changes. This has made it possible to revise operational procedures, since operators do not have to spend their time supervising the reactor temperature.

Pulp Dryer Control

Drying processes are common in the process industries. The mechanisms involved in drying are complex and poorly understood, and their dynamics depend on many changing factors. There are often significant benefits in improved regulation, since an even moisture content is an important quality factor. There are also significant potential energy savings. Drying processes are thus good candidates for adaptive control.

In pulp drying, a wet pulp sheet passes a steam-heated drying section and cooling section. A typical system is shown schematically in Fig. 12.8. The moisture content of the sheet entering the dryer is about 55%. At the exit, it is typically 10–20%. It takes about nine minutes to pass the dryer and about half a minute to pass the cooler. The dryer dynamics are complicated. It is influenced by many factors, such as the pH of the sheet. The measurements of the moisture content are obtained by a traversing microwave sensor that moves back and forth across the pulp sheet, describing a diagonal pattern on the sheet. When one traverse movement is complete, the mean value of the diagonal is stored in the computer, the mean value algorithm is reset, the sensor moves back, and the procedure repeats itself. It takes a little less than one minute

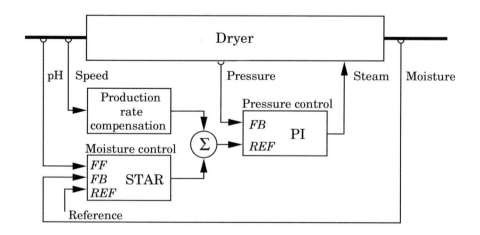

Figure 12.8   Schematic diagram of pulp drying and the control system.

for the sensor to move across the sheet. With manual control, the fluctuations in moisture content often exceed ±1%. The ABB adaptive controller was used in this application. The control system configuration is shown in Fig. 12.8. The moisture control is carried out by an adaptive software controller STAR. The moisture content measured by the traversing system is low-pass filtered and connected to the *FB* input of the STAR. The desired moisture content is chosen by the operator outside the ABB adaptive controller and software connected to the *REF* input of the STAR. The pH value, measured in an earlier process section, is used as the feedforward signal. This signal is connected to the *FF* input of the STAR. The control signal of the STAR defines the desired steam pressure, which is measured and controlled to the desired value by a conventional hardware PI controller. The control signal of this controller acts continuously on the steam flow valve.

The sampling period used in the adaptive controller was 3.5 minutes. A fourth-order Butterworth filter was used as an anti-aliasing filter. This was implemented by using the ABB adaptive controller tools. When the production rate was changed, large upsets were noticed, lasting for about 30 minutes, because it took 5–15 samples for the adaptive controller to settle. It was highly desirable to reduce these upsets, and this was done by introducing a special production rate compensation in the form of a pulse transfer function of the type

$$H(z) = \frac{b(z-1)}{z-a}$$

This gives a rapid change of the steam pressure when pulp speed changes. It was not necessary to make this filter adaptive. The system has been in operation since 1983 at a pulp mill at Mörrum's Bruk that produces 330,000 tons of paper pulp per year. The operational experiences have been very good. Fluctuations in moisture content have been reduced from 1% to 0.2%, which improves quality. It also allows the setpoint to be moved closer to the target value, resulting in significant energy savings.

Control of a Rolling Mill

The process control applications are typical steady-state regulation problems. The rolling mill control problem is much more batch-oriented. It illustrates the use of adaptive techniques in machine control. There are many types of rolling mills, each with its specific control problem. This particular application deals with a skin pass mill located at the end of the production line. The material processed by the mill may vary significantly in dimension and hardness.

The purpose of the mill is to influence quality variables such as hardness and yield limit. A schematic diagram of the process is shown in Fig. 12.9. Let v_1 be the speed of the strip entering the mill, and let v_2 be the speed of the strip at the exit. Because of the thickness reduction, the exit speed is larger

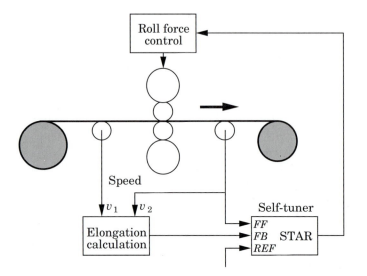

Figure 12.9   Schematic diagram of the rolling mill and the control system.

than the entrance speed. The elongation is defined as

$$\varepsilon = \frac{v_2 - v_1}{v_1}$$

The key control problem is to keep a constant elongation. There is a difficult measurement problem, since the velocity difference is so small. The process operates over a wide range of conditions; the following operating modes can be distinguished:

- Slow rolling at low speed during startup,
- Acceleration to fast rolling,
- Fast rolling at production speed,
- Intermediate decelerations to slow rolling or even to standstill,
- Deceleration to slow rolling at the end of the strip, and
- System at rest waiting for the next strip.

Transition from one mode to another is performed automatically on demand from the operator. It is essential that the control system handle these transitions well. The process dynamics relating elongation to roll force can be described as a high-order dynamical system with an open-loop response time of less than 0.05 s. Changes in production rate from 0 to 2000 m/min in less than 10 s are typical. The dynamics change drastically during the operation; the dynamics of rolling change because of variations in the speed, hardness, and dimension of the strip. There are also significant changes of the inertia of the coilers. All material starts on one coiler and ends up on the other. There are variations in the oil film on the roller bearings due to variations in speed and pressure. The dynamics of the hydraulic system vary with the operating point.

The changes in dynamics due to changing speed are predictable and can (in principle) be taken care of by gain scheduling. Variations in dimension can be handled similarly. The hardness cannot be measured directly on-line, so it must be handled by feedback and adaptation.

The ABB adaptive controller was used in this application. A block diagram of the control system is shown in Fig. 12.9. The speed variations are taken care of in an elegant way. In the ABB adaptive controller, sampling can be triggered by an arbitrary signal. In this case it is triggered by the pulse counters that measure strip speed. This means that sampling is related to the length of the strip, not to time. This is a simple way of making the control system invariant to strip speed (the same idea was used in the ship steering example in Section 9.5). The measurement of the velocity difference is implemented by using pulse generators and counters.

For each strip a saved model is loaded into the controller, and the adaptation is switched on with some delayed action (15 sampling intervals) to avoid adaptation during the first few steps, in which the measurement is irregular. The initial model is taken from a soft strip so that there will be no excessive control action at startup. Soon enough, the controller will adapt to the conditions of the new strip. Figure 12.10 illustrates a typical run of a strip. Notice

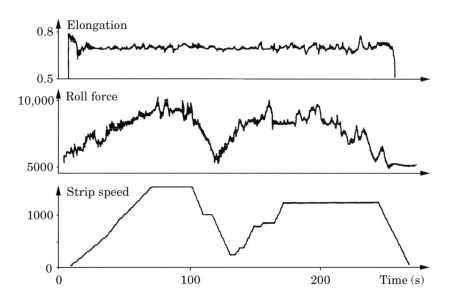

Figure 12.10 Elongation, roll force, and strip speed during a typical run with the system.

Figure 12.11   The cold rolling mill at Avesta-Sheffield is controlled by First Control's adaptive system. (With courtesy of Avesta-Sheffield, precision strip AB, Kloster.)

in particular how well the system copes with the velocity variations and with the mode changes. The installation of the system took about a week, mostly devoted to function and signal checking and tests. The controller functioned almost immediately when connected to the process. After that, approximately two days were devoted to checking and tuning performance. This involved experiments with different sampling rates.

A significant part of the installation time also involved other parts of the system, particularly the logic. Operational experiences with the adaptive control system have been very favorable. The variation in elongation was better than is found with a conventional system, and the adaptive system also settled faster during mode switches. The system has been in continuous operation since 1983.

Figure 12.11 shows a cold rolling mill at Avesta-Sheffield in Långshyttan, Sweden. The process is controlled by First Control's adaptive control system since 1990. The adaptive regulators keep the deviations in the strip thickness within 2–3 μm, which is considered to be very accurate for this kind of mill.

Pulp Digester

Control of the pulp digester is an important part in manufacturing of chemical pulp. The raw material is wood chips, which are broken down into fibers by processing in a liquor composed of sodium hydroxide and sodium sulfide (white

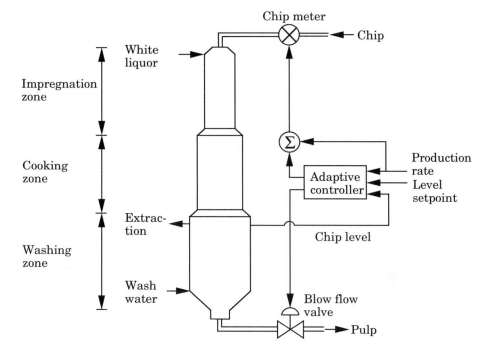

Figure 12.12   Schematic diagram of the chip level controller for a continuous Kamyr digester.

liquor). The process operates either in batches or, more commonly today, as a continuous process.

The Kamyr digester (see Fig. 12.12) is the standard continuous process. The production rate is determined by the chip meter, which feeds chips into the top of the digester. The flow of pulp from the digester is controlled by the blow flow at the bottom. The digester has three zones: impregnation, cooking, and washing. The dynamics that describe the material transport and the chemistry in the digester is very complicated. The total residence time in the digester is about 5 hours. An important control problem is the control of the chip level, which is controlled by the blow flow. The chip level signal is calculated from three strain gauges by using a scheme developed by MoDo Chemetics. The study reported here is a feasibility study made by Pulp and Paper Research Institute of Canada (Paprican) and the pulp company MacMillan Bloedel in Vancouver. The study has resulted in an adaptive controller for digester control developed in cooperation between MoDo Chemetics in Vancouver and Paprican. The commercial adaptive controller manipulates two inputs (blow flow and chip meter) as indicated in Fig. 12.12; in the feasibility study, only the blow flow was manipulated by the adaptive controller.

The industrial digester in the study produced 350 tons per day of kraft pulp. Two grades, R and K, are manufactured. From identification experiments

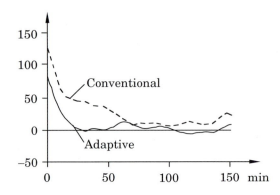

Figure 12.13   Autocovariance of the chip level under conventional and adaptive control. (With courtesy of Paprican.)

it was found that the digester can be described by the model

$$(1 + a_1q^{-1})\Delta y(t) = (b_0 + b_1q^{-1} + b_2q^{-2})\Delta u(t-2) + (1 + c_1q^{-1} + c_2q^{-2})e(t)$$

where $\Delta y(t)$ is the difference in level, $\Delta u(t)$ is the change in blow flow, and $e(t)$ is white noise. The sampling period is 5 minutes. The identification experiments also indicated that it would be feasible to use fixed values of all the parameters except the b_i's. The adaptive controller will thus be able to compensate for gain changes and changes in the time delay of the process. A GPC algorithm with $N_u = 1$, $N_1 = 1$, and $N_2 = 15$–20 is used (see Eq. 4.61). Figure 12.13 shows the autocovariance of the level when conventional PID control and adaptive control were used. The chip level signal is essentially uncorrelated after three lags (15 minutes). The standard deviation of the level decreased from 11.3% to 8.6%. This improvement of the chip level leads to direct improvements in pulp quality. Table 12.1 shows permanganate number (P-number), which is a standard laboratory test of the residual lignin in the pulp. The P-number is closely related to the kappa number. The P-numbers were measured on samples from the blow flow collected once every two hours. For both grades (R and K) the average P-numbers are closer to the target values, and the standard deviations are reduced.

   In summary, the advantages of the adaptive controller are

- Reduction of chip level and P-number variability,
- Reduced need for operator intervention,
- Elimination of manual retuning, and
- Prediction of potential problems with hang-ups in the chip column.

The pulp digester study is an example of a special-purpose adaptive controller. The process model is tailored to fit the specific application, and the parameters can be related to physical parts of the system.

Table 12.1   P-number variability under conventional and adaptive chip level control. (With courtesy of Paprican.)

Controller	Grade	Setpoint	Mean	Std. dev.	No days
Conventional	R	23.0	22.2	2.06	23
	K	21.0	19.9	1.91	6
Adaptive	R	23.0	22.4	1.76	18
	K	21.0	20.6	1.70	7

Pulp digesters have also been controlled by standard adaptive controllers. One example is the Vallvik mill at Assi Domän in Sweden where Novatune controllers in an ABB Master system are used extensively. Several Novatunes are used to control temperatures, flows and levels. The system was installed and commisioned by the regular mill staff with a core team of two enthusiastic engineers. The critical parameters in the Novatune were the sampling period and the prediction horizon; these values had to be selected individually for each application. Default values were used for the other parameters. The prediction sampling period is typically chosen to be 60% to 90% of the dead time, the predictions horizon is chosen as $KD = 2$ and the controller complexity as $NA = NB = NC = 3$. The standard procedure is to run the controllers in manual mode. The parameter estimation is switched on with restrictions on the control action, which are gradually removed.

The experience with adaptive control has been very good. Control performance is significantly better with adaptive control than with PID control. The systems have not required much attention after installation. The reason for improved performance is that tighter control is obtained with adaptive control. Experiments at the plant indicated that there was a good correlation between variations in chip level and the kappa-number. By introducing adaptive control of the chip level it was also possible to significantly reduce the variation in the kappa-number. The standard deviation was reduced from 0.52 to 0.30. It has also been observed that the adaptive controllers recover much faster from large upsets than the systems used previously.

12.6   AUTOMOBILE CONTROL

Microprocessor-based engine control systems were introduced in the automotive industry in the 1970s to address the demands of increased fuel economy and reduced emissions. Early electronic control systems had modest application. Today, the powertrain computer accomplishes a multitude of control tasks, including vehicle speed or "cruise" control, idle speed regulation, automatic transmission shift actuation, control of various emission-related systems,

Figure 12.14   This Ford Mustang has state-of-the-art adaptive power train controls. (With courtesy of Ford Motor Company.)

fuel control, and ignition timing, as well as many diagnostic functions. These highly I/0 intensive systems must be cost effective and function acceptably in many thousands of vehicles with attendant manufacturing variability over a wide range of operating conditions.

Many of the control functions in automobiles are open-loop look-up table oriented. Some automatic calibration methods have been developed to optimize table entries with respect to fuel economy, constrained by emissions. Typical closed-loop structures are comprised of individual operational loops, often PI or PID, and may contain several feedforward paths that are designed to reject measurable or predictable disturbances. Applications of adaptive control concepts can be found in many of those powertrain control functions where on-line self-tuning techniques are used to adjust controller parameters (in most cases, the feedforward parameters) to compensate for component and operating condition variability. One such adaptive control structure is the air-fuel ratio control method introduced by Ford in the mid-1980s to reduce sensitivity to component variability and calibration inaccuracy. Figure 12.14 shows a car with adaptive power train control.

Modern automobiles require precise control of air-fuel ratio to attain high catalytic converter efficiency and minimize tailpipe emissions. Air-fuel ratio control has two principal components: a closed-loop portion in which the fuel injectors are regulated in response to a signal fed back through a digital PI controller from an exhaust gas oxygen sensor located in the engine exhaust stream, and an open-loop or feedforward portion in which fuel flow is controlled in response to an estimate of the air charge entering the engine. (Compare Section 9.5.) This open-loop portion of the control is particularly important during engine transient when the inherent delay of the engine and exhaust system obviate the effectiveness of feedback, and during cold engine operation before the exhaust gas oxygen sensor has reached operational temperature. The purpose of the adaptive algorithm is to adjust the open-loop feedforward gain to reduce deviation from stoichiometric air-fuel ratio operation and improve emission performance under open- and closed-loop operation. This is essentially a gain scheduling process in which an adaptive multiplier is stored in a look-up table as a function of engine speed and load. Initially, all the table entries are unity. As the engine operates throughout its range, the appropriate cell values are increased or decreased to correct for parametric changes or inaccuracies in the initial calibration. In contrast to typical gain scheduling techniques, this adaptation is continuous throughout the vehicle's life.

Another application of adaptive control at Ford is in the area of automotive speed control or "cruise" control. These systems must provide acceptable steady-state error, excellent disturbance rejection, unnoticeable throttle movement, and must be robust to vehicle-to-vehicle variability and operating condition. An adaptive control design based on sensitivity analysis and gradient methods has been used to continuously tune the gains of a PI controller. This was accomplished by constructing a single quadratic cost function and adjusting the proportional and integral control gains to minimize this function. Additional modifications, such as projection and deadband together with slow adaptation, were used to avoid parameter drift and ensure robustness. In this manner, speed control performance is optimized for individual vehicles and operating conditions providing improved performance and reduced calibration effort compared to conventional fixed gain controllers.

12.7 SHIP STEERING

A conventional autopilot for ship steering is based on the PID algorithm. Such a controller has manual adjustment of the parameters of the PID controller and often also a dead zone called *weather adjust*—a simple version of a performance-related knob. Manual adjustments are necessary because the dynamics of a ship vary with speed, trim, and loading. It is also useful to change the autopilot settings when disturbances in terms of wind, waves, currents, and water depth are changed. Adjustment of an autopilot is a burden on the crew. A poor adjustment results in unnecessarily high fuel consumption. It is therefore of interest to have adaptive autopilots. A ship steering autopilot, Steermaster 2000 from Kockum Sonics AB in Sweden, and a roll damping equipment, Roll-Nix from SSPA Maritime Consulting AB in Sweden and Hyde Marine Systems in Ohio, are described in this section.

Ship Steering Dynamics

Simple ship steering dynamics were presented in connection with the discussion of gain scheduling in Section 9.5. That section detailed how the dynamics vary with the velocity of the ship and showed how the variations could be reduced by gain scheduling. It has been shown by hydrodynamic theory that the average increase in drag due to yawing and rudder motions can be approximately described by

$$\frac{\Delta R}{R} = k \left(\bar{\psi}^2 + \lambda \bar{\delta}^2 \right) \tag{12.2}$$

where R is the drag and $\bar{\psi}^2$ and $\bar{\delta}^2$ denote the mean square of heading error and rudder angle amplitude, respectively. The parameters k and λ will depend on the ship and its operating conditions. The following numerical values are typical for a tanker:

$$k = 0.014 \deg^{-2} \qquad \lambda = 1/12$$

It is thus natural to use the criterion

$$V = \frac{1}{T} \int_0^T \left(\left(\psi(t) - \psi_{\text{ref}} \right)^2 + \lambda \delta^2(t) \right) dt \tag{12.3}$$

as a basis for the design and evaluation of autopilots for steady-state course keeping. The disturbances acting on the system are due to wind, waves, and currents. A detailed characterization of the disturbances and their effect on the ship's motion is difficult. In a linearized model, disturbances appear as additive terms. It is common practice to describe them as random signals; the waves have a narrow band spectrum. The center frequency and the amplitude may vary significantly.

Autopilot Design

An autopilot has two main tasks: steady-state course keeping and turning. Minimization of drag induced by the steering is the important factor in course keeping, and steering precision is the important factor in turning. It is therefore natural to have a dual-mode operation. These two modes are described in the text that follows, together with the basic autopilot functions.

The influence of variations in the speed of the ship is handled by gain scheduling. The other disturbances are taken care of by feedback and adaptation. Implementation of the gain scheduling is discussed in Section 9.5. It requires a measurement of the forward velocity of the ship. If disturbances are regarded as stochastic processes, steady-state course keeping can be described as a linear quadratic Gaussian problem. It is then natural to estimate an ARMAX model (Eq. 2.38). The particular process model used is

$$\Delta\psi(t) - a\Delta\psi(t-h) = b_1\delta(t-h) + b_2\delta(t-2h) + b_3\delta(t-3h)$$
$$+ e(t) + c_1 e(t-h) + c_2 e(t-2h) \tag{12.4}$$

This model is built on Nomoto's approximation (compare Section 9.5). The additional b term was introduced to allow additional dynamics to be captured as an increased time delay. The difference occurs because there is a pure integration in the model from rate of turn to heading angle. A control law that minimizes the criterion of Eq. (12.3) is then computed by using the certainty equivalence principle. This approach requires the solution of a Riccati equation, which can be done analytically in the particular case. A straightforward minimum-variance control law was used in some early experiments. This was replaced by the LQG control law described previously, because there were significant advantages at short sampling intervals, which could not be used with the minimum-variance control law. The sampling interval in the model is set during commissioning.

Turning Controller

The major concern in turning is to keep tight control of the motion of the ship, even at the expense of rudder motions. For high turning rates the dynamics of many ships are nonlinear. The normal course-keeping controller can handle small changes in heading, but it cannot handle large maneuvers because of the nonlinearities discussed previously. A special turning controller was therefore designed. The controller is a high-gain controller in which the feedback is of PID type. (Compare Fig. 1.3.) Appropriate PID parameters are determined during commissioning. The model used is nonlinear. It is designed so that the command signal is *turning radius*. The turning rate is thus $r = u/R$, where u is the speed of the ship and R is the turning radius.

Human-Machine Interface

The fact that turning radius is used as a command signal instead of turning rate simplifies maneuvering considerably, because it is easy to determine an appropriate turning radius from the chart. It also improves path following, since the speed of the ship may change during a turn. This is then compensated for automatically. The man-machine interface is very simple. There is one joystick to increase and decrease the heading. An optional joystick provides override control; whenever this is moved, it gives direct control of the rudder angle. Control can be transferred to the autopilot by a reset button. In making a turn, the desired turning radius is set by increase-decrease buttons. The turn is initiated when the joystick is moved to the new desired course. The turn is then executed, and the ship turns until the desired course is reached. The fixed-gain controller is used during the turn, and the adaptive course-keeping controller is initiated when the turn is complete.

There are no adjustments on the course-keeping controller; everything is handled adaptively. Some default values are set during commissioning, but the fixed-gain controller can be activated when the operator pushes a switch labeled *fixed control*. This is typically used when there are heavy waves coming from behind (called a quartering sea). This condition makes steering difficult because the effective rudder forces are small and the disturbing wave forces are large.

Operational Experiences

Early versions of the autopilot were field-tested in 1973, and the product was announced in 1979. The product is used in various kinds of ships. One installation, in a ferry that navigates between Stockholm and Helsinki, has been in continuous operation since 1980. It uses adaptive control all the time. The ability to cope with large variations in speed has been found to be very useful, and the turning radius feature is particularly useful for navigation in archipelagos, where a lot of maneuvering is necessary. Figure 1.24 indicates the improvements in course-keeping that can be obtained through adaptation. The decreased drag with the data shown in the figure corresponds to a reduction in fuel consumption of 2.7%.

Rudder Roll Damping System

On many ships it is desirable to reduce the rolling motion. Conventional roll damping systems on large naval ships use active fins or active as

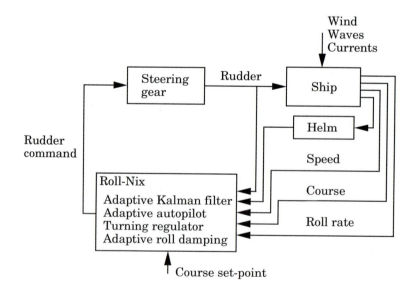

Figure 12.15 Block diagram of the Roll-Nix roll damping system. (With courtesy of SSPA Maritime Consulting AB.)

well as passive tanks. These systems are expensive to install, especially for retrofits. A third approach to roll damping is to use the rudder for roll damping as well as for maneuvering. High-frequency movements of the rudder damp the rolling without influencing the mean value of the heading of the ship. Such a system can be inexpensive, since it can easily be connected to the ordinary steering system. One such system, Roll-Nix, has been developed by SSPA Maritime Consulting in Gothenburg, Sweden. The system is also marketed by Hyde Marine Systems in Cleveland, Ohio. A block diagram of the system is shown in Fig. 12.15. Roll-Nix includes an adaptive Kalman filter, an adaptive course-keeping autopilot (optional), a high-gain turning controller (optional), and an adaptive roll damping controller. The first three parts are similar to those described for the Steermaster 2000 autopilot.

The system uses a roll rate sensor together with course gyro and speed log to determine rudder commands that are superimposed on the ordinary autopilot commands and fed into the steering engine. The operating principle is that the roll movements created by the rudder are opposite those of the roll movements caused by the waves. These counteractive moves damp the roll motions of the ship. In designing the roll damping system it is important to have quick rudder motions. Slow and large motions will influence the course

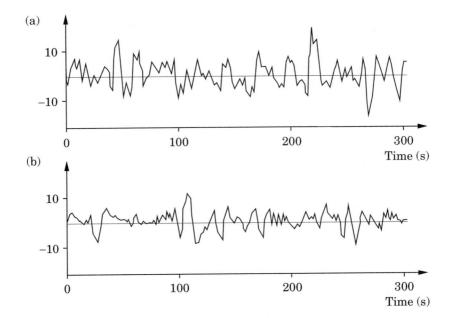

Figure 12.16 Results from sea trials with an attack craft at 27 knots and stern quartering seas (4 Beaufort): (a) without Roll-Nix; (b) with Roll-Nix. The significant roll angle was reduced by 58%, and the maximum roll angle was reduced by 53%. (With courtesy of SSPA Maritime Consulting AB.)

keeping. The Roll-Nix system is provided with an autopilot as an option. The adaptive feature of the roll damping system is necessary to handle different weather conditions and ship speed. The Kalman filter is used to obtain an accurate roll motion signal from the measured roll rate.

Roll-Nix has been tested on several types of ships. For instance, the system has been tested on two Royal Swedish Navy ships: one attack craft and one mine layer. The sea trials show that a significant roll reduction of 45–60% can be obtained for both the standard deviation and the maximum angle of the roll. The result from a sea trial with an attack craft is shown in Fig. 12.16. The roll reduction increases with increasing speed and rudder rates. Tests were also done on the mine layer HMS Carlskrona in September 1987. The following quote from the captain, Commander Hallin, gives an illustration of the performance of the system.

> This particular occasion was when the ship was off the Dutch coast, bound for Helder, with seas coming in from astern on the port quarter. I was resting in my cabin. The time was 04.00 hrs. Suddenly I sensed that the ship had started to roll perceptibly, and I wondered what was going on. At once, I went up on deck and asked the officer of the watch what on earth was happening, and what the reason was for this sudden increase in the ship's rolling motion. I was surprised to receive the reply, "We have just switched off the Roll-Nix. We need to have some data without Roll-Nix working, to see how much damping can be achieved." I think that that is the most illustrative experience I have had of the Roll-Nix system to date.

12.8   ULTRAFILTRATION

Patients with little or no renal function need some form of artificial blood purification to stay alive. In dialysis the blood is cleansed of waste products and excess water, and the electrolytes in the blood are normalized. More than 350,000 patients all over the world undergo this treatment a couple of times a week. In its most common form, hemodialysis, the blood flows past a semiper-meable membrane with a suitably composed dialysis fluid on the other side. Because of the large number of different dialyzers that are on the market, the control algorithm in the dialysis machine must be able to handle a wide span in process gain and other process characteristics.

An adaptive pole placement controller has been used in the fluid control monitor (FCM) developed by Gambro AB in Lund, Sweden. The system has been in use for many years and it has performed very well. This is probably one of the most widely used adaptive controllers in the world today. In this section we describe the system.

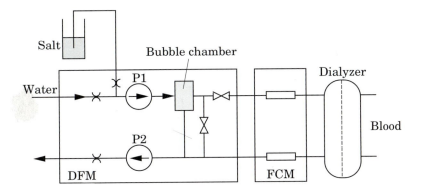

Figure 12.17 Schematic diagram of a dialysis system.

Process Description

A schematic view of the Gambro AK-10 dialysis system is shown in Fig. 12.17. Only the parts that are relevant to flow and pressure control are shown in detail. Clean water is heated to around 37°C, and salt is added to physiological concentration. A pressure drop in the restrictor is created by the first pump to degas the solution. The restrictor and the first pump (P1) determine the flow into the dialyzer. Because of the compressibility of the air in the bubble chamber, flow changes to the dialyzer will be slowed down by a time constant.

After passing a few measuring devices and a valve, the fluid leaves the dialysis fluid monitor (DFM) and passes the first flow-measuring channel of the FCM before entering the dialyzer. Before returning to the DFM, the second measuring channel of the FCM is passed. In the DFM a few more measuring devices and valves are passed before the second pump (P2). A restrictor is placed on the outlet to allow positive pressures in the dialyzer.

To maintain a specified transmembrane pressure, the DFM has a control system that is based on a conventional fixed-gain digital PI controller. (See the block diagram in Fig. 12.18.) This controller has a sampling period of 0.16 s and an integration time of about 30 s. The purpose of the fluid control module is to control weight loss during the treatment. This is done by the external control loop shown in Fig. 12.18, which has the flow difference Q_f as the measured variable and the setpoint to the pressure controller p_c as the control variable.

Process Dynamics

The dialyzer dynamics can be approximately described by the model

$$C \frac{dp}{dt} = Q_f - Bp$$

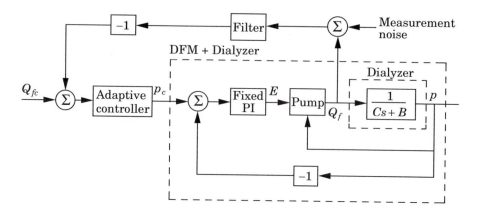

Figure 12.18 Block diagram of the system for controlling transmembrane pressure p and the flow difference Q_f control system.

where p is the transmembrane pressure and Q_f is the net fluid flow from the dialyzer. The constant C is the compliance. Parameter B, which represents the static gain, may, for example, vary from $1.6 \cdot 10^{-12}$ to $120 \cdot 10^{-12}$ (m^3 s^{-1} Pa^{-1}), that is, a gain variation by a factor of 75.

The complete dynamics of the pressure loop can be approximately described as a second-order transfer function. It has one pole associated with the dynamics of the ultrafiltration and another associated with the pressure control system. The PI controller is tuned conservatively so that both poles are real. The dominating time constant is 30–50 s. The transfer function from the pressure setpoint to the flow Q_f then also has the same poles, but it also has a zero corresponding to the pole $s = -B/C$ of the ultrafiltration (see Fig. 12.18). This zero can change significantly with the type of dialysis filter used. A consequence is that there is a drastic difference in the dynamics obtained for different filters.

The main function of the system is to control the total water removal V during the treatment. The water removal is given by

$$\frac{dV}{dt} = Q_f \tag{12.5}$$

An Earlier Control System

An earlier system used a PI controller in the outer loop. Because of the large gain variations, it was necessary to use a conservative setting with low gain. This resulted in very sluggish control of the weight loss. Experiments with various simple forms of gain adjustment did not solve the problem and it was decided to test if an adaptive controller was feasible.

Adaptive Control

The adaptive controller was designed as an indirect adaptive pole placement algorithm.

Parameter Estimation.   The dynamics can be expected to be of third order, representing the dynamics of the pressure loop and the dynamics of the filter introduced to filter the flow signal. This filter has a time constant of about 30 s. Experiments with system identification indicated, however, that data could be fitted adequately by

$$Q_f(t) = aQ_f(t-h) + b_1 p_c(t-h) + b_2 p_c(t-2h) \qquad (12.6)$$

where Q_f is the filtration flow and p_c is the setpoint of the pressure loop. This model represents first-order dynamics with a time delay. A sampling interval of 5 s was found to be suitable. The parameter estimation was made on differences to avoid problems with a constant level in the signals.

The estimated steady-state gain is an important parameter. With a low estimated gain, the gain in the controller will be large. It is therefore advantageous to have the sum of the b parameters as one of the estimated parameters so that it is easy to set a lower limit to the estimated gain. This has been done in the FCM by using the regression vector

$$\Big(\ Q_f(t) \quad p_c(t) \quad p_c(t) - p_c(t-h) \ \Big)$$

instead of

$$\Big(\ Q_f(t) \quad p_c(t) \quad p_c(t-h) \ \Big)$$

If the estimated gain becomes too small, the estimate is stopped at the limit.

A constant forgetting factor of 0.999 is used to track slowly time-varying parameters. To improve numerics, only the diagonal elements of the covariance matrix P are divided by this factor. It is well known that the equation for $P(t)$ may be sensitive to numerical precision when a forgetting factor is used. This is because the eigenvalues of the P-matrix may be widely separated. Several methods to handle this problem were described in Chapter 11 and in the discussion of the ABB adaptive controller and Firstloop in this chapter. In this case the problem was avoided by careful scaling, and an ordinary recursive least-squares method could be used.

Control Design.   A conventional pole placement algorithm and a design that guarantees integral action were used. (See Section 3.6.) Several factors influence the choice of desired closed-loop poles. If a smooth control is desired in the steady state, the speed of setpoint changes should not be set too high. Second, the first step response at startup must not be quicker than the time required to get a reasonable model. A reasonable response time in accumulated flow is one hour. The other closed-loop poles, which correspond to flow changes, were specified by time constants of 25 s, 15 s, and 15 s.

The controller can be reparameterized to correspond to a PID controller with a filtered derivative part. The structure was chosen so that the controller corresponds to a discrete-time PID controller in which the reference signal enters only the *P* and *I* parts. This corresponds to $\beta = 1$ in Eq. (8.3) in Chapter 8. A possible common factor in the estimated model was canceled before entering the design calculations.

Special Design Considerations

Control of fluid removal during dialysis has a direct influence on the patient's well-being. This imposes heavy demands on the control system. Several safety features have been included. Smooth performance from the first moment of control is essential. This can be achieved by a careful choice of certain parameters, as we discuss next.

Filtering.   The measured flow signal is corrupted by measurement noise. Since a new value is available every second, it is possible to filter the signal. With a sampling period for control of 5 s, it was found to be suitable to use a first-order filter with a time constant of 30 s to filter flow and accumulated flow before using the values in the control algorithm. This smooths the control signal considerably without preventing fairly quick setpoint changes.

Limits on Setpoint Changes.   Both the absolute level and the rates of setpoint changes were limited on the basis of physical constraints. The PID controller was provided with conventional anti-windup protection to avoid problems with saturation. Parameter updating is also interrupted when the pressure setpoint is kept constant at a limit. At startup, when the model parameters may be far from their best values, it is also wise to prevent the control algorithm from changing the control signal (i.e., the pressure setpoint) too rapidly. The rate limit on the pressure setpoint prevents this; experience has shown that this limit is hit only rarely.

Startup.   A critical moment for an adaptive controller is the start, before the model parameters have been accurately estimated. It was required that its step response be almost perfect from the beginning. For this reason, most of the development time was spent in adjusting the parameters to ensure a smooth start. The following parameters were then found to be important:

- Initial values of the parameter estimates,
- Initial values of the covariance matrix *P*,
- The desired closed-loop poles,
- The time allowed for signals to settle before estimation and control starts,
- Limits on the estimated parameters (especially the static gain), and
- The limit on control changes (and control).

The initial values of the parameter estimates are important, since they determine the initial controller parameters. They were chosen to model a high-gain dialyzer, with an extra time delay, to give a cautious low-gain controller. This is perfect for a highly permeable (i.e., high-gain) membrane, but for normal membranes the pressure changes will be too small, a situation that is soon detected by the parameter estimator.

It is important to choose the P matrix carefully. This determines the speed of parameter estimation. Values of P that are too large will make the estimates noisy, and there is a risk that the estimates may temporarily give bad controllers. Also, a value of P that is too large can quickly eliminate the carefully chosen initial parameters in the estimator. With values of P that are too small the time needed to find a good model can be very long, a situation that is not at all acceptable.

It was found to be advantageous to introduce a lower bound on the estimated gain in the model. With low-gain dialyzers there would otherwise be a tendency for the estimator to decrease the gain estimate too much, and the controller gain would be too high for a while. A suitable limit for the model gain could be determined from the known data of existing dialyzers. To facilitate the checking of the estimated gain, a special form of the process model was used. The estimated pole was also bounded away from a pure integrator, since this pole enters the expression for the gain limit.

The limit on the setpoint changes also helps to ensure a smooth startup. The desired closed-loop poles are important design parameters. The equivalent time constants should be chosen to be long enough to give the estimator time to find a good model before the setpoint is approached for the first time. They should also be as short as possible to give a rapid response to setpoint changes. The equivalent time constants of the closed-loop systems were chosen to be 720, five, and three sample intervals, which correspond to 1 hour, 25 s, and 15 s, respectively. Without the requirement of a smooth startup it would have been possible to speed up the desired closed-loop dynamics considerably. However, setpoint changes are not very frequent, and smooth startup is much more important than rapid setpoint changes.

If by chance the desired pressure were already set at startup, there would be no pressure change that would help to improve the estimates of model parameters. Therefore there is a period of forced small pressure changes for the first eight minutes after a reset. This is accomplished by periodic changes of the setpoint every 45 s.

With an adaptive controller it is very important to ensure that the estimated model is never destroyed. Therefore the estimator should always be given true values for control and measured signals. If for some reason, such as an alarm situation causing the DFM to bypass the dialysis fluid, the control signal is not allowed to do its job, the estimator must be turned off. The controller will then use the old estimates for a while.

After all such breaks and at startup, a settling period is allowed, during which correct signals are entered into all the vectors but no estimation is done. This settling period is very important, especially at startup, when the estimates are most sensitive to changes in the signals. Errors in the signals also force the *P*-matrix to decrease rapidly, so future learning is slowed down considerably.

Alarms.   Appropriate alarms are an important part of any useful control system. An alarm indicates if the volume control error is too large and also if something is wrong in the dialysis fluid monitor or with the pipes. If there is a stop in the blood pipe from the dialyzer to the drip chamber, the blood pressure within the dialyzer will rise, causing a large ultrafiltration rate and minimized pressure. The alarm in the FCM will then cause the DFM to enter a patient-safe condition.

Operational Experience

It has been possible to use the algorithm to handle ultrafiltration control for all kinds of dialyzers that are available today. Treatment modes such as single-needle or double-needle treatment or sequential dialysis with periods of isolated ultrafiltration have been tested. Dialyzers with variations in values of *B* by a factor of 75 have been tested in the laboratory without any problems. After a period of approximately five months of clinical trials at several clinics, full-scale production started in the autumn of 1986. Over 11,000 units had been

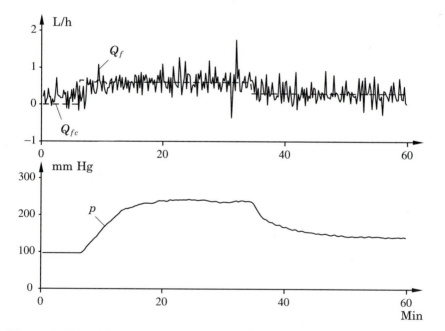

Figure 12.19   Adaptive control of a dialysis system. Responses in differential flow Q_f (solid line) and transmembrane pressure p to step changes in the setpoint Q_{fc} (dashed line) for a plate membrane. The adaptive control starts at $t = 6$. (With courtesy of Gambro AB.)

delivered as of December 1993. Since every machine may be used in several hundred treatments each year, there is now extensive practical experience with this algorithm, which seems to work well under all kinds of conditions.

Figure 12.19 shows responses in differential flow Q_f to step changes in the setpoint Q_{fc} when the system is under adaptive control. The pressure p is also shown. Despite the noisy flow measurement, the response of the closed-loop system is very good.

12.9   CONCLUSIONS

In this chapter we have tried to give an idea of how adaptive techniques are used in real control systems. A few general observations can be made.

Although there are many applications of adaptive control, it is clear that adaptive control is not a mature technology. The techniques were introduced in products in the early 1980s. Those in use today are mostly first-generation products; there are second-generation products in only a few cases.

The description of the products and the real applications show clearly that although the key principles are straightforward, many "fixes" must be done to make the system work well under all possible operating conditions. The need for safety nets, safety jackets, or supervision logic is not specific to adaptive control. Similar precautions must be taken in all real control systems, but since adaptive control systems are complex to start with, the safety nets that are required can be quite elaborate.

The examples clearly show that adaptive systems are not black box solutions that are a panacea. Rather, adaptive methods are useful in combination with other control design methods. Both in the rolling mill example and in the ship steering autopilot, adaptation was combined with gain scheduling. Another example is the use of a feedforward signal in the pulp dryer to improve the adaptation transient.

A third observation is that the human-machine interface is very important. A fourth observation is that some operating conditions are not conveniently handled by adaptive control. One example is the behavior of ship steering autopilots in a quartering sea.

There are unquestionably many different adaptive techniques, but so far, only a few of them have been used in industrial products. In many cases the choices have not been made by comparing several alternatives; one method has been chosen quite arbitrarily. This means that many alternatives have not been tried.

The computing power that is available has a significant influence on the type of control algorithms that can conveniently be implemented. The simple auto-tuners use simple 8-bit microprocessors, whereas some of the more advanced systems use full 32-bit architecture. In most process control applications there are no problems with computing time. The rolling mill applications, on the other hand, are quite demanding. The computing power that is available also has a significant impact on what human-machine interface can be implemented.

The applications also indicate the importance of the safety network. It is of interest to see the facilities provided in the toolbox and the specific solutions used in the dedicated systems. It is clearly much simpler to design a safety network for a dedicated system, in which good parameter bounds can be established.

The applications described in this chapter and elsewhere indicate that there are three cases in which it is very useful to use adaptive control:

- When the system has long time delays,
- When feedforward can be used, and
- When the character of the disturbances is changing.

In all these cases it is necessary to have a model of the process or the disturbances to effectively control the system. It is then beneficial to be able to estimate a model and to adapt to changes in the process.

REFERENCES

A number of applications are described in the books:

Narendra, K. S., and R. V. Monopoli, 1980. *Applications of Adaptive Control.* New York: Academic Press.

Unbehauen, H., ed., 1980. *Methods and Applications in Adaptive Control.* Berlin: Springer-Verlag.

Harris, C. J., and S. A. Billings, eds., 1981. *Self-Tuning and Adaptive Control: Theory and Applications.* London: Peter Peregrinus.

Narendra, K. S., ed., 1986. *Adaptive and Learning Systems: Theory and Applications.* New York: Plenum Press.

and in the survey papers:

Seborg, D. E., T. F. Edgar, and S. L. Shah, 1986. "Adaptive control strategies for process control: A survey." *AIChE Journal* **32**: 881–913.

Åström, K. J., 1987. "Adaptive feedback control." *Proc. IEEE* **75**: 185–217.

Proceedings of the IFAC, CDC, and ACC are also good reference sources. More details about the products are available in manuals, brochures, and application notes from the manufacturers.

The 3M PLC implementations is described in:

Alam, M. A., and K. K. Burhardt, 1979. "Further work on self-tuning regulators." *Proceedings of the 1979 IEEE Conference on Decision and Control,* pp. 616–620.

Alam, M. A., 1984. "A multivariable self-tuning controller for industrial application." *Preprints of the 9th IFAC World Congress*, pp. III:259–262. Budapest.

The Foxboro EXACT is described in:

Bristol, E. H., and T. W. Kraus, 1984. "Life with pattern adaptation." *Proceedings of the 1984 American Control Conference*, pp. 888–892. San Diego, Calif.

Kraus, T. W., and T. J. Myron, 1984. "Self-tuning PID controller uses pattern recognition approach." *Contr. Eng* June: 106–111.

The relay auto-tuning and the adaptive version are described in:

Åström, K. J., and T. Hägglund, 1984. "Automatic tuning of simple regulators with specifications on phase and amplitude margins." *Automatica* **20**: 645–651.

Åström, K. J., and T. Hägglund, 1988. *Automatic Tuning of PID Regulators.* Triangle Research Park, N.C.: Instrument Society of America.

Hägglund, T., and K. J. Åström, 1991. "Industrial adaptive controllers based on frequency response techniques." *Automatica* **27**: 599–609.

A survey of auto-tuners and adaptive PID controllers is given in:

Åström, K. J., T. Hägglund, C. C. Hang, and W. K. Ho, 1993. "Automatic tuning and adaptation for PID controllers: A survey." *Control Eng. Practice* **1**: 699-714.

The ABB adaptive controller/Master Piece is described in:

Bengtsson, G., and B. Egardt, 1984. "Experiences with self-tuning control in the process industry." *Proceedings of the 9th IFAC World Congress*, pp. XI:132–140. Budapest.

The description of the rolling mill example is based on:

Rudolph, W., H. Lefuel, and A. Rippel, 1984. "Regeln des Dressiergrades in einem Kaltbandwalzwerk." *Bänder Bleche Rohre* **25**(2): 36–37.

The description of the digester study is based on:

Allison, B. J., G. A. Dumont, L. H. Novak, and W. J. Cheetham, 1990. "Adaptive-predictive control of Kamyr digester chip level." *AIChE Journal* **36**(7): 1075–1086.

The adaptive systems used at Vallvik are described in:

Brattberg, Ö., 1994. "Adaptive control of a continuous digester." *Preprints of the Control Systems 94*, pp. 298–306. Swedish Pulp and Paper Research Institute, Stockholm.

The ship steering example is based on:

Källström, C. G., K. J. Åström, N. E. Thorell, J. Eriksson, and L. Sten, 1979. "Adaptive autopilots for tankers." *Automatica* **15**: 241–254.

Källström, C. G., P. Wessel, and S. Sjölander, 1988. "Roll reduction by rudder control." *Proceedings of the Spring Meeting, Society of Naval Architects and Marine Engineers*, pp. 67–76. Pittsburgh, Pa., June 8–10.

The particular control algorithm used in the product is described in:

Åström, K. J., 1980. "Design of fixed gain and adaptive ship steering autopilots based on the Nomoto model." *Proceedings of the Symposium on Ship Steering Automatic Control.* Instituto Internazionale delle Comunicazioni, June 25–27, Genoa, Italy.

The description of the control system for ultrafiltration is based on:

Sternby, J., 1995. "Adaptive control of ultrafiltration." Submitted, *Trans. Control Systems Technology* **3**.

CHAPTER 13

PERSPECTIVES ON ADAPTIVE CONTROL

13.1 INTRODUCTION

In this final chapter we attempt to give some perspective on the field of adaptive control. This is important but difficult because the field is in rapid development. The starting point is a short discussion of some closely related areas that are not covered in the book. These include adaptive signal processing in Section 13.2 and extremum control in Section 13.3. Particular attention is given to the field of adaptive signal processing, in which a cross-fertilization with adaptive control appears particularly natural.

Adaptive regulators and auto-tuning have complementary properties. Auto-tuners require very little prior information and give a robust ballpark estimate of gross system properties. Adaptive regulators require more prior knowledge, but they can give systems with much improved performance. It thus seems natural to combine auto-tuning with adaptive control in systems that combine several algorithms. Apart from algorithms for control, estimation, and design, it may also be useful to include supervision. It seems logical to use an expert system to monitor and control the operation of such a system. Systems of this type have been called *expert control systems* and are briefly discussed in Section 13.4. The use of expert systems also provides a natural way to separate algorithms from logic that occurs in all control systems.

Adaptation is related to learning; in Section 13.5 we discuss some early uses of learning in control systems and how it is related to adaptive control as we now understand it. In Section 13.6 we attempt to speculate on future directions in the theory and practice of adaptive control.

13.2   ADAPTIVE SIGNAL PROCESSING

Automatic control and signal processing have strong similarities; similar mathematical models and techniques are used in the two fields. However, there are also some significant differences. The time scales can be different. Signal processing often deals with rapidly varying signals, as in acoustics, in which sampling rates of tens of kilohertz are needed. In control applications it is often (but not always) possible to work with much slower sampling rates.

A more significant difference is that time delays play a minor role in signal processing. It is often permissible to delay a signal without any noticeable difficulty. Because control systems deal with feedback, even small time delays can result in drastic deterioration in performance. A third difference is in the industrial markets for the technologies. In signal processing, there are some standard problems that have a mass market, as in the field of telecommunications. The control market is more diversified and fragmented. Adaptive control is used to design control systems that work well in an unknown or changing environment. The environment is represented by process dynamics and disturbance signals. Adaptive signal processing is used to process signals whose characteristics are unknown or changing. More emphasis is given in signal processing to fast algorithms. Although there have been attempts to bring the fields closer together, much more effort is needed in this direction. To illustrate this, we will describe a few typical adaptive signal processing problems.

Prediction, Filtering, and Smoothing

Prediction, filtering, and smoothing are typical signal processing problems, which can all be described as follows: Given two signals x and y and a filter F, determine the filter such that the signals y and $\hat{y} = Fx$ are as close as possible. The problem can be illustrated by the block diagram in Fig. 13.1. In a typical case we have

$$x(t) = s(t) + v(t) \qquad \text{and} \qquad y(t) = s(t + \tau)$$

where s is the signal of interest and v is some undesirable disturbance. The problem is called *smoothing* if $\tau < 0$, *filtering* if $\tau = 0$, and *prediction* if $\tau > 0$. Solutions to such problems are well known for signals with known spectra and quadratic criteria. The corresponding adaptive problems are obtained when the

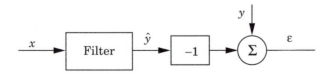

Figure 13.1   Illustration of filtering, prediction, and smoothing.

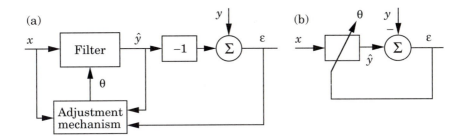

Figure 13.2   (a) An adaptive system for filtering, prediction, or smoothing and (b) its simplified representation.

signal properties are not known. All recursive parameter estimation methods can be applied to the adaptive signal processing problems. This is illustrated in Fig. 13.2, which gives a typical adaptive solution. The adjustment mechanism can be any recursive parameter estimator. The details depend on the structure of the filter and the particular estimation method chosen. An example illustrates the idea.

EXAMPLE 13.1   **Output error parameter estimation**

Assume that the filter is represented as an ordinary pulse transfer function

$$F(z) = \frac{b_0 z^{n-1} + b_1 z^{n-2} + \cdots + b_{n-1}}{z^n + a_1 z^{n-1} + \cdots + a_n}$$

To obtain a recursive estimator, the parameter vector

$$\theta = \begin{pmatrix} a_1 & \cdots & a_n & b_0 & \cdots & b_{n-1} \end{pmatrix}$$

and the regression vector

$$\varphi(t-1) = \begin{pmatrix} -\hat{y}(t-1) & \cdots & -\hat{y}(t-n) & x(t-1) & \cdots & x(t-n) \end{pmatrix}$$

are introduced. The error is then given by

$$\varepsilon(t) = y(t) - \hat{y}(t) = y(t) - \varphi^T(t-1)\hat{\theta}(t-1)$$

and the equation for updating the estimate is

$$\hat{\theta}(t) = \hat{\theta}(t-1) + P(t)\varphi(t-1)\varepsilon(t) \qquad\qquad \square$$

The special case of Example 13.1, obtained when the filter is an FIR filter and a gradient parameter estimation scheme is used, is particularly simple. This is the *LMS algorithm*.

Driver's microphone

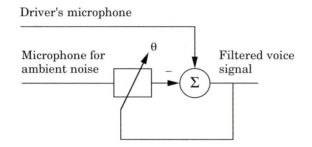

Figure 13.3   Use of an adaptive filter for adaptive noise cancellation.

Block Diagram Representation

The block diagram in Fig. 13.2 represents a solution to a generic signal processing problem. To make it easy to build large systems, it is convenient to consider this module as a building block that can be used for many different purposes. This is simpler if a proper representation is used. For that purpose it is convenient to represent the module as a block that receives signals x and y and delivers estimates $\hat{y}$ and $\hat{\theta}$. Such a representation, shown in Fig. 13.2(b), makes it possible to describe several adaptive signal processing problems.

Adaptive Noise Cancellation

Consider the situation of a mobile telephone in a car where there is a considerable ambient noise. Assume that two microphones are used. One is directional and picks up the driver's voice corrupted by noise; the other is directed away from the driver and picks up mostly the ambient noise. By connecting the microphones to an adaptive filter as shown in Fig. 13.3, it is possible to obtain a signal that is considerably improved. Removal of power frequency hum from measurement signals is another application at adaptive noise cancellation.

Adaptive Differential Pulse Code Modulation (ADPCM)

Digital signal transmission is becoming important because of the rapid development of new hardware. Its use in ordinary telephone communication is increasing. Pulse code modulation (PCM) is the standard method for converting analog signals to digital form. The analog signal is filtered and digitized by using an analog-to-digital (A-D) converter. The digitized signal is then transmitted in serial form. If the A-D converter has B bits and the sampling is f Hz, the transmission rate required is fB bits/s. For standard voice signals, a sampling rate of 8 kHz is typically used. A resolution of 12 bits in the A-D converter is required to get good-quality transmission. The bit rate required is thus 96 kbit/s. By having an A-D converter with a nonlinear characteristic it

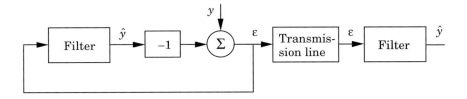

Figure 13.4   Block diagram of a differential pulse code modulation (DPCM) system.

is possible to reduce the bit rate to 64 kbit/s, which is the standard for digital voice transmission.

It is highly desirable to reduce the transmission rate, because more communication channels are then obtained with the same transmission equipment. The bit rate can be reduced significantly by using differential pulse code modulation (DPCM). In this technique the innovations of the signal are computed as $\varepsilon = y - \hat{y}$, where $\hat{y}$ is generated by filtering the innovations through a predictive filter. Only the innovations are transmitted (see Fig. 13.4). The receiver has a prediction filter with the same characteristics as the filter in the sender. The signal $\hat{y}$ can then be reconstructed in the receiver. The bit rate that is required is reduced significantly because fewer bits are required to represent the residual. It has been shown that for voice signals, a resolution of 4 bits is sufficient. This means that the bit rate required for the transmission can be reduced from 64 kbit to 32 kbit.

The prediction filter depends on the character of the transmitted signal. Substantial research into the characterization of speech has shown that it can be well predicted by linear filters. However, the properties of the filter will change with the particular sound that is spoken. To predict speech well, it is thus necessary to make the filters adaptive. The transmission scheme obtained is then called *adaptive differential pulse code modulation (ADPCM)*. Such a scheme, which uses an adaptive filter based on the output error method, is shown in Fig. 13.5. Notice that the adaptive filters at the transmitter and the receiver are driven by the residual only. If the filters in the receiver and the transmitter are identical, the filter parameters will automatically be the same. The adaptive filters have therefore been standardized by CCITT (*Comité Consultatif Internationale de Télégraphique et Téléphonique*). The filter that is used has the transfer function

$$H(z) = \frac{b_0 z^5 + b_1 z^4 + \cdots + b_5}{z^4(z^2 + a_1 z + a_2)}$$

The regression vector associated with the output error estimation is

$$\varphi(t) = \left(\begin{matrix} -x(t) & -x(t-1) & e(t) & \cdots & e(t-5) \end{matrix} \right)$$

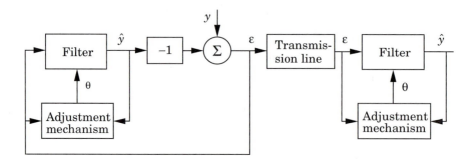

Figure 13.5   Block diagram of an adaptive differential pulse code modulation system.

and the associated parameter vector is

$$\theta = \begin{pmatrix} a_1 & a_2 & b_0 & \dots & b_5 \end{pmatrix}$$

The standard least-squares estimator is of the form

$$\hat{\theta}(t + 1) = \hat{\theta}(t) + P(t + 1)\varphi(t)\varepsilon(t + 1)$$

Several drastic modifications are made to simplify the calculations. A constant value of the gain is used. The multiplication is avoided by just using the signs of the signals. Leakage is also added to make sure that the estimator is stable. The updating of the parameters b_i is then given by the *sign-sign algorithm*

$$\hat{b}_i(t) = \left(1 - 2^{-8}\right)\hat{b}_i(t) + 2^{-7}\, \text{sign}\, (e(t - i))\, \text{sign}\, (e(t)) \qquad (13.1)$$

Similar approximations are made in the other equations. The computations in Eq. (13.1) are very simple. They can be done by shifts and the addition of a few bits, which can be accomplished with a small VLSI circuit. The CCITT ADCPM standard was achieved after significant experimentation. It is a good example of how drastic simplifications can be made with good engineering.

13.3   EXTREMUM CONTROL

The control strategies that have been discussed in the book have mainly been such that the reference value is assumed to be given. The reference value is often easily determined. It can be the desired altitude of an airplane, the desired concentration of a product, or the thickness at the output of a rolling mill. On other occasions it can be more difficult to find the suitable reference value or the best operating point of a process. For instance, the fuel consumption of a car depends, among other things, on the ignition angle. The mileage of the car

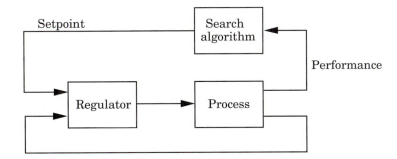

Figure 13.6   A simplified block diagram of an extremum control system.

can be improved by a proper adjustment, but the efficiency will depend on such conditions as the condition of the road and the load of the car. To maintain the optimal efficiency, it is necessary to change the ignition angle.

Tracking a varying maximum or minimum is called *extremum control*. The static response curve relating the inputs and the outputs in an extremum control system is nonlinear. The task of the controller is to find the optimum operating point and to track it if it is varying. Several processes have this kind of behavior. Control of the air-fuel ratio of combustion is one example. The optimum will change, for instance, with temperature and fuel quality. Another example is water turbines of the Kaplan type, in which the blade angle of the turbine is changed to give maximum output power. The same problem is encountered in wind power plants, in which pitch angle is changed depending on wind speed.

Extremum control is related to optimization techniques; many of the ideas have been transferred from numerical optimization. There was great interest in extremum control in the 1950s and 1960s, and some commercial products were put on the market. For instance, the first computer control systems installed in the process industry were motivated by the possibility of optimizing the setpoints of the controllers. The interest then declined, partly because of the difficulty of implementing the optimizing controllers. Furthermore, there is great difficulty in finding appropriate process models. Developments in computers have led to a renewed interest in extremum control and its combination with adaptive control. Improved efficiency of the process can result in large savings in the energy and raw material costs.

Figure 13.6 shows a simplified block diagram of an extremum control system. The process can work in open loop or in closed loop, as in the figure. The most important feature is that the process is assumed to be nonlinear in the sense that at least the performance is a nonlinear function of the reference signal. The goal of the search algorithm is to keep the output as close as possible to the extremum despite changes in the process or the influence of disturbances. The output used in the search algorithm is some measurement of the performance of the system—for instance, efficiency. The conventional regulator can also use this signal, but it is more common for the regulator to use some other output of the process.

Models

Extremum control systems are, by necessity, nonlinear. How the processes are modeled is therefore all-important. Many investigations of extremum control systems assume that the systems are static. This assumption can be justified if the time between the changes in the reference value is sufficiently long. For static systems it is possible to use many of the methods from numerical optimization. A typical description of the process is

$$y(t) = f(u(t), \theta, t) \tag{13.2}$$

where f is a nonlinear function and θ is a vector of unknown parameters that may change with time.

If there are dynamics in the process, the performance may not have settled at a new steady-state value before the next measurement is taken. This will give an interaction in the control system that can be difficult to handle. The dynamic influence will increase the complexity considerably.

In many applications it is not easy to find the appropriate models and to determine the exact nature of the nonlinearities involved. It can therefore be appropriate to combine adaptivity and extremum control. One way to simplify the identification of an unknown nonlinear model is to assume that the process can be divided into one nonlinear static part and one linear dynamic part. Models with different properties are obtained if the nonlinearity precedes or follows the linear part. The complexity of the problem will also depend on which of the variables in the process can be measured. One special type of model that has been used in extremum control systems is the *Hammerstein model*. A typical discrete-time model of this type is

$$A(q)y(t) = B(q)f(u(t)) \tag{13.3}$$

where f is a nonlinear function, typically a polynomial.

The main effect of an input nonlinearity is that it restricts the possible input values for the linear part. The nonlinear control problem can then be treated as a linear control problem with input constraints. The case with the output nonlinearity perhaps leads to more realistic problems but is also more difficult to solve.

Extremum Control of Static Systems

The first extremum control systems were based on analog implementation. One way to perform the optimization is the so-called perturbation method. The basic idea is to add a known time-varying signal to the input of the nonlinearity, then observe the effect on the output and make a correlation between these two signals. Depending on the phase between the two signals, the direction toward the extremum can be determined. The perturbation method has been used for extremum control of chemical reactors, combustion engines, and gas furnaces, for instance.

Extremum control of static systems as in Eq. (13.2) is in essence a problem of numerical optimization. With the analog implementations the possible methods were severely restricted. When a digital computer is available, standard algorithms for function minimization can be used. Usually, it is possible only to measure the function values, not its derivative. The function minimization then has to be done by using numerically computed derivatives. Some methods use only function comparisons. These methods can be used even for minimization of nonsmooth functions.

Performance measurements are typically corrupted by noise. It is then necessary to average out the influence of the noise. This implies that the gain in the optimization algorithm should go to zero. However, if the extremum is changing with time, the gain should not go to zero. This is the same compromise as is discussed in connection with tuning and adaptive control.

Most schemes for extremum control of static systems do not build up any information about the nonlinearity. The "states" of the algorithms are essentially the current estimate of the optimum point and some previous measurements. By using a model and system identification it is possible to utilize the measurements of the system better and to follow time variations in the process.

Extremum Control of Dynamic Systems

If there are dynamics in the process, it is necessary to take this into consideration in doing the optimization. The correlation and interaction between different measurements of the performance will otherwise confuse the optimization routine. One possibility, discussed previously, is to wait until the transients have vanished before the next change is made. Of course, this will increase the convergence time, especially if the process has long time constants. One way around the problem is to base the optimization on nonlinear dynamic models. An example is the Hammerstein model. A model of this type with an input nonlinearity of second order is

$$A(q)y(t) = b_0 + B_1(q)u(t) + B_2(q)u^2(t) + C(q)e(t) \tag{13.4}$$

The main reason for the popularity of the Hammerstein model is not that it is a good picture of the reality, but rather that it is linear in the parameters. The parameters can be estimated, for example, by using recursive least squares. The static response between the input and the output is given by

$$A(1)y_0 = b_0 + B_1(1)u_0 + B_2(1)u_0^2$$

The methods for static optimization discussed previously can now be used. Also note that the gradients and the Hessian are easily computed, a feature that will speed up the convergence.

Conclusions

The field of extremum control is far from mature. One crucial point is the modeling of the processes and the nonlinearities. It is generally very difficult to analyze nonlinear control problems and to derive optimal controllers, especially if there are stochastic disturbances acting on the system. The extremum control problem also has connections with the dual control problem discussed in Chapter 7. Extremum-seeking methods combined with adaptive control are of great practical interest, since even small improvements in the performance can lead to large savings in raw material and energy consumption. There are commercial extremum controllers.

13.4   EXPERT CONTROL SYSTEMS

All practical control systems contain heuristics. This appears as logic around the basic control algorithm. Adaptive systems have a lot of heuristics in the safety logic. Expert systems offer an interesting possibility of structuring the logic in a control system. If a good way to handle heuristic logic is available, it is also possible to introduce more complex control systems that contain several different algorithms. For example, it is possible to combine auto-tuners and adaptive algorithms that have complementary properties. The auto-tuner requires little prior information; it is very robust and can generate good parameters for a simple control law. Adaptive regulators can be more complex, with potentially better performance. Since they are based on local gradient procedures, they can adjust the regulator parameters to give a closed-loop system with very good performance, provided that reasonably good *a priori* guesses of system order, sampling period, and parameters are given. The algorithms will not work if the prior guesses are too far off. With poor prior data they may even give unstable closed-loop systems. This has led to the development of the safety logic discussed in Chapters 11 and 12.

Expert Systems

One objective of expert systems is to develop computer-based models for problem solving that are different from physical modeling and parameter estimation. An expert system attempts to model the knowledge and procedures used by a human expert in solving problems within a well-defined domain. Knowledge representation is a key issue in expert systems. Many different approaches have been attempted, such as first-order predicate calculus (logic), procedural representations, semantic networks, production systems or rules, and frames. A knowledge-based expert system consists of a knowledge base, an inference engine, and a user interface.

The Knowledge Base.   The knowledge base consists of data and rules. The data can be separated into *facts* and *goals*. Examples of facts are statements such as "The system appears to be stable," "PI control is adequate," and "Deviations are normal." Typical examples of goals are "Minimize the variations of the output," "Find out whether gain scheduling is necessary," and "Find a scheduling table." Data is introduced into the database by the user or via the real-time knowledge acquisition system. New facts can also be created by the rules. The rule base contains production rules of the type: "*If* premise *then* conclusion *do* action." The premise represents facts or goals from the database. The conclusion can result in the addition of a new fact to the database or modification of an existing fact. The action can be to activate an algorithm for diagnosis, control, or estimation. These actions are different from those found in conventional expert systems. The rule base is often structured in groups or knowledge sources that contain rules about the same subject. This simplifies the search. In the control application the rules represent knowledge about the control and estimation problem that are built into the system. This includes the appropriate characterization of the algorithms, judgmental knowledge about when to apply them, and supervision and diagnosis of the system. The rules are introduced by the knowledge engineer via the knowledge acquisition system, which assists in writing and testing rules.

Inference Engine.   The inference engine processes the rules to arrive at conclusions or to satisfy goals. It scans the rules according to a strategy, which decides from the context (current database of facts and goals) which production rules to select next. This can be done according to different strategies. In *forward chaining* the strategy is to find all conclusions from a given set of premises. This is typical for a data-driven operation. In *backward chaining* the rules are traced backward from a given goal to see whether the goal can be supported by the current premises. This is typical for a diagnosis problem. The search can be organized in many different ways, depth-first or breadth-first. There are also strategies that use the complexity of the rules to decide the order in which they are searched. To devise efficient search procedures, it is convenient to decompose the rule base into pieces that deal with related chunks of knowledge. If the rules are organized in that way, it is also possible for a system to focus its attention on a collection of rules in certain situations. This can make the search more efficient.

User Interface.   The user interface can be divided into two parts. The first part is the development support that the system gives. This contains tools such as rule editor and rule browser for development of the system knowledge base. The other part is the run-time user interface. This contains explanation facilities that make it possible to question how a certain fact was concluded, why a certain estimation algorithm is executing, and so on. It is also possible to trace the execution of the rules. The user interface can also contain facilities to deal with natural language.

Expert Control

The idea of expert control is to have a collection of algorithms for control, supervision, and adaptation that are orchestrated by an expert system. A block diagram of such a system is shown in Fig. 13.7. A comparison with Fig. 1.19 shows that the system is a natural extension of a self-tuning regulator. Instead of having one control algorithm and one estimation algorithm, the system has several algorithms. It also has algorithms for excitation and for diagnosis, as well as tables for storing data. Apart from this, the system also has an expert system, which decides when a particular algorithm should be used. The expert system contains knowledge about particular algorithms and the conditions under which they can be used.

In the special case in which there is only one algorithm of each category, Fig. 13.7 can be viewed as a well-structured way of implementing safety logic for an ordinary adaptive regulator. In that case the approach has the advantage that it separates the safety logic from the control algorithms. Another advantage is that the knowledge is explicit and can be investigated via the user interface.

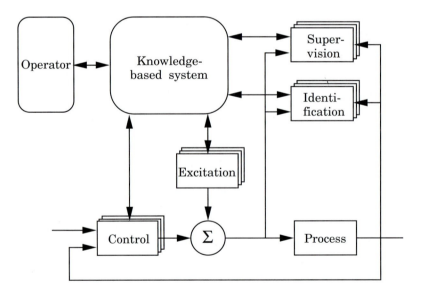

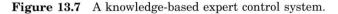

Figure 13.7   A knowledge-based expert control system.

13.5   LEARNING SYSTEMS

The notion of learning systems has been developed in the fields of artificial intelligence, cybernetics, and biology. In its most ambitious form, learning systems attempt to describe or mimic human learning ability. Attainment of this goal is still far away. The learning systems that have actually been implemented are simple systems that have strong relations to adaptive control. The systems have many names: neural nets, connectionist models, parallel distributed processing models, and so on.

Michie's Boxes

This system grew out of early work on artificial intelligence (see Michie and Chambers, 1968) and attempts to balance an inverted pendulum (see Fig. 13.8). The system has four state variables, φ, $\dot{\varphi}$, x, and $\dot{x}$, which are quantized in a crude way, with five levels for the position variables x and φ and three levels for the velocity variables $\dot{x}$ and $\dot{\varphi}$. The state space can thus be described by 225 discrete states. The control variable is quantized into two levels: force left (L) or force right (R). The control law can be represented by a binary table with 225 entries. In the experiment the table was initialized with randomly chosen L's and R's in the table. A simple scoring method was used to update the table entries as a result of experimental runs. Scoring was based on how long the pendulum stayed upright and the number of times the pendulum was in a discrete state. The system was able to balance the pendulum for about 25 minutes after a 60-hour training period. The table that defines the control action can be expressed in logic as:

> *If* cart *is* far left *and* cart *is* hardly moving *and* pendulum *is* hardly leaning *and* pendulum *is* swinging to right *then* apply force right.

For this reason the control law is also called *linguistic control*. When the logic is replaced by fuzzy logic, it is also called *fuzzy control*.

The training algorithm that is used in Michie's Boxes is similar to that used in programs for playing checkers and chess, but the pendulum problem is

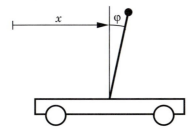

Figure 13.8   An inverted pendulum.

simpler than game playing. Training can be shortened by using a teacher, that is, by applying a scoring algorithm to an experiment in which the pendulum is balanced by an expert. A learning system of this type is obviously closely related to a model-reference adaptive system. The reference model can be viewed as a teacher.

The Perceptron

In a system such as Michie's Boxes the control law is a logic function that gives the control action as a function of sensor patterns. The function is adaptive in the sense that it will adjust itself automatically. The *perceptron* proposed by Rosenblatt (1962) is one way to obtain a learning function. To describe the perceptron, let u_i, $i = 1, 2, \ldots, n$, be inputs, and let y_i, $i = 1, 2, \ldots n$, be outputs. In the perceptron the output is formed as

$$y_i(t) = f\left(\sum_{j=1}^{n} w_{ij}(t)\, u_j(t) - b \right) \qquad i = 1, 2, \ldots, m \qquad (13.5)$$

where w_{ij} are weights, b is a bias, and f is a threshold function, for example,

$$f(x) = \begin{cases} 1 & \text{if } x \geq 0 \\ 0 & \text{if } x < 0 \end{cases}$$

To update the weights, the perceptron uses a very simple idea, which is called *Hebb's principle*: Apply a given pattern to the inputs and clamp the outputs to the desired response, then increase the weights between nodes that are simultaneously excited.

This principle was formulated in Hebb (1949), in an attempt to model neuron networks. Mathematically it can be expressed as follows:

$$w_{ij}(t + 1) = w_{ij}(t) + \gamma u_i(t) \left(y_j^0(t) - y_j(t) \right) \qquad (13.6)$$

where y_j^0 is the desired response and y_j is the response predicted by the model Eq. (13.5). By regarding the weights as parameters, it becomes clear that the updating formula of Eq. (13.6) is identical to a gradient method for parameter estimation.

Widrow and Hoff (1960) developed special-purpose hardware, called the Adaline, to implement perceptronlike devices. The learning algorithm used by Widrow was based on a simple gradient algorithm like Eq. (13.6). In devices like the perceptron and the Adaline, learning is interpreted as adjusting the coefficients in a network. From this point of view it can equally well be claimed that an adaptive system like the MRAS or the STR is learning. The mechanisms for determining the parameters are also similar.

A drawback of the perceptron is that it can recognize only patterns that can be separated linearly. It fell into disfavor because of exaggerated claims that could not be justified. It was heavily criticized in a book by Minsky and Papert (1969). The idea of designing learning networks did, however, persist.

The Boltzmann Machine

The Boltzmann Machine may be viewed as a generalization of a perceptron. It was designed to be a highly simplified model of a neural network. The machine consists of a collection of elements whose outputs are zero or one. The elements are linked by connections having different weights. The output of an element is determined by the outputs of the connecting elements and the weights of the interconnections. The firing is randomized in such a way that the probability of firing increases with the weighted sum of the inputs to an element. Some elements are connected to inputs and others to outputs, and there are also internal nodes. The connections in a Boltzmann Machine are assumed to be symmetric, which is a significant restriction.

In the perceptron there is a direct coupling between the inputs and the output. The Boltzmann Machine is much more complicated, because it can also have internal nodes. This implies that Hebb's principle cannot be applied directly. An extension called *back-propagation* has been suggested in Rumelhart and McClelland (1986).

There are many variations of neural networks. Dynamics can be introduced in the nodes. Hopfield observed that the weights could be chosen so that the network would solve specific optimization problems (Hopfield and Tank, 1986).

Hardware

An interesting feature of the neural networks is that they operate in parallel and that they can be implemented in silicon. Using such circuits may be a new way to implement adaptive control systems. A particularly interesting feature is that it is easy to integrate the networks with sensors.

13.6   FUTURE TRENDS

In this section we speculate on open research issues and the future of adaptive control. One interesting aspect of adaptive control is that it may be viewed as an automation of the modeling and design of control systems. To apply a technique automatically, it is necessary to have a very clear understanding of the conditions under which it can be applied. Ideally, this understanding should be formalized. Research on adaptive control will thus sharpen the understanding of control and parameter estimation.

Industrial Impact

There are adaptive systems that have been in continuous operation since the mid-1970s. Several products were announced in the early 1980s, and the development has accelerated since that time. Adaptation is used both in general-purpose controllers and in dedicated systems. Practically all PID controllers that are introduced today have some facility for tuning and adaptation. This applies even to simple temperature controllers. It has taken longer for adaptive techniques to appear in distributed systems for process control, but many distributed control systems are now provided with tuning and adaptation. There is a rich variety of special-purpose systems that use adaptation. For example, it has been shown that adaptation can provide improved riding quality in cars.

In summary, many adaptive algorithms are well understood. Our insight into how adaptive methods can be used to engineer better control systems is growing. Insight, understanding, and appropriate computing hardware are available. It seems likely that a large proportion of the control systems made in the future will have automatic tuning or adaptation. When adaptive control becomes more widely used, interesting phenomena that demand theoretical understanding will undoubtedly also be observed. For instance, what happens when many adaptive controllers are connected to one process? Will they interact? How should the system be initialized? We can thus look forward to interesting developments.

Algorithm Development

There are several important issues that relate to algorithm development. Current toolboxes for adaptive control use only a few of the algorithms that have been developed. It seems safe to guess that the toolboxes will be expanded, and it would also seem useful to include auto-tuners in the toolboxes to simplify initialization. Significant improvements can thus be achieved with tools that are already known, but there is also a need for improved techniques. Better methods for control system design are needed. Techniques that can explicitly handle actuator constraints and model uncertainties would be valuable contributions. It would be very useful to have methods for estimating the unstructured uncertainties.

Diagnostic routines that will tell whether a control algorithm is behaving as expected are needed. Such algorithms are well known for minimum-variance control, in which monitoring can be done simply by calculating covariances. It is straightforward to develop similar techniques for other design methods.

There is both theoretical and experimental evidence that probing signals are useful. It is also clear that it is not practical to introduce probing via stochastic control theory because of the excessive computational requirements. A significant challenge is therefore to find other ways to introduce probing. There are many who intuitively object to introducing probing signals intentionally. It must be remembered that a poorly tuned regulator will give larger than necessary deviations in controlled variables.

A systematic approach to design and implementation of safety networks is an issue of great practical relevance. Expert systems may be useful in this context.

Multivariable Adaptive Control

In this book we have focused on single-input, single-output systems, mainly to keep the presentation simple, but there has also been much research on multivariable adaptive control. Many of the results can be extended, but there is one large difficulty. For single-input, single-output systems it is possible to find a good canonical form to represent the systems in which the only parameter is the order of the system. For multivariable systems it is necessary also to know the Kronecker indices to obtain a canonical form. This is difficult both in theory and in practice. For special systems such as those found in robotics, a suitable structure can often be found by using prior knowledge of the system.

Most adaptive control systems used so far are single-loop control. Coupled systems can be obtained by interconnection via the feedforward connection. Interesting phenomena can occur when such regulators are used on multi-variable systems; analysis of the behavior of such systems is a fascinating problem.

Theoretical Issues

There are many unresolved theoretical problems in adaptive control. For example, we have no good results on the stability of schemes with gain scheduling. Much work is also needed on analysis of convergence rates. On a very fundamental level, there is a need for better averaging theorems. Many results apply only to periodic signals. This is natural, since the theory was originally developed for nonlinear oscillations. It would be highly desirable to have results for more general signal classes.

Several important problems have arisen in applications. The most important one is the design of proper safety logic; this is currently done in an ad hoc fashion. The development is also hampered by the fact that much of the information is proprietary, for competitive reasons.

13.7 CONCLUSIONS

In this book we have attempted to give our view of the complex field of adaptive control. There are many unresolved research issues and many white spots on the map of adaptive control. The field is developing rapidly, and new ideas are continually popping up.

Our opinion is that adaptive control is a good tool that a control engineer can use on many occasions. We hope that this book will help to spread the use of adaptive control and that it may inspire some of you to do research that will enhance our understanding of adaptive systems.

REFERENCES

There is an extensive literature on adaptive signal processing. A good treatment is given in:

Widrow, B., and S. D. Stearns, 1985. *Adaptive Signal Processing.* Englewood Cliffs, N.J.: Prentice-Hall.

There are strong international efforts by the IFAC and the IEEE to connect the fields of adaptive control and adaptive signal processing. A student would be well advised to pay attention to this effort. Adaptive filters are discussed in:

Treichler, J. R., C. R. Johnson, Jr., and M. G. Larimore, 1987. *Theory and Design of Adaptive Filters.* New York: John Wiley & Sons.

The CCITT standard on adaptive differential pulse code modulation is described in:

Jayant, N. S., and P. Noll, 1984. *Digital Coding of Waveforms: Principles and Applications to Speech and Video.* Englewood Cliffs, N.J.: Prentice-Hall.

Optimalizing control was introduced in:

Draper, C. S., and Y. T. Li, 1966. "Principles of optimalizing control systems and an application to the internal combustion engine." In *Optimal and Self-optimizing Control*, ed. R. Oldenburger. Cambridge, Mass.: MIT Press.

Extremum control problems are discussed in:

Blackman, P. F., 1962. "Extremum-seeking regulators." In *An Exposition of Adaptive Control*, ed. J. H. Westcott. Oxford, U.K.: Pergamon Press.

Sternby, J., 1980. "Extremum control systems: An area for adaptive control?" *Preprints of the Joint American Control Conference, JACC.* San Francisco. Paper WA2-A.

Lachmann, K.-H., 1982. "Parameter adaptive control of a class of nonlinear processes." *Proceedings of the 6th IFAC Symposium on Identification and System Parameter*, pp. 372–378, Washington, D.C.

Wittenmark, B., 1993. "Adaptive control of a stochastic nonlinear system: An example." *Int. J. Adapt. Control and Signal Processing* 7: 327–337.

The notion of expert control was introduced in:

Åström, K. J., J. J. Anton, and K. E. Årzén, 1986. "Expert control." *Automatica* **22**(3): 277–286.

A detailed description of a system based on this idea is given in:

Årzén, K.-E., 1987. "Realization of expert system based feedback control." Ph.D. thesis TFRT-1029, Department of Automatic Control, Lund Institute of Technology, Lund, Sweden.

Good sources for knowledge about expert systems are:

Barr, A., and E. A. Feigenbaum, eds., 1982. *The Handbook of Artificial Intelligence.* Los Altos, Calif.: William Kaufmann.

Hayes-Roth, F., D. Watermann, and D. Lenat, 1983. *Building Expert Systems.* Reading, Mass.: Addison-Wesley.

The program Boxes is described in:

Michie, D., and R. Chambers, 1968. "Boxes: An experiment in adaptive control." In *Proceedings of the 2nd Machine Intelligence Workshop*, eds. Dale, E., and D. Michie, pp. 137–152. Edinburgh, U.K.: Edinburgh University Press.

Fuzzy logic was introduced in:

Zadeh, L. A., 1973. "Outline of a new approach to the analysis of complex systems and decision processes." *IEEE Trans. Systems, Man and Cybernetics* **SMC-3**: 28–44.

Early examples of learning systems are given in:

Fu, K. S., 1968. *Sequential Methods in Pattern Recognition and Machine Learning.* New York: Academic Press.

Saridis, G. N., 1977. *Self-organizing Control of Stochastic Systems.* New York: Marcel Dekker.

The perceptron is described in:

Rosenblatt, F., 1962. *Principles of Neurodynamics.* New York: Spartan Books.

A critique of the perceptron is given in:

Minsky, M., and S. Papert, 1969. *Perceptrons: An Introduction to Computational Geometry.* Cambridge, Mass.: MIT Press.

Hebb's principle for adjusting the weights in a neural network is described in:

Hebb, D. O., 1949. *The Organization of Behavior.* New York: Wiley.

The Adaline is described in:

Widrow, B., and M. Hoff, 1960. "Adaptive switching circuits." *IRE WESCON Convention Record*, pp. 96–104, Pt 4.

This system was applied to many problems, such as how to stabilize an inverted pendulum. Examples of neural networks and some of their uses are found in:

Kohonen, T., 1984. *Self-organization and Associative Memory.* Berlin: Springer-Verlag.

Grossberg, S., 1986. *The Adaptive Brain. I: Cognition, Learning, Reinforcement, and Rhythm,* and *The Adaptive Brain. II: Vision, Speech, Language, and Motor Control.* Amsterdam: Elsevier/North-Holland.

Rumelhart, D. E., and J. L. McClelland, 1986. *Parallel Distributed Processing*, Vols. 1 and 2. Cambridge, Mass.: MIT Press.

These books contain many references and a detailed treatment of the Boltzmann machine. A spectacular application of the Boltzmann machine is given in:

Sejnowski, T., and C. R. Rosenberg, 1986. "NETtalk: A parallel network that learns to read aloud." JHU-EECS-8601, Johns Hopkins University.

Hopfield's network is described in:

Hopfield, J. J., and D. W. Tank, 1986. "Computing with neural circuits: A model." *Science* **233**: 625–633.

Methods for implementing neural networks in silicon and integrating them with sensors are found in:

Hecht-Nielsen, R., 1989. *The Technology of Non-Algorithmic Information Processing.* Reading, Mass.: Addison-Wesley.

Mead, C. A., 1989. *Analog VLSI and Neural Systems.* Reading, Mass.: Addison-Wesley.

Ideas on how to combine adaptation with neural networks and other techniques in control systems are discussed in:

White, D. A., and D. A. Sofge, 1992. *Handbook of Intelligent Control: Neural, Fuzzy and Adaptive Approaches.* New York: Van Nostrand, Reinhold.

INDEX